TEMPERATURE-FATIGUE INTERACTION

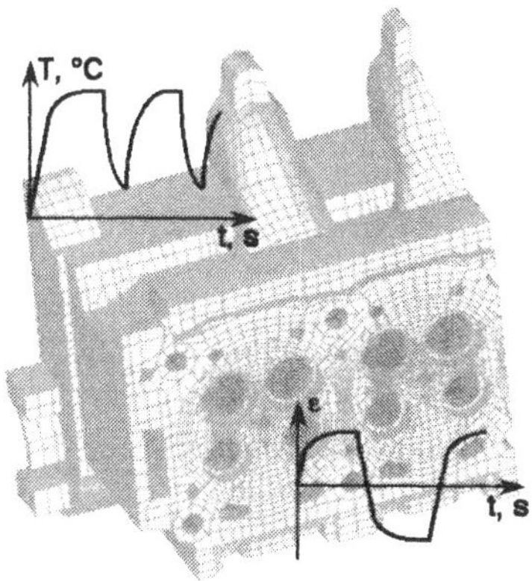

Other titles in the ESIS Series

EGF 1	*The Behaviour of Short Fatigue Cracks* Edited by K.J. Miller and E.R. de los Rios
EGF 2	*The Fracture Mechanics of Welds* Edited by J.G. Blauel and K.-H. Schwalbe
EGF 3	*Biaxial and Multiaxial Fatigue* Edited by M.W. Brown and K.J. Miller
EGF 4	*The Assessment of Cracked Components by Fracture Mechanics* Edited by L.H. Larsson
EGF 5	*Yielding, Damage, and Failure of Anisotropic Solids* Edited by J.P. Boehler
EGF 6	*High Temperature Fracture Mechanisms and Mechanics* Edited by P. Bensussan and J.P. Mascarell
EGF 7	*Environment Assisted Fatigue* Edited by P. Scott and R.A. Cottis
EGF/ESIS 8	*Fracture Mechanics Verification by Large Scale Testing* Edited by K. Kussmaul
ESIS/EGF 9	*Defect Assessment in Components -* *Fundamentals and Applications* Edited by J.G. Blauel and K.-H. Schwalbe
ESIS 10	*Fatigue under Biaxial and Multiaxial Loading* Edited by K. Kussmaul, D.L. McDiarmid and D.F. Socie
ESIS 11	*Mechanics and Mechanisms of Damage in Composites and Multi-Materials* Edited by D. Baptiste
ESIS 12	*High Temperature Structural Design* Edited by L.H. Larsson
ESIS 13	*Short Fatigue Cracks* Edited by K.J. Miller and E.R. de los Rios
ESIS 14	*Mixed-Mode Fatigue and Fracture* Edited by H.P. Rossmanith and K.J. Miller
ESIS 15	*Behaviour of Defects at High Temperatures* Edited by R.A. Ainsworth and R.P. Skelton
ESIS 16	*Fatigue Design* Edited by J. Solin, G. Marquis, A. Siljander and S. Sipilä
ESIS 17	*Mis-Matching of Welds* Edited by K.-H. Schwalbe and M. Koçak
ESIS 18	*Fretting Fatigue* Edited by R.B. Waterhouse and T.C. Lindley
ESIS 19	*Impact of Dynamic Fracture of Polymers and Composites* Edited by J.G. Williams and A. Pavan
ESIS 20	*Evaluating Material Properties by Dynamic Testing* Edited by E. van Walle
ESIS 21	*Multiaxial Fatigue & Design* Edited by A. Pineau, G. Gailletaud and T.C. Lindley
ESIS 22	*Fatigue Design of Components.* ISBN 008-043318-9 Edited by G. Marquis and J. Solin
ESIS 23	*Fatigue Design and Reliability.* ISBN 008-043329-4 Edited by G. Marquis and J. Solin
ESIS 24	*Minimum Reinforcement in Concrete Members.* ISBN 008-043022-8 Edited by Alberto Carpinteri
ESIS 25	*Multiaxial Fatigue and Fracture.* ISBN 008-043336-7 Edited by E. Macha, W. Będkowski and T. Łagoda
ESIS 26	*Fracture Mechanics: Applications and Challenges.* ISBN 008-043699-4 Edited by M. Fuentes, M. Elices, A. Martin-Meizoso and J.M. Martinez-Esnaola
ESIS 27	*Fracture of Polymers, Composites and Adhesives.* ISBN 008-043710-9 Edited by J.G. Williams and A. Pavan
ESIS 28	*Fracture Mechanics Testing Methods for Polymers Adhesives and Composites.* ISBN 008-043689-7 Edited by D.R. Moore, A. Pavan and J.G. Williams

For information on how to order titles 1-21, please contact MEP Ltd, Northgate Avenue, Bury St Edmonds, Suffolk, IP32 6BW, UK. Titles 22-28 can be ordered from Elsevier Science (http://www.elsevier.com).

TEMPERATURE-FATIGUE INTERACTION

Editors: L. Rémy and J. Petit

SF2M

ESIS Publication 29

This volume contains 37 papers, peer-reviewed from those presented at the International Conference on Temperature-Fatigue Interaction, Ninth International Spring Meeting organised by the Fatigue Committee of SF2M, held in Paris, France, 29-31 May 2001.

ESIS

2002

Elsevier

Amsterdam – London – New York – Oxford – Paris – Shannon – Tokyo

ELSEVIER SCIENCE Ltd.
The Boulevard, Langford Lane
Kidlington, Oxford OX5 IGB, UK

First edition 2002

Library of Congress Cataloging in Publication Data.
A catalog record from the Library of Congress has been applied for.

British Library Cataloguing in Publication Data.
A catalogue record from the British Library has been applied for.

ISBN : 0 08 043982 9
ISSN : 1566-1369

♾ The paper used in this publication meets the requirements of ANSI/NISO Z39.48-1992 (Permanence of Paper).
The papers presented in these proceedings have been reproduced directly from the authors' 'camera ready' manuscripts. As such, the presentation and reproduction quality may vary from paper to paper.

Transferred to digital printing 2006
Printed and bound by CPI Antony Rowe, Eastbourne

CONFERENCE COMMITTEES

International Scientific Committee

Canada: F. Ellyin. **Czech Republic**: P. Lukas. **Finland**: G. Marquis. **France**: J.-L. Chaboche, D. Fournier, F. Rezai-Aria. **Germany**: P. Dolabella Portella, D. Löhe, D. Munz. **Italy**: M. Marchionni. **Japan**: M. Igare,. R. Ohtani. **Poland**: K. Golos. **Spain**: M. Anglada. **The Netherlands**: J. Bressers. **United Kingdom**: P. Skelton. **USA**: C. Chu, A. Saxena, R. Schafrik, H. Sehitoglu.

Organisation Committee and National Scientific Committee

C. Amzallag (EDF), C. Bathias (CNAM), A. Bignonnet (PSA), S. Degallaix (EC Lille), J. F. Flavenot (CETIM), Y. Franchot (SF2M) A. Galtier (IRSID), J.Y. Guédou (SNECMA), G. Henaff (ENSMA), A. Koster (ENSMP), S. Kruch (ONERA), H.P. Lieurade (CETIM), P. Merrien (CETIM), J. Petit (ENSMA), P. Rabbe (INSA Lyon), L. Rémy (ENSMP), J. Renard (EMP/CDM), C. Sarrazin-Baudoux (ENSMA), G. Thauvin (SERMA Technologies).

Conference Chairmen

J. Petit (ENSMA), L. Rémy (ENSMP)

CONTENTS

Damage under Thermal-Mechanical Loading

Crack Growth

Design and Structures

PREFACE

The International Conference "Temperature-Fatigue Interaction", held at Paris in May 29-31, 2001, was organised by the Fatigue Committee of the Societé Française de Métallurgie et de Matériaux (SF2M, French Society for Metallurgy and Materials) under the auspices of the European Society for the Integrity of Structures (or European Structural Integrity Society). This meeting was sponsored by DGA/DSP of the French Ministry of Defence under contract MS/SC N° 160002/A000/DSP/SREA/SC/SR and the automotive manufacturer PSA.

This meeting was the 20th Spring Meeting organised by the Fatigue Committee of SF2M and the 9th International Edition. This series of meetings is the result of a long friendship between board members. This conference, like any other conference of the series, aimed to disseminate recent research results and promote the interaction and collaboration amongst materials scientists, mechanical engineers and design engineers.

Many engineering components and structures used in the automotive, aerospace, power generation and many other industries experience cyclic mechanical loads at high temperature or temperature transients causing thermally induced stresses. The increase of operating temperature and thermal mechanical loading trigger the interaction with time-dependent phenomena as creep and environment effects (oxidation, corrosion).

A large number of metallic materials were investigated including:

Aluminium alloys for the automotive industry

Steels and cast iron for the automotive industry and materials forming

Stainless steels for power plants

Titanium

Composites

Intermetallic alloys

Nickel base superalloys for aircraft industry

Polymers

Important progress was observed in testing practice for high temperature behaviour, including environment and thermo-mechanical loading as well as in observation techniques.

A large difficulty, which was emphasized upon, is to know precise service loading cycles under non isothermal conditions. This was considered critical for numerous thermal fatigue problems discussed in this conference. Thermo-mechanics of fluids and fluid-structure interaction, friction heating in brakes are to be analysed properly to estimate heat exchange coefficients and temperature transients : such transient thermal analyses are now carried out in numerous industries, due to advances in computer programming and performances.

Viscoplastic models which were implemented in simplified stress analyses software some 10 to 20 years ago are now used for a number of components under 3D cases. Impressive non linear computations were shown with a very high number of degrees of freedom (between ten thousand and 1 million) with or without parallel computers.

Experimental studies are more and more complex and point out the interactions between creep, oxidation and fatigue. The influence of gaseous species, oxygen versus water vapour, and that of hydrogen embrittlement, is still controversial. The global fracture mechanics concepts are still popular to analyse crack growth at high temperature. The stress analysis at crack tip is now often used to bring a clearer understanding of crack growth under high temperature loading but in many cases, much remains to be done.

In many cases, fairly simple damage models are still used by engineers for designing high temperature industrial components. Robust approaches are still to be developed which incorporate essential features of damage, as far as key physical mechanisms are concerned, when complex interactions exist between various forms of damage. Looking back at the progress achieved in the field of constitutive modelling since the 80's, one can be reasonably optimistic for the future progress of damage modelling in industry.

This International Conference brought together some 100 participants from ten countries including European countries, Japan and United States. 50 papers were presented and 32 were given orally. All the contributions, even oral ones, were exposed as posters in order to favour interaction between participants. The single session format and the poster sessions gave the opportunity for in depth discussion between delegates and for young doctorate students to interact with seniors. Lunches taken in a single room during the conference as well as an informal dinner on a boat trip on the River Seine brought in a warm atmosphere. Three overview lectures were given by R. Schafrik, from General Electric, A. Saxena from Georgia Tech and A. Bignonnet from PSA. Prof. H. Sehitoglu closed the symposium with an outline of the perspectives of research on Temperature-Fatigue Interaction.

The editors wish to thank all the authors and delegates for their contribution. After reviewing, 37 papers are finally presented in this volume which aims to become a helpful and valuable reference in the field of Temperature-Fatigue Interaction for scientists as well as for engineers.

The success of this event is due to the help of many people. We would like to thank the members of the International Committee and the Organising Committee, and the session chairmen: a number of them were really effective in the peer review process.

Special thanks are due to Mrs Veronique Matos, Dr Alain Köster, M. Yves Franchot, secretary of SF2M, and Mrs Chantal Ianarelli for their invaluable assistance in the preparation of the conference, including the web site, during the symposium, and for the editing of the proceedings.

Luc Rémy

Ecole des Mines de Paris, ARMINES, CNRS, Paris

Jean Petit,

ENSMA, CNRS, Poitiers/Futuroscope

Symposium Chairmen and Editors

Thermomechanical Behaviour

THERMO-MECHANICAL FATIGUE BEHAVIOR OF CAST 319 ALUMINUM ALLOYS

C.C. ENGLER-PINTO JR.[1], HUSEYIN SEHITOGLU[1], H.J. MAIER[2] and T.J. FOGLESONG[1]
[1] *Department of Industrial and Mechanical Engineering, University of Illinois, Urbana, IL 61801, U.S.A.*
[2] *Universität-GH Paderborn, Lehrstuhl für Werkstoffkunde, FB 10, Pohlweg 47-49, 33098, Paderborn, Germany*

ABSTRACT

Stress-strain behavior and durability of cast 319 aluminum-copper alloys are studied at high temperatures and under thermo-mechanical fatigue (TMF), exposing rate sensitivity and microstructural changes. The decrease in strength during cycling was attributed to the significant coarsening of the precipitates at high temperatures, which was confirmed with transmission electron microscopy. The results show that the stress-strain response is similar under out-of-phase (OP) and in-phase (IP) thermo-mechanical fatigue. However, TMF-IP fatigue lives are substantially lower compared to TMF-OP lives, which are very close to the isothermal low cycle fatigue (LCF) life obtained at a similar inelastic strain range. In fact, it is observed that TMF IP loading induces significant creep damage, while transgranular fracture predominates in all other testing conditions.

KEYWORDS

Thermo-mechanical fatigue, cast aluminum, fracture mechanism, stress-strain response, microstructural coarsening.

INTRODUCTION

The automotive industry has been facing the challenge to increase the engine efficiency and overall performance and at the same time deliver a vehicle that meets increased customer expectation for safety, fuel economy and price. The use of cast aluminum alloys has provided a significant reduction in weight, notably in the cylinder heads and blocks. To increase efficiency, however, the maximum operation temperature of these components has also increased from below 170°C in earlier engines to peak temperatures well above 200°C in recent engines [1].

Thermal gradients arising during transient regimes of start-up and shutdown operations produce a complex thermal and mechanical fatigue loading which limits the life of these components, especially on thinner sections, like the valve bridge area on the cylinder head. The isothermal low cycle fatigue (LCF) design philosophy is generally used for life prediction and residual lifetime assessment. However, the microstructure modifications and the crack initiation and propagation mechanisms may be different if the material is submitted to isothermal or non-isothermal fatigue. More accurate and reliable assessments under thermo-mechanical fatigue conditions are urgently needed to assist with the design and evaluation of components undergoing thermo-mechanical fatigue (TMF).

This paper investigates the stress-strain behavior and the fatigue life of the cast aluminum alloy Al319-T7B under thermo-mechanical fatigue and isothermal low cycle fatigue. This alloy is used in the fabrication of cylinder heads and blocks for automotive engines. Despite some investigations on the TMF stress-strain behavior [1-4] and on the room temperature fatigue behavior [5-7] of this class of alloys, a thorough analysis on the TMF and LCF lifetimes at high temperatures is still lacking.

MATERIALS AND EXPERIMENTAL METHODS

Material

The alloy investigated is a Al319 aluminum alloy, which presents an Al-Si-Cu microstructure and the nominal chemical composition given in Table 1. This is a secondary alloy (obtained by the remelting of aluminum alloys), which presents a higher iron (Fe) content – ~0.8% – as compared to a previously investigated primary alloy [1-4]– ~0.4%. Iron is an undesirable impurity, which decreases the feeding ability of the metal during casting and can reduce the ductility and toughness through the formation of brittle intermetallics. In order to differentiate both alloys, the secondary alloy in this paper is designated EAP319 and the primary alloy WAP319.

Both materials consist of precipitate hardenable alloys, where the primary strengthening phase is Al_2Cu. The alloys were submitted to the T7B heat treatment (solutionizing at 495°C for 8 hours followed by precipitating at 260°C for 4 hours) before testing. This treatment produces an overaged microstructure that confers thermal stability to the component.

Table 1. Nominal composition of EAP319 in weight percent.

Al	Si	Cu	Mg	Fe	Mn	Sr	Ti
Bal.	7.35	3.32	0.22	0.78	0.24	0.03	0.13

The samples were prepared from a sand-cast wedge with a copper chill positioned at the apex of the wedge. The wedge geometry results in different solidification rates, based upon a similar principle of the varied-cooling rate castings used in an earlier work [1]. This solidification control permits the machining of samples with controlled secondary dendrite arm spacing (SDAS) sizes, which replicates the solidification conditions and microstructure present in some critical locations of cast cylinder heads. All samples used in the present investigation were taken from the region where the SDAS is between 15–30 μm – solidification rate of approximately 2.5°C/sec. The samples were machined with a diameter of 7.6 mm and a gage length of 25.4 mm.

The TEM picture of the precipitates present after the T7B treatment for the EAP319 alloy is shown in Fig. 1(a). Note that the precipitates are mostly θ' and are located on [001] habit planes, as was previously observed for the WAP319 alloy [1, 2]. Fig. 1(b) shows the precipitate structure of the material after 45000 isothermal fatigue cycles (approximately 25 hours) at 300°C and 0.2% mechanical strain range ($\Delta\varepsilon_m$). The micrograph shows a much higher density of dislocations and that the precipitates have coarsened and approached an spheroidal morphology.

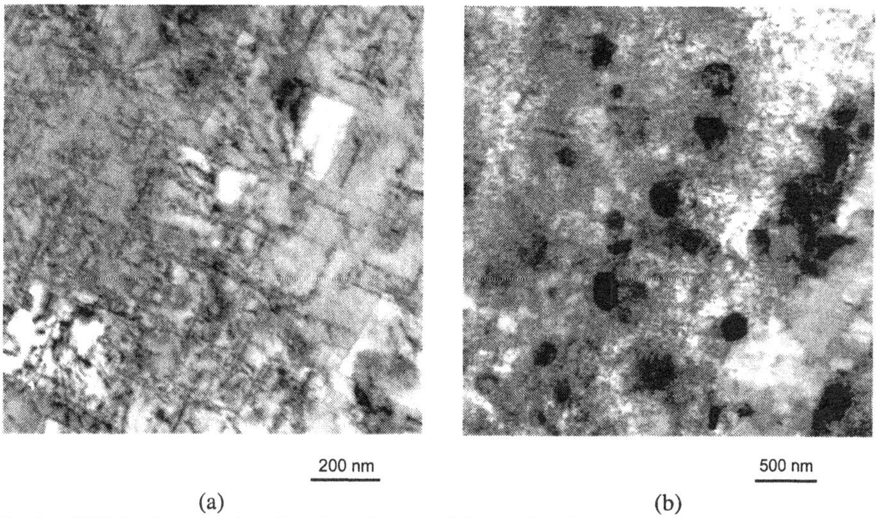

Fig. 1. TEM micrographs showing the precipitates in the EAP319-T7B microstructure: (a) untested sample and (b) after 45000 isothermal fatigue cycles (300°C, $\Delta\varepsilon_m = 0.2\%$).

Experimental Procedures

All isothermal and thermo-mechanical fatigue experiments were conducted under total strain control and constant strain rate. The isothermal fatigue experiments were performed at temperatures ranging from 20°C to 300°C with three different frequencies and strain rates: 40 Hz ($\sim 2\times10^{-1}$ s^{-1}), 0.5 Hz ($\sim 2\times10^{-3}$ s^{-1}) and 5×10^{-5} s^{-1}. A wide range of mechanical strain ranges ($\Delta\varepsilon_m$) was considered (0.2% to 2.0%) and a total of 51 isothermal tests were conducted, which include the room temperature tests.

The thermo-mechanical fatigue tests were conducted using a servo-hydraulic Instron testing machine in conjunction with a 15 kW Lepel induction heater. The temperature was measured using a Raytek non-contact infrared pyrometer. All thermo-mechanical fatigue experiments were conducted in total strain control. The strain and temperature waveforms followed a triangular wave-shape. Each TMF experiment was conducted at constant mechanical strain rate of 1.33×10^{-5} s^{-1} to 5×10^{-5} s^{-1}. The temperature range for all tests was 100-300°C, with a constant temperature rate during heating and cooling of 0.5°C/s to 1.33°C/s, depending on the applied mechanical strain range. Two TMF cycle types were considered in the present study: out-of-phase (OP), where the maximum mechanical strain occurs at the minimum temperature of the cycle, and in-phase (IP), where the maximum mechanical strain occurs at the maximum temperature. A total of 22 TMF tests were conducted with both EAP319 and WAP319 alloys.

RESULTS AND DISCUSSION

Cyclic Behavior

The stress-strain hysteresis lops for the out-of-phase and in-phase TMF tests, at different portions of the observed fatigue life, are presented in Fig. 2(a) and 2(b), respectively. The tests shown in Fig. 2 were performed at similar strain ranges (0.60% for TMF-OP and 0.54% for

TMF-IP), resulting in similar stress ranges for both OP and IP tests. However, because of the differences on the strain-temperature phasing, the alloy response is different in tension and in compression, resulting on a positive mean stress for the TMF-OP loading and on a negative mean stress for TMF-IP loading. Figure 2 also shows that the material softens cyclically, which is explained by the coarsening of precipitates, as shown previously on Fig. 1.

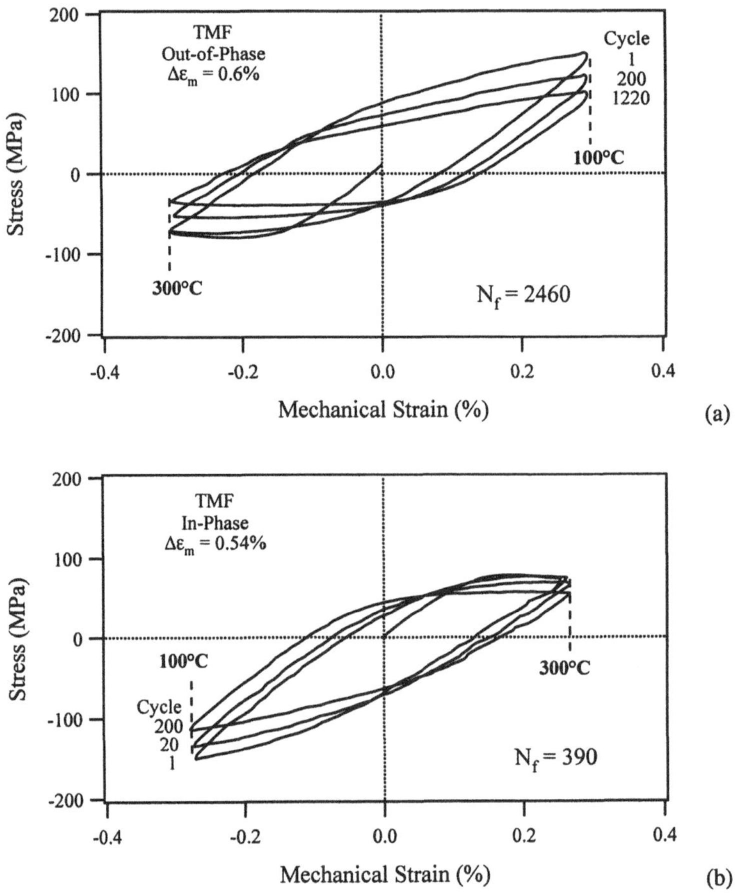

Fig. 2. Typical hysteresis loops obtained from TMF tests: (a) out-of-phase and (b) in-phase.

The TMF OP and IP loops may be plotted in a way to eliminate the difference between both phasing conditions, as shown in Fig. 3. In fact, by inverting the axes of the IP hysteresis (dashed lines) and plotting them together with the OP stress-strain loops, it is observed that the behavior is nearly the same under either OP or IP, which demonstrates that the deformation mechanisms are very similar for this alloy under either tension or compression. This observation is confirmed by the graph in Fig. 4, which plots the peak stresses in tension and in compression for all TMF tests. In fact, it is observed that the peak stresses occurring at the same temperature (100°C or 300°C) follow the same trend for both OP and IP tests.

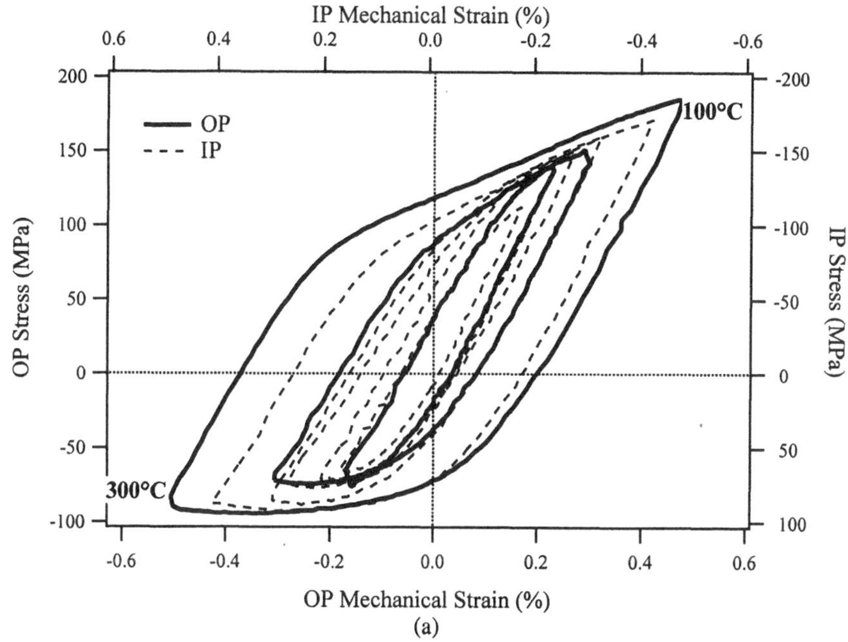

Fig. 3. Initial TMF hysteresis loops for different mechanical strain ranges (a) and inelastic strain ranges (b). Note that the axes for the in-phase loops are inverted to allow a direct comparison between them.

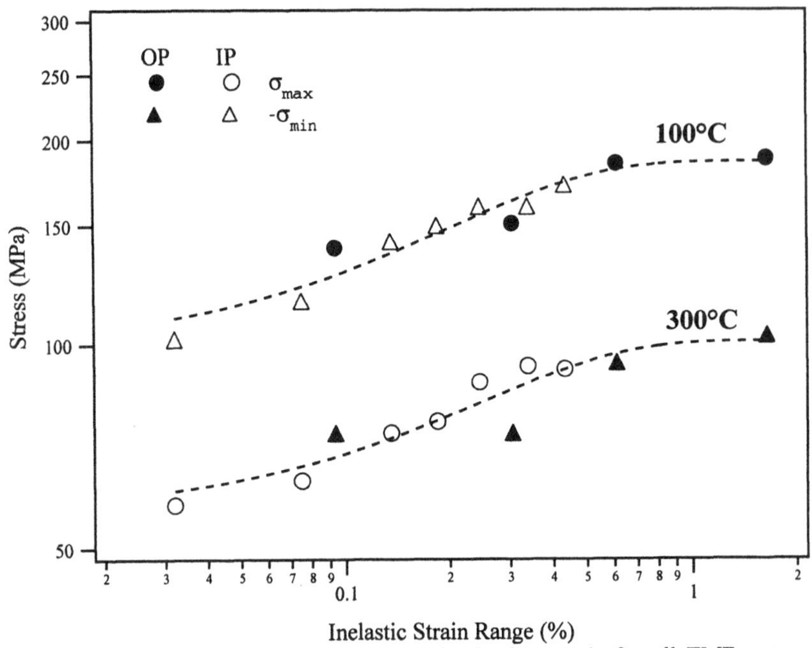

Fig. 4. Peak tensile and compressive stresses in the first cycle for all TMF tests with the EAP319-T7B alloy.

LCF and TMF Life

Figure 5 summarizes the TMF and isothermal LCF life results obtained for the EAP319-T7B alloy. The average room temperature fatigue life, which represents the maximum life, is also included for reference (dashed line). The high temperature results fall below the room temperature curve because damage by oxidation and/or creep diminishes the fatigue life. It can be observed from this plot that the TMF IP corresponds to the most critical loading condition for this class of alloys. We also note that the data can be more easily visualized with different plots presented below.

The stress-life plot, Fig. 6, clearly shows the effect of different temperatures and strain rates. The stress range for each test is taken at half-life, because there is no stabilized cycle for this material for the conditions investigated, as shown on Fig. 7. It can also be observed from Fig. 7 that the stress range for the IP test at half-life is higher than for the OP test performed at a similar mechanical strain range. That occurs just because the IP life is shorter, and it also explains why the lives for the IP are closer to the OP on a stress range basis.

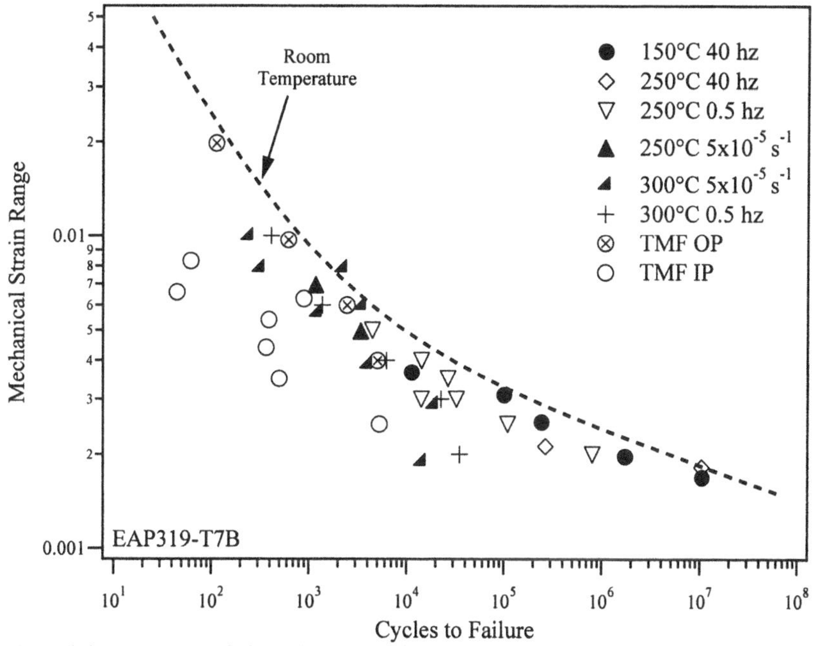

Fig. 5. High temperature fatigue life results for all tests with the EAP319-T7B alloy.

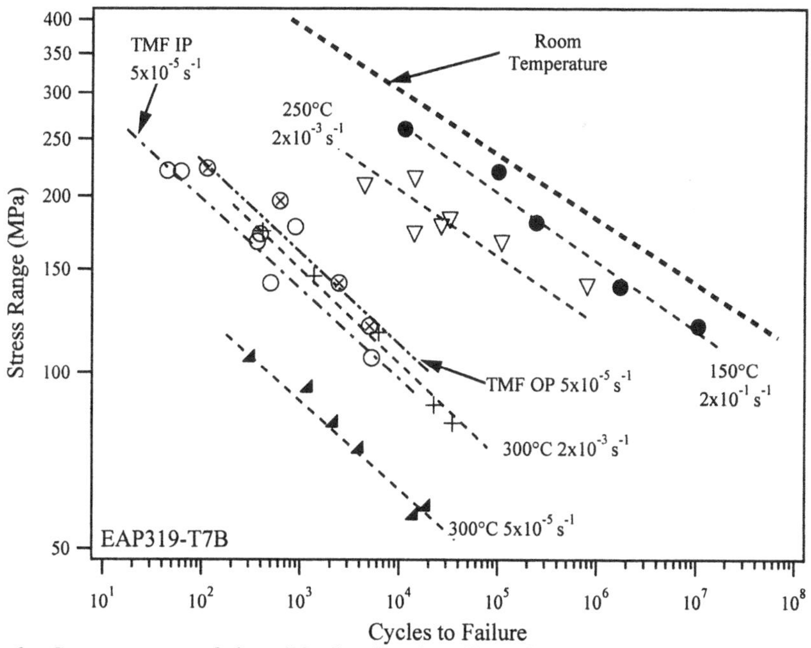

Fig. 6. Stress range *vs.* fatigue life showing the effect of temperature and strain rate for the EAP319-T7B alloy.

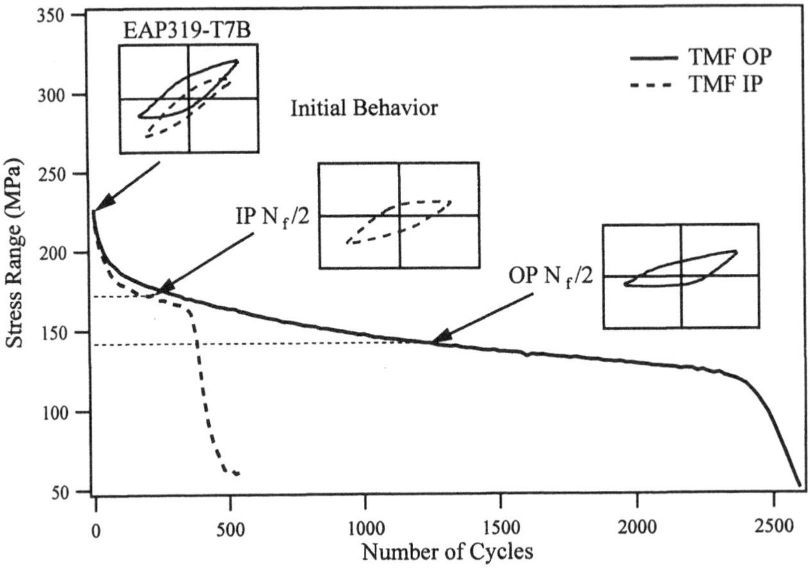

Fig. 7. Evolution of the stress ranges during out-of-phase and in-phase TMF tests under similar mechanical strain ranges for the EAP319-T7B alloy. Note that at half-life the stress ranges for OP and IP differ, while for the first cycle they are nearly identical.

The effect of strain rate at 300°C is also observed on cyclic stress-inelastic strain curve presented in Fig. 8. The higher stress range for the TMF is due to the variable temperature in this case. This figure also shows that the secondary alloy investigated in this paper, EAP319 is softer than its correspondent primary alloy, WAP319 under the same thermal treatment (T7B).

A better correlation for all fatigue lives obtained from different test conditions is presented in Fig. 9, which plots the inelastic strain range *vs.* the fatigue life. In fact, with the exception of the TMF-IP tests, all fatigue lives seems to be on a single Manson-Coffin curve. This leads to the conclusion that, under the test conditions reported here, the damage introduced by increasing the temperature or decreasing the strain rate consist mainly in increasing the inelastic strain range for the TMF-OP and LCF tests. The TMF-IP lives, however, are significantly shorter, which indicates that the inelastic strain range is not the only factor affecting the fatigue life. A more complete investigation on the fatigue life for these alloys, including fatigue life modeling, is under work and will be the subject of a future publication.

The analysis of the fracture surface of the tested samples revealed that transgranular fracture predominated for all TMF-OP and LCF tests, while the fracture surfaces for the TMF-IP tests appear to be intergranular. In addition, under TMF-IP loading, cracks may reside at the boundaries of large eutectic particles. This requires further investigation. Figures 10(a) and 10(b) show the fracture surfaces of two samples tested under TMF-OP and TMF-IP, respectively. The intergranular-like fracture on the TMF-IP tested samples is an indication that substantial creep damage occurs for this type of loading, what explains the shorter lives observed for the TMF-IP tests.

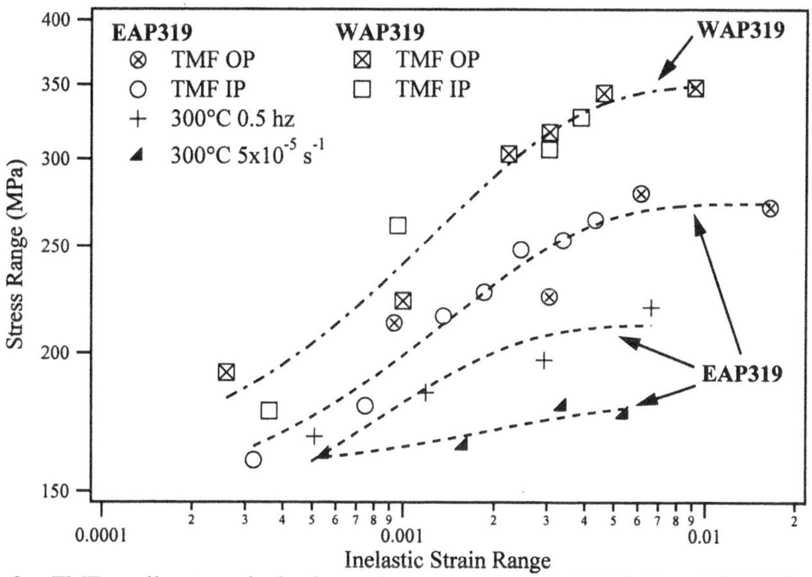

Fig. 8. TMF cyclic stress-inelastic strain curves for the WAP319 and EAP319 alloys compared to the isothermal behavior of the EAP319 alloy.

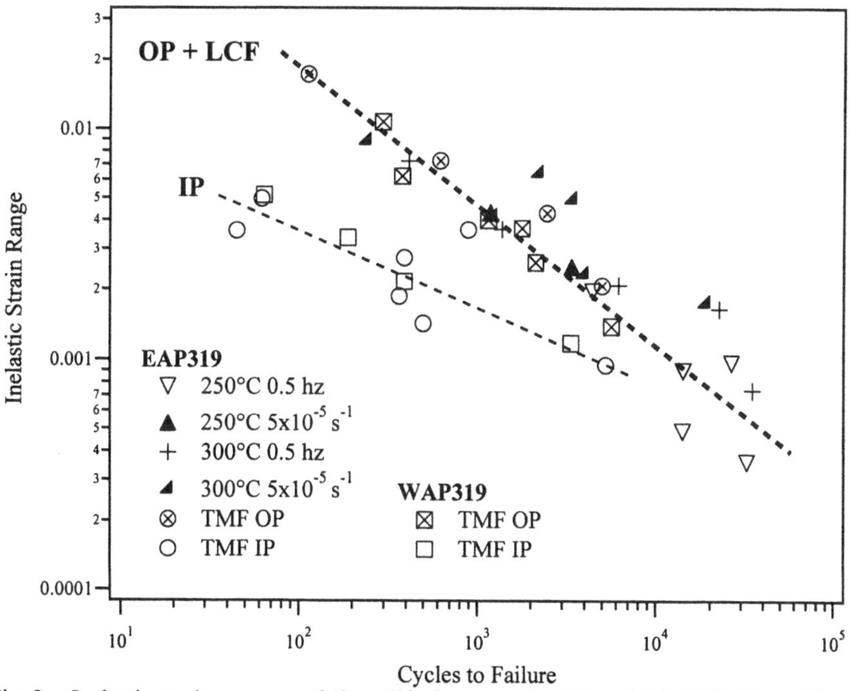

Fig. 9. Inelastic strain range *vs.* fatigue life for the WAP319 and EAP319 alloys. Note that TMF-IP lives are significantly shorter than TMF-OP lives.

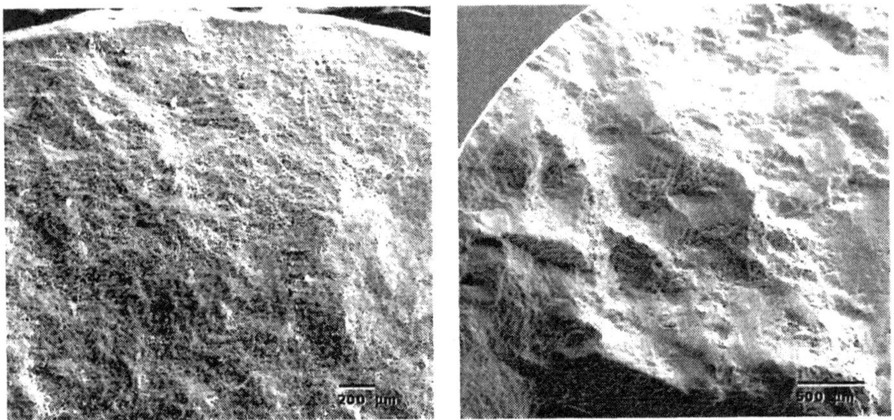

(a) (b)

Fig. 10. Fracture surface of EAP319 tested samples: (a) TMF-OP, $\Delta\varepsilon_m = 0.60\%$, $N_f = 2460$ cycles and (b) TMF-IP, $\Delta\varepsilon_m = 0.54\%$, $N_f = 390$ cycles.

SUMMARY

An extensive investigation of the high temperature fatigue behavior of the cast aluminum alloy A319-T7B has been carried out, including SEM and TEM analysis of the fractured samples. Isothermal low cycle fatigue and thermo-mechanical fatigue tests were conducted at temperatures up to 300°C. The results of this work can be summarized as follows:

1. The TMF stress-strain behavior is identical for both IP and OP loading conditions. However, based on the mechanical or inelastic strain range, the TMF-IP lives are substantially shorter than the TMF-OP lives.
2. The secondary alloy (EAP319) is softer than the primary alloy (WAP), but the TMF lives are very similar on mechanical or inelastic strain basis.
3. Creep damage dominates for TMF-IP loading and large strain range, providing a partial explanation for the shorter lives in TMF-IP compared to the other loading conditions.

ACKNOWLEDGEMENTS

The authors would like to thank Dr. John E. Allison and John V. Lasecki, from Ford Research Laboratory, Dearborn, for their support of this research. Dr. Carlos Engler-Pinto is also grateful to *Fundação de Amparo à Pesquisa do Estado de São Paulo* – FAPESP, São Paulo, Brazil, for the partial post-doctoral fellowship granted. Some of the isothermal tests were conducted at Westmoreland Laboratory and AMTEL. The SEM investigations were completed at the Center for Microanalysis of Materials at the University of Illinois, which is supported by the United States Department of Energy under Grant No. DEFG02-91-ER45439.

REFERENCES

1. Smith, T.J., Maier, H.J., Sehitoglu, H., Fleury, E., and Allison, J. (1999) *Metall. & Mater. Trans. A* **30**, pp.133-146.
2. Sehitoglu, H., Smith, T.J., and Maier, H.J. (2000) In: *Thermo-Mechanical Fatigue Behavior of Materials: Third Volume, ASTM STP* **1371**, pp. 53-68, Sehitoglu, H., and Maier, H.J. (Eds.), American Society for Testing and Materials, West Conshohocken, PA, USA.
3. Sehitoglu, H., Qing, X., Smith, T.J., Maier, H.J., and Allison, J.A. (2000) *Metall. & Mater. Trans. A* **31**, pp.139-151.
4. Smith, T.J., Sehitoglu, H., Qing, X., and Maier, H.J. (1998) In: *Low Cycle fatigue and Elasto-Plastic Behaviour of Materials*, pp. 167-172, Rie, K.-T., and Portella, P.D. (Eds.), Elsevier.
5. Ting, J.C., and Lawrence, F.V., Jr. (1993) *Fatigue Fract. Engng. Mater. Struct.* **16**, No. 6, pp. 631-647.
6. Dabayeh, A.A., and Topper, T.H. (2000) *Fatigue Fract. Engng. Mater. Struct.* **23**, pp. 993-1006.
7. Caton, M.J., Jone, J.W., Boileau, J.M., and Allison, J.E. (1999) *Metall. & Mater. Trans. A* **30**, pp. 3055-3068.

LOW CYCLE FATIGUE BEHAVIOUR OF DUPLEX STAINLESS STEELS AT HIGH TEMPERATURES

S. HEREÑÚ[1], I. ALVAREZ – ARMAS[1], A. ARMAS[1],
A. GIRONÈS[2], L. LLANES[2], A. MATEO[2] and M. ANGLADA[2]

1. Instituto de Física Rosario, CONICET, Universidad Nacional de Rosario, Rosario, (Argentina)
2. Dept. de Ciència dels Materials i Enginyeria Metal·lúrgica, Universitat Politècnica de Catalunya, Barcelona, (Spain)

ABSTRACT

Cyclic hardening-softening response has been investigated for two duplex stainless steels (DSSs), with different nitrogen content, in the range of temperatures from 200 °C to 500 °C. It is reported the existence of an inverse strain rate sensitivity as well as an abnormal cyclic stress hardening and both effects are associated to DSA (Dynamic Strain Ageing). Analysis of the substructural evolution suggests that the ferritic phase develops a preponderant role in the cyclic response of DSSs in the DSA temperature range. Finally, a comparison of the cyclic response of the two DSSs shows that the level of nitrogen influences the magnitude of cyclic hardening but not the qualitative behaviour of DSSs in the range of temperature investigated.

KEYWORDS

Duplex stainless steels, austenite, ferrite, nitrogen alloying, dynamic strain ageing, low cycle fatigue

INTRODUCTION

Since the early 1980s, the usage of duplex stainless steels (DSSs) has experienced a rapid growth due to the outstanding combination of strength and corrosion resistance that they exhibit as compared with standard austenitic steels [1-2]. The main evolution in the chemical composition of DSSs has been the progressive increase in the nitrogen content which leads to improvements in yield stress, corrosion resistance and weldability. Thus, nowadays DSSs may be classified in three commercial generations: a first one given by DSSs whose nitrogen content is lower than 0.10 wt%; a second one consisting of DSSs with a percentage in nitrogen between 0.10 and 0.20 wt%; and a third one, usually referred to as superduplex, corresponding to materials with more than 0.20 wt% in nitrogen.

Cyclic deformation response at elevated temperatures is an important aspect for materials used in industrial equipments such as gas turbines, boilers and nuclear reactors. In the case of DSSs, it is well known that when these materials operate at temperatures between 250 °C and

500 °C, they are susceptible to experience "475 °C embrittlement". As a consequence, mechanical strength and hardness increase, but at the same time ductility and toughness are reduced, limiting the temperature range of safe operation. Thermal ageing effects on the cyclic response of DSSs at room temperature have been extensively studied [3]. Systematic studies on the cyclic properties at relatively high temperatures (up to 600 °C) for ferritic and austenitic stainless steels are also available in the literature [4-7]. However, information about cyclic behaviour of DSSs at high temperatures is scarce.

The behavior exhibited by single phase stainless steels in this range of temperatures is usually related with the operation of DSA (Dynamic Strain Ageing) phenomena. Due to the austeno-ferritic nature of the duplex alloys, it was expected that they exhibit this phenomenon too. Recent works of the present authors confirm this hypothesis [8-10]. In this sense, tensile tests at temperatures ranging from 275 to 475 °C were performed in a superduplex stainless steel type UNS S32750. The dependence of yield stress and ultimate tensile strength on temperature indicates a significant effect of DSA. In order to evaluate the influence of strain rate on this phenomenon, tests at two different strain rates were conducted, both at 325 °C, temperature at which DSA was found to be maximum for this material. The results showed that the flow stress had an inverse strain rate sensitivity which established the existence of DSA in the steel under study [10].

On the basis of the above mentioned results, the present investigation is devoted to analyze the cyclic deformation behaviour of DSSs at temperatures from 200 °C to 500 °C. In doing so, special attention is paid to clarify the effects of DSA on the mechanical response of DSSs at elevated temperatures.

EXPERIMENTAL PROCEDURE

The materials used in this investigation are a first generation DSS type DIN W. Nr 1.4460 and a superduplex stainless steel UNS S32750 (commercial designation SAF 2507). They were supplied in the form of cylindrical bars of 20 mm in diameter, by Metales Villares S.A., Brasil, and Sandvik AB, Sweden, respectively. Table 1 gives their chemical composition in weight percent. The microstructure of DIN W. Nr 1.4460 consisted of islands of austenite within a ferritic matrix (72% volume fraction), while the superduplex one consisted of 56% austenitic grains within a ferritic matrix and the grain shape is highly elongated in the rolling direction. This direction corresponds with the loading axis in all the specimens tested.

Table 1. Chemical composition in weight percent and volume fractions of the constitutive phases for each of the DSS studied.

	Cr	Ni	Mo	Mn	Si	C	N	α/γ (%)
DIN W. Nr 1.4460	24.1	4.8	1.49	1.29	0.43	0.03	0.04	72/28
UNS S32750	25.0	7.0	3.79	0.4	0.34	0.01	0.26	44/56

Cylindrical specimens were machined from the bars. Low cycle fatigue tests were conducted under total strain control at a frequency of 0.1 Hz. Two different total strain ranges were considered: $\Delta\varepsilon_T = 0.57\%$ and $\Delta\varepsilon_T = 0.85\%$. The experimental procedure followed consisted of heating from room temperature to the working temperature and holding at this temperature for 30 minutes before starting the test, just to ensure the thermal stability of the system. After this period of thermal stabilization, the temperature of the bulk of the specimen was initially controlled with a thermocouple, which was directly connected to the specimen but outside the gauge length.

Thin foils of 3 mm diameter were prepared using double jet-polishing technique at 0 °C with a 90% vol. ethanol – 10% vol. perchloric acid electrolyte. Dislocation structures developed during cyclic loading within each of the constitutive phases were examined in a transmission electron microscope (TEM).

RESULTS AND DISCUSSION

Mechanical response

In Fig. 1, the cyclic stress response of DIN W. Nr 1.4460 in the temperature range 200 – 500 °C is represented for a total strain range of 0.57%. Three different cyclic response trends can be distinguished. At 200 °C, a rapid hardening is followed by a saturation stage that remains during the main part of the fatigue test. Finally, a secondary hardening stage occurs before fracture. At 300 °C and 400 °C, the cyclic stress response is characterized by a pronounced cyclic hardening rate during the early stage of fatigue life (about 100 cycles) which is followed by a slow hardening. At 500 °C, the initial hardening is less pronounced and it gives way to a noticeable softening stage.

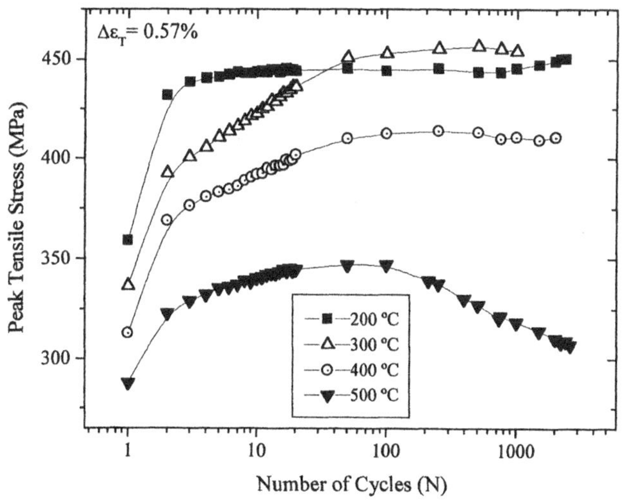

Fig. 1. Influence of temperature on the cyclic stress response at a total strain range of 0.57% for the duplex stainless steel type 1.4460.

Qualitatively similar responses were observed for fatigue tests at a total strain range of 0.85%, as can be seen in Fig. 2. Nevertheless, it is worthwhile to remark that the cyclic hardening rate developed at 300 and 400 °C is more pronounced at this higher imposed strain.

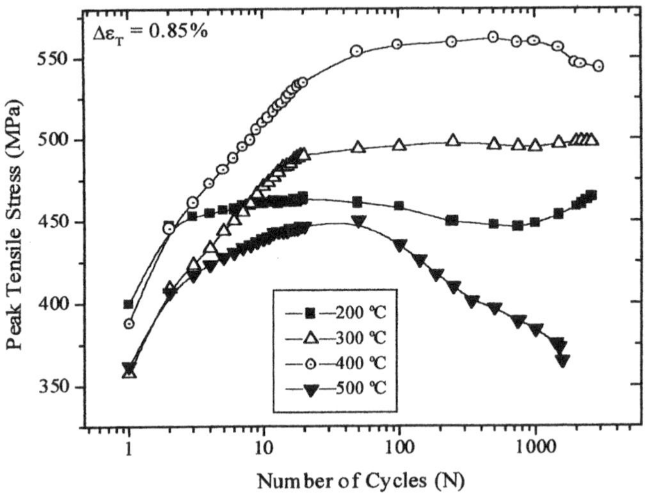

Fig. 2. Influence of temperature on the cyclic stress response at a total strain range of 0.85% for the duplex stainless steel type 1.4460.

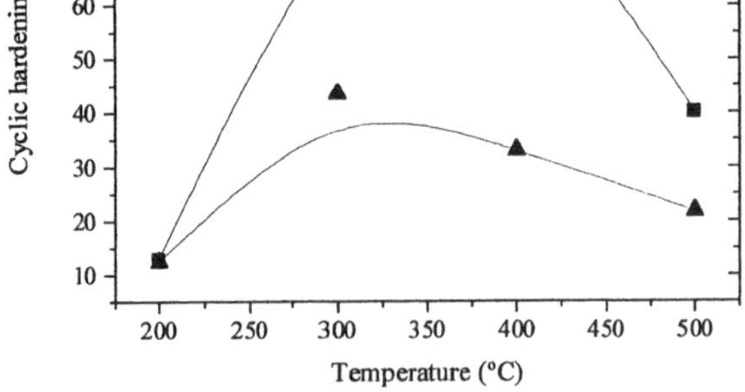

Fig. 3. Effect of temperature on cyclic hardening of DIN W. Nr 1.4460 tested at total strain ranges of 0.57% and 0.85%, respectively.

The initial cyclic hardening response of DIN W. Nr 1.4460 tested at the previously referred total strain ranges ($\Delta\varepsilon_T$ = 0.57% and $\Delta\varepsilon_T$ = 0.85%) has been evaluated for the different temperatures (200 – 500 °C) in terms of $\Delta\sigma$, which is defined as the difference of stresses reached during the cycles number 2 and 20 (Fig. 3). The value of N = 2 was considered because it is the first cycle immediately after a reversal, whereas N = 20 has been chosen because, in all tests performed, a maximum hardening is attained at this value of fatigue life.

From Fig. 3, it is clear that cyclic hardening is more pronounced at the highest imposed strain, as previously shown. At the same time, it can be noted that both curves present a maximum within the range 300 - 400 °C. This range should correspond to the temperature at which this material exhibits maximum DSA effect. This fact can be explained in terms of density of dislocations necessary to accommodate the strain imposed during the test. When increasing the strain imposed, the density of dislocations increases and therefore a higher stress is necessary to unblock them while DSA mechanisms are active.

In order to study the sensitivity of the material to the strain rate, cyclic tests at a total strain range of 1.14% were performed within the range of temperatures from 20 °C to 500 °C. The experimental procedure was the following: the sample was allowed to reach approximately 300 cycles with a strain rate of 2.3×10^{-3} s^{-1}. This number of cycles corresponds to an almost stabilized peak tensile stress (σ_1) in most of the tests performed at different temperatures. Then, without interrupting the test, the strain rate was lowered to 2.3×10^{-4} s^{-1}. The specimen was submitted to cyclic loading at this strain rate until reaching a saturation state, whose stress was designed as σ_2. Afterwards, again without stopping the test, the strain rate was changed to 2.3×10^{-5} s^{-1} to obtain a correspondent stress value σ_3.

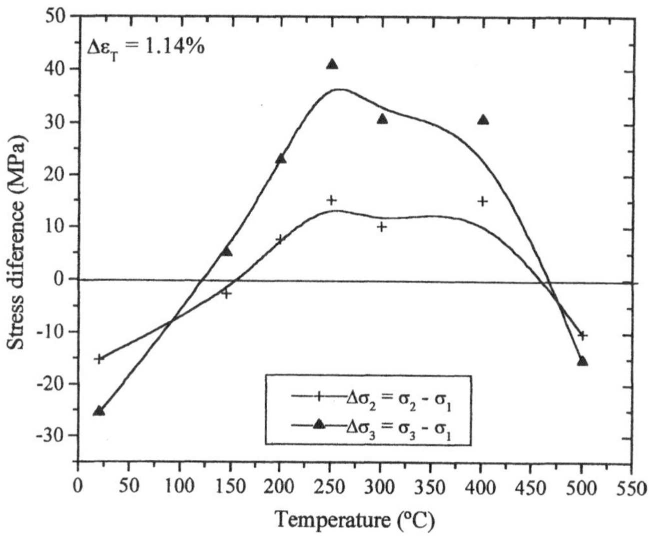

Fig. 4. Strain rate sensitivity measured as the stress difference of DIN W. Nr 1.4460 tested at a total strain range of 1.14%.

In Fig. 4, the difference of stresses was represented ($\Delta\sigma_2 = \sigma_2 - \sigma_1$ and $\Delta\sigma_3 = \sigma_3 - \sigma_1$). At the lowest temperatures (25 – 150 °C), the material exhibits a normal behaviour, i.e. lower stresses for lower strain rates. At temperatures between 150 and 450 °C, the difference of stresses plotted is positive which reveals that the steel presents inverse strain rate sensitivity (higher stresses correspond to lower strain rates). Here it is also worthwhile to remark that the inverse strain rate sensitivity and the abnormal initial cyclic hardening behaviour occur within the same range of temperatures.

On the other hand, cyclic stress response at 400 °C and 475 °C of DIN W. Nr 1.4460 and UNS S32750 were compared (Fig. 5). The hardening-softening curves are qualitatively the same for both steels, but it can be noted that stress levels are higher for UNS S32750. In order to explain this fact, it must be kept in mind that the DSSs under comparison contain different nitrogen levels. Higher nitrogen content induces a marked strengthening in austenite. Therefore, the material with the higher nitrogen content will exhibit higher strength properties, i.e. higher stresses will be necessary to accommodate the strain imposed.

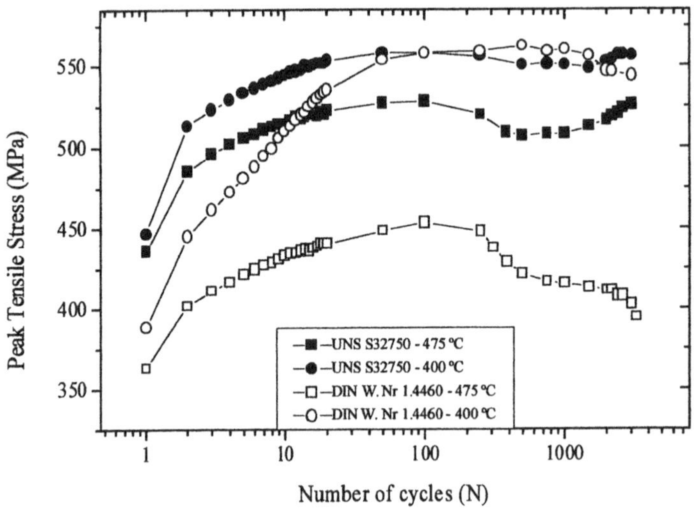

Fig. 5. Cyclic behaviours of DIN W. Nr 1.4460 and UNS S32750 fatigued at total strain range of 0.85% and for two temperatures (400 - 475 °C).

Both materials experience similar cyclic responses at 400 °C, which can be described as strong hardening during the first cycles of life. However, an interesting point to remark is that the first generation DSS (DIN W. Nr 1.4460) exhibits higher cyclic hardening at all temperatures studied. For rationalizing this finding it must be taking into account the higher volume fraction of ferrite in this duplex (72%), because ferrite is the constituent phase which develops the most important role in the DSA phenomenon at these temperatures [7].

At 475 °C, the cyclic behaviours of these two alloys are different in the final part (more than 300 cycles) of the tests: the superduplex exhibits a period of secondary hardening, whereas DIN W. Nr 1.4460 cyclically softens.

Substructural features

DSA phenomena can be described as a discontinuous gliding of dislocations, so in order to evaluate the effect of DSA within each of the constitutive phases of the DSSs, the dislocation substructures developed during fatigue testing at a total strain range of 0.85% and for two different temperatures (400 and 475 °C) have been studied.

Fig. 6. Dislocation structure in a ferritic grain of DIN W. Nr 1.4460 cycled at $\Delta\varepsilon_T = 0.85\%$ and at 400 °C.

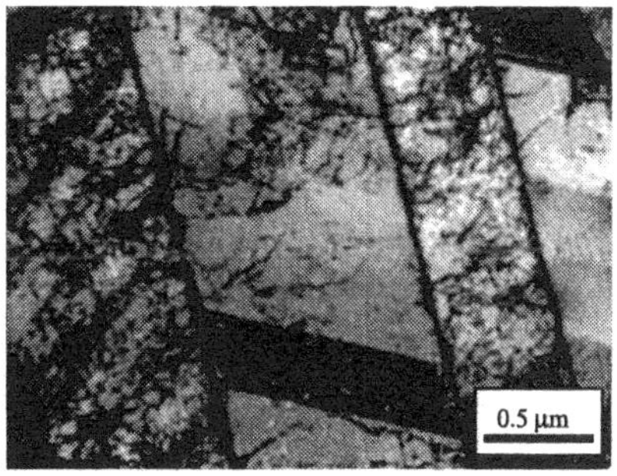

Fig. 7. Dislocation structure in an austenitic grain of DIN W. Nr 1.4460 cycled at $\Delta\varepsilon_T = 0.85\%$ and at 400 °C.

Figures 6 and 7 show the typical dislocation arrangement in ferritic and austenitic grains of DIN W. Nr 1.4460 tested at a total strain range of 0.85% at 400 °C. The dislocation density is considerably high and it assumes a configuration of uncondensed and poorly defined walls in the ferritic grains (Fig. 6). However, neither cell structures nor walls and channels are observed, whilst they are the most common situation in ferrite when DSSs are cyclically deformed at room temperature [11-12]. On the other hand, stacking faults and arrangements of entangled dislocations coexist in austenitic grains (Fig. 7). It is worthwhile to point out that the fatigue-related dislocation substructure in the austenitic phase does not vary noticeable with temperature in the range from 200 °C to 400 °C.

For the same strain range and temperatures, the substructural features have been analyzed for UNS S32750 alloy (Figs. 8 and 9). Despite the enhanced planar-slip promoted by its high nitrogen content, a tendency to a wavy glide is observed in austenitic grains (Fig. 9). This behaviour may be attributed to the DSA phenomenon that mainly affects the ferritic phase and therefore, once the ferrite is plastically saturated, austenite has to contribute to the accommodation of the plastic strain. So, austenite is forced to change its mode of plastic deformation from a planar to a wavier glide.

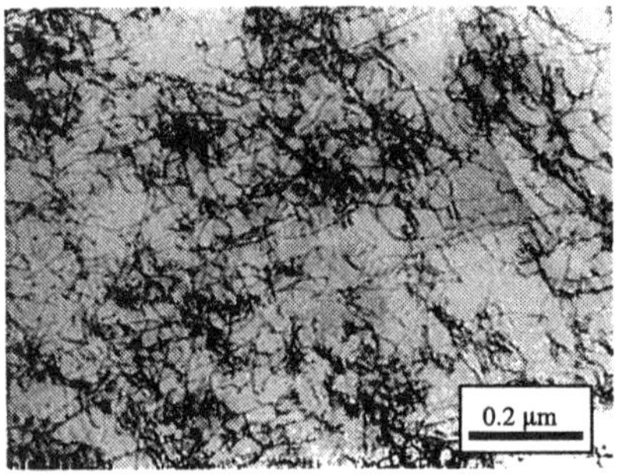

Fig. 8. Dislocation structure in a ferritic grain of UNS S32750 cycled at $\Delta\varepsilon_T = 0.85\%$ and at 400 °C.

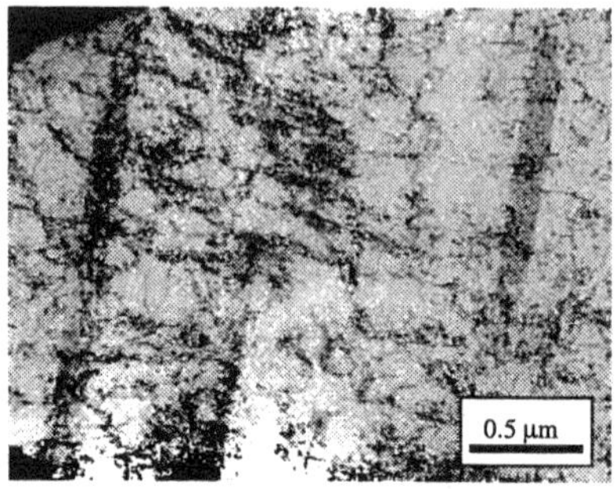

Fig. 9. Dislocation structure in an austenitic grain of UNS S32750 cycled at $\Delta\varepsilon_T = 0.85\%$ and at 400 °C.

CONCLUSIONS

On the basis of the results reported in this work, the following conclusions are drawn:

- The cyclic hardening-softening response of duplex stainless steels strongly depends on temperature in the range 200 - 500 °C.
- The pronounced cyclic hardening is an experimental indication of the operation of dynamic strain ageing processes. The maximum effect of this phenomenon is observed at temperatures between 300 and 400 °C.
- The substructural evolution in both constitutive phases has been analyzed and the observed features sustained that ferrite is mainly affected by the DSA.

Acknowledgements

The financial support of Spanish CICYT (Grant MAT99-0781) and *Comissionat per a Universitats i Recerca de Catalunya* (ACI-2000-26) is kindly acknowledged for the Spanish group. The authors owe their gratitude to Sandvik AB (Sweden) for supplying SAF 2507.
The Argentinean group wants to acknowledge to *Consejo Nacional de Investigaciones Científicas y Técnicas* CONICET- PIP N° 197/98 and *Agencia Nacional de Promoción Científica Y Tecnológica* – ANPCYT – PICT N° 12 – 03287.

REFERENCES

1. Gunn, R.N. (1997). In: *Duplex Stainless Steels*, Abington Publishing, Cambridge, p. 175.
2. Johansson, K. (2000). In: *Proc. World Duplex 2000 Conference*, Ed. Associazione Italiana di Metallurgia, p. 18.
3. Iturgoyen, L. and Anglada, M. (1997). *Fat. Fract. Engng. Mater. Struct.*, **20**, (5), p. 645.
4. Christ, H.-J.; Wamukwamba, C.K. and Mughrabi, H. (1999). In: *Proc. 7th Int. Fatigue Congress*, Ed.: X.R. Wu and Z.G. Wang, **4**, p. 2165.

5. Tjong, S.C. and Zhu, S.M. (1997). *Metallurgical and Materials Transactions A*, **28 A**, p. 1347.
6. Ilola, R.; Kemppainen, M. and Hänninen, H. (1999), *Materials Science Forum*, **318-320**, p. 407.
7. Srinivasan, V.S.; Sandhya, R.; Valsan, M.; Bhanu Sankara Rao, K.; Mannan, S.L. and Sastry, D.H. (1997). *Scripta Materialia*, **37**, (10), p. 1593.
8. Hereñú, S.; Alvarez-Armas, I. and Armas, A.F. (2000). In: *Proc. World Duplex 2000 Conference*, Ed. Associazione Italiana di Metallurgia, p. 681.
9. Hereñú, S.; Gironès, A.; Alvarez-Armas, I.; Armas, A.; Mateo, A. and Anglada, M. (2001). *Anales de Mecánica de la Fractura*, **18**, p. 73.
10. Gironès, A.; Mateo, A.; Llanes, L. and Anglada, M. (2001). *Rev. Metal. Madrid,* **37**, p. 150.
11. Mateo, A.; Gironès, A.; Keichel, J.; Llanes, L.; Akdut, N. and Anglada, M. (2001). *Mater. Sci. Engng. A314*, p. 176.
12. Mateo, A.; Llanes, L.; Iturgoyen, L. and Anglada, M. (1996). *Acta Materialia*, **44**, p. 1143.

Temperature-Fatigue Interaction
L. Rémy and J. Petit (Eds.)

VALIDATING THE PREDICTIVE CAPABILITIES: A KEY ISSUE IN MODELLING THERMOMECHANICAL FATIGUE LIFE

H.J. MAIER[1] and H.-J. CHRIST[2]

[1] *Lehrstuhl für Werkstoffkunde, Universität Paderborn, Pohlweg 47-49, D-33098 Paderborn, Germany*

[2] *Institut für Werkstofftechnik, Universität Siegen, Paul-Bonatz-Str. 9-11, D-57068 Siegen, Germany*

ABSTRACT

Virtually all models developed to predict thermomechanical fatigue (TMF) behaviour are limited by the fact that life prediction is based on experimental data generated under conditions significantly different from the actual service conditions of most high-temperature components. The current research strongly supports the idea that life models that are closely related to the relevant microstructural processes provide a more reliable basis for life prediction. It will be discussed how microstructural arguments can be used to extend a given life model to loading situations that cannot be simulated satisfactorily in the laboratory and/or estimate the limits within which reliable life predictions can be made.

Data are presented that demonstrate that non-conservative life prediction may result despite a seemingly excellent correlation of model predictions with experimentally obtained TMF life. It is emphasized that TMF tests designed to validate a life model should focus on revealing the presence of potential couplings between the various damage mechanisms such as creep, environmental degradation and cyclic plasticity.

KEYWORDS

crack initiation, crack growth, cyclic stress-strain response, damage evolution, environmental effects, life prediction, microstructure, modelling, thermomechanical fatigue

INTRODUCTION

Operating conditions of many high-temperature components include severe thermal transients and mechanical strain cycles, and thus, thermomechanical fatigue (TMF) often is the life-limiting degradation mode. This has triggered considerable research effort directed towards predicting life under such conditions, and various life prediction methodologies have evolved from this.

TMF tests are time consuming and require expensive test equipment. Consequently, these tests usually have to be conducted under conditions much more severe than those encountered by an actual component. It is, however, long known that cyclic stress-strain response, crack initiation and fatigue crack growth can vary significantly depending on the actual loading conditions, see *e.g.* Refs [1, 2]. From this it should be clear that the predictive capabilities of any TMF life model that relies on data generated under laboratory conditions are a key issue. Still, few studies [3–6] have addressed this topic due to the complexity involved.

In a related study [6] five widely used life models were modified to allow for life prediction under TMF loading conditions, and their predictive capabilities were then evaluated. Figure 1 taken from this study was chosen as it nicely illustrates the problems encountered in validating a life model. The TMF life data shown in Fig. 1 were obtained on a particulate-reinforced aluminium alloy. The hot extruded material was produced via a powder-metallurgy route from rapidly solidified gas-atomized powders. Nominal composition of the material was Al-8Fe-4Ce, in wt pct. In the reinforced material 12.5 vol. pct SiC particles with an average diameter of 3 μm were present. For life prediction input data from isothermal low-cycle fatigue (LCF) and creep tests were used. Creep-damage governs life in all *in-phase* TMF tests analysed, and thus, models such as the Linear Accumulation of Time Dependent Damage (AC) [7, 8] or the Strain Rate-Modified Accumulation of Creep-Damage (SRM) [9] can be expected to yield reliable life prediction results. In fact, as seen in Fig. 1 the *in-phase* TMF tests marked #1, #2, #4, #6 and #8, are all predicted reasonably well both by the AC and the SRM approach.

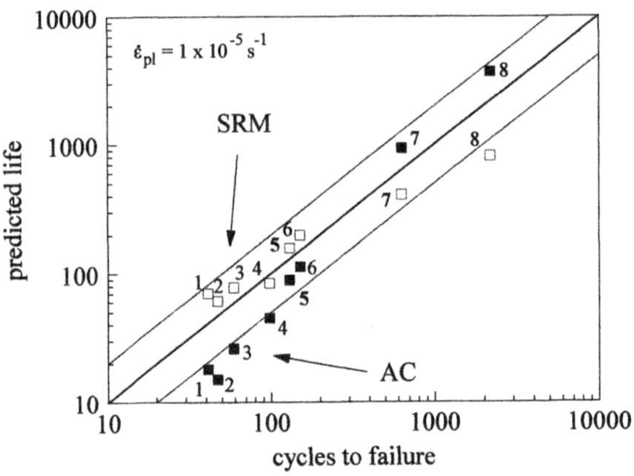

Figure 1: TMF life prediction results obtained on an SiC-reinforced aluminium alloy [6]. See main text for details.

Careful metallography has shown that creep-damage is negligible in the *out-of-phase* TMF tests shown in Fig. 1. Still, life prediction results obtained for the *out-of-phase* TMF tests labelled #3, #5, #7 appear to be satisfactory for both models. This is a result of the fact that for the specific loading conditions used, the amount of creep damage calculated for these tests by chance almost equals the additional amount of damage that resulted from mean stress effects and a coupling between environmental degradation and cyclic plasticity in the out-of-phase TMF tests. Clearly, life prediction for a real component could be drastically non-conservative if the damage mechanism assumed in the model is different from the dominating one in the real component. It should also be clear that this is not a criticism of the specific models employed to predict life in Fig. 1. The data shown in Fig. 1 were selected only in order to demonstrate that a TMF test *per se* does not always allow to validate a life model. As discussed in detail in Ref. [6] analysis of life prediction data obtained for AISI 304 L stainless steel has revealed that similar problems exist for this material as well.

It has long been argued that models that are closely linked with microstructure should be superior to purely empirical ones. In recent years the present authors and their co-workers have used various microstructure-related models in order to predict TMF life for a variety of materials [10–13]. The objective of the current study is *not* to promote a specific model, but to stimulate further research in this field.

Life prediction relies both on accurate knowledge of the cyclic stress-strain response for given loading conditions and on some kind of damage criterion that evolves with the number of cycles. The approach taken in the present paper is to first address both aspects separately. In a later section, however, coupling effects between cyclic plasticity and certain damage mechanism will be discussed, which can drastically affect damage evolution. Finally, first suggestions for designing TMF tests intended to validate life models will be given.

MICROSTRUCTURAL APPROACHES TO MODELLING OF THERMO-MECHANICAL FATIGUE LIFE

An in-depth discussion of all aspects that govern cyclic stress-strain response under TMF loading conditions is clearly beyond the scope of the present paper. In the following only the major effects will be focused upon. In this section materials where stress-strain response is dominated by coarsening of microstructural constituents will be discussed first. Then, alloys where a change in slip mode governs cyclic stress-strain response will be addressed. Finally, both aspects will be combined in order to demonstrate how microstructural arguments can be used to extrapolate a stress-strain model to long-term service conditions that cannot be simulated accurately in the laboratory.

Cyclic stress-strain response

Various approaches have been proposed to model the cyclic stress-strain response under TMF loading conditions, and models that employ constitutive equations have made substantial progress in the last decade. The major drawback of such models is the large number of parameters involved. Usually, these parameters are established from tension tests and isothermal low-cycle fatigue tests, respectively. Often TMF tests are then used as means to verify the predictive capabilities of the model.

In a recent study [11] a stress-strain model that is based on unified equations was used to predict the stress-strain response of a cast 319-T6 aluminium alloy under TMF loading. Specimens were machined from a sand-cast wedge that allowed for machining of samples with a pretermined secondary dendrite arm spacing. The material, which had a nominal composition of Al-7.4Si-3.3Cu-0.22Mg-0.38Fe-0.24Mn (in wt pct), was first solutionized for 8 hours at 495° C and then given a T6 heat treatment (5h/190° C) in order to achieve a peak-aged condition. In this material substantial microstructural changes occur during TMF as the precipitates present are not thermally stable at elevated temperatures. This in turn affects the cyclic stress-strain response, and drastic cyclic softening is observed after few cycles. Figure 2 demonstrates that the cyclic softening is predicted quite accurately with the model employed. As explained in detail in Ref. [11], the good correlation between predictions and experimentally obtained data results largely from the incorporation of microstructural observations into the model.

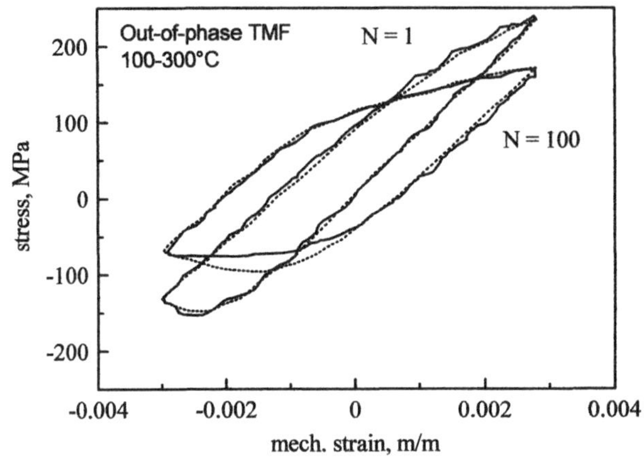

Figure 2: Evolution of cyclic stress-strain response in a TMF test on Al319-T6 at a strain rate of 5×10^{-5} s^{-1}. The material had a secondary dendrite arm spacing of ≈ 30 μm. Solid and dashed lines refer to experiment and simulation, respectively. For details see Ref. [11].

In the present paper, only one example of this will be presented briefly. One key parameter in the model is the drag stress K, which is a state variable and represents the size of the stress surface. This parameter can vary in a complex fashion with time, temperature and number of cycles. In alloys such as Al319, the parameter K is mostly affected by changes in precipitate spacing. Figure 3 illustrates the microstructural effects that can be achieved by extended ageing at elevated temperatures. Obviously, the lowest value of K is obtained in an almost dislocation free sample with a drastically overaged microstructure. This data can then be used to model the temperature and cycle dependence of the drag stress under long-term elevated temperature service conditions in a reliable and numerically stable manner [11, 14]. It should be clear that a microstructure such as the one shown in Fig. 3b provides a suitable means for extrapolating the model but needs not to occur at all in the actual component during service loading. In a similar manner the other parameters used in the model were correlated with microstructural observations. Pronounced microstructural changes during high-temperature exposure dominate stress-strain response not only in aluminium alloys but other materials as well, and can be modelled in a similar fashion through changes in state variables [15].

It is emphasized that such a microstructural approach to modelling does not depend on the actual stress-strain model employed. However, microstructural observations are easier to incorporate into certain types of stress-strain models. Multi-component models are a case in point as these models are inherently linked closely to microstructure [16–19]. For the two materials discussed next a multi-component model was employed for modelling of stress-strain response [10]. Again, focus will not be on the details of the model but on the use of microstructural observations for modelling in general.

If the TMF cycle is not dominated by high homologous temperatures, there are still situations were isothermal and thermomechanical loading can produce significantly different

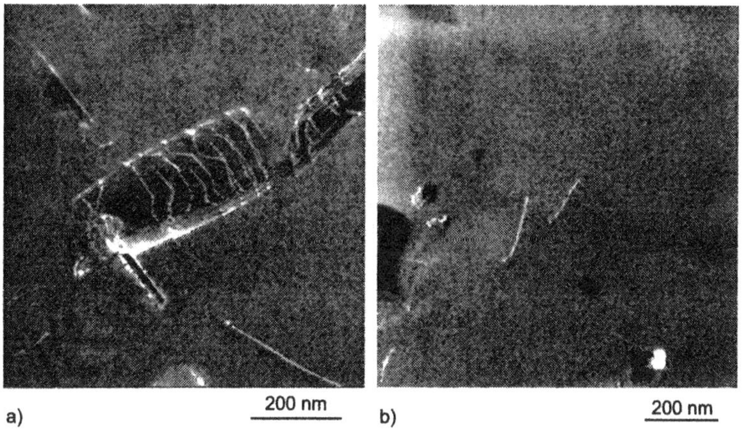

Figure 3: TEM g/3g weak-beam dark-field images showing (a) dislocation particle interaction in peak-aged Al319 aluminium alloy and (b) dislocations that move almost unhindered by precipitates after the material has been drastically overaged at 250° C for 1000 h [14].

stress-strain responses. An example are changes in dislocation slip mode. One of the best characterized materials in this respect probably is AISI 304 stainless steel. Zauter *et al.* [20] have shown that in isothermal tests planar dislocation slip dominates at test temperatures around 400° C due to dynamic strain ageing. If the test temperature is significantly below or above this temperature, dislocation slip mode becomes wavy. The simplest approach to modelling is to assume that the stress-strain response at any given temperature along a TMF hysteresis loop is identical to that of the corresponding isothermal fatigue test. Dislocation rearrangements are, however, not very rapid. Consequently the dislocation arrangement in an individual TMF cycle will not change back and forth between truly planar dislocation slip and wavy dislocation slip. Instead, a unique dislocation arrangement will often be established in the TMF test that does not change significantly within the cycle [20, 21].

These microstructural findings can be used to predict stress-strain response for loading situations not covered by the actual TMF experiment. In the following this will be demonstrated for high-temperature titanium alloy IMI 834 (nominal composition Ti-5.8Al-4.0Sn-3.5Zr-0.7Nb-0.5Mo-0.35Si-0.06C, in wt pct). The material had a bimodal microstructure consisting of 15 vol. pct primary α phase embedded in a lamellar transformed β matrix. Average grain size of the primary α phase was about 14 μm. In this material critical resolved shear stress (CRSS) is quite different for the various slip systems at low temperatures, and thus, planar dislocation slip prevails for test temperature up to about 600° C. Above this transition temperature the differences in CRSS are diminished and dislocation slip becomes more wavy. Indeed, transmission electron microscopy (TEM) [22] has revealed that *independent of the actual test mode*, planar dislocation slip dominates as long as the test temperature does not exceed 600° C. As expected a TMF test conducted with a maximum temperature of the cycle ($T_{\max}$) that does not significantly exceed the transition temperature is satisfactorily predicted based on isothermal LCF data only [10]. The idea of an almost constant microstructure within an individual TMF cycle can then be used to predict the cyclic stress-strain response for cases were $T_{\max}$ significantly exceeds the transition temperature. For such cases the high-temperature part of the TMF cycle should dominate the microstructural arrangement, *i.e.* wavy dislocation slip is expected and microstructures should be established that cannot convert back to planar arrangements during the low temperature part of the cycle. Figure 4a shows the cyclic stress-strain response observed during an in-phase TMF test devised to test this hypothesis. In this specific test the sample was cycled until a stabilised stress-strain response was obtained. Then, temperature cycling was interrupted at the minimum temperature of the TMF cycle ($T = 400°$ C) while strain cycling continued. As seen in Fig. 4a the stress-strain response is drastically different from the one obtained in a conventional isothermal LCF test run at 400° C. It is also evident from Fig. 4a that the dislocation rearrangement is indeed a slow process.

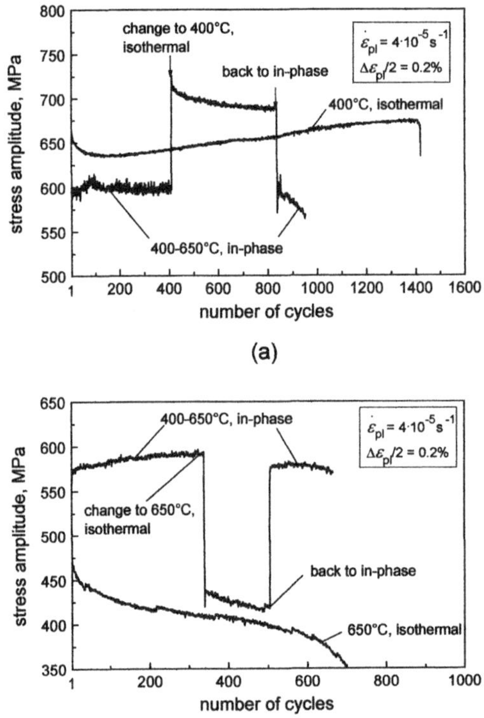

Figure 4: Cyclic stress-strain response in interrupted in-phase TMF tests as compared to isothermal LCF tests [22]. See main text for details.

In another TMF test temperature cycling was interrupted at 650° C while strain cycling continued. In this case the stress-strain response after temperature cycling was interrupted closely resembled that from a conventional isothermal LCF test run at 650° C, *cf.* Fig. 4b. This indicates that wavy dislocation slip does indeed dominate stress-strain response in such a TMF test. In the following it will briefly be demonstrated how such microstructural observations can be used to predict cyclic stress-strain response under long-term service conditions not covered in the actual TMF tests. For an extended discussion the reader is referred to Ref. [12].

Firstly, high-temperature titanium alloy IMI 834 normally is used with a bimodal microstructure that represents the best compromise between fatigue and creep properties [23]. During long-term high-temperature exposure, however, the bimodal microstructure displays significant microstructural instability [24]. The main effects are coarsening of the silicides and deterioration of the lamella boundaries. Similar to the case of the Al319 aluminium alloy, cyclic stress-strain response of a suitably degraded microstructure can be used to obtain model parameters characteristic of material that has experienced long-term high-temperature service conditions. For IMI 834 such a microstructure is an equiaxed one, *i.e.* one without lamella boundaries [12].

Another concern in predicting cyclic stress-strain response from data generated in the laboratory are stress state effects. In a TMF test one usually tries to obtain an uniaxial stress state. In real components, however, multiaxial loading conditions are often dominant. Again, microstructural arguments can be used to assess the validity of the model predictions. Doong *et al.* [25] have demonstrated that materials that deform by wavy dislocation slip will form identical microstructures both under proportional and nonproportional multiaxial loading. By contrast, materials that exhibit planar slip in a uniaxial test will display significant extra hardening in multiaxial fatigue tests as the formation of multislip dislocation arrangements is forced under such conditions. From Fig. 4 it is evident that depending on the actual dislocation arrangement stress amplitude can vary by as much as 100 MPa.

Clearly, any model that predicts cyclic stress-strain response for IMI 834 under multiaxial loading conditions will fail if the model is based on uniaxial data generated at temperatures where planar slip prevails. Based on the microstructural observations a solution to this would be to first force formation of multislip dislocation arrangements by conducting a conventional LCF test at temperatures where wavy dislocation slip dominates ($T \gtrapprox 650°$ C). This will introduce a microstructure similar to the one present under multiaxial loading conditions. Next, cyclic stress-strain response of the specimen with the modified microstructure is recorded for the actual test conditions of interest. From this, the model parameters relevant for multiaxial loading are then established from the cyclic stress-strain response in the first few cycles that followed the change of test conditions.

Damage evolution

Life models that incorporate a physically measurable quantity of damage appear to provide the most versatile and reliable basis for life prediction under complex loading situations [26]. Crack length provides such a physically-based definition of damage and microcrack propagation models have been applied successfully to predict TMF life for a variety of materials, see *e.g.* Refs [12, 13, 26]. The preceding section has shown that actual loading mode can have substantial effects on the microstructural evolution, and both crack initiation and crack growth depend on the microstructural state. Consequently, a close correlation between the model and the relevant microstructural processes and damage mechanisms should allow for a better assessment of the predictive capabilities and/or help to define the limits of applicability of the life model.

In the following focus will be once more on high-temperature titanium alloy IMI 834 as the difference between short-term TMF tests and long-term service loading conditions can be illustrated quite easily in this system. In a related study [12], a microcrack propagation model has been proposed to predict TMF life for IMI 834. Basically, the model predicts life by integrating a crack propagation law. For the purpose of the current paper it is sufficient to note that all such models are very sensitive to initial crack length. Hence, it is most important to know how the initial crack length depends on the actual loading conditions.

IMI 834 has a bimodal microstructure, *i.e.* one that consists of small primary α grains embedded in a lamellar transformed-β matrix. In *isothermal* LCF tests cracks initiate preferentially within the primary α grains as long as planar dislocation slip dominates [12]. Within the planar slip band crack initiation is rapid, and thus, initial crack size a_0 equals the primary α grain size ($\approx$ 14 μm), *cf.* Fig. 5a. For an in-phase TMF test conducted in the temperature regime where planar dislocation slip dominates one would expect that the cracks also initiate within the primary α grains. However, as seen in Fig. 5b the fatal cracks tend to form within the lamellar transformed-β matrix, and thus, the initial crack size is significantly larger ($\approx 25 - 30$ μm). Note that crack like features are evident in Fig. 5b within the primary α grains as well, but these do not extend across the grain boundaries. For a detailed discussion see Ref. [12]. Within the scope of the present paper it is sufficient to emphasize that the loading condition can affect crack initiation mechanism. Clearly, one needs this microstructural information from the TMF tests to accurately incorporate the crack initiation mechanisms into the life model. As already discussed, multiaxial loading tends to suppress planar slip. Thus, similar to TMF, multiaxial loading should favour crack initiation in the larger lamellar transformed-β grains despite the fact that uniaxial LCF tests indicate the opposite for $T \lessapprox 600°$ C.

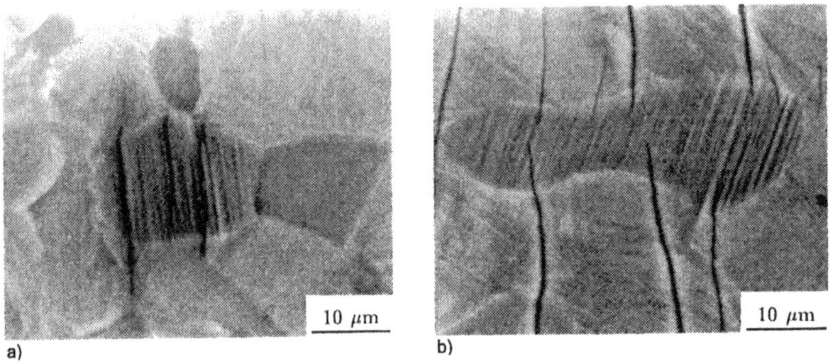

Figure 5: SEM micrographs [12] showing crack initiation representative for (a) isothermal LCF tests performed at $T \leq 400°$ C and (b) an in-phase TMF test conducted between 400° C and 600° C. Plastic strain amplitude was 2×10^{-3} in both cases, and external stress axis is horizontal.

The second issue that is relevant in this context is the fact that titanium alloys do form a non-protective oxide layer. In high-temperature tests ($T \gtrapprox 600°$ C) oxygen can diffuse rapidly into the material and cause severe embrittlement of the subsurface layer. Consequently, both of the crack initiation mechanisms shown in Fig. 5 may be irrelevant if the actual component is exposed to high temperatures for an extended period of time. In this case initial crack size will be given by the thickness of the oxygen-embrittled subsurface layer.

DISCUSSION

In isothermal high-temperature fatigue and even more so in TMF the experiment is rarely performed under conditions that closely resemble those of the actual component. TMF is different in this respect from many other fields. Clearly, a good correlation between model predictions and experimental data alone is not a sufficient indication of the predictive capabilities of a life model, *cf.* Fig. 1.

The examples given in the present paper have shown that knowledge of the relevant microstructural processes can be used in various ways in order to improve prediction of cyclic stress-strain response under conditions not covered by the actual experiments. Modelling of cyclic stress-strain response has made substantial progress over the last decades, and it appears that nowadays an increasing number of researchers successfully correlate their models with microstructural observations. From the results presented for high-temperature titanium alloy IMI 834 it is clear that deformation behaviour, *e.g.* planar *vs* wavy dislocation slip, can also effect damage evolution through changes in crack initiation mechanism or effects on crack growth rate. This effect of cyclic stress-strain response on damage evolution can be implemented in a life model in a rather straightforward manner. Similarly, enhancement of oxygen-uptake by cyclic plasticity that is observed for high-temperature titanium alloys such as IMI 834 [12] can be accounted for as the effect can be analysed easily by microhardness measurements [27]. It is noted that a coupling between cyclic plasticity and oxidation has been reported for many materials, see *e.g.* Refs [12,13,28–31]. For most materials, however, it is difficult to analyse this experimentally.

It appears that the real challenge in validating life prediction models is to describe the complex couplings that may be present between the various damage mechanisms. A detailed analysis of the TMF tests conducted on the particulate-reinforced aluminium alloy [13] revealed that the damage mechanisms operating under TMF conditions are strongly coupled. However, a quantitative assessment of the coupling itself was not yet possible. Analysis of the TMF life data available for AISI 304 L-type stainless steel indicated that for certain loading conditions similar problems do exist for this material as well [6].

Whenever a strong coupling between the various damage mechanisms occurs, non-conservative life prediction may result if this is not accounted for in the life model. At present, however, accurate physical modelling of such effects is lacking. It is proposed that whenever TMF tests are conducted to validate a certain life model these tests should be designed such as to reveal potential couplings. From Fig. 1 it is obvious that the test parameters used in the specific TMF tests shown were not suitable to validate the models employed for life prediction. Clearly, the actual test parameters of a validation test are most important. As a first step, it is suggested that TMF tests should be designed such that no single damage mechanism dominates. For such test conditions coupling between the various damage mechanisms may become apparent. If current life models are to be improved significantly, a close link of the models with microstructural observations appears to be indispensable. At present, however, there is no broad consensus on this issue, and it is hoped that the current study stimulates more research in this field.

At least for some cases it has been demonstrated in this study that a close correlation with microstructure does allow for extending the models to situations not covered in the actual experiments. The additional amount of microstructural characterization needed may appear as a major drawback of this approach. Within a certain class of materials, however, similar approaches may be applied. Indeed, related work indicated that a great many of the relevant processes identified in high-temperature titanium alloy IMI 834 also operate in a titanium aluminide.

ACKNOWLEDGEMENT

Financial support of this study by Deutsche Forschungsgemeinschaft is gratefully acknowledged.

References

[1] Miller, D. A. and Priest, R. H. (1987). In: *High Temperature Fatigue: Properties and Life Prediction*, pp. 113-175, Skelton, R.P. (Ed.). Elsevier Applied Science, London.

[2] Ellison, E.G. and Al-Zamily, A. (1994). *Fatigue Fract. Engng. Mater. Struct.* **17**, 39.

[3] Danzer, R. (1988). *Lebensdauerprognose hochfester metallischer Werkstoffe im Bereich hoher Temperaturen*, Gebr. Borntraeger, Berlin.

[4] Ellyin, F. (1997). *Fatigue damage, crack growth and life prediction*, Chapman & Hall, London.

[5] Bernstein, H. L. (1982). In: *Low-cycle fatigue and life prediction, ASTM STP 770*, pp. 105-134, Amzallag, C., Leis, B.N. and Rabbe, P. (Eds). ASTM, PA.

[6] Maier, H.J., Teteruk, R.G. and Christ, H.-J. (2000). *Mater. at High Temp.*, in review.

[7] Taira, S., Fujino, M. and Ohtani, R. (1979). *Fatigue Eng. Mater. Struct.* **1**, 495.

[8] Spera, D. A. (1969). *The calculation of elevated temperature cyclic life considering low cycle fatigue and creep*, NASA TN D-5317.

[9] Danzer, R. and Bressers, J. (1986). *Fatigue Fract. Engng Mater. Struct.* **9**, 151.

[10] Maier, H.J. and Christ, H.-J. (1997). *Int. J. Fatigue* **19**, Supp. 1, 267.

[11] Sehitoglu, H., Qing, X., Smith, T., Maier, H.J. and Allison, J.A. (2000). *Metall. Mater. Trans. A* **31A**, 139.

[12] Maier, H.J., Teteruk, R.G. and Christ, H.-J. (2000). *Metall. Mater. Trans. A* **31A**, 431.

[13] Jung, A., Maier, H.J. and Christ, H.-J. (2000). In: *Thermo-mechanical Fatigue Behavior of Materials: Third Volume, ASTM STP 1371*, pp. 167-185, Sehitoglu, H. and Maier, H.J. (Eds). ASTM, West Conshohocken.

[14] Smith, T., Maier, H.J., Sehitoglu, H., Fleury, E. and Allison, J. (1999). *Metall. Mater. Trans. A* **30A**, 133.

[15] Sehitoglu, H. (1989). *Trans. ASME, J. Eng. Mater. Techn.* **111**, 192.

[16] Mughrabi, H. (1987). *Mater. Sci. Eng.* **85**, 15.

[17] Holste, C. and Burmeister, H.-J. (1980). *phys. stat. sol. (a)* **57**, 269.

[18] Polak, J. and Klesnil, M. (1982). *Fatigue Eng. Mater. Struct.* **5**, 19.

[19] Maier, H.J. and Christ, H.-J. (1996). *Scripta Mater.* **34**, 609.

[20] Zauter, R., Petry, F., Christ, H.-J. and Mughrabi, H. (1993). In: *Thermomechanical Fatigue Behavior of Materials, ASTM STP 1186*, pp. 70-90. Sehitoglu, H. (Ed.). ASTM, Philadelphia.

[21] Zauter, R., Christ, H.-J. and Mughrabi, H. (1994). *Metall. Mater. Trans. A* **25A** 407.

[22] Pototzky, P., Maier, H.J. and Christ, H.-J. (1998). *Metall. Mater. Trans. A* **29A**, 2995.

[23] Neal, D.F. (1985). In: *Titanium Science and Technology*, pp. 2419-2424, Lütjering, G., Zwicker, U. and Bunk, W. (Eds). DGM, Oberursel.

[24] Borchert, B. and Daeubler, M.A. (1988). In: *Proc. 6th World Conf. on Titanium*, pp. 467-472, Lacombe. P., Tricot, R. and Béranger, G. (Eds). Les éditions de physique, Les Ulis.

[25] Doong, S.-H., Socie, D.F. and Robertson, I.M. (1990). *Trans. ASME, J. Eng. Mater. Techn.* **112**, 456.

[26] Miller, M.P., McDowell, D.L., Oehmke, R.L.T and Antolovich, S.D. (1993). In: *Thermomechanical Fatigue Behavior of Materials, ASTM STP 1186*, pp. 35-49, Sehitoglu, H. (Ed.). ASTM, Philadelphia.

[27] Liu, Z. and Welsch, G. (1988). *Metall. Trans. A* **19A**, 527.

[28] Antolovich, S.D. (1982). In: *Pressure vessels and piping: design technology–1982 A decade of progress*, pp. 533-540, Zamrik, S.Y. and Dietrich, D. (Eds). ASME, New York, NY.

[29] Reger, M. and Rémy, L. (1988). *Metall. Trans. A* **19A**, 2259.

[30] Karayaka, M. and Sehitoglu, H. (1991). *Metall. Trans. A* **22A** , 697.

[31] Esmaeili, S., Engler-Pinto Jr, C.C., Ilschner, B. and Rézai-Aria, F. (1995). *Scripta Metall. Mater.* **32**, 1777.

Temperature-Fatigue Interaction
L. Rémy and J. Petit (Eds.)

HIGH TEMPERATURE FATIGUE AND CYCLIC CREEP OF P91 STEEL

L. KUNZ and P. LUKÁŠ
Institute of Physics of Materials, Academy of Sciences of the Czech Republic,
Žižkova 22, 616 62 Brno, Czech Republic

ABSTRACT

Cyclic plastic response of martensite ferritic 9%Cr steel was investigated at 600°C under symmetrical and asymmetrical loading. Cyclic softening was found to be the main characteristic cyclic deformation feature. The tensile mean stress promotes the cyclic plasticity and results in cyclic creep. Fatigue loading with 2Hz frequency reduces the creep rate when compared to sustained loading at maximum stress. An exponential relation between the instantaneous plastic strain amplitude and the cyclic creep rate for constant mean stress was experimentally determined.

KEYWORDS
9%Cr steel, fatigue, cyclic creep, cyclic plasticity, mean stress, dislocation structure.

INTRODUCTION

Tempered martensite ferritic 9% Cr steels with excellent creep resistance were developed for high temperature applications mainly in power generating industry. Their creep properties has been thoroughly studied and well documented [1]. Outstanding creep resistance is determined by characteristic microstructure of tempered martensite. The microstructure consists of elongated cell or subgrains within the former austenite grains. Carbides situated along the prior austenite grain and tempered lath boundaries effectively stabilize this structure. Under sustained constant loading at high temperature this microstructure exhibits excellent long-term stability.

High temperature fatigue resistance of 9%Cr steels has been studied to a less extent than the creep behaviour, though the fatigue is frequent in many industrial applications. Similarly to the creep studies the main attention was paid to the degeneration of the strengthened microstructure due to mechanical loading. It was found that combined fatigue with creep loading (performed as hold periods inserted into fatigue loading) produces more expressive microstructural changes than the pure creep or than the isothermal aging for equivalent times [2]. The typical fatigue behaviour under strain controlled loading is rapid initial softening followed by gradual softening until macroscopic crack appears [3, 4]. The cyclic softening is more pronounced for high strain amplitudes. The dislocation density generally decreases with increasing number of cycles and dislocations move to the cell boundaries. Both the increase of temperature and the stress amplitude result in an increase of plastic strain amplitude and shorter lifetime. Cyclic stress strain curve, i.e. the dependence of the stress amplitude on plastic strain amplitude for 50 percent of cycles to failure, $N_{f/2}$, was shown to obey the power-law relationship [5].

Cyclic plastic deformation depends on broad variety of loading parameters, among others on loading history. Generally simplifications are made when studying the complex relation between material response to the cyclic loading and changes of microstructure. Because the majority of studies of the cyclic plasticity was performed at constant amplitude loading (usually of sine or triangle waveform) with constant frequency and at symmetrical stress or strain loading the knowledge concerning the cyclic plastic behaviour and microstructural changes is limited mainly to the fatigue loading characterized by the above mentioned conditions. One of the important fatigue parameters strongly influencing the fatigue life is the mean stress. For stress controlled tests with tensile mean stress evolution of unidirectional strain, i.e. the cyclic creep was observed [6]. Though it is generally known that the tensile mean stress shortens the fatigue life and the compression mean stress has an opposite effect, the cyclic plastic behaviour under mean stress was studied rather scarcely. The majority of papers report a decrease of cyclic plastic amplitude with increasing mean stress. On the other hand a week influence or even an increase of cyclic plasticity with increasing mean stress was also reported [7].

The aim of this paper is to clarify the relation between unidirectional and cyclic plastic behaviour of a 9%Cr steel at high temperature under asymmetrical cyclic loading.

MATERIAL AND EXPERIMENTAL PROCEDURE

Experiments were performed on martensite ferritic 9%Cr1Mo steel P91. The chemical composition in wt. percent is given in Tab.1. The heat treatment consisting of austenitization at 1060°C for 1 h and air-cooling followed by tempering at 760°C for 2 h resulted in ultimate tensile strength of 730 MPa and the yield stress $\sigma_{0.2} = 550$ MPa at room temperature. Round bar

Tab.1. The chemical composition of 9%Cr steel in wt.%.

C	Mn	Si	P	S	Cu	Ni	Cr	Mo	V	Nb	N	Al	Fe
0.09	0.56	0.021	0.009	0.009	0.05	0.46	8.36	0.86	0.20	0.06	0.065	0.007	bal.

type specimens with gauge section of 6 mm in diameter and 20 mm gauge length were machined from rods in longitudinal direction. Load controlled uniaxial fatigue tests were carried out in a 10 kN servohydraulic fatigue machine equipped with a temperature chamber at the temperature 600°C in laboratory air. The frequency of loading was 2Hz with the exception of hysteresis loop recording. This was performed at a frequency lowered by a factor 100. Fatigue tests were started at least 1 h after the temperature stabilization. Full load amplitude was applied from the first loading cycle. Two sets of tests with different initial stress amplitudes were performed. Load symmetrical tests with initial mean load zero and asymmetrical tests with initial tensile mean stress 71 MPa. Strain was detected by a clip gauge extensometer outside the temperature chamber attached to drawbars connected with the specimen. The accuracy of the plastic strain measurement was 2×10^{-5}. The stress and strain were detected continuously and hysteresis loops were recorded in suitable intervals. The cyclic creep strain was determined from the shift of hysteresis loops along the strain axis. Repeated replacement of the clip gauge was necessary in the case of higher cyclic creep strains.

RESULTS

Cyclic deformation curves for symmetrical loading for three stress amplitudes are shown in Fig.1. At the very beginning of the cyclic loading, more precisely for the first five to ten cycles, an inexpressive transition hardening can be detected. A short region of nearly stable cyclic plastic behaviour follows. Continuous cyclic softening is the typical cyclic deformation feature for the decisive fraction of lifetime. This is better documented in Fig.2 where the plastic strain amplitude is presented in dependence on the relative number of cycles to fracture N/N_f. All specimens failed by fatigue crack propagation.

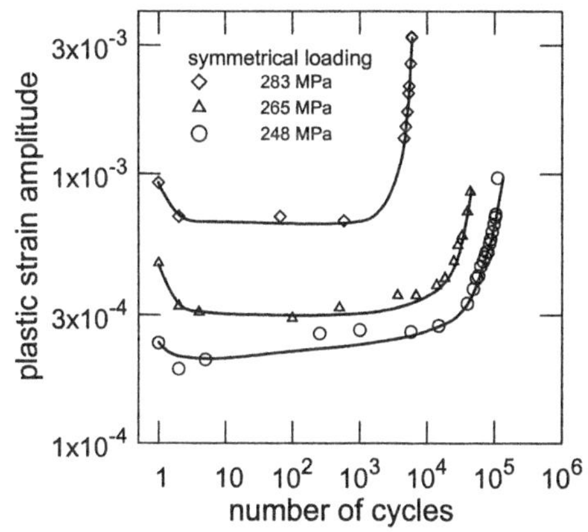

Fig.1. Cyclic deformation curves for symmetrical loading.

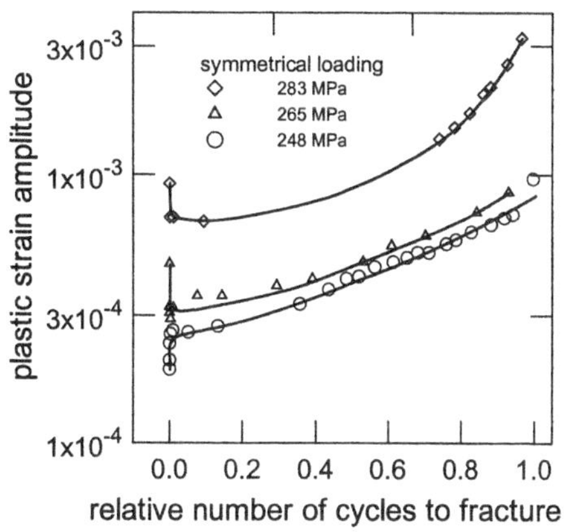

Fig.2. Cyclic deformation curves in relative co-ordinates.

The main difference between the cyclic plastic response under tensile mean stress and under symmetrical loading is the development of unidirectional strain; moreover, a change of the mechanism of fracture can be stated. The increase of unidirectional strain due to cycling i.e. the cyclic creep strain in dependence on the number of cycles for the tensile mean stress 71 MPa and different stress amplitudes is presented in Fig.3. A clear knee on the cyclic creep curves presented in semi-log coordinates is observed for higher stress amplitudes. The type of final fracture was either ductile for higher stress amplitudes or fatigue initiated for lower stress amplitudes. Multiple crack initiation was observed.

Cyclic deformation curves corresponding to the tensile mean stress 71 MPa are shown in Fig.4. From comparison with Fig.1 it can be concluded that the application of the tensile mean stress does not qualitatively change the cyclic plastic behaviour. Only small differences can be claimed. The inexpressive hardening in the first cycles observed for symmetrical loading for higher stress amplitudes disappears. For low stress amplitudes a weak hardening take place before the onset of cyclic softening. Nevertheless, the cyclic softening remains the main characteristic feature of the cyclic plastic deformation. A quantitative comparison of cyclic softening curves for symmetrical loading and loading with tensile mean stress for two stress amplitudes is depicted in Fig.5. It can be seen that application of tensile mean stress results in

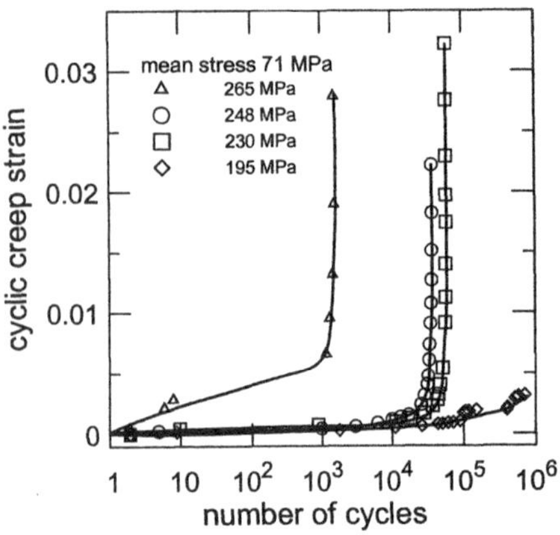

Fig.3. Cyclic creep curves.

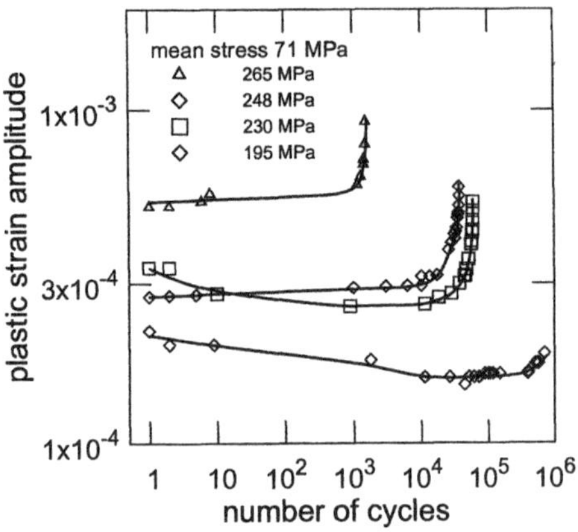

Fig.4. Cyclic deformation curves for loading with tensile mean stress.

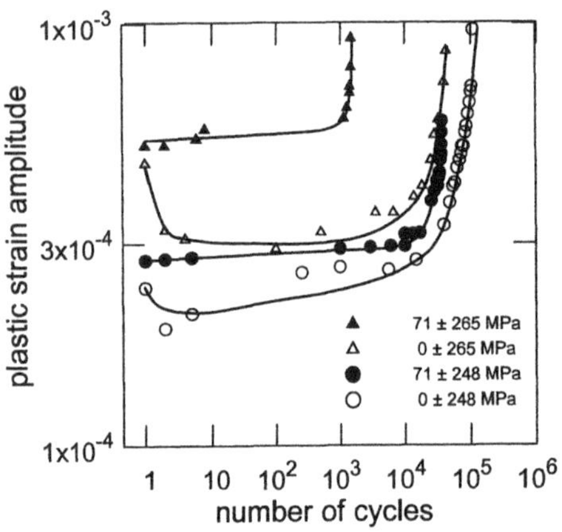

Fig.5. Comparison of cyclic deformation curves for symmetrical loading and loading with tensile mean stress.

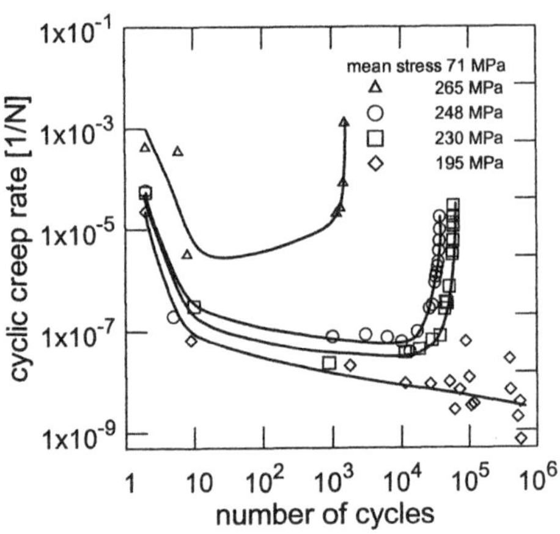

Fig.6. Differentiated cyclic creep curves.

higher plastic strain amplitudes for the whole lifetime. Moreover, it can be seen that the tensile mean stress reduces the lifetime.

Differentiated cyclic creep curves are shown in Fig.6. An initial drop of the cyclic creep rate at the beginning of loading is replaced by an increase after a minimum creep rate is reached. The number of cycles to the onset of the cyclic creep rate increase depends on stress amplitude, being higher for lower stress amplitudes. For the lowest amplitude 195 MPa the test was interrupted after $7x10^5$ cycles at a cyclic creep rate of the order of 10^{-8}/cycle.

DISCUSSION

From comparison of experimental results for symmetrical loading and loading with mean stress 71 MPa it can be concluded that the mean stress does not qualitatively change the main cyclic deformation behaviour, namely the cyclic softening. This finding is in full agreement with results published for P91 steel for temperature 530°C [6]. Simultaneously, the tensile mean stress applied promotes the cyclic plasticity. For given stress amplitude the plastic strain amplitude defined as a half-width of a hysteresis loop is larger for loading with tensile mean stress. Because the similar behaviour was found for this steel also at room temperature [7], it can be concluded that this behaviour is characteristic for broad temperature range. Analogical results were reported for mild steel at room and at elevated temperatures [8].

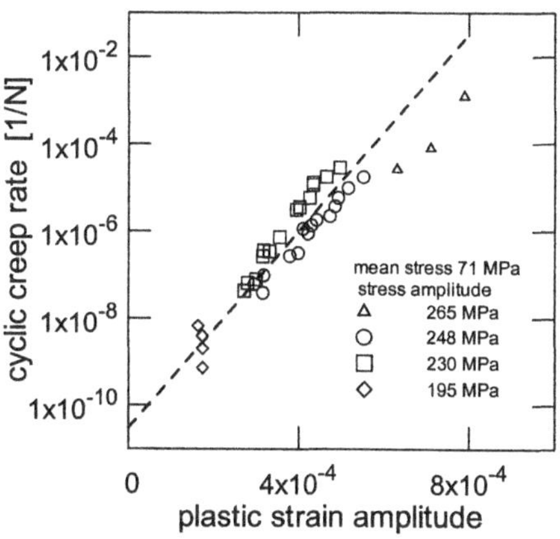

Fig.7. Dependence of instantaneous cyclic creep rate on instantaneous plastic strain amplitude.

The relation between the cyclic plasticity and creep is far from being satisfactorily described and understood. A limited data are available for materials exhibiting initial rapid hardening followed by saturated cyclic plastic behaviour. Eckert at al. [9] found for copper the relationship in the form of a power-law dependence of the steady state creep rate on saturated plastic strain range for constant stress ratio R. The P91 steel studied does not exhibit cyclic saturation or some kind of cyclic stable behaviour. Both the plastic strain amplitude and cyclic creep strain develops with increasing number of cycles. Experimental results indicate qualitatively similar behaviour of the cyclic softening and cyclic creep rate. From the comparison of corresponding cyclic deformation curves and cyclic creep curves for given mean stress it follows that the onset of acceleration of cyclic creep corresponds to the onset of the cyclic softening. In other words, there seems to be a causal relation between the cyclic and unidirectional plasticity. Experimental data corresponding to the increasing branches of cyclic creep curves, i.e. "above the knee" (which means for the decisive part of lifetime) are shown in Fig.7. An exponential relation between the instantaneous cyclic creep rate and instantaneous plastic strain amplitude can approximate the experimental data for all stress amplitudes used. The best fit in Fig.7 depicted by the dashed line and extrapolated to zero plastic strain amplitude yields the cyclic creep rate $3x10^{-11}$/cycle. The stress dependence of time to fracture and minimum creep rate in the stress interval from 100 to 300 MPa was determined on the same batch of steel previously [10]. The stress dependence of minimum creep rate is shown in Fig.8. Extrapolation of the power-law dependence down to the stress value 71MPa, which is the mean stress of the cyclic creep experiment, yields the creep rate $4.3x10^{-12}$/s. This value is in reasonable agreement with the

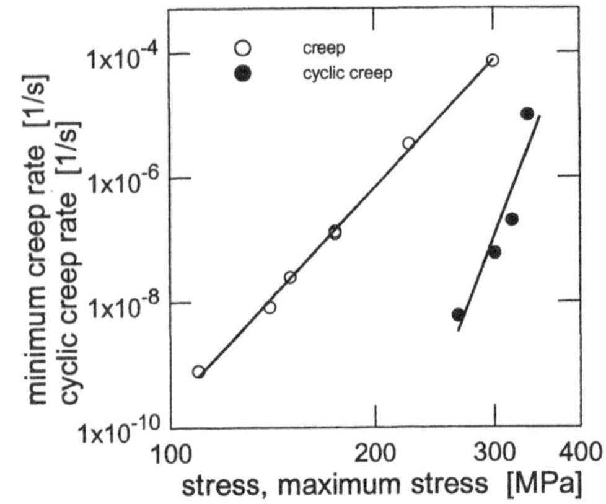

Fig.8. Comparison of stress dependence of minimum creep rate and the dependence of the minimum cyclic creep rate on maximum stress in loading cycle.

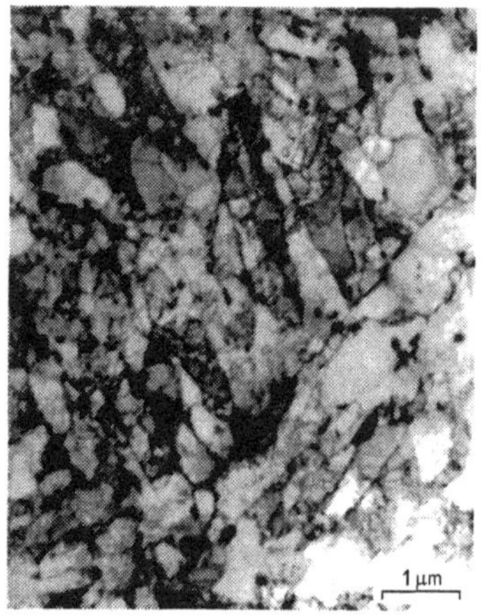

Fig.9. Microstructure of as received steel.

cyclic creep rate $6x10^{-11}$/s determined from Fig.7 for zero plastic strain amplitude, taking into account the loading frequency 2 Hz.

The stress dependence of minimum creep rate determined for pure creep is compared with the dependence of the minimum cyclic creep rate on maximum stress in loading cycle in Fig.8. For the cyclic loading parameters and temperature used a strong retardation of creep rate due to cycling is obvious. This is in contradiction to the observed enhancement of creep due to periodical unloading and reloading e.g. in Cu, Al and in C-Mn steel at room temperature [11] or for Ni-base alloy at 750°C [12]. The recovery rate after unloading was found to be generally low and strongly dependent on material microstructure [12]. This indicates the dependence of the cyclic creep behaviour on loading frequency. For low enough frequency, which is usually the case of dwell periods inserted into creep tests, the creep is enhanced. Contrary to this, conventional fatigue tests characterized by loading reversals which are usually substantially shorter than the time necessary for full recovery after unloading or reloading result in a decrease of the creep rate when compared to the pure creep.

A balance between work hardening and recovery determines the cyclic creep rate. The recovery rate decreases with decreasing temperature. For low temperatures the unidirectional strain was found to increase with time after the start of cycling in an approximately parabolic manner [11]. The strain produced by each cycle diminishes with increasing strain, which means that the cyclic creep rate decreases. For material exhibiting cyclic softening the structure developed by work hardening becomes unstable when cycling proceeds and the strain amplitude increases. The microstructure of P91 steel before cycling consists of elongated subgrains and/or transformed martensite laths with high dislocation density due to martensitic transformation, Fig.9. Cyclic loading results partially in annihilation, partially in entanglement of dislocations in the walls. An example of dislocation structure after symmetrical loading with the stress amplitude 283 MPa is shown in Fig.10. Lower dislocation density in subgrains and a tendency to more equiaxed subgrains in comparison to Fig.9 can be felt here. Fatigue loading with the stress amplitude 265 MPa and mean stress 71 MPa results in the microstructure

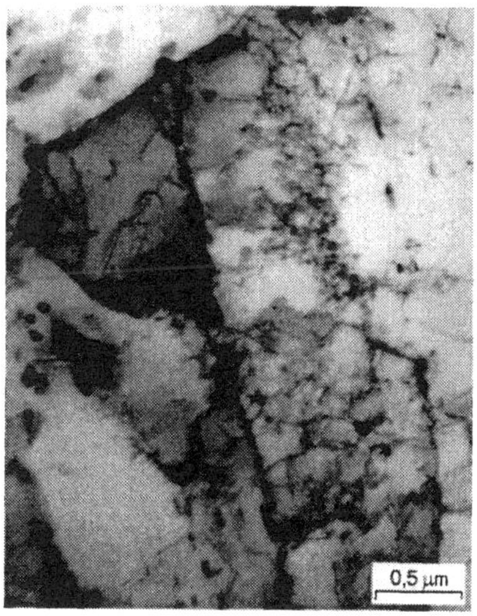

Fig.10. Dislocation microstructure after symmetrical loading at the stress amplitude 283 MPa.

Fig.11. Dislocation microstructure after loading with mean stress 71 MPa and stress amplitude 265 MPa.

presented in Fig. 11. It can bee seen that this microstructure resembles more the original microstructure than the microstructure after symmetrical fatigue loading. The shape and the dimension of subgrains or laths seems to be unchanged whereas dislocation density within the subgrains is lower. The cyclic plastic behaviour corresponding to symmetrical and asymmetrical loading is similar. No clear relation between the subgrain size and the cyclic softening can be stated. The high temperature creep behaviour is determined both by changes of the dislocation microstructure and by carbide coarsening. The lifetime of fatigued specimens was short when compared to standard creep tests. Comparison of the microstructure of as received steel with the microstructure after symmetrical fatigue and fatigue with tensile mean stress did not reveal any substantial difference in distribution and size of carbides situated along prior austenite grain and subgrain boundaries. The decrease of the dislocation density within subgrains or laths seems to be general, i.e. both for symmetrical and asymmetrical loading. The dislocation microstructure observed by transmission electron microscopy does not reveal all the details of deformation process. The observed dislocation distribution itself does not say much about the stress fields, which have to be overcome by the moving dislocation. This is due to the fact, that the transmission electron micrographs do not show (without special analysis) the sign of the dislocations. Without this any information about the stress fields can be obtained. Thus the practically undistinguishable details of the dislocation structure in subgrains and/or martensite laths boundaries can determine the motion of dislocations.

CONCLUSIONS

Symmetrical load controlled fatigue of 9%Cr steel P91 at 600°C results in cyclic softening. Tensile mean stress 71 MPa does not change qualitatively this behaviour, but it promotes the cyclic plasticity and the development of cyclic creep. Instantaneous cyclic creep rate depends on the instantaneous plastic strain amplitude. Fatigue loading with frequency 2 Hz decelerates the creep when compared to constant loading on maximum stress.

Acknowledgement. This work was supported by the Czech Academy of Sciences under the grant S2041001.

REFERENCES

1. Lecomte-Beckers, J., Schubert, F. and Ennis, P. J., Eds. (1998). *Materials for Advanced Power Engineering, Proc. of the 6th Liege Conference.* Forschungszentrum Jülich GmbH.
2. Jones, W. B. (1983). In: *Proc. of ASM Int. Conf. on Ferritic Steels for High Temperature Appl.*, pp. 221-235, Khare A. K. (Ed.) Warrendale (1981).
3. Ebi, G. and McEvily A. J. (1984). *Fat. Engng Mater. Struct.* **7**, 299.
4. Choudhary, B. K., Bhanu Sankra Rao, K. and Mannan, S. L. (1991). *Mat. Sci. Eng.* **A148**, 267.
5. Petersmeier, Th., Martin, U., Eifler, D. and Oettel, H. (1998). *Int. J. Fatigue* **20**, 251.
6. Röttger, D. and Eifler, D. (1998). In: *Proc. of Low Cycle Fatigue and Elasto-Plastic Behaviour of Materials,* pp.57-62, Rie, K-T. and Portella P.D. (Eds.) Garmisch-Partenkirchen (1998).
7. Kunz, L. and Lukáš, P. (2001). *Mat. Sci. Eng.* in print.
8. Christ, H.-J., Wamukwamba C. K. and Mughrabi, H. (1997) *Mat. Sci. Eng.* **A234-236**, 382.
9. Eckert, R., Laird, C. and Bassani, J. (1987). *Mat. Sci. Eng.* **91**, 81.
10. Knésl, Z., Kunz, L., Sklenička V. and Vilímek D. (1996). In: *Localized Damage IV. Computer Aided Assessment and Control*, pp.465-472. Nisitani, H. (Ed.), Computational Mechanical Publ. Southampton, Fukuoka (1996)
11. Evans, J. T. and Parkins R. N. (1976). *Acta Met.* **24**, 511.
12. Šustek, V., Pahutová, M. and Čadek, J. (1995). *Mat. Sci. Eng.* **A201**, 127.

Temperature-Fatigue Interaction
L. Rémy and J. Petit (Eds.)

INTERNAL AND EFFECTIVE STRESS ANALYSIS DURING CYCLIC SOFTENING OF F82H mod. MARTENSITIC STAINLESS STEEL

A.F. ARMAS[1], I. ALVAREZ-ARMAS[1], C. PETERSEN[2], M. AVALOS[1] and R. SCHMITT[2]
[1]*Instituto de Física Rosario, CONICET,*
Universidad Nacional de Rosario, Rosario, (Argentina)
[2]*Institut für Materialforschung II,*
Forschungszentrum Karlsruhe, (Germany)

ABSTRACT

Low cycle fatigue tests were performed on samples of F82H mod. stainless steel with two different initial microstructures: "ferritic-pearlitic" and "tempered martensite". Samples of both microstructures show, after the first tensile ramp, an exponential cyclic softening that continues up to failure. Whereas samples with "tempered martensite" structure display the most striking softening behaviour. The observed differences and the evolution in flow stress at different temperatures were analysed by studying internal and effective stresses obtained from the hysteresis loops after the method originally proposed by Cottrell. From this analysis during cycling of the steel F82H mod. at 723K it can be concluded that the strong softening obtained in samples with "tempered martensite"-structures are produced by a decrease in internal stresses. A different behaviour of internal and effective stresses was revealed during the softening of samples with "ferritic-pearlitic"-structure. The internal stress remains almost constant from the beginning of the test up to the failure of the specimen but the effective stress behaves like the applied stress. These differences in behaviour correlated with observations from electron microscopy are discussed in the paper.

KEYWORDS

Low cycle fatigue, transformable stainless steels, internal and effective stresses, dislocation structure.

INTRODUCTION

High-chromium ferritic (-martensitic) steels containing 8 to 14 wt % chromium have received extensive use at elevated temperatures because of their economical combination of good mechanical and corrosion properties. Nevertheless, the development of low activation steels has involved the substitution of principal alloying elements by elements that exhibit a lower radiological impact. A reduced activation ferritic/martensitic steel, the Japanese F82H mod., is one of the candidate alloys in first wall applications for fusion reactor systems [1]. This steel attracted much attention because of low ductile to brittle transition temperature (DBTT), adequate weldability and high creep rupture strength [2]. The first wall of such a reactor is assumed to be operated under the conditions of surface heat load, coolant pressure and bombardment from energetic particles. The structural material would be subjected to inelastic deformation and cyclic loading at high temperatures during operation.

Together with creep, low cycle fatigue behavior needs to be better understood in order to use this steel as structural material for the first wall of a fusion reactor. The synergetic effect of temperature and cyclic straining in ferritic/martensitic steels usually causes softening that could lead to some cumulative damage [3]. Although many studies have been made on low cycle fatigue behavior of ferritic/martensitic steels, very limited studies have been performed on F82H mod. steel [4, 5]. Ishii et al. [4] have investigated the influence of precipitations on the fatigue softening of F82H mod. but the evolution of dislocation structure during the test was not studied. The main purpose of the present study is to evaluate the cyclic behavior of the steel for both possible microstructures: Ferritic-pearlitic and tempered martensite and to analyze such behavior correlating it with the dislocation structure and its evolution during cycling.

Since the early work of Seeger [6] it was assumed that the flow stress to produce plastic deformation during monotonic tensile tests is composed by two contributions. One is a thermal component, named effective stress, caused by short range obstacles and the other, the internal stress, is almost independent of temperature and originated by barriers too large, as dislocations, precipitates and grain boundaries, for thermal activation to be significant.

For cyclic tests, the analysis of the flow stress was originally suggested by Cottrell [7] and employed by Kuhlmann-Wilsdorf and Laird [8] and Handfield and Dickson [9]. Upon this method, the flow stress obtained from the hysteresis loops is also the result of the same two kinds of resistance to plastic deformation: The short distance interaction stress was named "friction stress" and the long distance interaction stress received the name "back stress". This well-known method is illustrated in Figure 1. The scheme is used in the assertion that the hysteresis loops may be considered reversible.

At the peak stress, the applied stress σ_{max}, is the sum of the friction stress and the back stress. On lowering the applied stress, the friction stress will oppose to the backward motion of dislocations. Reverse plasticity will be obtained when the applied stress, σ_y, aided by the back stress, can overcome the friction stress.

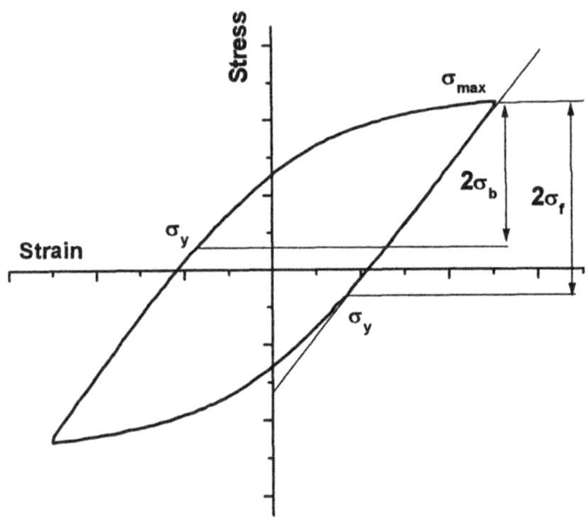

Figure 1. Schematic representation of the "effective" (friction) stress and the "internal" (back) stress.

In this work the stress originated by short range obstacles will be called "effective" stress and represented by σ_f. The stress originated by long range obstacles is the "internal" stress and represented by σ_b. Information on the types of obstacles to dislocation movement can be obtained from the measure of these stresses as shown in Figure 1.

The main purpose of this work is to obtain additional information about the cyclic behaviour of F82H mod. stainless steel at high temperature. The paper aims to report different cyclic softening behaviour observed in this steel and to analyse such manifestations on studying the evolution of the internal and effective stresses and correlating this evolution with the observed dislocation structure.

MATERIAL AND EXPERIMENTAL PROCEDURES

The chemical composition of the material investigated is given in Table 1. Cylindrical fatigue specimens were machined with a diameter of 5 mm and a gauge length of 18.4 mm. All specimens were austenitized at 1313 K for thirty minutes in vacuum. After this initial treatment, the samples were divided in two groups. One of them was cooled down from the 1313 K to 1040 K, annealed for 50 hs at this temperature and than air-cooled [10]. This microstructure is designated as "ferritic–pearlitic" microstructure and consists of a chromium ferrite matrix with pearlite (Figure 2). Another group of samples was air cooled from 1313 K, tempered for two hours at 990 K and finally air cooled. The fully martensitic structure transforms during tempering into the so-called "tempered martensite" microstructure (Figure 3).

Table1. Chemical composition (in wt%):

	C	Si	Mn	P	S	Ni	Cr	Mo	Al	V	Ti
F82H mod.	0.089	0.11	0.16	0.002	0.002	0.019	8.16	0.002	0.023	0.16	0.002

Nb	Cu	N	W	Ta	B
0.0001	0.006	0.007	0.09	0.02	< 0.0002

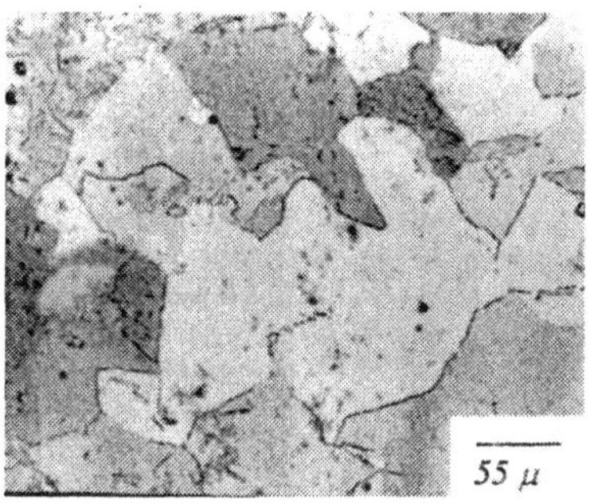

Figure 2. Ferritic-pearlitic structure

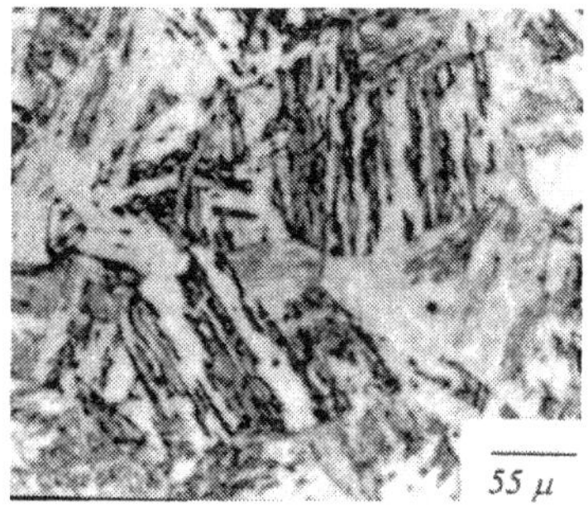

Figure 3. Tempered martensite structure.

Low cycle fatigue tests were performed in air at 723 K on the above described specimens under total axial strain control with a completely reversed triangular wave form. Specimens were tested at a total strain range of 0.6 % and total strain rate of $2 \times 10^{-3}\ s^{-1}$.

A data acquisition system including a personal computer and A/D converter was used to record the strain and stress values of the hysteresis loops. Two hundred and fifty points per cycle were taken, from which one hundred correspond to the unloading part of the hysteresis loop that runs from the maximum strain up to zero strain. A computer programme using a linear regression method fit the elastic portion of this part of the loop. Despite the considered number of cycles and the used microplasticity criterion to determine the flow stress, σ_y, at which the reverse plasticity starts, the accuracy of the measurement could not be better than 5%. The values of the effective stress, σ_f, and the internal stress, σ_b, were obtained from σ_{max} and σ_y according to Figure 1.

Thin foils of 3-mm diameter were prepared using double jet-polishing technique at 273K with a 90% vol. ethanol-10% vol. perchloric acid electrolyte. Dislocation structures were examined in a transmission electron microscope (TEM), Philips EM 300, operating at 100kV. Owing to the ferromagnetic character of the specimens, it was necessary to minimize the mass of the thin foils to reduce the influence of the astigmatism in the TEM. With a 1-mm punch the thin area of the thin etched area was punched out. The 1-mm diameter samples were then fixed between a sandwich Cu grid.

RESULTS

Mechanical behaviour

Figure 4 shows the cyclic curves obtained on F82H mod. cycled at 723 K with ferritic-pearlitic and tempered marten-

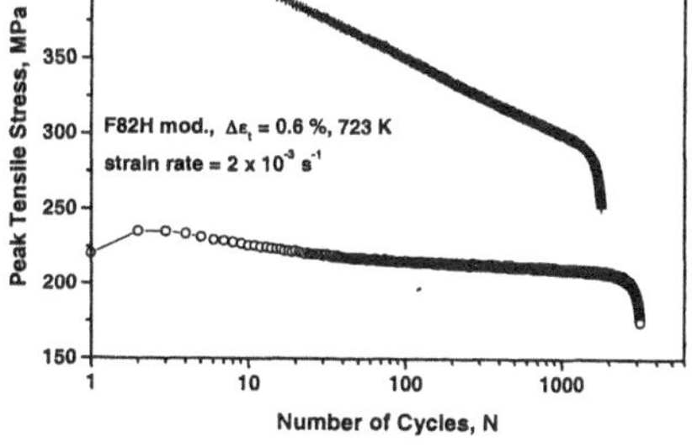

Figure 4. Cyclic softening curves obtained for both microstructures.

site structures. Only cyclic softening, that continues up to failure, was observed for both structures after the small hardening occurring after the first stress reversal. Samples with "tempered martensite" structure show the most striking softening behaviour: Thirty percent of stress reduction from the beginning of the test up to the onset of the stress reduction caused by the fracture of the specimen. Only ten percent of stress reduction was observed on samples with "ferritic-pearlitic" structures.

Figures 5(a) and 5(b) show in a linear scale the evolution of the effective and internal stresses at 723K of samples with both microstructures. The peak tensile stress curve is also shown for comparison. The temperature 723K was selected as representative of the temperature range where dynamic strain ageing phenomena were observed in ferritic stainless steels [11]. Anomalous cyclic hardening is one of the typical manifestations of strain ageing phenomena during cyclic tests. However, no cyclic hardening was observed in F82H mod. with both microstructures and in the middle temperature range. As a matter of clarity, only some points are represented of each curve but error bars are also included in the set of curves showing the scatter band observed in the points.

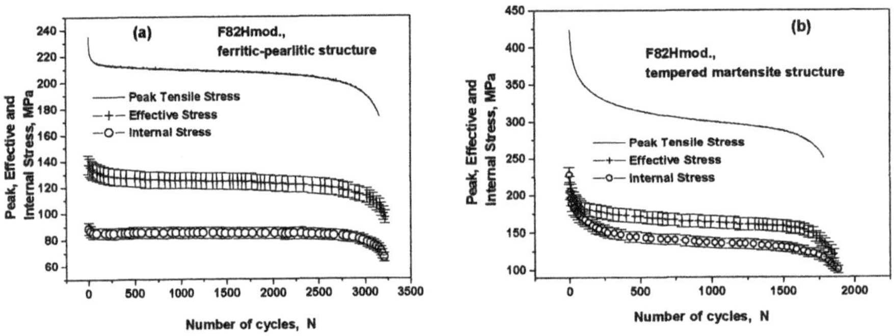

Figures 5(a) and 5(b). Peak tensile, effective and internal stress evolution of F82H mod. with ferritic-pearlitic and tempered martensite structure.

The small cyclic softening occurring at the beginning of the tests in samples with ferritic-pearlitic structure obeys to the softening observed on both: The effective and the internal stresses. The internal stress presents a rapid softening at the beginning and saturates up to the failure of the specimen. The effective stress shows a more pronounced cyclic softening. The evolution of the applied stress in samples with tempered martensite structure is also a consequence of the evolution of both, the effective and the internal stress. From Figure 5(b) it can be concluded that in this case the internal stress is responsible for most of the softening observed in the applied stress. The effective stress evolution is similar to the observed in samples with ferritic-pearlitic structure.

Dislocation Substructures

In order to correlate the observed mechanical behaviour with the dislocation substructure, microstructural observations with TEM were performed.

Figure 6(a) shows the dislocation structure observed in samples of F82H mod. with ferritic-pearlitic structure in a sample obtained from the specimen head. The observed microstructure would correspond to the material exposed at the same temperature than the gauge section but undeformed. Only grains with a low dislocation density were observed. A two-dimensional cell structure (Figure 6(b)) with dense dislocation walls was observed in the present steel cycled up to fracture. The interiors of the cells are almost free of dislocations and precipitates.

The general TEM microstructure of tempered martensite F82H mod. samples is given in Figure 7(a). The main structure is tempered martensite that contains tempered martensite laths with some precipitates along the grain boundaries. The decoration of the grain boundaries by precipitates is very poor after quenching. The microstructure observed in Figure 7(b), corresponding to a sample fatigued up to 100 cycles, shows that the original lath structure is transforming to an equiaxed subgrain structure. This structure resembles a partial recrystalized structure where the lath structure begins to dissolve. The presence of carbides inside expanded laths, which are lined up parallel to the lath boundaries, clearly indicates that former martensite lath boundaries migrate during testing. A striking structure is observed in Figure 7(c) corresponding to the steel F82H mod. cycled up to fracture.

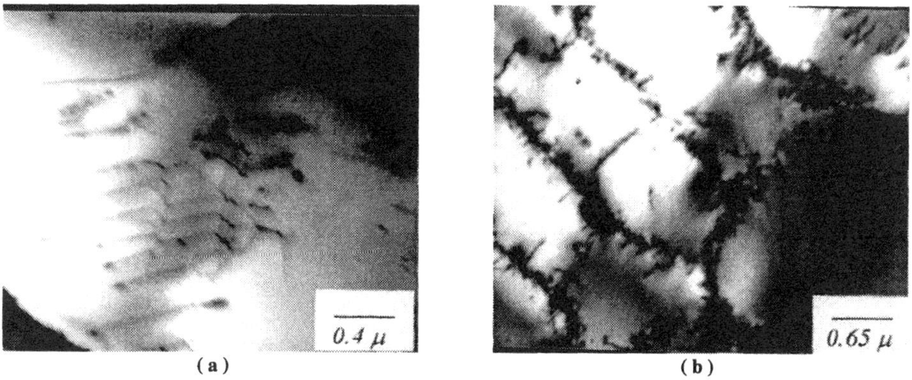

Figures 6(a) and 6(b). Dislocation arrangement observed in F82H mod. with ferritic-pearlitic microstructure in the undeformed section (6(a)) and the gauge section (6(b)) of a sample cycled up to fracture.

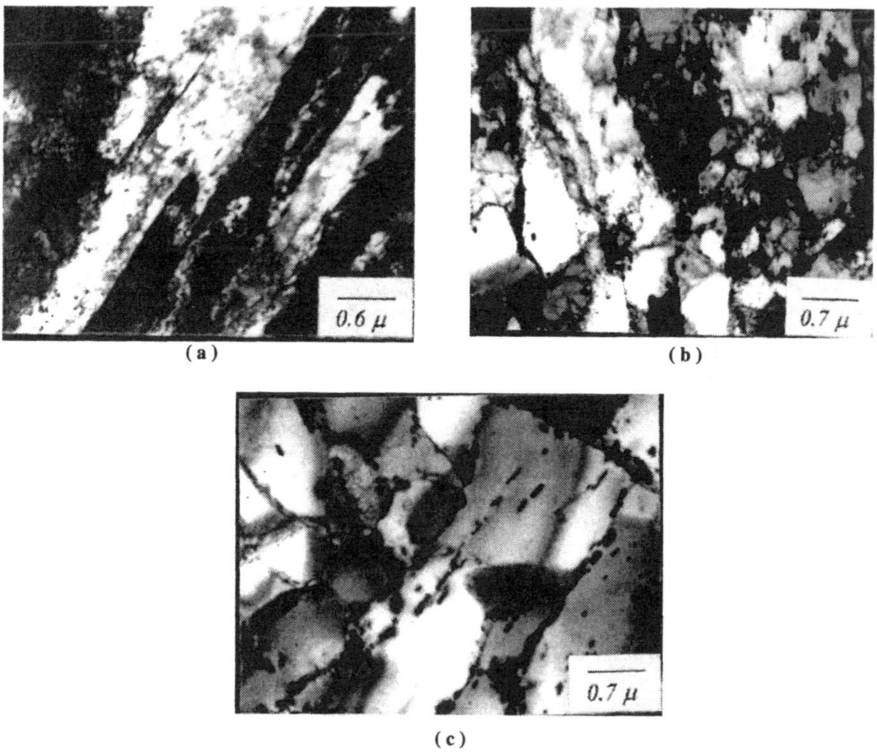

Figures 7(a), 7(b) and 7(c). Dislocation Substructure observed in F82H mod. with tempered martensite microstructure at different stages of the fatigue life.

No tempered martensite structure was observed but growing equiaxed ferrite subgrains. The dislocation density was drastically reduced and condensed in the subgrain boundaries. The interiors of the subgrains are almost free of dislocations but decorated with large $M_{23}C_6$ - carbides. Much carbide is also distributed along the subgrain boundaries that were formed by dynamic recovery.

DISCUSSION

Although ferritic stainless steels [11, 12] are prone to exhibit anomalous cyclic behaviour due to dynamic strain ageing phenomena in a middle temperature range, 523-773 K, only cyclic softening was observed in F82H mod. with ferritic-pearlitic and tempered martensite structure.

The evolution of the internal and effective stresses in the steel with ferritic-pearlitic structure indicates that most important changes occur at the beginning of the test. The rapid softening observed in the internal stress is an evidence that all dislocations produced after the first cycles evolves to a softer substructure that does not change appreciably during the cycles. Figure 6(b) shows the cell structure formed at the end of the test. Microstructural observations realized at the beginning of the test, e.g. 50 cycles, show that the cell structure is already formed at this stage. On the other side, from the evolution of the effective stress it can be said that short-range obstacles are becoming less effective to the dislocation movement. It can be speculated that, during cycling, interstitial solute atoms become less effective in causing solid solution hardening [13].

The internal stress is the main responsible component for the cyclic softening observed at this temperature range in F82H mod. with tempered martensite structure(Figure 5(b)). The figure shows that the level of both stresses, effective and internal, are higher than those corresponding to the steel with ferritic-pearlitic structure. As a consequence of the harder tempered martensite structure the initial value of the internal stress is also higher than that for the effective stress. After the Cottrell model [7], the internal stress is representative of long-range obstacles hiding the movement of dislocations. The striking exponential cyclic softening observed during the evolution of the internal stress for the tempered martensite samples implies that the original hardened material will evolve to a softened condition. High dislocation density and carbides are considered long-range obstacles [7]. Ishii et al. [4] reported the $M_{23}C_6$ - coarsening and Laves phase precipitations as a consequence of cyclic deformation of F82H mod. at high temperatures. These authors concluded that these are the dominant factors for the cyclic softening observed in this steel. But no TEM observations of the dislocation structure were reported.

The present work reveals the importance of the dislocation substructure evolution during cycling of the alloy with tempered martensite structure. After cyclic straining at 723K, the lath structure was evolved toward an equiaxed subgrain structure and the dislocation structure reduced even further (Figures 7(a), 7(b) and 7(c)). The process of transformation of tempered martensite with high dislocation density (strong obstacles) to a subgrain structure with a rather low dislocation density (soft obstacles) happened in a high temperature fatigue test should be related to dynamic recovery (Figure 7(c)).

The mobility of the lath interfaces in 8-12% Cr steels is expected to increase as the coarsening and spheroidization of carbides proceed, especially if these particles are primary responsible for the pinning of the lath boundaries. However, the lath morphology in steels containing vanadium remained intact for long periods at temperatures up to 923K. This effect was attributed to the interfacial pinning by vanadium carbide precipitates, which coarsen very slowly [14]. In the case of the present F82H mod. steel, with vanadium content, the observed results show a drastic recovery of the lath structure when cycled at 723K. These and earlier results in other 8-12% Cr ferritic stainless steels alloys [15] indicate that the martensitic lath structure of the steel is destabilized under cyclic strain conditions. It is proposed in the present work that lath interfaces of the tempered martensite structure, pinned by carbide particles, would be stable under prolonged annealing at higher temperatures than 723K. However, the coarsening of the carbide particles, also those containing vanadium, becomes accelerated as a consequence of the cyclic strain. Consequently, such coarsening would be expected to promote the breakdown of the lath structure evolving to an equiaxed subgrain structure.

CONCLUSIONS

From the present results it can be concluded that the cyclic behaviour of F82H mod. is strongly structure dependent. From the evolution of the effective and internal stress it can be concluded that cyclic softening of ferritic-pearlitic samples is attributed to the rapid small softening occurring at the beginning of the test in both stresses. Dislocations produced during the first cycles of the test evolve rapidly to a softened dislocation cells structure. This structure remains almost stable up to the failure of the specimen. A large exponential softening process is observed in samples with tempered martensite structure. The cause of this pronounced softening is attributed principally to the internal stress evolution. Long range obstacles like dislocation lath structure, formed during quenching, appears to evolve to an equiaxed subgrain structure. Accelerated carbide coarsening due to strain cycling destabilizes the lath structure leading to a dynamic recovered subgrain structure.

ACKNOWLEDGEMENTS

The authors acknowledge the financial support of Consejo Nacional de Investigaciones Científicas (CONICET) y Técnicas and Agencia Nacional de Promoción Científica y Tecnológica (ANPCYT) of Argentina. This work was performed within the Special Intergovernmental Agreement between Germany and Argentina, sponsored by Forschungszentrum Karlsruhe, Germany, within the framework of the Nuclear Fusion Programme.

REFERENCES

1. Shiba, K.. Suzuki, M.and Hishinuma, A. (1996) J. Nucl. Mater. **237**, 309.
2. Tamura, M., Hayakawa, H., Yoshitake, A., Hishinuma, A. and Kondo, T., (1988) J. Nucl. Mater. **155-157**, 620.
3. Alvarez-Armas, I., Armas, A. and Petersen C., (1994) Fatigue Fract. Engng Mater. Struct., **17**, 671.
4. Ishii, T., Fukaya, K., Nishiyama, Y., Suzuki, M. and Eto, M.,. (1998) J. Nucl. Mater. **258-263**, 1183.
5. Stubbins, J. and Gelles, D., (1996) J. Nucl. Mater. **233-237**, 331.
6. Seeger, A. (1955). In: *Defects in Crystalline Solids*, p. 328, The Phys. Soc., London.
7. Cottrell, A. H., *Dislocations and Plastic Flow in Crystals.* (1953) Oxford University Press, London.
8. Kuhlmann-Wilsdorf, D. and LAIRD, C., (1979) Mater. Sci. Engng. **37**, 111.
9. Dickson J., Boutin J., and Handfield, L., (1984), Mater. Sci. Engng., **64**, L7-L11.
10. Schirra, M. and Finkler H., FZKA-Report, No. 5607, 1995.
11. Armas, A., Alvarez-Armas, I., Avalos, M., Petersen, C. and Schmitt, R. (1997) Fusion Technology 1996. C. Varandas and F. Serra (editors). Elsevier Science B.V. 1359.
12. Avalos, M., Moscato, M., Alvarez-Armas, I., Petersen, C., Schmitt R. and Armas A., (1998) J. Nucl. Mater. **258-263**, 1204.
13. Munz, D., Scripta Met. (1972), **6**, 815.
14. Jones, W. Hills, C. and Polonis, D., (1991), **22A**, Met. Trans. A, 1049.
15. Jones, W. (1981). pp. 221-235. Proc. ASM Intern. Conf. on Ferritic Steels for High Temperature Applications, Ashok Khare, ed. Warrendale, PA., ASM, Metals Park, OH.

Damage under Isothermal Loading

Temperature-Fatigue Interaction
L. Rémy and J. Petit (Eds.)

EFFECT OF NOTCHES ON HIGH TEMPERATURE FATIGUE/CREEP BEHAVIOUR OF CMSX-4 SUPERALLOY SINGLE CRYSTALS

P. LUKÁŠ, P. PRECLÍK, L. KUNZ, J. ČADEK and M. SVOBODA
Institute of Physics of Materials, Academy of Sciences of the Czech Republic,
Žižkova 22, 61662 Brno, Czech Republic

ABSTRACT

Effect of notches on high temperature fatigue/creep strength of CMSX-4 single crystals has been investigated. Cylindrical bars of the orientation <001> with circumferential notches were tested at 850 °C under constant loads both without and with superimposed high frequency cyclic loads. Under creep conditions, the notched specimens exhibit a longer creep lifetime than the smooth specimens for the same net-section stress. Stress-strain analysis of the notched specimens subjected to constant loads was performed by an elastic-plastic FEM procedure; the experimentally determined creep data of smooth specimens were used as input data. An excellent correlation was found between the creep lifetime of the notched specimens and the average value of the calculated steady-state creep strain rate. The creep life curves of notched specimens were found to be identical with the creep life curves of smooth specimens when expressed in terms of Monkman-Grant diagram. Cyclic load components superimposed on static load reduce the time to failure. This reduction increases with increasing stress amplitude. Moreover, the mode of failure is changed from the ductile creep mode to the sharply localised fatigue mode.

KEYWORDS

Fatigue/creep, notch effect, superalloys, stress-strain analysis, Monkman-Grant relationship, fatigue slip bands.

INTRODUCTION

Presence of notches in components operating at high temperatures is often inevitable. The notches cause stress concentration and change the stress state from uniaxial to multiaxial even in the case of uniaxial remote loading. Moreover, the components can be subjected to a complex stress system varying from simple uniaxial fatigue or creep loading to multiaxial combined fatigue/creep loading. The uniaxial fatigue and creep data generally do not suffice for the description of the component behaviour under more complex loading. Therefore it is necessary to seek the methods of laboratory measurements and theoretical calculations which are relevant for the case of the complex loading and at the same time remain reasonably economical.

While the notches are always detrimental in the case of cyclic loading (fatigue), they can be both detrimental and beneficial in the case of constant load (creep). Notch effect in fatigue has been studied quite extensively and there are reliable procedures for its estimation; see e.g. [1]. Notch effect in creep has been also studied by a number of investigators, nevertheless there is still no generally accepted procedure for its estimation. Quantitatively, the degree of malignity or benignity of a notch can be best seen when the lifetime data of both the smooth and notched specimens are plotted in dependence on the net section stress. For a given net section stress, the notch strengthening (beneficial effect) means that the lifetime of the notched specimen is higher than that of the smooth specimen. The opposite is true for the notch softening (detrimental effect). The available data show both notch strengthening [2-11] and notch softening [3,4,5,8,10] in dependence on the material, notch geometry and testing conditions (temperature, applied load, surrounding environment). The constraint of axial plastic deformation in the triaxial stress-state field caused by a notch leads to the notch strengthening. On the other hand, for very sharp notches and cracks the damage is highly localised, the mode of failure is changed from the global to the local-crack mode and thus such notches lead to the notch softening in spite of their high geometrical constraint. The behaviour of real notched components can represent mixture of both these modes and is thus dependent on a number of variables listed above. The notch effect in conditions of interaction of high cycle fatigue with high temperature creep was studied in our preceding papers [12,13] for two creep-ductile steels.

The experimental results show that superimposed cycling with very small cyclic stress amplitudes has either no or even a slightly beneficial effect. At mild cyclic stress amplitudes the beneficial effect of notches is removed and at high enough stress amplitudes expressive notch softening occurs. This proves that vibration of sufficient amplitude is one of the factors contributing to the transition from the global to the localized mode of failure.

This paper deals with one of the hi-tech high temperature materials, namely with single crystals of superalloy CMSX-4. The aim of the paper is (i) to offer a general procedure for the evaluation of the notch effect under creep conditions, and (ii) to determine experimentally the effect of notches under fatigue/creep conditions.

MATERIAL AND METHODS OF TESTING

Testing was carried out on <001> - oriented CMSX-4 single crystals. The chemical composition of the superalloy CMSX-4 is given in Table 1.

Table 1. Chemical composition of CMSX-4 (wt. %).

Cr	Mo	W	Co	Ta	Re	Hf	Al	Ti	Ni
6.5	0.6	6.4	9.7	6.5	2.9	0.1	5.7	1.0	bal.

Single crystals were delivered as cast rods in fully heat treated condition. The microstructure of the as-received rods consists of cuboidal γ' precipitates embedded in a γ matrix. The γ' particle size lies between 0.4 and 0.5 μm and the volume fraction of the γ' phase is about 70%.

Three types of specimens were machined from the tests bars, namely:
(1) Smooth cylindrical specimens with gauge length of 50 mm and gauge diameter of 3 mm.
(2) Cylindrical specimens with circumferential V-notch. The depth of the V-notch was 0.5 mm, the opening angle was 60° and the radius of the notch root was 0.2 mm. The diameter of the net section was 2 mm and the diameter of the gross section was 3 mm. For this notch geometry the theoretical (elastic) stress concentration factor is $K_t = 2.54$.
(3) Cylindrical specimens with circumferential semicircular U-notch. The depth and the radius of the U-notch was 1 mm. The diameter of the net section was 3 mm and the diameter of the gross section was 5 mm. The theoretical (elastic) stress concentration factor is K_t = 1.61.

All the creep tests on smooth and notched specimens were performed in air at a temperature of 850°C under constant load regime in tension using standard creep machines. The fatigue/creep tests on notched specimens were performed in a modified resonant pulsator also in air at a temperature of 850°C. The specimens were subjected to a static load corresponding to the net stress of 600 MPa and to superimposed cyclic loads with amplitudes varying from specimen to specimen. The start-up procedure of the fatigue/creep tests was the following. The static load was applied first (after the specimen had reached the desired temperature), then the cyclic load (frequency 90 to 95 Hz) was applied by switching on the resonant loading system. The full amplitude was reached within 500 cycles. In both the types of tests (i.e. creep and fatigue/creep) the elongation was continuously measured by means of linear variable differential transformers coupled with a digital data acquisition system.

EXPERIMENTAL RESULTS

Basic creep data

Four specimens were tested. One of the creep curves is shown in differentiated representation (creep rate versus time) in Fig.1. It is clear that there is only a very short steady-state stage. Nevertheless the minimum creep rate can be deducted from the creep curve very easily in the case of this specimen as well as in the case of the other specimens. This makes it possible to determine stress dependence of the minimum creep strain rate $\dot{\varepsilon}_{min}$. The results show that this dependence can be best described by the exponential function of the type

$$\dot{\varepsilon}_{min} = a \exp(b\sigma), \qquad (1)$$

where $a = 1.67 \times 10^{-13}\ s^{-1}$ and $b = 2.12 \times 10^{-2}\ MPa^{-1}$. The creep life data are presented (together with the data for notched specimens) in Fig.2.

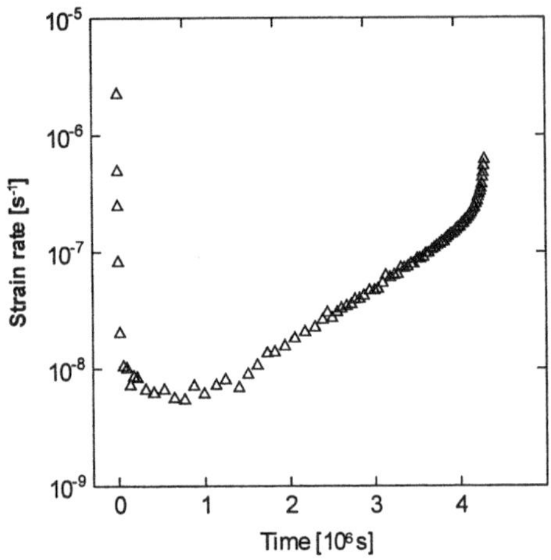

Fig. 1. Differentiated creep curve of smooth specimen tested at 500MPa and 850 °C.

Creep life of notched specimens

Inspection of Fig.2 shows that the creep lifetime curves for notched specimens (net section stress, i.e. load divided by the minimal cross-section, versus time to complete failure) are shifted towards higher stresses with respect to the curve for smooth specimens. For example, for the same net stress of 600 MPa, the creep life of the notched specimens is by almost exactly one order of magnitude longer than that of the smooth specimen. We can thus state a strong notch strengthening effect which is slightly more expressive for the V-notch than for the U-notch.

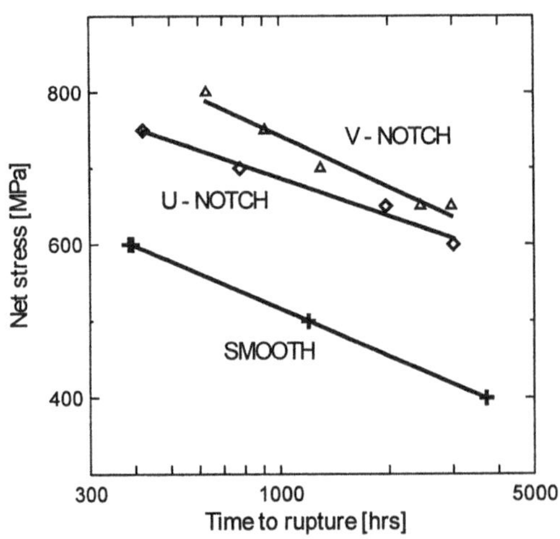

Fig. 2. Creep life curves of smooth and notched specimens at 850 °C.

Life of notched specimens under fatigue/creep loading

Effect of vibrations on the life of notched specimens under fatigue/creep loading can be seen in Fig.3. Here the time to rupture is plotted in dependence on the stress amplitude. The numbers attached to the experimental points are the numbers of elapsed cycles at the given stress amplitude. It is interesting to note that the number of cycles needed to bring the specimen to failure at the lowest stress amplitude used (120 MPa) approaches the gigacycle fatigue region (4.2×10^8 cycles). Cyclic stress component generally shortens the life. In comparison to pure creep, a very small cyclic stress component superimposed on a large static stress component has no substantial harmful effect, higher stress amplitudes reverse the notch strengthening effect into the notch softening effect.

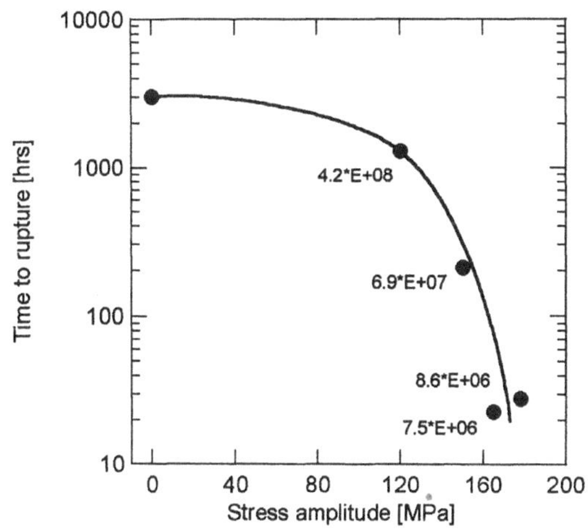

Fig. 3. Effect of vibrations superimposed on static stress of 600 MPa on life of the notched bars (U-notch) at 850 °C.

STRESS-STRAIN ANALYSIS

Stress and strain distribution in crept notched specimens was determined by time-dependent elastic-plastic calculations. The finite element method applied in this paper is basically of the same kind as the method used earlier (Hayhurst et al. [14], Eggeler et al. [15]). The distributions of stress, strain and displacement were computed using finite element program ANSYS. Formulation of the boundary conditions corresponded to the experimental set-up. Due to the rotational symmetry the stress-strain analysis could be taken as a two-dimensional problem. Material was assumed to be isotropic. The influence of the length of the specimen and its grip in the testing machine was also taken into account. Directly after loading, the stress distribution in the notched specimens corresponds to the elastic situation. As creep occurs, the initial stress field redistributes at a rate given by geometry of the notched body, applied stress and material properties. The "steady-state" material properties are given by equation (1) with the above presented values of the constants. This equation was used in the computation. It is important that the rate of redistribution decreases and the computed stress and strain distribution reaches relatively shortly after the application of the load its steady state. Example of the distribution of one of the stress components just after application of the external load and in the steady state is shown in Fig.4. Fig.4a shows the stress component parallel with the load axis, σ_{zz}, immediately after loading, Fig.4b shows the corresponding distribution in the steady state. The stress component σ_{zz} is in these diagrams normalized by the nominal stress σ_{nom} (net section stress) and is presented as a function of the distance from the notch root. Stress distribution in Fig.4a corresponds to the elastic solution and - as expected – the value of the σ_{zz}/σ_{nom} at the notch root is equal to the K_t –value. It can be seen that the peak at the notch root is relaxed out in the steady state (Fig.4b) and that the maximum value of the σ_{zz}/σ_{nom} lies below the specimen surface.

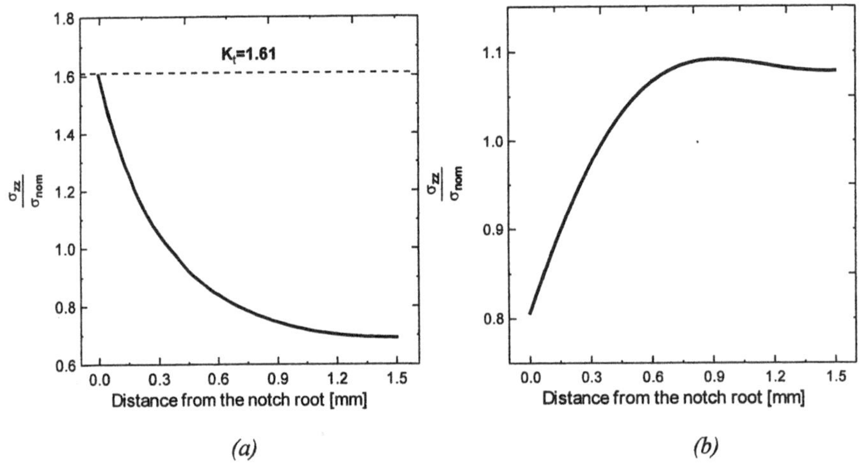

Fig. 4. Stress component σ_{zz} as a function of the distance from the notch root, σ_{nom}= 600MPa , T = 850 °C. (a) Immediately after loading (elastic case); (b) In the steady state.

The creep behaviour of the notched specimens can be characterised by the creep strain rate component in the direction of the applied stress, $\dot{\varepsilon}_{zz}$. Its value averaged over the net section, $\bar{\dot{\varepsilon}}_{zz,notch}$, will be used in the following for the presentation of the lifetime data of the notched specimens. This quantity is defined as the mean value of the $\dot{\varepsilon}_{zz}$ over the net-section plane with the initial area S, i.e. as

$$\bar{\dot{\varepsilon}}_{zz} = \frac{\iint \dot{\varepsilon}_{zz}}{S}. \quad (2)$$

DISCUSSION

Creep life of notched specimens

Already in the fifties of the last century Monkman and Grant [16] related the creep life of smooth specimens with the minimum creep rate. Their relation can be written as

$$\dot{\varepsilon}_{min}\, t_f^n = \text{constant}, \tag{3}$$

where the exponent n lies near to 1. Let us try to modify the Monkman-Grant relation for the case of notched specimens. Instead of $\dot{\varepsilon}_{min}$ it is necessary to use a value of the creep rate characterizing the notched specimen. For that purpose we shall use the value defined by equation (2), i.e. $\bar{\dot{\varepsilon}}_{zz}$. Fig.5 shows the Monkman-Grant plot both for the notched specimens and for the smooth specimens. Not only that the Monkman-Grant relation can be used also for the notched specimen, but moreover the experimental points for smooth specimens and for notched specimens with different notch geometries fall into one scatter band. This offers a good possibility for the estimation of the effect of notches on the creep life solely on the basis of smooth creep data (creep life and minimum creep rate) combined with the above outlined computation.

The above computation does not take into account crystal anisotropy. In spite of that the agreement between the smooth and the notched data is very good. This is probably due to the fact that the effect of anisotropy in <100>-oriented cylindrical crystals with circumferential notch is masked by the simultaneous activity of eight equally stressed {111}<110> slip systems. The experiments on sheet crystals [10] indicate that the anisotropy must be taken into account in less symmetrical cases.

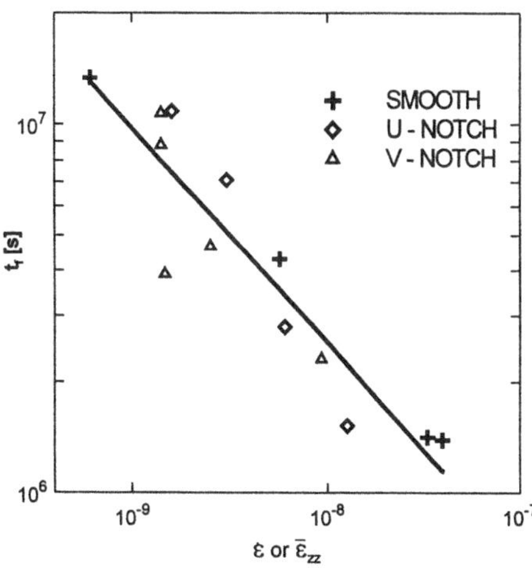

Fig. 5. Monkman – Grant plot for notched and smooth specimens at 850 °C.

Life of notched specimens under fatigue/creep loading

The above presented treatment of creep life of notched specimens is valid in the range of creep-ductile behaviour, i.e. in the range where the damage of the notched specimens is not localised and the failure is of global, i.e. of ductile type. The transition from the global type of failure to the localised failure depends on a number of factors. To the main factors enhancing the local crack type failure belong (i) low uniaxial creep ductility, (ii) too a high degree of plastic flow constraint caused by too a high triaxiality and sharpness of the notch (this is true especially for cracks) and (iii) low applied stresses. The last point is the most important one as it concerns the question of the possibility to extrapolate the short term and medium term laboratory data to the long-term behaviour of components loaded at high temperatures.

Superposition of vibrations on the static load can be understood as the strongest factor enhancing the local crack type failure. This is confirmed by the TEM observation of the structure in the nearest vicinity of the fracture surfaces after the tests were completed. Fig.6 shows the structure in the crept notched specimen. Both the γ/γ' structure and the dislocation structure can be seen. The dislocations are seen in the γ channels. Fig.7 shows the structure as

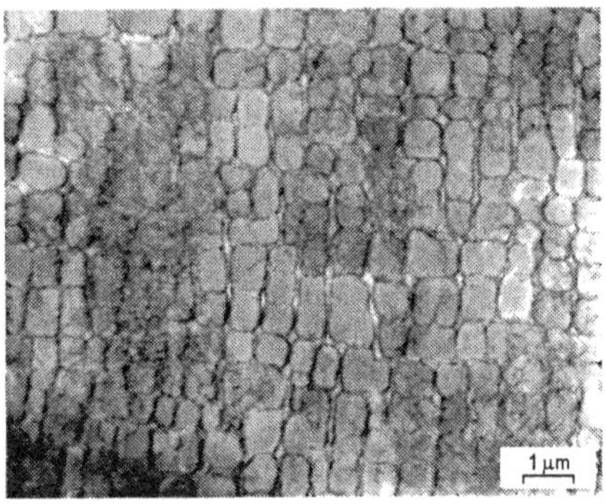

Fig. 6. γ/γ' structure and dislocation configuration in notched specimen subjected to creep loading. Foil perpendicular to load axis prepared from the vicinity of fracture surface.

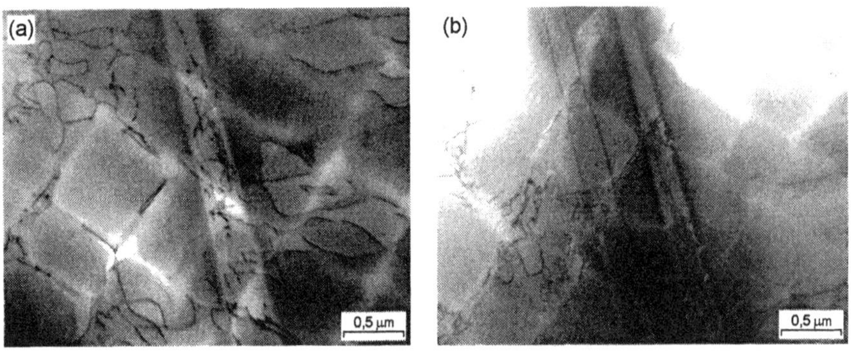

Fig. 7. Examples of fatigue slip bands in the notched specimen subjected to fatigue/creep loading. Foil perpendicular to load axis prepared from the vicinity of fracture surface.

seen in the specimen subjected to static load with superimposed vibrations. Fatigue slip bands going right through both the γ matrix and the γ' precipitates can be well seen. The bands lie along the {111} slip planes. The angle between the {111} slip planes and the plane of the foil (001) is 54.7 degrees. That is why the fatigue slip bands appear in the foil (thickness about 200 nm) as broad bands. In reality they are extremely thin slabs. Nevertheless they represent the sites for the nucleation of fatigue cracks. Thus the reason for the shortening of life due to the cyclic stress component lies in the transition from the ductile type failure to the fatigue type, i.e. fatigue crack, type failure.

CONCLUSIONS

Cylindrical bars of the orientation <001> with circumferential notches were tested at 850 °C under constant loads and under constant loads with superimposed high frequency (90 Hz) vibrations. Under creep conditions, the notched specimens exhibit a longer creep lifetime than the smooth specimens for the same net-section stress. This can be attributed to the strain constraint caused by the stress triaxiality. Stress-strain analysis of the notched specimens subjected to constant loads was performed by an elastic-plastic FEM procedure; the creep data of smooth specimens were used as input data. The distributions of stress, strain and displacement in dependence on time were computed. All the named parameters reach their steady-state values. An excellent correlation was found between the creep lifetime of the notched specimens and the average value of the calculated steady-state creep strain rate. For both the notches the creep life curves of this type were found to be identical with the life curve for smooth specimens expressed in terms lifetime vs. steady state creep rate. Thus a modified Monkman-Grant relationship is valid both for smooth and notched specimens. This offers a basis for the evaluation of the notched creep life solely on the basis of the smooth creep data.

In comparison to pure creep loading, cyclic stress component generally shortens the life. The vibrations of sufficiently high amplitude contribute to the transition from the global type of failure to the localised type of failure. Fatigue slip bands were found in the notched specimens subjected to fatigue/creep loading.

ACKNOWLEDGEMENTS

This research was supported by the Academy of Sciences of the Czech Republic under contracts nos. A2041002 and K1010104. This support is gratefully acknowledged.

REFERENCES

1. Taylor, D. and Wang, G. (2000) *Fatigue Fract. Engng. Mater. Struct.* **23**, 387.
2. Ellison, E.G. and Wu, D. (1983) *Res Mechanica* **7**, 37.
3. Lloyd, G.J., Barker, E. and Pilkington, R. (1986) *Engng. Fract. Mech.* **23**, 359.
4. Curbishley, I., Pilkington, R. and Lloyd, G.J. (1986) *Engng. Fract. Mech.* **23**, 383.

5. Konish, H.J. (1988) *J. Pressure Vessel Tech.* **110**, 314

6. K. H. Wu, F. A. Leckie, (1990) *Fatigue Fract. Engng. Mater. Struct.* **13**, 155.

7. Eggeler, G., Tato, W., Jemmely, P. and deMestral, B. (1992) *Scripta Metall. Mater.* **27,** 1091.

8. Muller, J.F. and Donachie, M.J. (1975) *Metall. Trans. A* **6**, 2221.

9. M. C. Pandey, A. K. Mukherjee, D. M. R. Taplin, J. Mater. Sci. 20 (1985) 1201.

10. Sugimoto, K., Sakaki, T., Horie, T., Kuramoto, K. and Miyagawa, O. (1985) *Metall. Trans. A* **16**, 1457.

11. Lukáš, P., Preclík, P. and Čadek, J. (2001) *Mat. Sci. Eng.* **A298**, 84.

12. Lukáš, P., Kunz, L., Knésl, Z. and Kuna, M. (1994). In: *Proc. 4th Int. Conf. Biaxial/Multiaxial Fatigue*, Vol. 1, pp. 171-180, Societe Francaise de Metallurgie et de Materiaux, St. Germain en Laye.

13. Lukáš, P., Knésl, Z., Kunz, L. and Preclík, P. (1999). In: *Progress in Mechanical Behaviour of Materials, Proc. ICM 8*, Vol.1, pp. 412-417, Ellyin, F. and Provan, J.W. (Eds.). Fleming Printing Ltd., Victoria.

14. Hayhurst, D.R. and Henderson, J.T. (1977) *Int. J. Mech. Sci.* **19,** 133.

15. Eggeler, G. and Wiesner, C. (1993) *J. Strain Analysis* **28,** 13.

16. Monkman, F.C. and Grant, N.J. (1956) *Proc. ASTM* **56**, 593.

Temperature-Fatigue Interaction
L. Rémy and J. Petit (Eds.)

CREEP-FATIGUE LIFE PREDICTION OF AGED 13CrMo44 STEEL USING THE TENSILE PLASTIC STRAIN ENERGY

GEEWOOK,SONG, JUNGSEOB,HYUN, JEONGSOO,HA
Korea Electric Power Research Institute (KEPRI),
103-16,Daejeon, Korea

ABSTRACT

Low cycle creep-fatigue tests of 13CrMo44 steel used for boiler header of fossil power plant for 185,000 hr are conducted at 515°C with triangular and trapezoidal strains wave. Trapezoidal wave is considered about hold time for creep effects. The relationship between the tensile hysteresis energy and number of cycles to failure is examined to predict the low cycle creep-fatigue life of 13CrMo44 steel. The life, predicted by the tensile hysteresis energy method, is found to coincide with experimental results and analytical results obtained from models such as Coffin-Manson method.

KEYWORDS

13CrMo44 steel, plastic strain energy, creep-fatigue interaction, tensile hysteresis energy, boiler header, life prediction.

INTRODUCTION

The power plants use high temperature and high pressure for good thermal efficiency and these components of a power plant such as boiler header, steam pipe, turbine rotor and casing, etc. may undergo low cycle fatigue at high temperature. These critical components are subjected to transient loadings resulting from the large number of start-up and loading changes induced by the daily and seasonal variations in the electricity demand. Thus, most materials of components used in conditions of high temperature and high pressure suffer from thermal stresses as well as mechanical stresses. Although actual low cycle fatigue fracture is caused by thermal stresses due to the temperature changing, the behavior of materials is customarily investigated at constant temperature under mechanical loading.

Previously, most of the evaluation of low cycle fatigue life of materials has been done at uninterrupted (triangular) cycling. Correlation with actual service experience using this data is not encouraging. Therefore, introducing hold times makes a simulation closer to the actual service conditions. Since the fatigue damage is generally caused by the cyclic plastic strain, the plastic strain energy plays an important role in the damage process. Therefore, the idea of relating fatigue life to the plastic work during a load cycle has been proposed [1-5]. In this approach, plastic strain energy was proposed by Morrow. Ellyin proposed a fatigue failure criterion based on the strain energy density damage law. Ostergren proposed the net tensile hysteresis energy that can be used as the damage function. The objective of this paper is the investigation of the effect of hold time in the creep-fatigue tests and evaluate remaining life of the 185,000 hr, 439 start-up, aged 13CrMo44 steel by using the tensile hysteresis energy as the measure of creep-fatigue damage per cycle.

EXPERIMENTAL PROCEDURE

Material, specimens and device

The material used in this study was 13CrMo44 steel. The test specimens were cut out from boiler header of fossil power plant that was manufactured by MAN during the early 1960's. The boiler header had been in service for 185,000 hours at a temperature of 515°C. Serviced conditions of the boiler head are summarized in Table 1. The chemical composition and the mechanical properties of the testing material at room temperature and 515°C are shown in Tables 2 and 3, respectively.

Fig. 1. shows the metallurgical microstructure of testing material. The microstructure consisted of ferrite and pearlite. The change of microstructure during high temperature service mainly results from morphological of pearlite. It is seen from this figure that carbides in pearlite are dispersed leaving little trace of the original pearlite area.

The creep rupture tests carried out ex-service testing material and new material. The test results are shown in Fig. 2. The creep life reduction of ex-service testing material is larger than new material. From the results of microstructure investigations and creep rupture tests, we knew that the condition of the testing material had been high degradation. Fig. 3 shows the shape and size of the specimens.

The tests were conducted in a closed-loop, electro hydraulic servo-controlled fatigue testing machine INSTRON 8521 with an induction heating equipment. The temperature was measured by the optical pyrometer, and the strain is measured by the 12.5mm extensometer in the gauge length, and real-time hysteresis loop was obtained by I/O terminal board from load, strain and temperature controller.

Table 1. Service conditions

Pressure (Kg/cm^2)	Temperature (°).	Operation Time (hour)	Number of starts
100	515	185,000	439

Table 2. Chemical composition of the testing material

Composition (wt%)	C	S	Si	P	Mn	Cr	Mo
	0.18	0.02	0.33	-	0.58	0.74	0.69

Table 3. Mechanical properties of the testing material at room temperature and 515°C

Temp. (°C)	Young's Modulus (GPa)	Ultimate Tensile Strength (MPa)	0.2% Yield Stress (MPa)	Elongation (%)
25	205.1	484.1	280.7	33.8
515	190.2	315.9	198.6	35.1

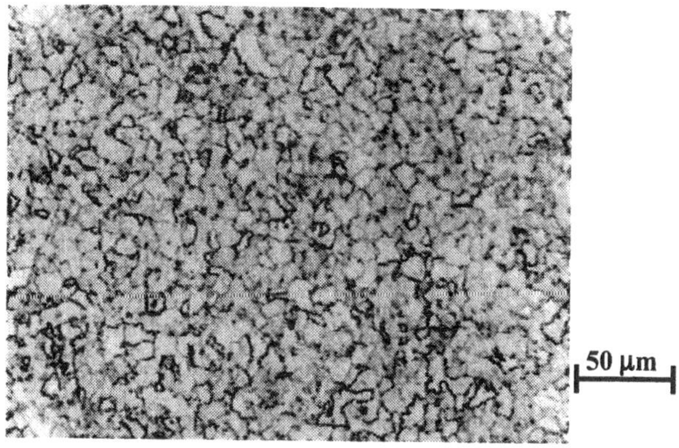

Fig. 1. Metallurgical microstructure of testing material

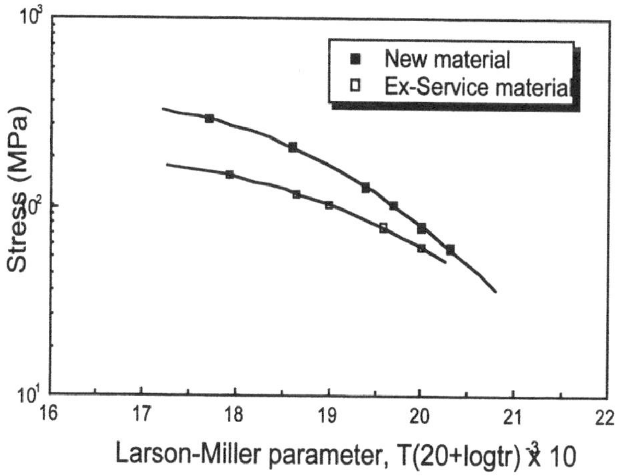

Fig. 2. Microstructure of testing material

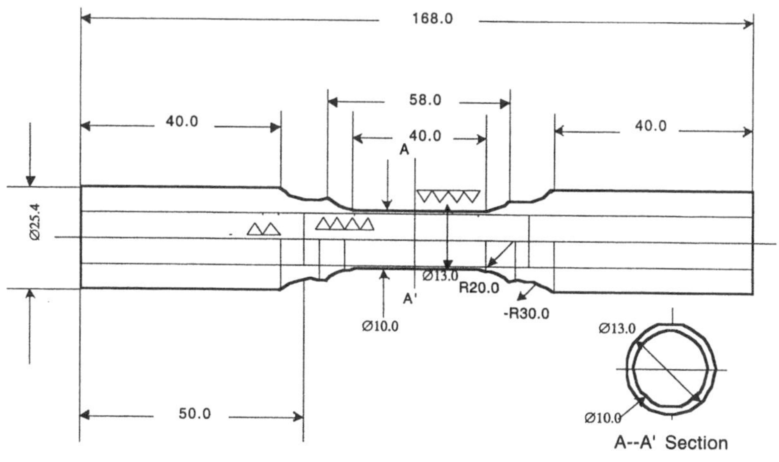

Fig. 3. Geometry of creep-fatigue test specimen

Experimental test conditions

Isothermal creep-fatigue experiments were performed in uniaxial tension-comprehension under total strain control, each condition. In all experiments, the strain ratio is held constant at R=-1, and the strain rate is maintained at 0.004/s, and total strain amplitude is 0.2~0.6%. The tests were carried out at 515°C under triangular without hold time and trapezoidal strain waveform with 60, 600, 1800 s hold time in tension, as shown in Fig. 4. All tests were continued until a 25% load drop of maximum tensile stress, although several specimens were fractured before 25% load drop.

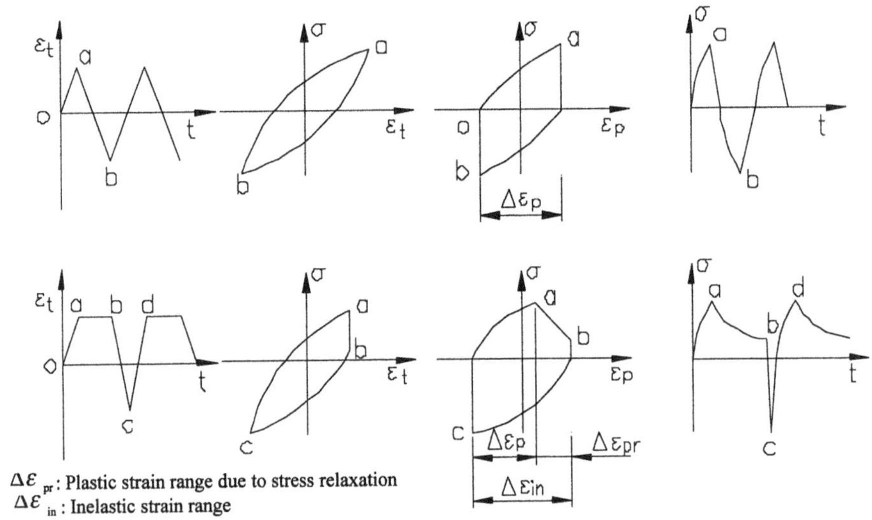

Fig 4. Schematic stress-strain diagram for strain control cycling under triangular and trapezoidal wave

THE TENSILE HYSTERESIS ENERGY

The low cycle fatigue damage is generally caused by the cyclic plastic strain. At high cyclic stresses, plastic strain is the predominant cause of energy dissipation in metals. Part of this energy is transformed into heat, the rest is absorbed by the material. Thus, the plastic strain energy plays an important role in the low cycle fatigue damage process.

In general it has been found that small cracks do initiate very early in the total life and that low cycle fatigue is primarily a crack propagation phenomenon. If we consider that low cycle fatigue is essentially a problem of crack propagation, the deformation occurring in the portion of the cycle with the open crack contributes to the state of damage by propagating the crack. Elber has calculated for a plate specimen that crack closure occurs for $\sigma < \sigma_o$, where σ equals the gross stress and σo equals positive constant. The deformation damage occurs only for $\sigma > \sigma_o$. The net tensile hysteresis energy can be considered as the measure of fatigue damage. [6] Ostergren develops this approach for predicting strain controlled low cycle fatigue life at elevated temperature. This physical quantity is a measure of the energy going into the material during crack propagation and has the advantage of including stress and strain parameters. The tensile hysteric energy ΔW_T may be calculated at half life. Similar to the total energy of a hysteresis loop, a first approximation to ΔW_T may be computed as

$$\Delta W_T = \alpha\,(\sigma_T - \sigma_O)\,\Delta\varepsilon_P \qquad (1)\ [6]$$

where α is a numerical constant, σ_T is maximum tensile stress, and σ_O is the stress level for crack closure.

For the nominally plastic situation of low cycle fatigue the local and gross stresses are likely to be in phase, and the assumption of crack closure for $\sigma < \sigma_O$ with $\sigma_O = 0$ is reasonable.

$$\Delta W_T N_f^{\beta} = C_1 \qquad (2)\ [6]$$

$$\alpha\,\sigma_T\,\Delta\varepsilon_P N_f^{\beta} = C_1 \qquad (3)\ [6]$$

$$\sigma_T\,\Delta\varepsilon_P N_f^{\beta} = (C_1/\alpha) = C \qquad (4)\ [6]$$

where C and β are material constants.

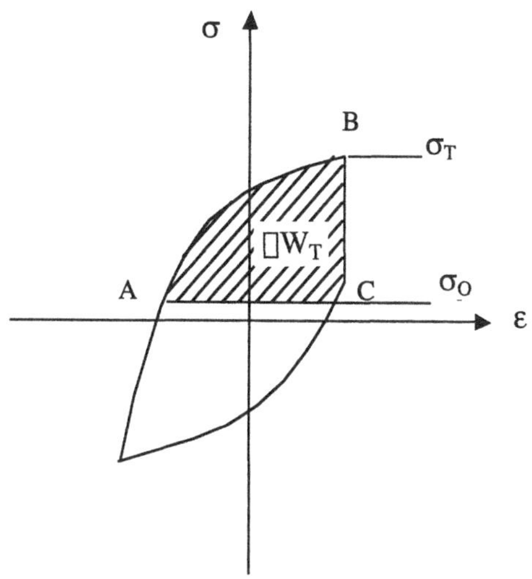

Fig. 5. Schematic hysteresis loop defining net tensile hysteresis energy

RESULTS and DISCUSSION

Influence of hold time on low cycle creep-fatigue life

The hysteresis loops at half life without and with hold time are shown in Fig. 6. The shape of the stress relaxation curve due to creep effect will depend on the number of cycles. Therefore, on the amounts of stress relaxation range a~b per cycle depend on the number of cycles and increases with the hold time.

Generally, these hold times are shown to have a life-reducing effect at elevated temperature when compared without hold time cycling under the same condition. The stress behavior for the fatigue cycles is shown in Fig. 7. Since the strain range limits are fixed, the stresses and plastic strain range will vary from cycle to cycle. The ratio between number of cycles to failure with 1800 s hold time and without hold time is found to be 1/2.7 at $\Delta\varepsilon_t = 0.6\%$. Besides, during the initial cycling, there is a cyclic softening as indicated by Fig. 7. It is seen that there is quite a noticeable decreasing of stress amplitude during the softening phase. For example, after one hundred cycles (end of softening process) the stress amplitude drops to 15% of its initial value. The strain amplitude, $\Delta\varepsilon_a/2$, versus number of cycles to failure, N_f, with hold time are plotted on a log-log scale in Fig. 8. The mathematical expressions for the low cycle fatigue curves in Fig. 8 may be written as:

$$\frac{\Delta\varepsilon_a}{2}=\frac{\Delta\varepsilon_a^e}{2}+\frac{\Delta\varepsilon_a^p}{2}=\frac{\sigma_f'}{E}(2N_f)^b+\varepsilon_f'(2N_f)^c \qquad (5)$$

where (σ_f'/E) and ε_f' are the strain amplitudes corresponding to the elastic and plastic intercept for the first half cycle.

At a given strain range, life reduction increases until hold time is 600 s by the creep-fatigue interaction effects, but there seems to be a saturation effect, no further reduction of fatigue life above 600 s hold time by the long creep effects.

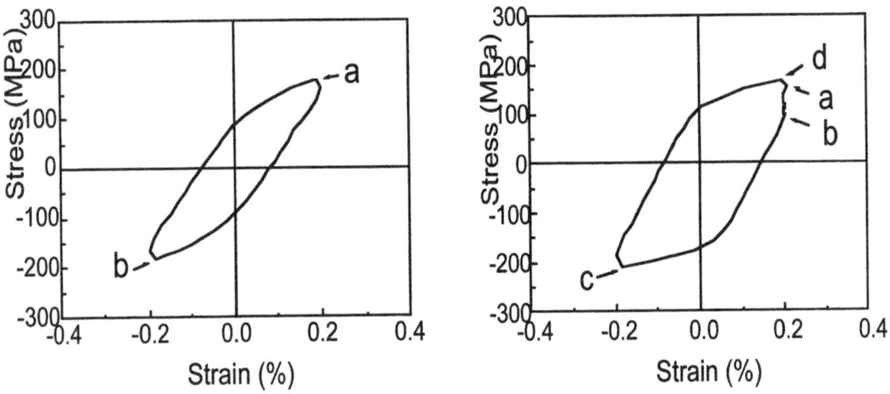

a) Hold time = 0s b) Hold time = 600s

Fig. 6. Hysteresis loops at half life without and with hold time

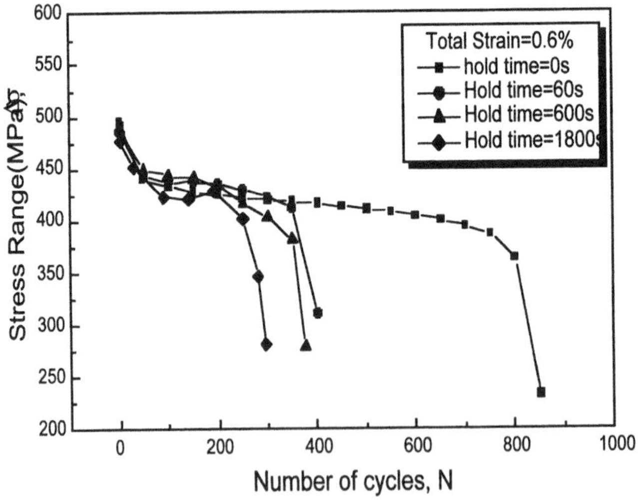

Fig. 7. The stress behavior for the fatigue cycles

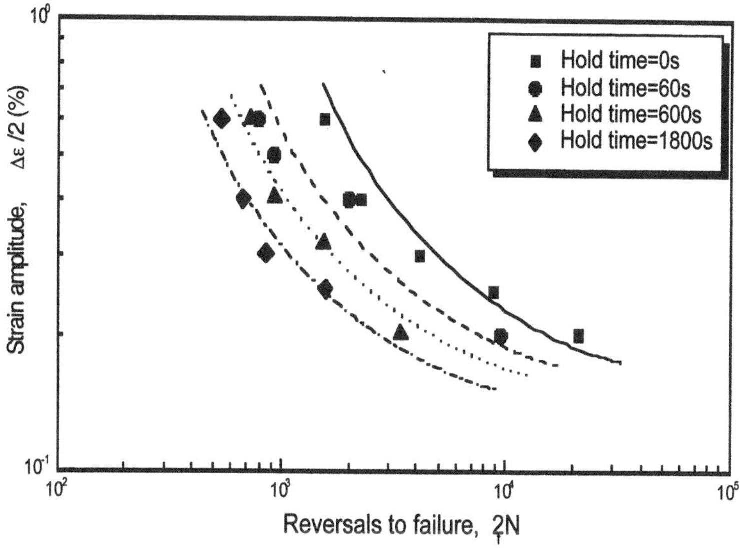

Fig. 8. Total strain versus number of cycles to failure

Life prediction by the tensile hysteresis energy

The tensile hysteresis energy at half life is plotted against number of cycles to failure in Fig. 9. It is seen that the tensile hysteresis energy each hold time have close relation to the fatigue life. The relation equations between tensile hysteresis energy and fatigue life may be shown in Table 4. The values of material constants, C and β, are obtained by the least-square errors technique. The straight lines leading to equations for each hold time in Table 4 are shown solid, dash, dot and dot-dash lines in Fig. 9. The predicted life obtained by the tensile hysteresis energy equation in Table 4 and the actual life from experimental data are compared in Fig. 10 indicating a range within a factor of 2. The results are compared with those obtained by Coffin-Manson method. Thus, it may be possible to estimate the creep-fatigue life at high temperature by the calculation of the tensile hysteresis energy, so that this result can be used for the quantitative life assessment of components in power plants.

Table 4. The relationship between the Tensile hysteresis energy and fatigue life each hold time

Hold time	Equations (MPa)
0 sec	$\sigma_T \Delta\varepsilon_P = 121.78 \times (N_f)^{-0.738}$
60 sec	$\sigma_T \Delta\varepsilon_P = 166.47 \times (N_f)^{-0.824}$
600 sec	$\sigma_T \Delta\varepsilon_P = 134.61 \times (N_f)^{-0.858}$
1800 sec	$\sigma_T \Delta\varepsilon_P = 521.87 \times (N_f)^{-1.153}$

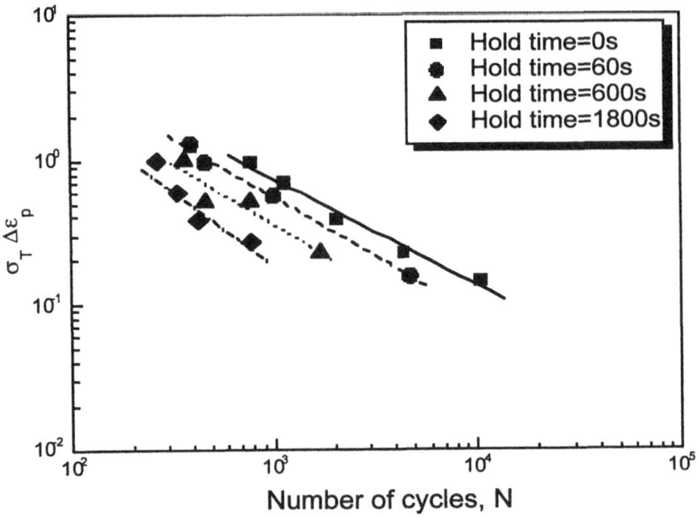

Fig. 9. The tensile hysteresis energy versus number of cycles to failure

CONCLUSION

In this study, with specimens extracted from aged boiler header in a power plant, the isothermal low cycle creep-fatigue tests are performed. The results are as follows.

1. At a given strain range, the fatigue life decreases with increasing hold time. There seems to be a saturation effect, no further reduction of the fatigue life above 600 s hold time is apparent by the long creep effects.

2. The tensile hysteresis energy approach for in-serviced material tests is introduced. This correlated to the number of cycles to failure by the relationship, $\sigma_T \Delta\varepsilon_P N_f^{\beta} = C$. Comparison of calculated values with experimental results shows good agreement.

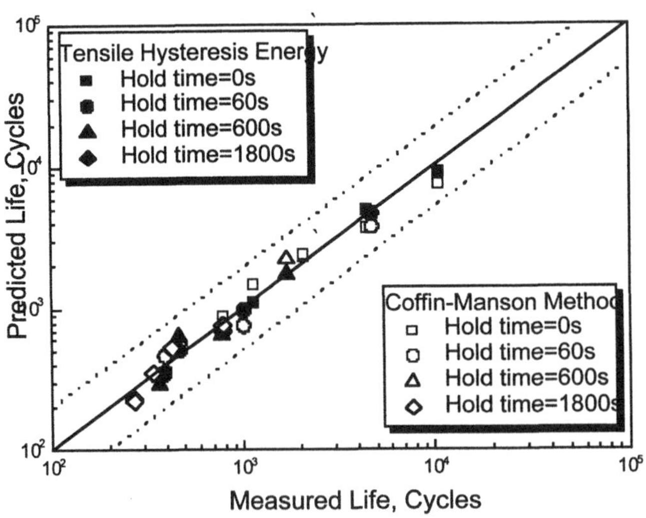

Fig. 10. Measured life versus predicted life

REFERENCES

1. Morrow, J. D., (1965) *ASTM STP 378: Internal Friction, Damping and Cyclic Plasticity*, pp.45~84.
2. Ellyin, F. and Kujawski, D., (1984) *Journal of Pressure Vessel Technology* **106**, 342.
3. Lefebvre, D. and Ellyin, F., (1984) *Int. Journal of Fatigue* **6**, 9.
4. Ellyin, F., (1985) *Journal of Engineering Materials and Technology* **107**, 119.
5. Giglio, M. and Vergani, L., (1995) *Trans. of the ASME* **117**, 50.
6. W. J. Ostergren, (1976) *Journal of Testing and Evaluation* **4**, 327.

Temperature-Fatigue Interaction
L. Rémy and J. Petit (Eds.)

THERMOMECHANICAL FATIGUE AND AGING OF CAST ALUMINUM ALLOY : A LINK BETWEEN NUMERICAL MODELING AND MICROSTRUCTURAL APPROACH

I. Guillot*, B. Barlas**, G. Cailletaud**, M. Clavel*, D. Massinon***

* Université de Technologie de Compiègne, Laboratoire Roberval, UMR CNRS 6066,
BP 20529 - 60205 Compiègne cedex France
** Centre des Matériaux de l'Ecole Nationale Supérieure des Mines de Paris, UMR CNRS 7633
BP 87 - 91003 Evry France
*** Fonderie Montupet, 67 rue J. de la Fontaine, 60181 Nogent-sur-Oise France

ABSTRACT

The present paper is devoted to the modeling of the stress-strain behaviour of a cast aluminum alloy for cylinder heads (AISI 319), which is studied from the initial state T5 to saturated aging (320°C). The evolution of the microstructure during heating corresponds to the sequence : $\theta' \rightarrow \theta$ (Al_2Cu). The effect of transformations and coarsening can be characterized either by mechanical testing or by hardness measurements and by coarsening model using TEM image analysis. The observed coarsening of spherical θ precipitates is in good agreement with the Lifshitz-Slyozov-Wagner (LSW) theory. The use of these data in precipitation hardening theories provides a "microstructural" evaluation of the strength evolution. A mechanical model is written in a viscoplastic framework, since viscous effects play an important role in this temperature range. Constitutive equations include the description of Bauschinger effect and of aging, through a scalar internal variable a. A careful comparison is made between the two approaches, allowing us to present a physically supported mechanical behaviour, which can be extrapolated to other metallurgical compositions.

KEYWORDS

Aluminum alloy, viscoplastic modeling, aging effect, transmission electron microscopy (TEM), image analysis, particle coarsening.

INTRODUCTION

Over the past decade, the automotive industry has increasingly employed cast aluminum alloys as a replacement for cast iron in the production of engine components. Improvements in engine performance have caused the temperatures in aluminum cylinder heads to increase, especially in the inter valve zone, from below 170°C in earlier engines to peak temperatures above 300°C in recent engines [1]. The 319 aluminum family is commonly used in casting cylinder heads, due to its low density, good thermal conductivity and good casting properties. This class of material exhibits mechanical properties that depend strongly on microstructural features such as secondary dendritic arm spacing (SDAS), porosity, intermetallic compounds, hardening precipitates [2, 3, 4] ... The effects of porosity [5, 6, 7], intermetallics [8, 9, 10, 11], grain refinement [12, 13], silicon concentration [14, 15] are well established and the performance of these alloys has also been improved by modeling casting processes and the thermal treatments [12, 16]. Previous studies have been performed on the same subject by some of the present authors [17, 18] or other groups [3, 1]. The purpose of these works was to obtain a mechanical model to define viscoplastic behavior in presence of a microstructural evolution. Well–selected mechanical experiments, which expose the material behavior under various loading and temperature histories, are an essential requirement to identify the coefficients of these models, which are purely phenomenological, so that a new experimental data base must be built if a new chemical composition is considered. On the other hand, the use of structural hardening models, including the effect of precipitate size distribution is still rare for the aluminum casting alloys. One of the reasons is that a good knowledge of the precipitate coarsening law during aging is needed in order to build a physically supported model of the mechanical behavior. The purpose of the present paper, which wants to bridge the gap between mechanics and metallurgical models, is therefore to gather careful mechanical testing and microstructural observations.

MATERIALS AND EXPERIMENTAL PROCEDURE

The material of the study is a cast aluminum 319 alloy with a T5 heat treatment (24 hours at room temperature and 5 hours at 210°C). Its chemical composition is shown in table 1. All specimens used for microstructural characterization and mechanical testing were cast in a metallic die and provided by Montupet S.A. as 20 mm diameter rods. The T5 specimens were then aged at 100, 200, 250, 280 and 320°C, under various aging time up to 1000 hours and air–cooled in order to obtain the corresponding hardness values. The samples for TEM observations and mechanical testing were 100 h heat–treated at various temperatures.

Element	Si	Cu	Mg	Fe	Mn	Zn	Ti	Ca	Sr
Wt (pct)	8.2	3.3	0.3	0.47	0.24	0.24	0.2	0.0005	0.010

Table 1: Material composition in weight percent.

The samples used for TEM observations were cut with a low–speed cutting saw, mechanically rounded and thinned down to 3 mm and 120 μm respectively. The thin disks were electropolished in a double-jet Tenupol using a 33% nitric acid in methanol solution maintained at -30°C and 12 V. Thin foils were observed using a TEM operated at 100 kV.

The particle dimensions were measured in two ways. In the case of the θ' precipitates, the length and thickness were measured from dark-field TEM micrographs of the particles viewed edge-on in the $[001]_{\rm Al}$ zone axis. The measurements were made on 3.5–time magnification blow–ups of the negatives. The measured lengths of the θ' precipitates are the projected lengths in the $\langle 001 \rangle_{\rm Al}$ directions. In fact, the precipitate plates have a somewhat irregular shape and the measured length, close to the plate diameter, $(\overline{d} = \overline{L})$, corresponds to the big axis of the ellipse with the same surface of the circled precipitate. Therefore, the thickness is equal to the small ellipse diameter. The non–coherent θ precipitates exhibit quite a globular morphology and can be considered as spheres. The *radii*, ($\bar{r} = \bar{d}/2$), were measured from bright–field TEM micrographs of the particles performed on zero tilt. The measurements were made on 2.5–time magnification blow–ups of the negatives. The geometrical parameters of the precipitates, θ' or θ, are obtained with populations of about 1000 particles. All data are obtained from a two–dimensional projection of three–dimensional volume elements in a thin foil of finite thickness, z. The conversion of the particle size distributions from planar to volumetric distributions was performed according to Shah and Altstetter analysis [19] with the Schwartz-Saltykov method [20, 21].

Vickers macrohardness was carried out with a 10 kg weight at room temperature. Each data point represents an average of five measurements. The scatter is approximately 2 Hv. Vickers microhardness measurements were performed on the α–phase with a 3 g weight. Each data point represents an average value of twenty measurements. The low cycle fatigue experiments were conducted until failure under strain control on a servohydraulic machine using cylindrical specimens, the strain being adjusted at each cycle to keep the plastic strain amplitude constant. The axial strain was determined using an extensometer located on the gauge length. Thermomecanical fatigue tests are also performed by adapting a setup previously developed at ONERA [22]. This experiment is described in detail elsewhere [23].

MICROSTRUCTURAL OBSERVATIONS

The microstructure of the as–received materials consists of a dendritic aluminum structure containing Cu and an aluminum–silicon eutectic with intermetallic compounds between dendritic arms (fig. 1). The cooling rate control during solidification gives a SDAS (fig. 1a) around 20 μm and a grain size around 0.4 mm. Figure 1b depicts silicon particles and intermetallic compounds. The amount and morphology of silicon particles were determined using image analysis. The value of the elongation factor, corresponding to the ratio of the maximum/minimum segments of the silicon particles, is close to 2. Therefore silicon eutectic can be described as pseudo–globular particles resulting from the eutectic structure modification related to Sr additions.

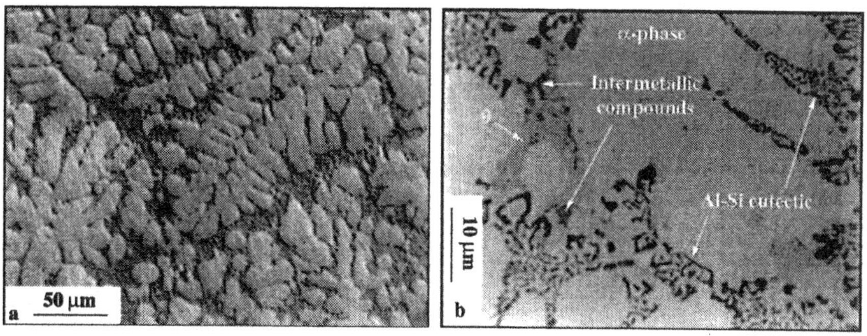

Figure 1: (a) SEM micrograph showing 319–T5 with SDAS = 20 μm. (b) Microstructure of the as–received 319–T5 alloy.

Figure 2a shows a dark–field TEM image of the precipitates (T5) viewed edge–on in the $[001]_{Al}$

zone axis. From the analysis of diffraction patterns, it was found that the disc shaped precipitates were all θ'. As shown in figure 2, the evolution of the precipitate features in the annealed specimens corresponds to the sequence : $\theta' \longrightarrow \theta$ (Al_2Cu). Below 200°C, the only microstructural change is the coarsening of the θ' phase, which is well known for the Al-4wt% Cu binary alloy [24, 25, 26, 27, 28]. The subsequent growth of non–coherent θ precipitates takes place for higher temperatures. As previously indicated, the particle dimensions were determined using image analysis. These results are summarized in table 2. Following the Lifshitz-Slyozov-Wagner [29, 30] theory (LSW theory), the volume fraction of particles is expected to remain constant during coarsening under steady-state condition. The mean value obtained is close to 3.8%, in agreement with the results due to Cerri *et al.* [31].

Aging conditions	Precipitates	$\bar{d} = \bar{L}$ (nm)	$\bar{d} = 2\bar{r}$ (nm)
T5	θ'	18.2 ± 3	–
T5 + 100 h at 100°C	θ'	20.8 ± 3.5	–
T5 + 100 h at 200°C	θ'	30.9 ± 4.5	–
T5 + 100 h at 250°C	θ	–	39.4 ± 5.5
T5 + 100 h at 320°C	θ	–	107.0 ± 8

Table 2: Mean volumetric precipitate diameters from TEM image analysis.

MECHANICAL BEHAVIOR

The precipitate coarsening effect can be characterized either by mechanical testing (fig. 3) or by means of hardness measurements (fig. 4). Figure 3 describes the cyclic behavior at room temperature of the alloy after several aging conditions in plastic strain controlled tests at $\Delta\varepsilon^p/2 = 0.3\%$. The evolution of the peak stress amplitude, $\Delta\sigma/2$, (the average of the tensile and the compressive peak stresses) *versus* cumulated plastic strain clearly shows the effect of aging on mechanical properties. The total amount of softening represents about 100% of the initial stress (T5) for specimen aged at 320°C.

Figure 4 illustrates the evolution of hardening *versus* time. Macrohardness measurements (fig. 4a) were found to be correlated with yield stress values, and give a good view of the transformation rate. The measurements confirm that the maximum softening occurs for aging at 320°C. A two–level test at 200°C then 320°C demonstrates that the asymptotic value obtained during the preheating at 200°C is forgotten at the second level, and that the asymptotic value at 320°C is reached. Other tests [18] show that, in a complex temperature history, the final asymptotic value depends first on the maximum temperature. Microhardness measurements (fig. 4b) show that the softening is supported by the α-phase due to the coarsening of the Al_2Cu precipitates.

PRESENTATION OF THE MACROSCOPIC MODEL

The model is fully described in a companion paper in this conference [33]. Since the purpose of this work is to show that the microstructural evolution of the alloy determines its mechanical response, a short summary is given here.

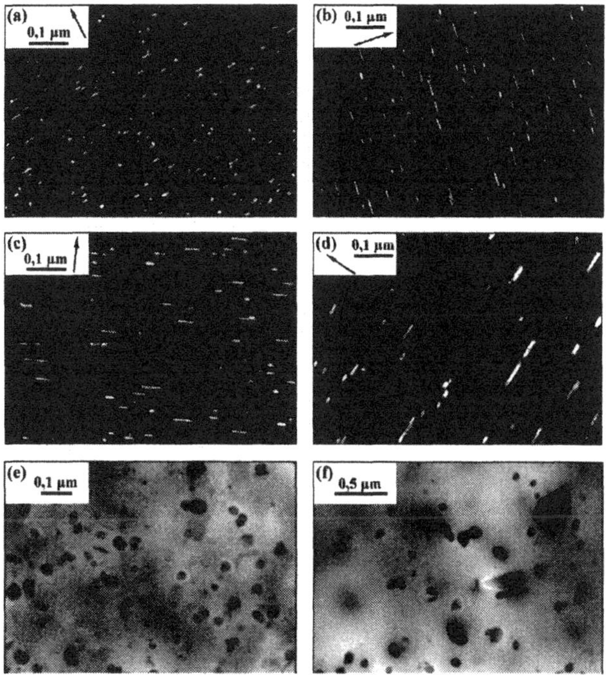

Figure 2: Growth of Al_2Cu precipitates during 100 h aging from the T5 Treatment (a) T5 (b) T5 + 100 h at 100°C (c) T5 + 100 h at 175°C (d) T5 + 100 h at 200°C (e) T5 + 100 h at 250°C (f) T5 + 100 h at 320°C . (a-d) dark field TEM micrographs, $\vec{g} = [002]_{\theta'}$, (e,f) bright field TEM image performed on zero tilt.

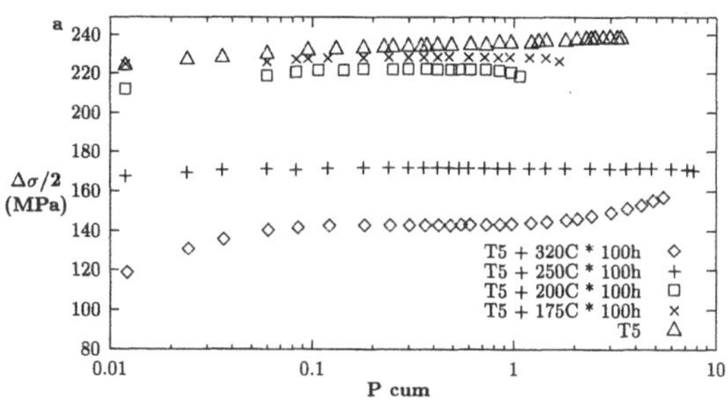

Figure 3: Effect of aging on mechanical properties : stress amplitude at room temperature after several aging conditions versus cumulated plastic strain rate.

The model is an extension of a classical unified viscoplastic model for cyclic loadings [34],

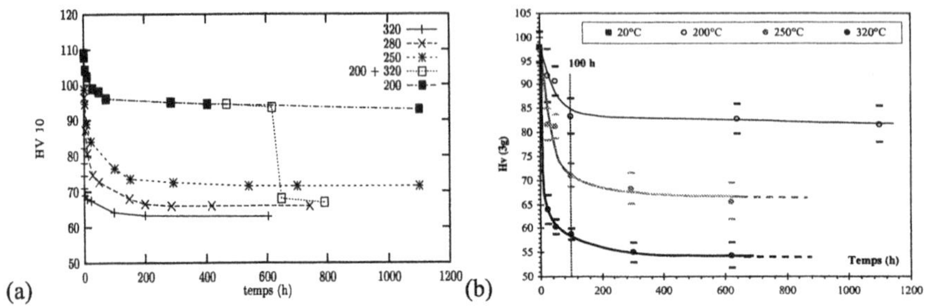

Figure 4: Aging at different temperatures (a) Macrovickers hardness evolution of the alloy (10 kg) (b) Microvickers hardness evolution of the α-phase (3 g).

with additional terms to take into account the aging of the alloy. Indeed, the physical properties of this material are very sensitive to the thermomechanical loading history. The kinematic variable is the tensor $\underset{\sim}{\mathbf{X}}$ and the isotropic variable is the scalar R. The yield criterion is defined with a von Mises function :

$$f = J(\underset{\sim}{\sigma} - \underset{\sim}{\mathbf{X}}) - R - k \tag{1}$$

$$\text{with} \quad J(\underset{\sim}{\mathbf{A}}) = \left((3/2)\underset{\sim}{\mathbf{A}}^d : \underset{\sim}{\mathbf{A}}^d\right)^{0.5} \quad \text{and} \quad \underset{\sim}{\mathbf{A}}^d \ \text{deviatoric part of} \quad \underset{\sim}{\mathbf{A}} \tag{2}$$

The normality rule is applied to compute viscoplastic strain rate as a power function of f as :

$$\dot{p} = \left\langle \frac{f}{K} \right\rangle^n \quad \text{and} \quad \dot{\underset{\sim}{\varepsilon}}^p = \dot{p}\frac{\partial f}{\partial \underset{\sim}{\sigma}} \tag{3}$$

where $\dot{p}$ is the cumulated plastic strain rate, $\dot{\underset{\sim}{\varepsilon}}^p$ the plastic strain rate and $\langle x \rangle$ the positive part of x. The aging part is represented by a scalar internal variable a, starting from zero, and tending to an asymptotic value a_∞ (when the maximum aging state is reached, $a_\infty = 1$), depending on temperature and describing the precipitate growth. Its evolution is exponential, with a temperature dependent time constant τ :

$$\dot{a} = \left\langle \frac{a_\infty - a}{\tau} \right\rangle \tag{4}$$

The model can thus describe softening related to overheating. It has been identified on isothermal LCF tests and validated on anisothermal TMF tests using an original device given elsewhere [18]. For monotonic onedimensional loadings, explicit expressions are available for X, R, the elastic part k and the viscous part σ_v :

$$X = \tfrac{C_1}{D_1}(1 - e^{-D_1\varepsilon^p}) + \tfrac{C_2}{D_2}(1 - e^{-D_2\varepsilon^p})(1 - a) \tag{5}$$

$$R = Q(1 - e^{-b\varepsilon^p}) \tag{6}$$

$$k = R_0 + R_0^*(1 - a) \tag{7}$$

$$\sigma_v = K\,(\dot{\varepsilon}^p)^{1/n} \tag{8}$$

Finally, in a tension test: $\sigma(\varepsilon^p, \dot{\varepsilon}^p, a, T) = X + R + k + \sigma_v$, where each parameter is a function of temperature except D_1 and D_2 which are chosen to be constant. These parameters are identified with experimental data and define the material file to be implemented in the code. The model has been implemented in the general finite element code ZeBuLon [35] , having the numerical simulation of automotive cylinder head in view.

DISCUSSION

Following [36], it is now interesting to compare the macroscopic stress model developed above with the stress resulting from microscopic investigations. Indeed, both models are supposed to be linked by a classical relation between the apparent yield strength variation and the microscopic strength variation due to precipitate coarsening.

Macroscopic level

The stress amplitude can be obtained analytically [34] for a model using isotropic and nonlinear hardening variables. Since aging only affects the isotropic part and the second kinematic hardening, its variation will be simply :

$$\Delta\sigma/2 = \left(R_0^* + \frac{C_2}{D_2}\text{th}(D_2\frac{\Delta\varepsilon^p}{2})\right)(1-a) \tag{9}$$

Microscopic level

Particles are big enough to be looped following the well-known Orowan looping mechanism. For impenetrable hard particles, the Orowan expression is given by :

$$\tau_{Oro} = \frac{2\Gamma}{bL} = \frac{2\beta\mu b}{L} \tag{10}$$

with $\Gamma = \beta\mu b^2$, the tension line where β is a parameter close to 0.5, μ the alloy shear modulus at room temperature and b the Burgers vector lenght ($2.86.10^{-10}$ m). The previous expression as to be corrected by α-phase volume fraction containing the precipitates ($f_\alpha = 73\%$) as follows :

$$\tau_{Oro} = \frac{2\beta\mu b}{L} f_\alpha \tag{11}$$

According to [36], the interparticle spacing is assumed to be :

$$L = \bar{r}\left[\left(\frac{\pi}{f_\theta}\right)^{1/2} - 2\right] \tag{12}$$

where $\bar{r}$ is the mean *radius* of the precipitates (cf tab 2) and f_θ the volume fraction of particles in the α-matrix (3.8%). Since the macroscopic model reference state corresponds to the alloy at maximum aging (100 hours at 320°C and $a = 1$), the variation $\Delta\tau$ to be considered for the microscopic model is :

$$\Delta\tau_{Oro} = \tau_{(T)} - \tau_{(320°C)} \tag{13}$$

Comparison

A transition rule must be introduced to go from the granular scale in the microscopic models to the macroscopic scale. Two crude assumptions may be first considered for that purpose. The static model assumes that all the grains have the same stress (no intergranular residual stresses), and provides a lower bound of the solution. On the other hand, Taylor's model [37] assumes that each grain will present the same plastic strain. The result of both models can be written as :

$$\Delta\sigma_{macro} = M\Delta\tau_{Oro} \tag{14}$$

In texture–free FCC material, the static model gives $M = 2.24$, and Taylor's model predicts a value of 3.07. Self–consistent approaches provide more precise descriptions, which are valid for disordered microstructures, with varying values of M. The result of this study is given in figure 5. Each value is presented with its error bar, keeping in mind that the larger error comes from the measurement of $\bar{r}$ in equation (12). The point (0,0) obtained for maximum aging has to belong to the lines. The value of M is found to be close to 2.5, which means that the present model is intermediate but tend to a static model. This value could change with plastic strain range.

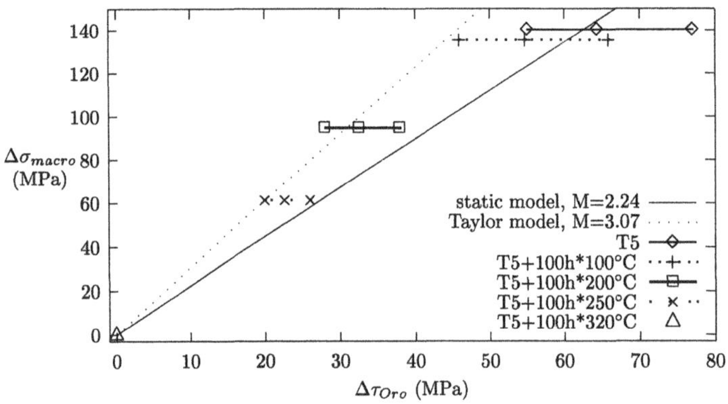

Figure 5: Comparison between macroscopic and microscopic models.

CONCLUSION

The behaviour of a cast aluminum alloy for cylinder head (AISI 319) has been investigated between its initial state T5 and saturated aging (320°C). Variations of physical properties due to microstructure evolution during heating have been exhibited, using micro and macro hardness measurements, TEM image analysis and mechanical testing. It has been found that the coarsening of precipitates follows the Lifshitz-Slyozov-Wagner theory. A numerical macroscopic model, written in a viscoplastic framework and taking into account the description of aging and Bauschinger effect, has been developed. In this model, aging is represented by a scalar internal variable a depending on temperature and time. A comparison can be made between the macroscopic mechanical model and the microscopic approach (Orowan theory). There is a good agreement between the two classes of theories, since the value of the apparent factor between the shear variation in the microscopic models and the variation of the macroscopic yield limit is close to 2.5.

References

[1] SMITH T.J., MAIER H.J., SEHITOGLU H., FLEURY E., ALLISON J. *Metall. Mater. Trans.*, **30A**(1):133–146, 1999.

[2] CATON M.J., JONES J.W., BOILEAU J.M., ALLISON J.E. *Metall. Mater. Trans.*, **30A**(12):3055–3068, 1999.

[3] SEHITOGLU H., QING X., SMITH T., MAIER H.J., ALLISON J.A. *Metall. Mater. Trans.*, **31A**(1):139–151, 2000.

[4] STOLARZ J., MADELAINE-DUPUICH O., MAGNIN T. *Mater. Sci. Eng.*, **A299**:275–286, 2001.

[5] CÁCERES C.H., DJURDJEVIC M.B., STOCKWELL T.J., SOKOLOWSKI J.H. *Scripta metall. mater.*, **40**(5):631–637, 1999.

[6] ROY N., SAMUEL A.M., SAMUEL F.H. *Metall. Mater. Trans.*, **27A**(2):415–429, 1996.

[7] SAMUEL A.M., SAMUEL F.H. *Metall. Mater. Trans.*, **26A**(9):2359–2372, 1995.

[8] CÁCERES C.H., DAVIDSON C.J., GRIFFITHS J.R., WANG Q.G. *Metall. Mater. Trans.*, **30A**(10):2611–2618, 1995.

[9] GUSTAFSSON G., THORVALDSSON T., DUNLOP G.L. *Metall. Trans.*, **17A**(1):45–52, 1986.

[10] HIROSAWA S., SATO T., KAMIO A., FLOWER H.M. *Acta mater.*, **48**:1797–1806, 2000.

[11] SAMUEL A.M., GAUTHIER J., SAMUEL F.H. *Metall. Trans.*, **27A**:1785–1798, 1996.

[12] GREER A.L., BUNN A.M., TRONCHE A., EVANS P.V., BRISTOW D.J. *Acta mater.*, **47**(17):4253–4262, 1999.

[13] MOHANTY P.S., GRUSZLESKI J.E. *Acta mater.*, **44**(9):3749–3760, 1996.

[14] PLAZA D., ASENSIO J., PERO-SANZ J.A., VERDEJA J.I. *Materials Characterization*, **40**:145–158, 1998.

[15] VELDMAN N.L.M., DAHLE A.K, ST. JOHN D.H., ARNBERG L. *Metall. Mater. Trans.*, **32A**(1):147–155, 2001.

[16] NASTAC L. *Acta mater.*, **47**(17):4253–4262, 1999.

[17] CAILLETAUD G., DEPOID C., MASSINON D., NICOULEAU-BOURLES E. In MAUGIN *et al.*, editor, *Continuum thermodynamics : the art and science of modeling material behaviour*. Klüwer, 2000.

[18] NICOULEAU-BOURLES E. Thèse de doctorat de l'Ecole Nationale Supérieure de Mines de Paris, 1999.

[19] SHAH D., ALTSTETTER C. *Mater. Sci. Eng.*, **26**:175–183, 1976.

[20] DE HOFF R.T., RHINES F.N. *Microscopie quantitative*. Masson, Paris, 1972.

[21] UNDERWOOD E.E. *Quantitative stereology*. Addison-Wesley Publishing Co., Reading, MA, 1970.

[22] CAILLETAUD G., CULIÉ J.P., KACZMAREK H. *La Recherche Aérospatiale*, **2**:85–97, 1981.

[23] NICOULEAU-BOURLES E., EL-MAYAS N., MASSINON D., CAILLETAUD G. In SKRYPEK J.J., HETNARSKI R.B., editor, *Thermal stress '99*, pages 241–244, Cracow, June 2000.

[24] AARONSON H.I., CLARK J.B., LAIRD C. *Met. Sc. J.*, **2**:155–158, 1968.

[25] BOYD J.D., NICHOLSON R.B. *Acta Metall.*, **19**(10):1101–1109, 1971.

[26] MERLE P., FOUQUET F. *Acta Metall.*, **29**:1919–1927, 1981.

[27] MERLE P., FOUQUET F., MERLIN J., GOBIN P.F. *Phys. Stat. Sol.*, **35**:213–222, 1976.

[28] SANKARAN R., LAIRD C. *Acta mater.*, **22**(8):957–969, 1974.

[29] LIFSCHITZ I.M., SLYOZOV V.V. *J. phys. Chem. Solids*, **19**(1/2):35–50, 1961.

[30] WAGNER C. *Z. Elektrochem.*, **65**(7/8):581–591, 1961.

[31] CERRI E., EVANGELISTA E., RYUM N. *Metall. Mater. Trans.*, **27A**(2):257–263, 1997.

[32] DICKSON J.I., BOUTIN J., HANDFIELD L. *Mater. Sci. Engng*, **64**:L7–L11, 1984.

[33] NICOULEAU-BOURLES E., FEYEL F., QUILICI S., CAILLETAUD G. In *Temperature-Fatigue interaction*, Paris, 29-31 May 2001. ESIS-Elsevier.

[34] LEMAITRE J., CHABOCHE J.L. *Mécanique des matériaux solides*. Dunod, Paris, 1988.

[35] BESSON, J. AND LE RICHE, R. AND FOERCH, R. AND CAILLETAUD, G. *Revue Européenne des Eléments Finis*, **7**(5):567, 1998.

[36] CULIÉ J.P., CAILLETAUD G., LASALMONIE A. *La Recherche Aérospatiale*, **2**:109–119, 1982.

[37] TAYLOR, G. I. *J. Inst. Metals*, **62**:307, 1938.

Temperature-Fatigue Interaction
L. Rémy and J. Petit (Eds.)

CYCLIC DEFORMATION AND LIFE TIME BEHAVIOUR OF NICR22CO12MO9 AT ISOTHERMAL AND THERMAL-MECHANICAL FATIGUE

M. Moalla, K.-H. Lang, and D. Löhe
Institut für Werkstoffkunde I, Universität Karlsruhe (TH)
Kaiserstr. 12, D-76131 Karlsruhe, FRG

ABSTRACT

In the present study, the materials reaction and the microstructural changes during isothermal and thermal-mechanical fatigue are presented. In total strain controlled isothermal fatigue tests at temperatures between 1123 and 1473K and a frequency of 10^{-2} Hz the cyclic deformation behaviour is influenced by thermally activated recovery and a neutral cyclic deformation behaviour is found. At this condition the life time behaviour is determined by creep-fatigue interactions. In total strain controlled in-phase and out-of-phase thermal-mechanical fatigue tests the initial values of the induced stress amplitudes and plastic strain amplitudes are the higher and the cyclic hardening is the more pronounced, the higher the total mechanical strain amplitude is. The observed cyclic hardening is on the one hand caused by the development of high dislocation densities due to plastic deformation at lower temperatures, and on the other hand by the precipitation of small semi-coherent carbides at higher temperatures. At high total mechanical strain amplitudes with the same magnitude, in-phase tests yield smaller lifetimes than out-of-phase tests. At low total mechanical strain amplitudes the contrary is true. This is the result of competitive processes: creep damage favoured by high tensile stresses at high temperatures under in-phase loading and tensile mean stresses developing during out-of-phase loading.

KEYWORDS

Isothermal fatigue, thermal-mechanical fatigue, Nickel base superalloy, cyclic deformation behaviour, microstructure.

INTRODUCTION

Components operating at high temperatures are subjected to both thermal and mechanical loadings. Due to repeated start-ups, load changes and shut-downs transient temperature fields induce complex stress and strain fields which may cause damage. This phenomenon is called thermal fatigue. In laboratory thermal-mechanical fatigue tests, the internal constraint acting in a component during thermal fatigue is replaced by an external constraint applied in a testing system. During stationary service the components are subjected to mechanical loadings at high temperatures. In this service phase damage may occur due to high temperature isothermal fatigue. Therefore, the cyclic deformation and life time behaviour of the used materials in both isothermal and thermal-mechanical fatigue is of high interest. A typical example of a thermally and mechanically high loaded component is the combustion chamber of a gas turbine. Combustion chambers of stationary gas turbines are commonly made from the Nickel base superalloy NiCr22Co12Mo9. Therefore, isothermal fatigue tests at different temperatures and out-of phase as well as in-phase thermal-mechanical fatigue tests at different maximum temperatures were carried out with specimens made from this material.

MATERIAL AND EXPERIMENTAL SETUP

Material

The material investigated is the solid solution and carbide precipitation hardened Nickel base superalloy NiCr22Co12Mo9 (Inconel Alloy 617, Nicrofer 5520 Co9). The chemical composition is 22.25 Cr, 11.45 Co, 8.88 Mo, 1.28 Al, 0.56 Fe, 0.04 Ti, 0.11 Si, 0.06 C, balance Ni (all quantities in wt. %). It was supplied by Krupp VDM as round bars with a diameter of 19 mm. From the supplier it was solution annealed at 1475K and water quenched. The microstructure of the test material shows grains with a high density of twin boundaries and a relatively high number of uniformly distributed primary M_6C carbides. The mean diameter of the grains is about 180μm. From the supplied bars, solid round specimens with a cylindrical gauge length of 10mm and a diameter of 7mm within the gauge length were machined. The material was investigated in the as received state.

Experimental Details

The isothermal and thermal-mechanical experiments were carried out in a servohydraulic fatigue testing machine with a maximum loading capacity of 100kN. For strain measurement, a high temperature capacitive extensometer was used. The specimens were heated up in an induction furnace with closed loop control. The temperature was measured with a Ni-CrNi thermocouple, which was spot welded close to the gauge length of the specimens. During the thermal-mechanical fatigue (TMF) tests the specimens were cooled by thermal conduction to the water cooled grips and, if necessary, additionally by a proportionally controlled air jet. All experiments were performed under total strain control. The isothermal fatigue tests were carried out at temperatures between 1123 and 1473K using a triangle shaped loading cycle at a frequency of f = 10^{-2}Hz. Thus, the deformation rates in the isothermal fatigue tests are comparable to the deformation rates in the TMF tests.

In the TMF tests T_{min} was always 473K and T_{max} was varied between 1023 and 1473K. The heating and cooling rate was 14K/s resulting in cycle periods ranging from 79 to 143s and frequencies between $1.3 \cdot 10^{-2}$ and $7 \cdot 10^{-3}$Hz, respectively. At the beginning of a TMF test, the specimen is first heated up to the mean temperature T_m. Then it is subjected to three triangle-shaped temperature cycles without any mechanical loading. To determine the thermal expansion and contraction, the total strain of each specimen is measured during these cycles. After that, the testing machine is switched to total strain control and the TMF loading is started. In-phase (IP) and out-of-phase (OP) thermal-mechanical fatigue tests with constant total strain amplitudes $\varepsilon_{a,t}$ were performed. As shown in Fig.1, $\varepsilon_{a,t}$ is the sum of the total mechanical strain amplitude $\varepsilon^{me}_{a,t}$ and the thermal strain amplitude $\varepsilon^{th}_{a,t}$ ($\varepsilon_{a,t} = \varepsilon^{me}_{a,t} + \varepsilon^{th}_{a,t}$). Therefore, in OP tests a total mechanical strain amplitude equal to the thermal strain amplitude is induced and the phase shift between the temperature and the mechanical strain is 180°. Thus, tensile stresses are acting at low temperatures and compressive stresses at high temperatures. In IP tests there is no phase shift

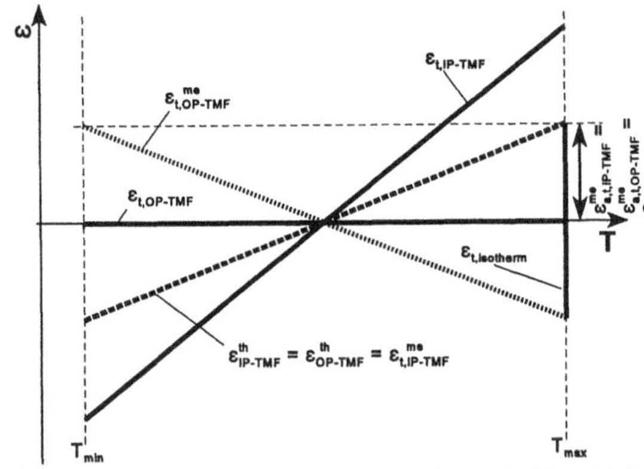

Fig.1: Strain versus temperature in isothermal, IP and OP thermal-mechnical fatigue tests

between mechanical strain and temperature and due to $\varepsilon^{me}_{a,t} = \varepsilon^{th}_{a,t}$ the total strain amplitude $\varepsilon_{a,t}$ is twice the thermal strain amplitude. Therefore, in IP TMF tests tensile stresses are acting at high temperatures and compressive stresses at low temperatures.

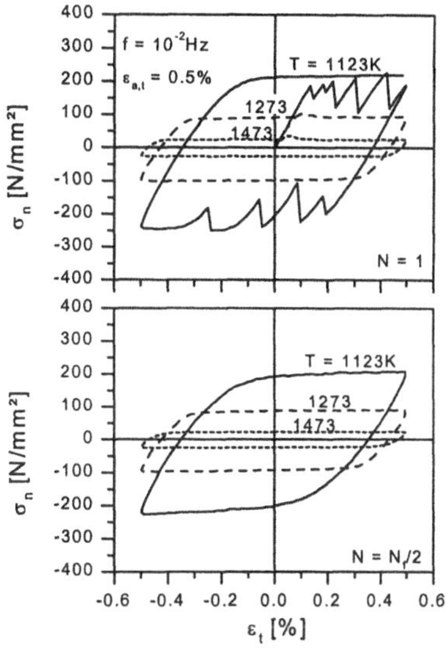

ISOTHERMAL FATIGUE

Cyclic deformation behaviour

In Fig. 2 the stress - total strain course at a total strain amplitude $\varepsilon_{a,t} = 0.5\%$ and test temperatures of 1123, 1273 and 1473K are shown for the first loading cycle (top) and at half of the life time (bottom). With increasing temperature the magnitude of the induced maximum and minimum stresses decreases. The particular values are determined by the material's resistance against deformation at the given deformation rate which is relatively low due to the low test frequency. The plastic strain amplitude which is indicated by the breadth of the hysteresis loops at mean stress increases significantly with increasing temperature because the thermal activation of the dislocation movement and the effectiveness of recovery processes rises. At T = 1123K distinctive fluctuations of the stresses appear in the first cycle. These stress drops are caused by dynamic strain ageing effects as interactions between gliding dislocations and diffusing alloying atoms [1,2]. Such irregularities do not appear any more at half of the number of cycles to failure. Apart from this finding there are no significant changes in the appearance of the hysteresis loops between the first cycle and $N_f/2$.

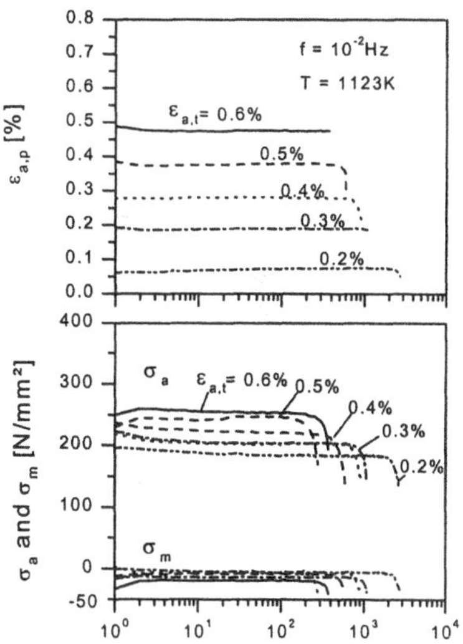

Fig. 3: cyclic deformation curves from isothermal fatigue tests

The plastic strain amplitude in Fig. 3 (top) and the stress amplitude and mean stress (bottom) plotted as a function of the number of cycles for T = 1123K show a neutral cyclic deformation behaviour of the material at all total strain amplitudes investigated. From the second cycle the induced values of $\varepsilon_{a,p}$, σ_a and σ_m remain practically constant up to macroscopic crack formation and crack propagation. Generally, with increasing $\varepsilon_{a,t}$ the plastic strain amplitude increases strongly and the stress amplitude increases slightly. In the first cycle low compressive mean stresses are produced which remain approximately constant during the complete life time. Additional experiments [3] show that at a given total strain amplitude an increase of the test temperature to T = 1273K increases the plastic strain amplitude and reduces the induced stresses. The neutral cyclic deformation behaviour remains, i.e. the plastic strain amplitudes and stress amplitudes produced in the first cycles remain constant up to macroscopic crack initiation. If the test temperature is increased to 1473K the induced stress amplitudes are reduced strongly and only reach magnitudes between 20 and 30 N/mm^2. At this temperature the cyclic deformation curves show a small decrease of σ_a at constant plastic strain amplitudes which presumably has to be put down to creep damage. Not only at 1273K but also at 1473K there are almost no mean stresses observed during the complete life time.

Life time behaviour

Fig. 4 shows the total strain Wöhler curves for the selected temperatures and the examined total strain amplitudes. The total strain amplitude is plotted double logarithmically over the number of cycles to failure. The lines plotted in the figure were calculated with the combination of the Coffin-Manson and Basquin relations [3]. The effect of the temperature on fatigue life is almost negligible at small total strain amplitudes. At high $\varepsilon_{a,t}$ values the number of cycles to failure is reduced with increasing temperature.

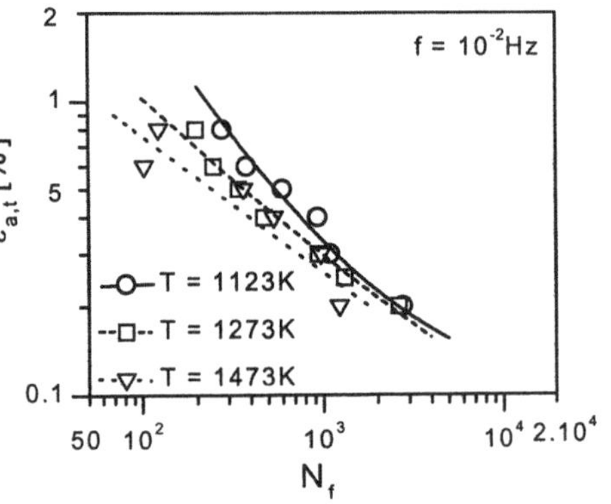

Fig. 4: Total strain Wöhler curves

Microstructure

For selected total strain amplitudes the microstructure of the broken specimens was examined. Fig. 5 shows TEM photographs of specimens loaded at T = 1123K at $\varepsilon_{a,t}$ = 0.2% (left) and 0.5% (right). For $\varepsilon_{a,t}$ = 0.2% first subgrains are formed in areas nearby grain boundaries which are impoverished in alloying atoms. Far from grain boundaries there are homogeneous dislocation networks indicating viscous gliding and dislocations which are diffusely distributed and bent. Regarding the carbide morphology the microstructure is characterised by homogeneously distributed fine secondary precipitations of the type $M_{23}C_6$ within the grains and coarser carbides at the grain boundaries. At $\varepsilon_{a,t}$ = 0.5% also inside the grains areas with a distinctive subgrain structure with numerous dislocations homogeneously distributed within the subgrains are observed. The carbide population can be compared with the one at $\varepsilon_{a,t}$ = 0.2%. Further investigations [3] show that with an increase of the temperature to T = 1273 K and 1473K the development of subgrains is more pronounced and the dislocation density within the subgrains decreases. Beyond that, only relatively coarse secondary carbides are found occasionally within the grains but the grain boundaries show a thick occupancy of carbide precipitations.

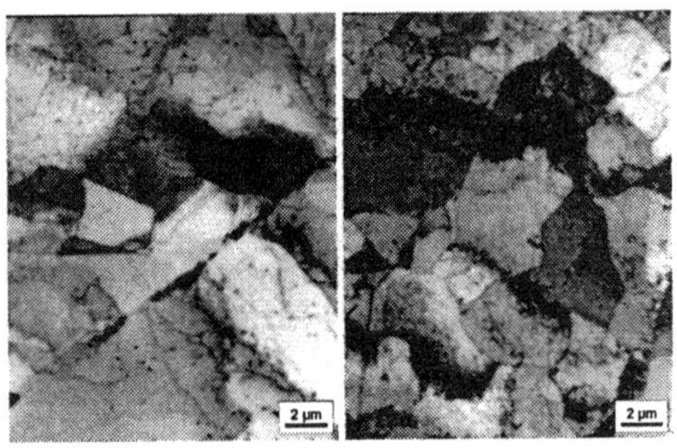

Fig. 5: TEM photos after isothermal fatigue at 1123K at $\varepsilon_{a,t}$ = 0.2% (left) and 0.5% (right)

THERMAL-MECHANICAL FATIGUE

Cyclic deformation behaviour

The σ_a, $\varepsilon^{me}_{a,t}$ hysteresis loops represented in Fig. 6 for the first loading cycle (top) and at the half number of cycles to failure (bottom) were taken from OP-TMF tests at different maximum temperatures T_{max}. During the first heating up of the specimens to T_{max} compressive stresses are induced by the suppression of the thermal expansion. At all maximum temperatures investigated these stresses lead to an elastic-plastic deformation. Due to this plastic compression tensile stresses are produced when cooling down the specimens. At the minimum temperature the tensile stresses reach approximately the same maximum value in the first cycle at all T_{max} investigated. For $T \geq$ 1123K stress drops are result of dynamic strain ageing processes which appear in the first cycle. At T_{max} = 1273K and 1473K the magnitude of the stresses decreases strongly before the maximum temperature and the minimum mechanical strain is reached. This effect is called dynamic stress relaxation in the following. At half of the life time the magnitude of the induced maximum and minimum stresses increases with decreasing T_{max}. A comparison with the material response at N = 1 shows that there is cyclic hardening at $T_{max} \leq$ 1273K which is the more pronounced the lower T_{max} is. The hysteresis loops for all T_{max} investigated exhibit dynamic relaxation effects which become more distinctive the higher T_{max} is. At T_{max} = 1473K typical irregularities caused by dynamic strain ageing effects still occur at $N_f/2$. From the first cycle the hysteresis loops are shifted to tensile stresses due to the different material resistance to plastic deformation at T_{min} and T_{max}. The width of the hysteresis loops indicates that the mechanical plastic strain amplitude $\varepsilon^{me}_{a,p}$ increases noticeably with increasing T_{max} because of the enhancement of thermally activated and diffusion controlled dislocation movement.

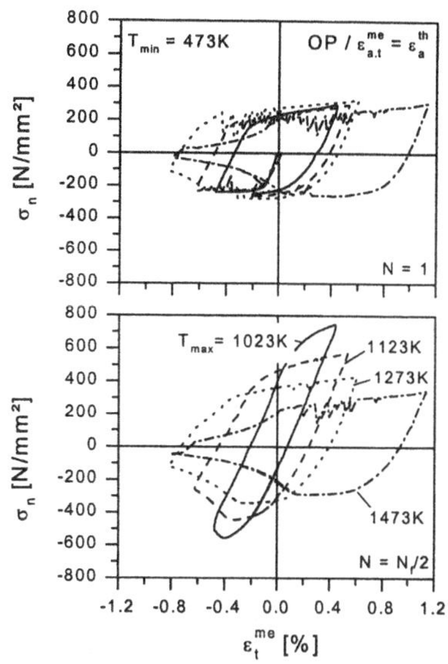

Fig. 6: Hysteresis loops of OP-TMF tests at N = 1 (top) and N = $N_f/2$ (bottom)

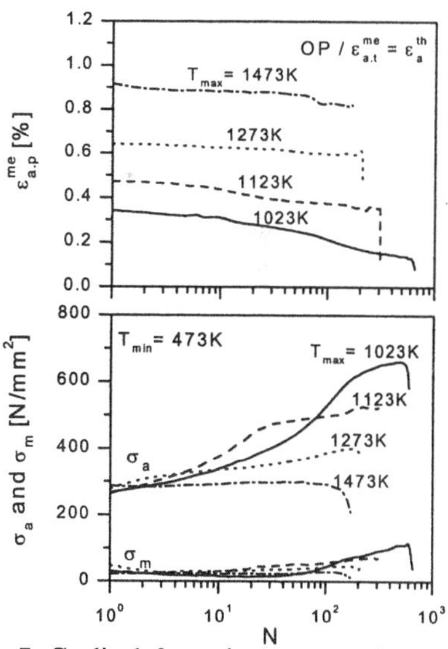

Fig. 7: Cyclic deformation curves at OP loading

Fig. 7 shows the development of $\varepsilon^{me}_{a,p}$ (top) and σ_a as well as σ_m (bottom) for OP TMF tests at different maximum temperatures. For T_{max} between 1023K and 1273K $\varepsilon^{me}_{a,p}$ decreases and σ_a increases with increasing number of cycles. As already mentioned, at this loading conditions the material shows cyclic hardening which is the more pronounced the lower T_{max} is. At T_{max} = 1473K $\varepsilon^{me}_{a,p}$ decreases slightly and σ_a remain nearly constant during the test. At all T_{max} investigated tensile mean stresses are induced in the first cycle which remains constant within the 10 first cycles at about 50 N/mm². During further OP loading at $T_{max} \leq$1273K a continuous increase of σ_m is measured which is most distinct at T_{max} = 1023K and leads to a maximum value of 130 N/mm². At T_{max} = 1473K the mean stress changes only insignificantly during the complete experiment.

In Fig. 8 the hysteresis loops determined at N = 1 (top) and $N_f/2$ (bottom) during in-phase thermal-mechanical fatigue tests at maximum temperatures between 1023 and 1473K are represented. According to the in-phase loading conditions tensile stresses are induced when heating up the specimens in the first cycle. At all T_{max} investigated elastic-plastic deformation occurs during heating up. For T_{max} = 1273 and 1473K dynamic relaxation processes are observed when the combination of temperature and induced stress exceeds a certain value. When cooling down the specimens to T_{min} compressive stresses arise which increase with increasing T_{max}. Between N = 1 and $N_f/2$ the magnitude of the induced maximum und minimum stresses rises at $T_{max} \leq$ 1273K. At the same time the width of the hysteresis loops decrease. This effect is the more pronounced, the lower T_{max} is. At $N_f/2$ dynamic relaxation effects appear in all of the hysteresis loops. In the first cycle all hysteresis loops show, both during heating up as well as during cooling down, considerable irregularities in the stress-strain course which are caused by distinct dynamic strain ageing processes. Such effects occur at N = $N_f/2$ only at T_{max} = 1273 and 1473K. Due to the dynamic stress relaxation at higher temperatures and stresses as well as the different material properties at high and low temperatures the hysteresis loops are moved to compressive mean stresses and positive mean strains in the first cycle. A further shift to higher compressive mean stresses occurs up to N = $N_f/2$.

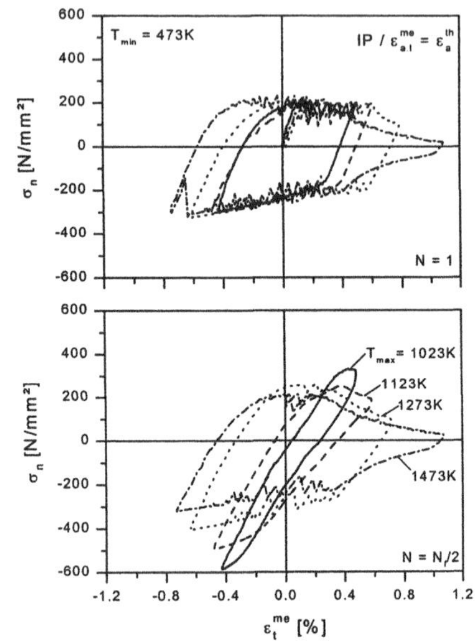

Fig. 8: Hysteresis loops of IP-TMF tests at N = 1 (top) and N = $N_f/2$ (bottom)

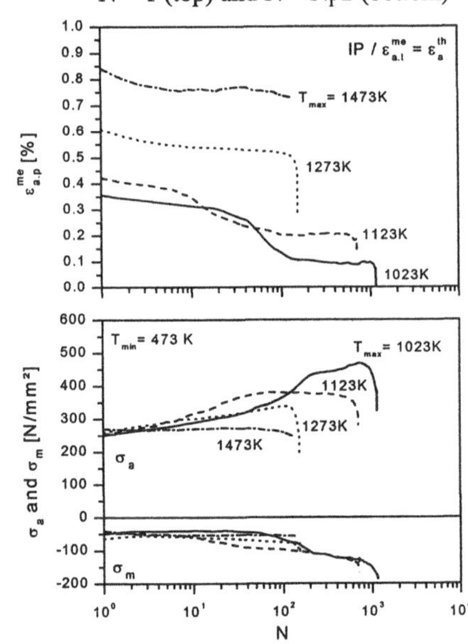

Fig.9: Cyclic deformation curves in IP loading

Fig. 9 shows the dependency of $\varepsilon^{me}_{a,p}$ (top) and σ_a as well as σ_m (bottom) from the number of cycles in IP-TMF with T_{max} between 1023 and 1473K. In all experiments a cyclic hardening appears at the beginning of the test. With increasing T_{max} the cyclic hardening becomes weaker and shorter. It is combined with a decrease of the plastic strain amplitude and an increase of the stress amplitude. After that, the cyclic deformation behaviour is neutral during the largest fraction of the life time where the plastic strain amplitude and the stress amplitude remain almost constant up to the appearance of macroscopic crack initiation. The induced mean stresses which are produced in the first cycle are approximately of the same magnitude for all T_{max} investigated and amount to about -50N/mm^2. Within the first cycles these compressive mean stresses remain nearly constant. Then, at T_{max} between 1023 and 1273 they increase continuously and reach e.g. at T_{max} = 1023 K a magnitude of about 180 N/mm^2. At T_{max} = 1473K the mean stress remains almost constant during the whole experiment.

Microstructure

The TEM photographs in Fig. 10 give a summary of the microstructure of the specimens after failure in OP loading with T_{max} between 1023 and 1473K. After IP-TMF with the same maximum temperature similar microstructures are found. At T_{max} = 1023 and 1123K there is a very high dislocation density which makes it difficult to make a clear correlation to the microstructure. Due to the relatively high mechanical total strain amplitudes and low maximum temperatures, planar dislocation structures dominate with numerous bent dislocation lines as result of viscous gliding. At T_{max} = 1123K first subgrain formation sets in near grain boundaries. While at T_{max} = 1023K the carbide precipitations are hardly visible because of the extremely high dislocation density, at T_{max} = 1123K a relatively high density of the fine secondary carbide precipitations appear which are often arranged along slip traces. At T_{max} = 1273K diffusion controlled relaxation processes at high temperatures increasingly contribute to the deformation processes. Therefore, beside areas with distinctive planar dislocation structures with a very high number of bent dislocations, also areas with numerous subgrains are noticed. These subgrains contain a relatively high dislocation density but only occasionally fine secondary $M_{23}C_6$ carbides. The increase of the maximum temperature to T_{max} = 1473K inreases the plastic strain amplitude which promotes planar dislocation gliding at low temperatures wihtin the TMF cycle on the one hand and enforces, on the other hand, creep controlled relaxation processes with spatial dislocation movements at high temperatures which favours subgrain formation. Furthermore, dissolving and coarsening of carbide precipitations is enhanced. Therefore, at T_{max} = 1473K a clear subgrain structure with planar dislocation arrangements within the cells is observed. The subgrains include high dislocation densities which however are qualitatively smaller than those at the lower maximum temperatures.

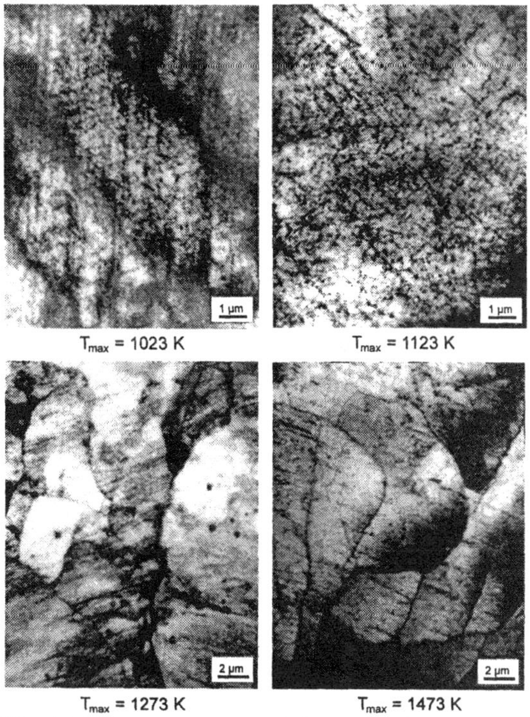

Fig. 10: Microstructure after failure under OP-TMF at different maximum temperatures

Life time behaviour

Fig. 11 shows the total mechanical strain Wöhler curves. The experimental data are well described by the combination of the Coffin-Manson and Basquin relations [3]. At higher $\varepsilon^{me}_{a,t}$ values which are combined with higher T_{max} values, IP loading leads to lower life times than OP loading. At lower $\varepsilon^{me}_{a,t}$ (T_{max}) values the contrary is true. This effect is caused by competitive processes: Tensile stresses at high maximum temperatures acting under IP loading produce creep damage which reduces the life time. At lower T_{max} the development of tensile mean stresses under OP loading is regarded as the dominating effect which influences the life time.

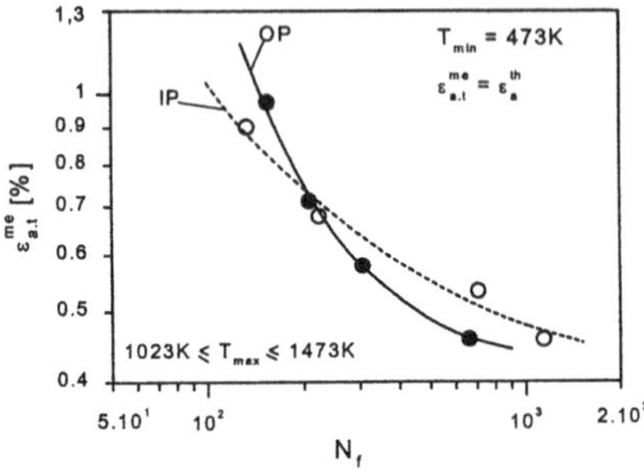

Fig. 11: Total mechanical strain Wöhler curves at OP- and IP-TMF

DISCUSSION

The examined alloy shows maximum dynamic strain ageing processes at a test frequency of 5Hz at temperatures between 673 K and 873 K [1,2,4,5]. A decrease of the frequency shifts this temperature range to lower temperatures. Hence, under the chosen isothermal fatigue conditions the remaining effect of dynamic strain aging is balanced by recovery processes even at 1123K, and a neutral cyclic deformation behaviour is found (see Fig. 3). At the chosen TMF loadings with T_{min} = 473K and T_{max} between 1023 and 1473K the temperature range of dynamic strain ageing effects is passed twice in every temperature cycle. Therefore, not only work hardening by plastic deformation at temperatures close to T_{min} but also dynamic strain ageing effects influence the material reactions during TMF loading. Fig. 12 shows cyclic deformation curves from isothermal tests at T = 1023K and from OP as well as IP TMF tests at T_{max} = 1123K. At all test procedures, with similar mechanical strain amplitudes at the beginning of the experiments, the magnitudes of the induced plastic strain amplitudes and of the induced stress amplitudes are similar. However, under TMF conditions the material shows pronounced cyclic hardening and, hence, much higher stress amplitudes and significantly smaller plastic strain amplitudes at higher number of cycles.

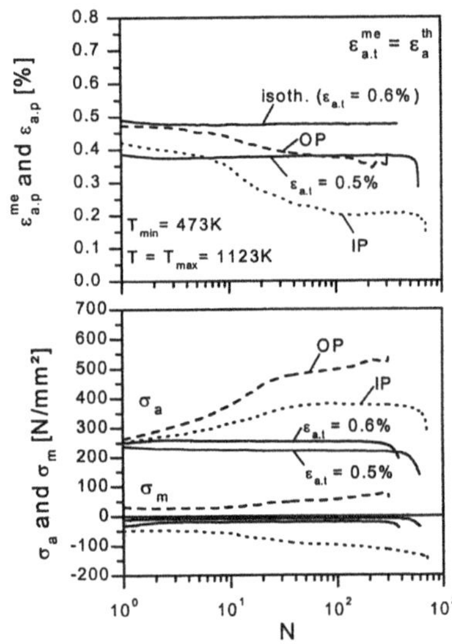

Fig. 12: Cyclic deformation curves under isothermal, IP and OP thermal-mechanical fatigue loading at $T = T_{max}$ = 1123K

Thermal-mechanical fatigued specimens exhibit after failure at lower T_{max} a diffuse planar dislocation structure with a high dislocation and carbide density and at higher T_{max} a subgrain structure with a relatively low dislocation density in the interior of the subgrains (see Fig. 10). This is attributed to the increase of the effectiveness of recovery processes like cross slip of screw dislocations and diffusion controlled climbing of edge dislocations with increasing T_{max}. At a given mechanical strain amplitude the appearing dislocation and precipitation structure is essentially determined by the maximum and minimum temperature of the thermal cycle [6]. Therefore, almost the same microstructures develop in OP and IP loading at the same $\varepsilon^{me}_{a,t}$ value. In contrast, isothermal fatigue at all temperatures and total strain amplitudes investigated results in the formation of subgrain structures (see Fig. 5). Here, obviously a balance exists between hardening processes which are caused by dislocation movements and production of new dislocations and recovery processes due to diffusion controlled climbing and thermally activated cross slip effects.

In Fig. 13 the numbers of cycles to failure in isothermal and thermal-mechanical fatigue are compared in a total mechanical strain Wöhler diagram. It is obvious that an estimation of the fatigue life under thermal-mechanical loading on the base of results from isothermal tests is not possible. Depending on total strain amplitude and temperature in the isothermal experiment, or on maximum temperature in the TMF experiment either too high or too low life times are assessed. Only at T = T_{max} = 1273K, N_f of isothermal and thermal-mechanical fatigue tests at a total (mechanical) strain amplitude of about 0.7% lies in the same life time range. The above mentioned different deformation mechanisms as well as the different damage developments under isothermal and thermal-mechanical loading are primarily regarded as reasons. In OP-TMF e.g. a distinct cyclic hardening results in high stress ranges and enforces the formation of high tensile stresses at lower temperatures. Therefore, the protective oxide layer is damaged and numerous surface cracks develop [7]. Such a process which is not observed under isothermal total strain controlled fatigue loading contributes to an earlier failure of the specimen because of the large stress ranges occurring during OP-TMF providing high driving forces for crack propagation. Under IP conditions the combination of high temperatures and high tensile stresses at higher temperatures as well as an asymmetrical gliding of grain boundaries produces distinct intergranular volume damage in the form of creep pores and creep cracks [8]. Comparable effects were not observed after isothermal fatigue not even at very high temperatures even though under the given loading conditions the deformation behaviour is influenced by thermally activated recovery processes and a neutral cyclic deformation behaviour is found. The isothermal life time behaviour is determined by creep-fatigue interactions. Therefore, at higher total strain amplitudes the increase of the plastic strain amplitude with increasing temperature results in a decrease of the number of cycles to failure.

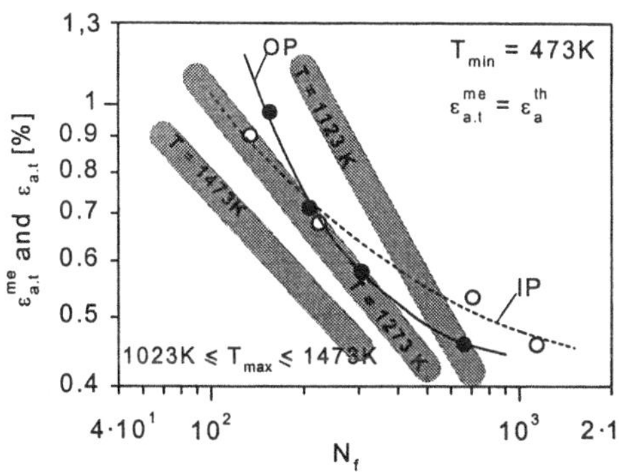

Fig. 13: Total mechanical strain Wöhler curves in isothermal, OP- and IP thermal-mechanical fatigue

SUMMARY

In the present study, the reaction of NiCr22C012Mo9 and the microstructural changes during total strain controlled isothermal and thermal-mechanical out-of-phase and in-phase fatigue tests are presented. In total strain controlled isothermal fatigue tests at temperatures between 1123 and 1473K and a frequency of 10^{-2} Hz the cyclic deformation and life time behaviour is determined by creep-fatigue interactions. In total strain controlled in-phase and out-of-phase thermal-mechanical fatigue tests, the initial values of the induced stress amplitudes and plastic strain amplitudes are the higher and the cyclic hardening is the more pronounced, the higher the total mechanical strain amplitude is. The observed cyclic hardening is on the one hand caused by the development of high dislocation densities due to plastic deformation at lower temperatures, and on the other hand by the precipitation of small semi-coherent carbides at higher temperatures. At high total mechanical strain amplitudes with the same magnitude, in-phase tests yield smaller lifetimes than out-of-phase tests because of creep damage favoured by high tensile stresses acting at high temperatures under in-phase loading. At low total mechanical strain amplitudes out-of-phase loading results in lower life times. Under these conditions tensile mean stresses developing during out-of-phase loading are regarded as the dominating effect which influences the life time.

AKNOWLEDGEMENT

The financial support of the investigation by the „Deutsche Forschungsgemeinschaft (DFG)" within the „Sonderforschungsbereich (SFB) 167 " is gratefully acknowledged.

LITERATURE

[1] G. Merckling, K.-H. Lang, D. Eifler, O. Vöhringer: Creep-fatigue behaviour of solid-solution hardened superalloys at temperatures up to 1473K. In: D.G. Brandond, R. Chaim, A. Rosen (Eds.) Proc. ICSMA 9, Haifa, Isreal, London (1991), Vol. I, 443-450

[2] G. Merckling, K.-H. Lang, E. Macherauch: Das Ermüdungsverhalten von NiCr22Co12Mo9 im Temperaturbereich 295K $\leq$ T $\leq$ 1473K. Z. Metallkd. 84 (1993) 12, 844-853

[3] M. Moalla: Isothermes und thermisch-mechanisches Ermüdungsverhalten bei überlagerter LCF/HCF-Beanspruchungen sowie isothermes Ermüdungsrißausbreitungsverhalten von NiCr22Co12Mo9 und CoCr22Ni22W14: Dr.-Ing. Thesis, Universität Karlsruhe (TH), (2000)

[4] K.-H. Lang, D. Eifler, E. Macherauch: Fatigue behaviour of Ni-base alloys up to 1273 K; In: P.O. Kettunen, T.K. Lepistö, M.E. Lehtonen (Eds.), Proc. ICSMA 8, Tampere, Finnland, Pergamon Press Oxford (1988), Vol. 2, 1245-1250

[5] G. Merckling: Kriech und Ermüdungsverhalten ausgewählter metallischer Werkstoffe bei höheren Temperaturen. Dr.-Ing. Thesis, Universität Karlsruhe (TH), (1989)

[6] B. Kleinpaß, K.-H. Lang, D. Löhe, E. Macherauch: Influence of the Minimal Cycle Temperature on the Thermal-Mechanical Fatigue Behaviour of NiCro22Co12Mo9. In: J. Lecomte-Beckers, F. Schubert, P.J. Ennis (Eds.) Materials for Advanced Power Engineering, Schriften des FZ Jülich, Reihe Energietechnik/Energy Technology; ISBN 3-89336-228-2, Vol. 5, Part III(1998) 1369-1377

[7] D.A. Boismier, H. Sehitoglu: Thermo-Mechanical Fatigue of MAR-M247: Part I - Experiments. J. Engineering Materials and Technology, Vol 112 (1990), 68-89

[8] B. Kleinpaß, K.-H. Lang, D. Löhe, E. Macherauch: Thermal-Mechanical Fatigue Behaviour of NiCr22Co12Mo9. In: J. Bressers, L. Rémy (Eds.), Fatigue under Thermal and Mechanical Loading, Kluwer Academic Publishers, Dordrecht/Boston/London (1996), 327-337

Temperature-Fatigue Interaction
L. Rémy and J. Petit (Eds.)

TEMPERATURE AND ENVIRONMENTAL EFFECTS ON LOW CYCLE FATIGUE RESISTANCE OF TITANIUM ALLOYS

J. Mendez, S. Mailly* and P.Villechaise
LMPM / ENSMA, 1 Avenue C. Ader BP 40109
F - 86961 FUTUROSCOPE CHASSENEUIL France

ABSTRACT

The fatigue behaviour of the titanium alloy Ti 6246 (6Al-2Sn-4Zr-6Mo) has been compared in air, considered as an aggressive environment, and in vacuum as an inert reference environment in the temperature range 20°C – 500°C. In this way the intrinsic effect of temperature and environment – temperature coupled effects has been identified. The environment effect on fatigue lifetime is relatively small at room temperature but becomes dramatic above 300°C. Moreover it has been shown that whatever the temperature is, below a value σ_{max}/σ_{YS} of 0.72 a transition in crack initiation mechanisms occurs from the surface to the interior of the specimen. In the last case no difference in the fatigue life in air and in vacuum is observed. The role of surface damage for the high stress levels has also been underlined by investigating the influence of high strength thin NiTi coatings. A significant increase of fatigue lifetimes has been obtained with this treatment, the fatigue lifetime in air of coated specimens exceeding the fatigue resistance in vacuum of the untreated material.

KEYWORDS

titanium alloys, temperature effects, environment effects, fatigue lifetime, coatings

INTRODUCTION

High resistant titanium alloys have been developed for applications in modern turbo engines in which these materials are commonly used for manufacturing turbine blades or compressor disks. Even if most of these components work at temperatures below 300°C, some of them can reach temperatures close to 500°C. Moreover, starts-stops sequences of the engine induce fatigue solicitations. This cyclic loading can initiate cracks which propagation can lead to the components failure.

The aim of this work was to determine the influence of temperature, in the range of 20°C - 500°C, on the low cycle fatigue behaviour, of an $\alpha+\beta$ titanium alloy Ti 6246 (6Al-2Sn-4Zr-6Mo). Tests were conducted in air considered as an aggressive environment and in vacuum as an inert reference environment. By this way the intrinsic effects of temperature and the environment-temperature coupled effects have been identified.

EXPERIMENTAL PROCEDURE

The chemical composition of the Ti 6246 is indicated in Table 1. The thermomechanical treatment applied to the Ti 6246 was: a “Hot Die” type forge at 950°C, a first solution treatment at 930°C during 2 hours with water-quenching, a second solution treatment at 900°C for 1 hour with air cooling and finally an annealing at 595°C during 8 hours with air

* present address : CECM-CNRS, Vitry sur Seine

cooling. This material is an (α+β) alloy with an α lamellar Widmanstätten type microstructure (lamellae length varying from 1 to 50µm) in a β transformed matrix hardened by fine secondary α lamellae (less than 1 µm in length). The microstructure morphology is illustrated in figure 1.

element	Al	Sn	Zr	Mo	C	O	Fe	H	Si	Cu
Wgt. %	5.68	1.98	3.96	6.25	0.01	<0.15	0.05	<0.015	0.04	<0.01

Table 1: Chemical composition of the Ti 6Al-2Sn-4Zr-6Mo alloy (wgt. %).

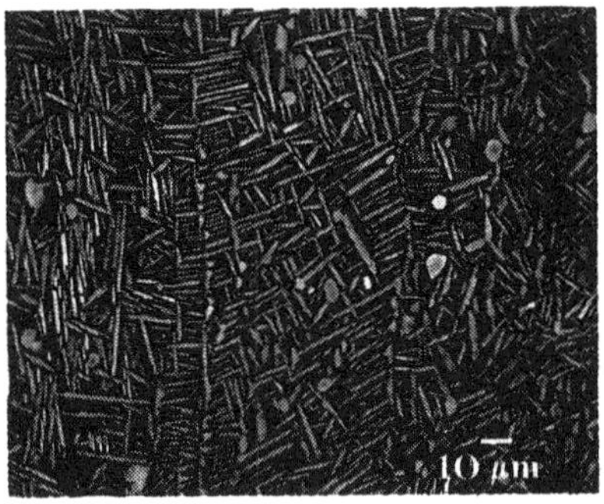

Fig. 1: microstructure of the Ti 6Al-2Sn-4Zr-6Mo alloy.

The mechanical characteristics for each temperature are summarised in Table 2. Young's modulus (E), determined by a resonant frequency method, is observed to decrease linearly with the temperature [1].

Ti6246	20°C	300°C	425°C	500°C
E (Gpa)	125.4	112.5	105.5	103
σ_y (MPa)	993	716	717	680
σ_m (MPa)	1140	980	930	900
A (%)	10.1	10.9	13.7	10.5

Table 2: Mechanical properties of Ti 6246.

In contrast, the yield stress evolution exhibits a plateau between 300°C and 425°C which can be explained by dynamic strain ageing due to interactions between solute atoms and dislocations [2].

Fatigue tests were performed on smooth cylindrical specimens having a gauge length and diameter of 13 mm and 4.4 mm respectively. After polishing up to a 1 μm diamond paste, tests were conducted under stress controlled push-pull mode (R_σ = -1) with a frequency of 0.15 Hz and a triangular wave form. Some additional tests, at the lowest stress levels, were performed with a frequency of 15 Hz. Tests were conducted in air or in vacuum (~5×10^{-4} Pa).

The role of thin NiTi coatings elaborated by a dynamic ion mixing PVD technique was investigated. Fatigue tests on coated specimens were performed at room temperature in stress control (R_σ=-1) or at 300°C and 425°C in total strain control (R_ε=-1).

RESULTS

INTRINSIC BEHAVIOR: FATIGUE LIFETIME IN VACCUM

The $\Delta\sigma/2$ - N_f curves (Stress amplitude versus the Number of cycles to failure) were established in vacuum for four temperature levels (20°C, 300°C, 425°C and 500°C). They are reported in figure 2. These results correspond to a testing frequency of 0.15 Hz for fatigue lifetimes lower than 10^5 cycles and of 15 Hz for higher fatigue lifetimes. Fatigue resistance is observed to decrease with temperature. A continuous evolution of the fatigue life with the stress amplitude is noted. For a given temperature, this evolution can be represented by a single relation $\Delta\sigma/2 = A.Ln\ N_f + B$.

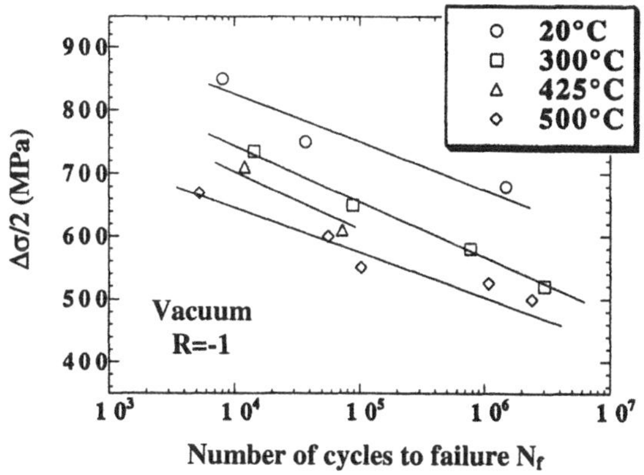

Fig. 2: Stress amplitude versus the Number of cycles to failure N_f

However, whatever the testing temperature, a modification in crack initiation mode was observed for lowest stress amplitudes. At high stress levels (fatigue life below 10^5 cycles) initiation takes place at the specimen surface. In contrast, at lower stress levels an internal initiation is observed. Such transition from the surface to the bulk takes place for stress amplitudes relatively high: about 80 % of the conventional yield stress (i.e. 580 MPa for 300°C and 550 MPa for 500°C). Moreover the normalisation of the results presented in figure 1 by the Young's modulus leads to a quasi-single curve (Figure 3). This shows that in vacuum for the stress amplitude range investigated in the pseudo-elastic regime, the elastic strain governs fatigue lifetime. This is in accordance with the fact that for such a high strength material having relatively low Young's modulus, cyclic plastic strain amplitude remains very low compared to the elastic strain amplitude in the low cycle fatigue range (for $\Delta\varepsilon_p/2=2.10^{-3}$, the cyclic elastic strain $\Delta\varepsilon_e/2=8.10^{-3}$ at 20°C and $\Delta\varepsilon_e/2=6.6\ 10^{-3}$ at 500°C). Therefore, the intrinsic effect of temperature in an inert environment is only related to the variation of the material elastic characteristics. Previous investigations have shown that in the range of 10^4 to 10^5 cycles, the total fatigue lifetime is mainly concerned by crack growth [3]. Now it has been shown that in vacuum crack growth rate is governed by $\Delta K/E$ [4]. It can be noted that for the highest fatigue lifetimes at which crack initiation stage becomes predominant, the rationalisation of the results by the Young's modulus is not so effective.

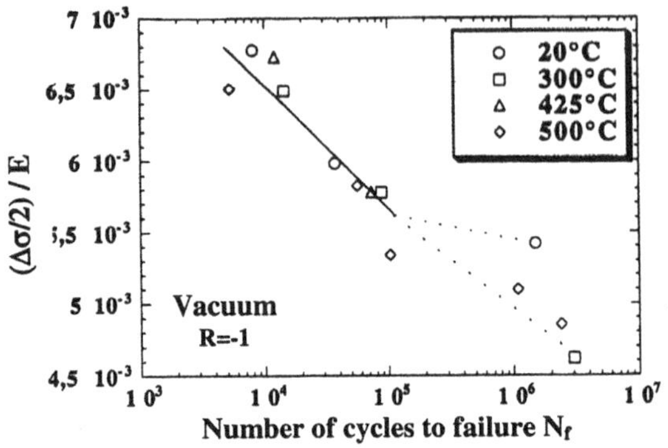

Fig. 3: ($\Delta\sigma/2$)/E – N_f curves for tests performed in vacuum from 20°C to 500°C.

FATIGUE LIFETIME IN AIR

The figure 4 presents fatigue lifetimes obtained in air at 20°C, 300°C, 425°C and 500°C. As for tests performed in vacuum, for the highest stress amplitudes the frequency was 0.15 Hz (fatigue lifetime less than 10^5 cycles) but for the lowest stress amplitudes the frequency was 15 Hz. In this figure the arrows indicate a non-failure of specimens tested at 0.15 Hz. It is clear that the detrimental effect of the increase of testing temperature in air is more marked than in vacuum. For instance, to keep a fatigue lifetime of 10^4 cycles by increasing the testing temperature from 20°C to 500°C, a decrease of 80 MPa is sufficient in vacuum. In contrast 200 MPa are required in air.

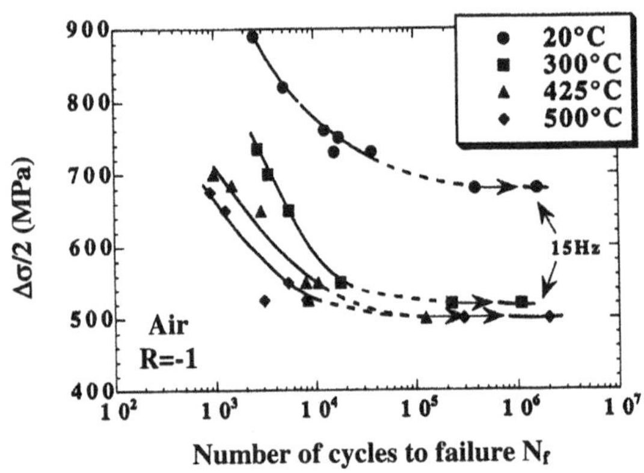

Fig. 4: Fatigue lives of Ti 6246 in air from 20°C to 500°C.

The most remarkable point in this figure is the discontinuity in the fatigue life evolution which appears at low stress levels. Beyond 10^4 - 5.10^4 cycles, a very small decrease in the stress level produces an increase in fatigue lifetime of two order of magnitude. This behaviour is noted whatever the temperature is for an amplitude of approximately 0.72 σ_{ys}. This is associated with a transition in crack initiation process from the surface to the bulk. In vacuum such a transition was observed at slightly higher stress levels (0.81 σ_{ys}). However in vacuum this transition does not imply a modification in fatigue lifetime. In air a strong increase in fatigue life is associated with this change in crack initiation mechanisms. This is explained by the fact that initiation and crack growth from the surface is highly assisted by the environment whereas the initiation within the bulk occurs without any influence of the environment. Thus, for the lowest stress levels no difference in fatigue lives is observed between tests performed in air and in vacuum. Figure 5 illustrates air-vacuum differences at 300°C, temperature at which environmental effects are particularly marked. In contrast at room temperature the effect of environment remains limited (Figure 6).

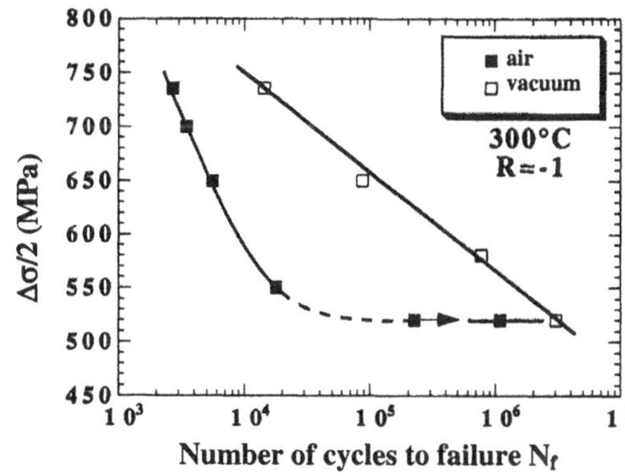

Fig. 5: $\Delta\sigma/2 - N_f$ curves of Ti 6246 in air and in vacuum at 300°C.

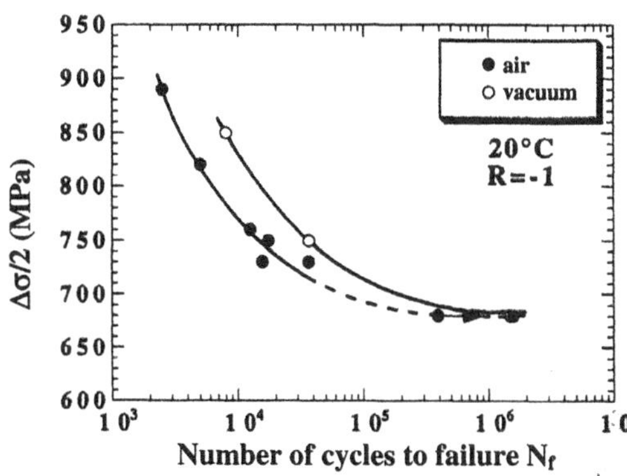

Fig. 6: $\Delta\sigma/2 - N_f$ curves of Ti 6246 in air and in vacuum at 20°C.

In the case of air tests the influence of temperature can not be rationalised by the evolution of the Young's modulus as in vacuum (see Figure 7). This is due to the higher effect of the environment at high temperature compared to room temperature. At 300°C, 425°C and 500°C the acceleration of damage by the aggressive species is of the same order and therefore fatigue lives are quite similar as a function of $\Delta\sigma/E$.

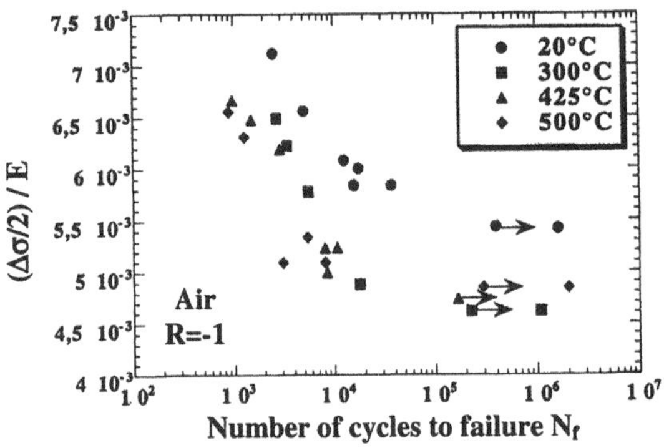

Fig. 7: $(\Delta\sigma/2)/E - N_f$ curves for tests performed in air from 20°C to 500°C.

DAMAGE MECHANISMS

At low stress amplitude cracks initiate within the bulk as illustrated in figure 8a. At the origin of this cracking process any particular microstructural defect as porosity or inclusion has been detected. The crack is transgranular with regard to ex-β grains. Further investigations are necessary to determine precisely the nature of microstructure favouring this damage process. Nevertheless, it is clear that this mechanism is not assisted by the environment.

In the stress amplitude range for which crack initiation takes place at the specimen surface, initiation sites are the α lamellae - β transformed matrix interfaces (see figure 8b), whatever the stress level and the temperature are [5].

(a)

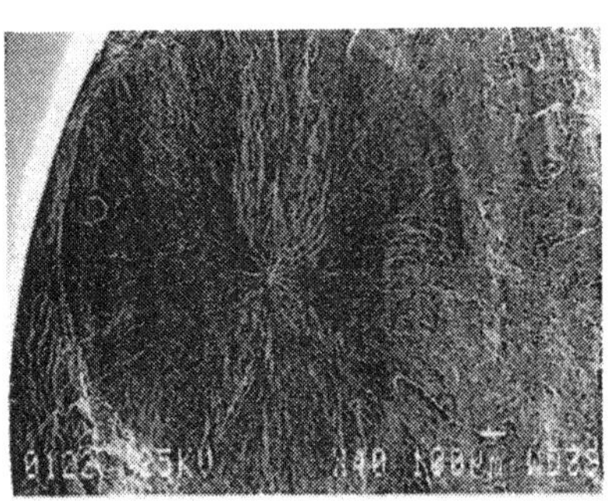

(b)

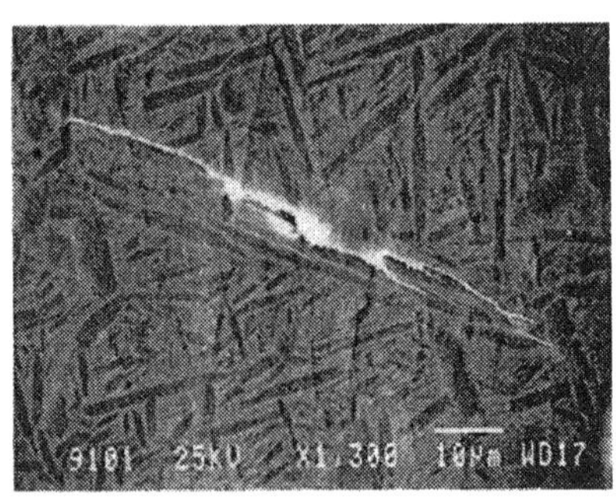

Fig. 8: a: crack initiation within the bulk.
b: crack initiation at the α lamellae - β matrix interfaces.

This fatigue damage mechanism is highly influenced by the environment due to adsorption and / or diffusion of active species (hydrogen, oxygen). Such process are favoured by the local stress gradient at the α/β interfaces caused by deformation incompatibilities between the two phases. Our tests do not allow to separate the respective role of oxidation and water vapour. However the intrinsic effect of an oxide layer has been investigated by performing fatigue tests in vacuum with pre-oxidised specimens. Pre-oxidation was performed by keeping the specimen at a constant temperature (300°C or 500°C) in air for a period corresponding to the duration of a fatigue test in this environment. Pre-oxidised samples were subsequently tested at 300°C or 500°C in vacuum. These specimens have presented the same fatigue lifetime than the reference non-oxidised ones. Therefore, the reduction of fatigue lifetimes in air clearly results of dynamical interaction between water vapour and/or oxygen and cyclic plastic deformation and can not be attributed to a damaging effects due to the oxide layer properties. Such environment coupling effect does not operate anymore at low stress levels at which cracks initiate within the bulk.

INFLUENCE OF THIN NiTi FILMS

In the low cycle fatigue range the free surface plays a major role in crack initiation processes. Consequently the application of surface treatments is a potential way for limiting or suppressing fatigue crack nucleation sites by modifying the mechanical, physical and chemical properties of surface layers. Over the past few years, the influence of thin films elaborated by dynamic ion mixing (DIM) on the fatigue resistance of different metallic materials has been studied [6,7]. This treatment has been developed at the Laboratoire de Métallurgie Physique of the University of Poitiers. It involves a coating deposition method combined with a simultaneous high energy ion implantation which permits the growth of amorphous or nanocrystalline films with thickness on the order of one micrometer [8]. Such coatings exhibit a gradual and diffuse interface with the substrate favouring a great adherence.

It has been found that DIM nanocrystalline or amorphous coatings, specially NiTi films, have considerable beneficial effects on the fatigue resistance of titanium alloys and austenitic stainless steels tested at 20°C and 300°C in the low cycle fatigue range [9]. For Ti6246 samples coated with NiTi films 0.2 μm and 1.0 μm in thickness have cycled in air in tension-compression ($\Delta\sigma/2$ = 850 MPa) up to 5,600 and 12,600 cycles respectively, the fatigue lifetime of the untreated material being 3,900 cycles for these conditions. Lower the cyclic stress amplitude, greater is the improvement: at $\Delta\sigma/2$ = 750 MPa, A 0.2 μm thick NiTi coated sample has been cycled up to 161,000 cycles without developing any surface damage, the reference fatigue lifetime being in this case 12,000 cycles. The fatigue resistance is then improved at least by a factor 18 [10]. Such improvement is not only due to a chemical protection since the fatigue lifetime of the untreated material tested in vacuum at $\Delta\sigma/2$ = 750 MPa is only 35,000 cycles. Ti6246 samples coated with NiTi 1 μm thick films have also been tested at higher temperatures in air. For a total strain amplitude of 6.8 10^{-3}, coated samples have been cycled up to 174,000 and 67,000 cycles, without any damage at 300°C and 425°C respectively. Both results correspond to an improvement by, at least, a factor 8. Therefore in the low cycle fatigue range at which crack initiation takes place at the specimen surface, strong improvement of fatigue resistance of Ti6246 can be obtained with high strength thin coatings. Such a result cannot be expected at low stress amplitudes (below 0.81 σ_{YS} in vacuum or 0.72 σ_{YS} in air for which crack initiation occurs within the specimen's bulk).

CONCLUSION

- The intrinsic effects of temperature on the fatigue resistance of Ti6246 determined by tests performed in vacuum is directly related to the evolution of elastic characteristics.
- In air, the effects of temperature is higher than in vacuum due to the increase of environmental effects beyond 300°C.
- The difference of fatigue behaviour in air and in vacuum is only observed for stress levels higher than 0.72 σ_{YS}. In this stress range, crack initiation takes place at the surface and damage processes are highly assisted by environmental effects. For lower stress levels, crack initiation appears in the bulk and fatigue lifetime is the same in air and in vacuum.
- In the stress range in which cracks initiate at the surface, a considerable improvement of the fatigue life can be obtained at 20°C as at higher temperatures with the application of thin NiTi coatings elaborated by dynamical ion mixing. Thus, this treatment permits the fatigue lifetime to exceed the intrinsic resistance obtained in vacuum.

REFERENCES

1. Péraud S., Pautrot S., Villechaise P., Mazot P., Mendez J. *Thin Solid Films* 292 (1997) 55-60
2. Kubin L.P., Estrin Y., *J.Phys. III.1* pp.929
3. Demulsant X., Mendez J. Fatigue and Fracture of Engineering Materials and Structures, 18 (1995), n°12, 1483-1497
4. Petit J., Henaff G., *Proc. of Fatigue'93* (1993), pp.503
5. Demulsant X., Mendez J., *Materials Science and Engineering* 219 (1996) 202-211
6. Villechaise P., Mendez J., Violan P. Delafond J., *Nucl. Instrum. Methods* B59/60 (1991) 837
7. Péraud S., Villechaise P., Mendez J., *Proc. of Titanium'95* (1995) 2015-2022
8. Jaulin M., Laplanche J.,Delafond J., Pimbert S., *Surf. Coat. Technol.* 37 (1984) 225
9. Péraud S., Villechaise P., Mendez J., *Materials Science and Engineering* A225 (1997) 162-172
10. Péraud S., Villechaise P., Mendez J., Delafond J., *Journal of Matererials Science* 34 (1999) 1003-1008

Temperature-Fatigue Interaction
L. Rémy and J. Petit (Eds.)

INFLUENCE OF TEMPERATURE ON THE LOW CYCLE FATIGUE BEHAVIOUR OF A GAMMA-TITANIUM-ALUMINIDE ALLOY

A.-L. Gloanec, G. Hénaff, D.Bertheau
Laboratoire de Mécanique et Physique des Matériaux
UMR CNRS n°6617, ENSMA- téléport 2,BP 40109
86961 FUTUROCOPE Cédex, France

Abstract

In this study, the influence of temperature on the low cycle fatigue behaviour of a cast alloy Ti- 48Al- 2Cr- 2Nb (atomic %) with a lamellar structure is examined. Two temperatures are considered: room temperature and 750°C, which is close to the brittle-to-ductile transition. Fatigue tests are conducted at different total-strain-amplitude values ranging from ±0.2% to 0.8% in air. The strain ratio ($R_\varepsilon = \varepsilon_{min} / \varepsilon_{max}$) values of –1 and 0 are used. All tests are carried out at a strain rate of 10^{-3} s^{-1}. For a same total-strain-amplitude, fatigue lives are nearly similar at room temperature and at 750°C. However the cyclic stress-strain behaviour at elevated temperature is different from that observed at room temperature: indeed a strain hardening is observed at room temperature during the whole life while at 750°C a stabilization of the stress amplitude is rapidly reached. A Bauschinger effect is detected at room temperature as well as at elevated temperature.

Keywords

TiAl intermetallics; temperature; cyclic stress-strain behaviour; Bauschinger effect; fracture.

INTRODUCTION

Alloys based on the intermetallic compound TiAl have received a considerable attention during the recent years. Indeed, thanks to their excellent combination of properties (low density, excellent high temperature strength of modulus retention, thermal expansion comparable to current alloys, good oxidation and hot corrosion resistance) these alloys are presented as potential competitors to steels and to superalloys in the range 600°C-800°C for applications in aeronautical engines. Nevertheless some aspects concerning their mechanical behaviour under cyclic loading and their durability in service still require thorough investigations. In particular some parts have to withstand low cycle fatigue. However, while many studies have been dedicated to the fatigue crack propagation [1-7], up to now only few data are available on the low cycle fatigue behaviour of these alloys [8,9]. The present study aims to characterize and understand the cyclic stress-strain behaviour of a cast Ti- 48Al- 2Cr- 2Nb alloy with a lamellar microstructure.

EXPERIMENTAL PROCEDURE

MATERIAL

The material investigated is a quaternary alloy (Ti- 48Al- 2Cr- 2Nb), provided in the form of cast bars of 20 mm in diameter and 100 mm in length.
These pieces are hipped under the following conditions: 1260°C / 172 MPa / 4h under argon, during one cycle and followed by a furnace cooling.
The material is then heat-treated as follow: 1300°C /20h under vacuum with gas fan cool.

The grains are columnar with the elongation axis oriented along the radial direction. Those grains have an average size of about 200-500µm. The centre of the bars contains equiaxed grains (their size is around 50-100 µm) due to the slower cooling rate. The size of this central zone decreases while coming close to the bar foot (Fig.1). The grains, either columnar or equiaxed, have a lamellar structure (Fig.2). The microstructure is nearly fully lamellar with coarse and thin lamellae.

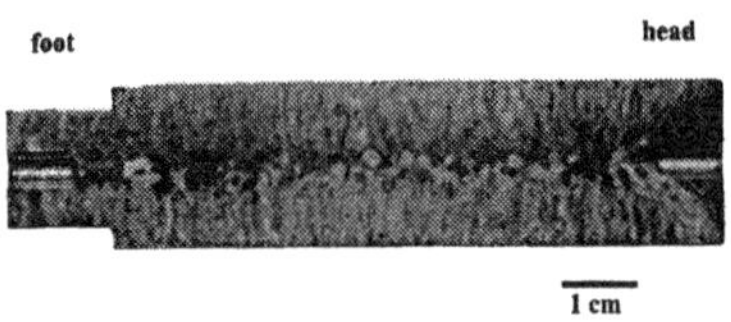

Fig.1: Macrography of a bar

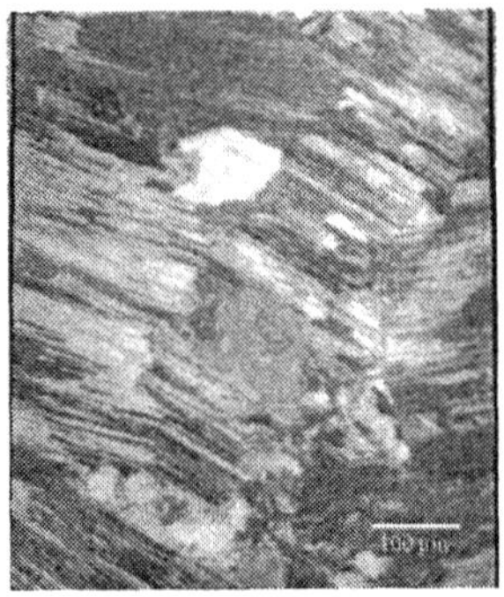

Fig.2: Microstructure of pieces

TEST SPECIMEN

Test specimens are cylindrical with a gage section of 6mm in diameter and 30mm in length and a total length of 92mm. To minimize the effects of surface irregularities on fatigue lives, final surface preparation is achieved by mechanically polishing the gage length of the test specimen with emery paper until grade 4000, to remove all circumferential and surface machining marks.

MECHANICAL TESTING

Low cycle fatigue tests are conducted on an electro-mechanical test system (INSTRON 1362) at different total-strain-amplitudes ranging from ±0.2% to ±0.8% in air. Two temperatures are considered: room temperature and 750°C. Total-strain-amplitude is controlled by an INSTRON extensometer, with a root of 15±0.5mm, placed on the gage of test specimen. Total-strain-amplitude is initiated either in tension or in compression. The test command signal is triangular. Tests are conducted with a constant strain rate of $10^{-3}s^{-1}$ in total-strain-amplitude control with a strain ratio ($R_\varepsilon = \varepsilon_{min}/\varepsilon_{max}$) of –1 and 0. The stress-strain hysteresis loops are periodically recorded by means of an analogic plotter and a computer. The recording of the stress as a function of time is carried out with an X-t recorder.

MONOTONIC STRESS-STRAIN BEHAVIOUR

Information about the monotonic stress-strain behaviour of the material can be derived from the first cycle. Average values of Young's modulus and yield stress at room temperature and at 750°C are given in table1.

	Room temperature	750°C
Young's modulus	166 ± 11 GPa	139 ± 2 GPa
Yield stress	420 ± 16 MPa	349 ± 8 MPa

Table1: Mechanical properties of the material

The Young's modulus and the yield stress slightly decrease with increasing temperature, about 14% for the Young's modulus and around 20% for the yield stress. Similar trends have been observed by Recina [10] and Jones [11].

RESULTS

CYCLIC STRESS-STRAIN BEHAVIOUR AT ROOM TEMPERATURE AND AT 750°C

AT ROOM TEMPERATURE. Figures 3 and 4 offer an outline of the variation of adimensionnal stress-amplitude (σ/σ_i where σ_i is the initial stress-amplitude) with the normalized lifespan (N/N_f where N_f is the fatigue life, i.e the number of cycle when the specimen breaks) with different strain ratios ($R_\varepsilon = -1$ in Fig.3 and $R_\varepsilon = 0$ in Fig.4).

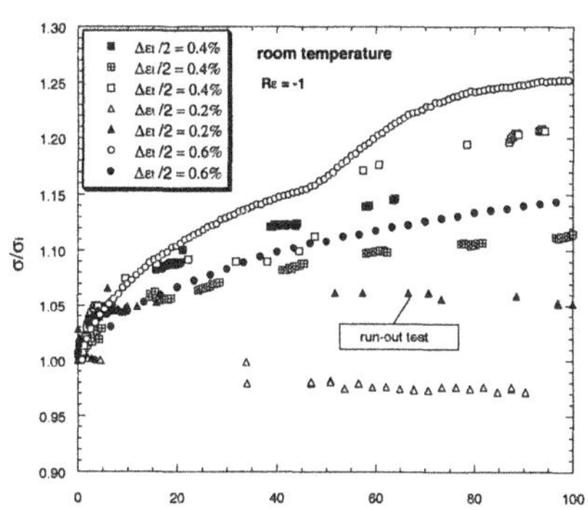

Fig.3: Variation of adimensionnal stress-amplitude versus normalized lifespan at room temperature and with $R_\varepsilon = -1$

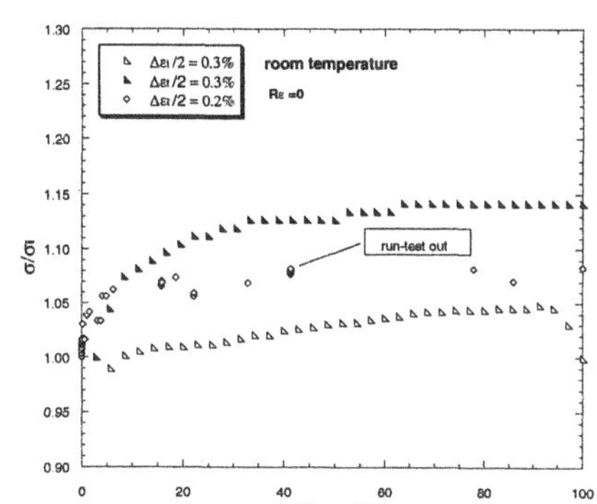

Fig.4: Variation of adimensionnal stress-amplitude versus normalized lifespan at room temperature and with $R_\varepsilon = 0$

For the highest total-strain-amplitudes (0.4% and 0.6%), the stress-amplitude increases continuously during cycling until failure. This increase is however more important during the first ten cycles. In contrast, for the lowest total-strain-amplitudes, the stress-amplitude increases only during the first ten cycles and then stabilises until failure. Umakoshi [12] and Srivatsan [13] have also observed a similar dependence of the CSS behaviour with the strain-amplitude. Besides, the comparison of Fig.3 and Fig.4 indicates that the hardening rate does not seem to be influenced by the strain ratio. The material hardens cyclically, whatever the total-strain-amplitude and the strain ratio are.

AT 750°C. The CSS behaviour at 750°C is presented in Fig.5.

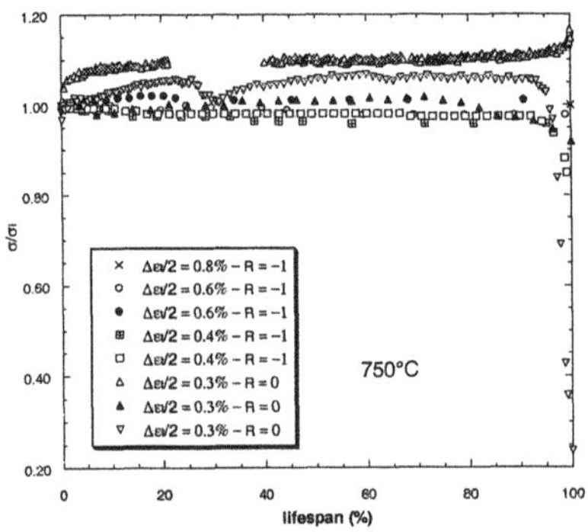

Fig.5: Variation of adimensionnal stress-amplitude versus normalized lifespan at 750°C

In contrast with room temperature, the standardized stress-amplitude, for specimens cycled with a strain ratio R_ε = -1, does not evolve during cycling. Actually, at this elevated temperature this alloy exhibits no hardening and no softening during cycling, but a cyclic stability of the strain-amplitude at all the total-stress-amplitude levels investigated. Such a cyclic stability behaviour was also observed by Malakondaiah et al. [14] on a Ti-46Al-2Nb-2Cr with a fully lamellar structure at 650°C and at 800°C.

At the low total-strain-amplitude (0.4%), a rapid drop in stress-amplitude is noticed at the end of the fatigue life. This drop is associated with a stable crack propagation stage.

On the other hand, for the specimens tested with a strain ratio R_ε = 0, the CSS behaviour seems to be similar at room temperature and at 750°C, i.e. a slightly hardening during the first ten cycles is followed by a stability of the strain-amplitude. In the same way that for specimens cycled with a strain ratio R_ε = -1, a rapid drop is observed just before failure, due to crack propagation.

Finally, it should noticed that the normalised stress-amplitude is more important for specimens cycled with a strain ratio R_ε = 0 than for those cycled with R_ε = -1.

BAUSCHINGER EFFECT.

A systematic examination of the hysteresis loops was conducted in order to analyze the hardening mechanisms. An example is described on the Fig.6, for each loop the variation of the size of the elastic domain and the variation of the elastic zone center in unloading or in loading is researched. If the size of the elastic domain changes during cycling the hardening presents an isotropic components and if the elastic zone center moves during cycling the hardening presents an kinematic components, indicating a Bauschinger effect. As a result, the observed hardening presents an isotropic and a kinematic components. It is also noteworthy that this Bauschinger effect is still present at elevated temperature, consistently with the observations reported by Hardy [15]. According to Recina [10], the Bauschinger effect would be related to the microstructure. The Bauschinger effect is larger in a material with a lamellar structure than in a duplex material. This difference leads to larger inelastic strains in each cycle causing irreversible damage in the material with a lamellar structure.

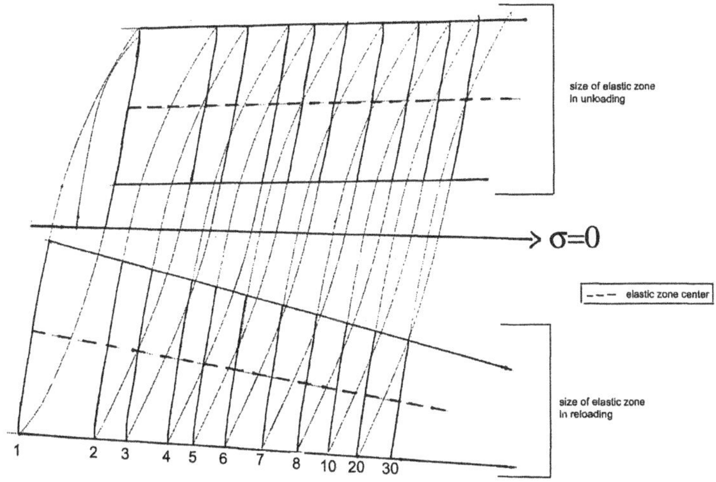

Fig.6: hysteresis loop analysis – bauschinger effect

FATIGUE LIVES AT ROOM TEMPERATURE AND AT 750°C

AT ROOM TEMPERATURE. The experiments showed that the fatigue life is strongly dependent on the total–strain-amplitude. For instance, the fatigue life of specimen cycled with 0.6% of total-strain-amplitude is approximately 100 cycles, 200 cycles for a total-strain-amplitude of 0.4% and did not fail in 50000 cycles for total-strain-amplitude of 0.2%. In one case an early rupture was observed, due to the presence of an internal defect.
In order to estimate a Coffin-Manson law the plastic-strain-amplitude at half of the fatigue life was considered. This relationship between plastic-strain-amplitude and number of reversals to failure $2N_f$ is: $\Delta\varepsilon_p/2 = 0.06*(2N_f)^{-0.7}$. The fatigue-ductility exponent (-0.7) is in agreement with the value derived from Recina and Karlson works [10].

AT 750°C. It seems that temperature does not influence the fatigue life of specimens cycled with a strain ratio $R_\varepsilon = -1$. Indeed, for comparable total-strain-amplitude values, the number of cycles is roughly the same at room temperature and at 750°C. The fatigue life decreases with the increase of the total-strain-amplitude (indeed the fatigue life of specimens cycled with 0.6% of total-strain-amplitude is approximately 100 cycles and 300 cycles for specimens cycled with a total-strain-amplitude equal to 0.4%). On the other hand, fatigue life of specimens cycled with $R_\varepsilon = 0$ seems to be influenced by temperature; fatigue life is here approximately 3000 cycles whereas at room temperature fatigue life is 40 cycles.

In a similar manner as at room temperature, a Coffin-Manson relationship can be determined. This relationship between plastic-strain-amplitude and fatigue life is: $\Delta\varepsilon_p/2 = 0.02*(2N_f)^{-0.4}$. At 650°C Malakondaiah et al. [14] found similar values for the fatigue-ductility exponent and the fatigue-ductility coefficient. The fatigue-ductility exponent is lower at 750°C than at room temperature.

FRACTURE CHARACTERISTICS AT ROOM TEMPERATURE AND AT 750°C

AT ROOM TEMPERATURE. Observations of the fracture surface features of the cyclically deformed TiAl intermetallic specimens were carried out with a JEOL scanning electron microscope (SEM). Macrographs of specimens fractured at different total-strain-amplitudes are presented in Fig.7 a to d.

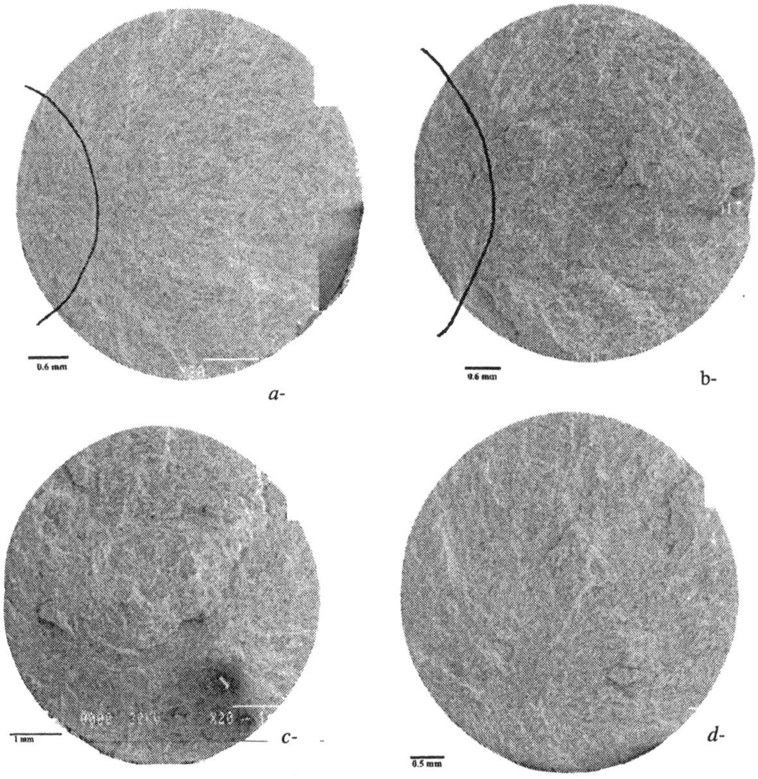

Fig.7: Macrographs of specimens fractured at different total-strain-amplitude at room temperature:

a- $\Delta\varepsilon_t=\pm 0.6\%$
b- $\Delta\varepsilon_t=\pm 0.4\%$
c- $\Delta\varepsilon_t=\pm 0.3\%$
d- $\Delta\varepsilon_t=\pm 0.2\%$

For all total-strain-amplitudes, the fracture surface presents the same aspect typical of a brittle fracture. It seems however that at high total-strain-amplitude the fracture surface is smoother and does not present secondary cracking. At the highest total-strain-amplitude, the initiation site is identified by the presence of river patterns converging towards the initiation site as indicated in Fig.7a and 7b. On the opposite, at low total-strain-amplitude, the initiation site, propagation site and fast fracture are scarcely perceptible. Observations at higher magnification indicate that the propagation is either translamellar or interlamellar (Fig.8).

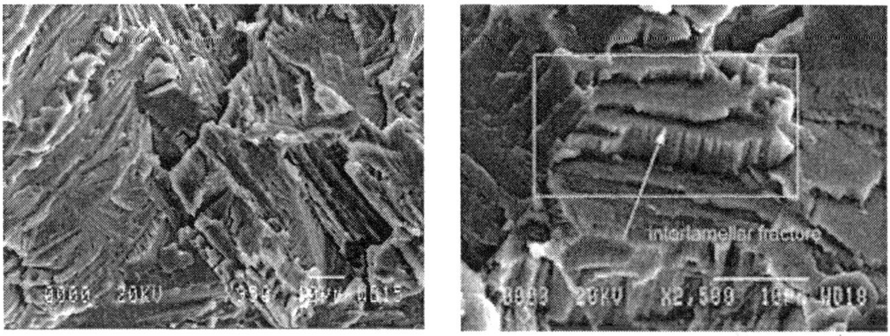

a- *b-*

Fig.8: Observations of specimens fractured at room temperature:
a- $\Delta\varepsilon_t = \pm 0.4\%$ and $R_\varepsilon = -1$, translamellar fracture
b- $\Delta\varepsilon_t = \pm 0.2\%$ and $R_\varepsilon = -1$, translamellar and interlamellar fracture

AT 750°C. Figures 9 and 10 present typical fracture surface feature. For this total-strain-amplitude ($\Delta\varepsilon_t = \pm 0.3\%$), the fractures topographies look more ductile than at room temperature.

At higher magnification the propagation is translamellar with extensive interlamellar splitting. In a previous study Chan [16] remarked that the deformation modes during fatigue are (111) slip on planes parallel to the lamellar interface for the easy-slip orientation and slip and twinning on translamellar (111) planes for the hard-slip orientation. Some fine markings can also be noticed in lamellae. The influence of temperature on fracture surfaces is indeed comparable to that noticed in fatigue crack growth by Mabru et al. [7].

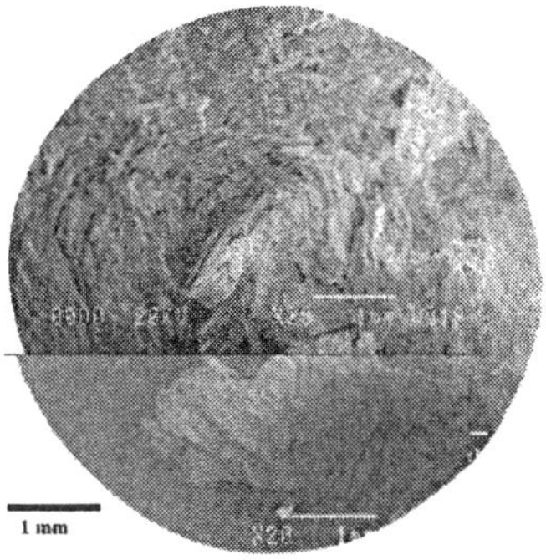

Fig.9: Macrography of the fracture surface at 750°C and with $\Delta\varepsilon_t = \pm 0.3\%$

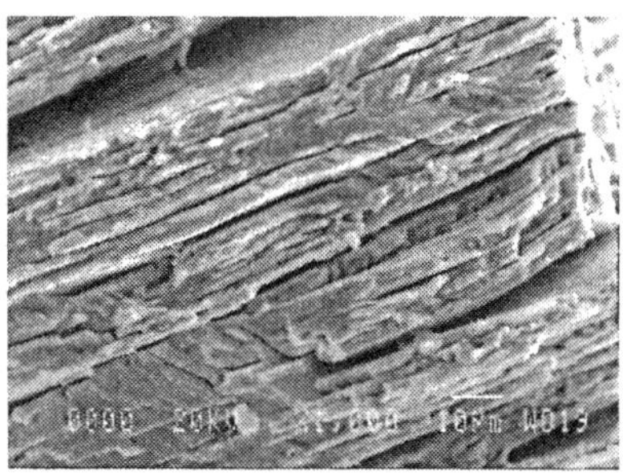

Fig.10: Higher magnification observation of the specimen presented in Fig.9

DISCUSSION

The results reported here indicate an influence of temperature on the CSS behaviour of a cast Ti-48Al-2Cr-2Nb alloy (Fig.11 and Fig.12) but not on the fatigue lives. At room temperature a continuous hardening is observed although at elevated temperature a stabilized regime is rapidly reached. This difference is specially marked at high total-strain-amplitude values ($\Delta\varepsilon_t=\pm0.6\%$ and $\Delta\varepsilon_t=\pm0.4\%$).

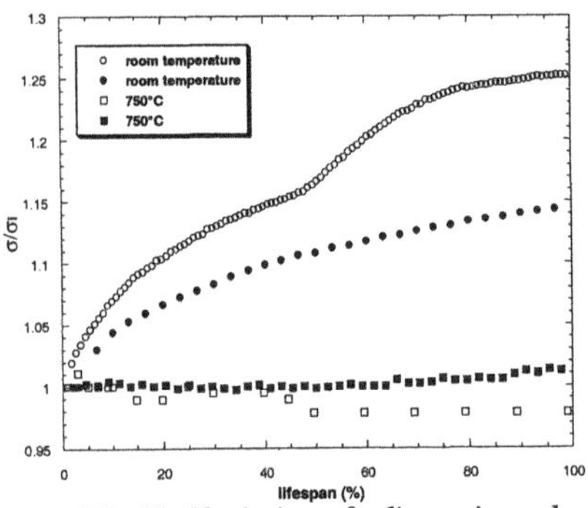

Fig.11: Variation of adimensionnal stress-amplitude versus normalized lifespan for a high total-strain-

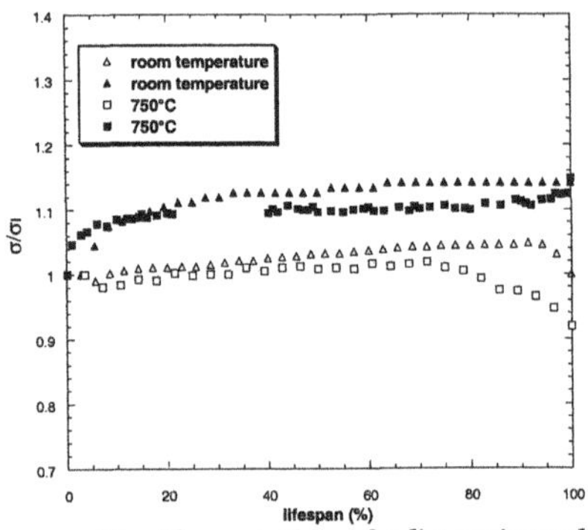

Fig.12: variation of adimensionnal stress-amplitude versus normalized lifespan for a low total-strain-

As mentioned in the introduction there are only very few data available on the low cycle fatigue behaviour of TiAl alloys. The most extensive work has been carried out by Umakoshi and co-workers [17,18] on polysynthetically twinned (PST) crystals (PST crystals are constituted of a single lamellar and colony as a consequence contain one lamella orientation) Their results reveal a strong plastic anisotropy under cyclic loading in relationship with lamellae orientation with respect to the load ratio. In the case of the alloy under investigation here, the lamellae orientation changes from one colony to another, so that the observed behaviour can be regarded as a mixture of different orientations. Nevertheless it is remarkable that the CSS behaviour as well as the fracture surfaces observed at room temperature with $R_\varepsilon=-1$ is extremely similar to that observed by Umakoshi and co-workers in case where the load axis is parallel to the lamellae interface. Their TEM observations indicate that at low temperature the primary deformation mode is related to twinning and slip by ordinary dislocation motion. Dislocations form a vein-like structure during their to-and-fro motion although twinning would operate during the early stage of cyclic loading. The respective contributions of these deformation modes would depend on the applied strain amplitude, but also presumably on the strain ratio. At higher temperatures the activation of twinning would be suppressed, resulting in a change in the CSS behaviour. This is consistent with observations of near-crack-tip plasticity carried out by Mercer [19] and Zghal [20]. Indeed they observed the presence of slip and twins at room temperature and a lower activity of twinning at 750°C. TEM observations of deformation substructures of specimens tested in the present study are currently under progress.

Sastry et al. [21] also noted a strong influence of temperature on fatigue lives. Clearly such an influence is not observed in the present study. The lowest fatigue lives at room temperature would be attributed to twinning which enhance the formation of extrusion and stress concentration at lamellae or domain interfaces. However the effect of temperature on fatigue lives results from a balance between various processes, namely deformation mechanisms, as described here above, but also environmental attack and oxidation. Clearly the respective contribution of those processes in the present alloy have to be investigated by conducting the same kind of low cycle fatigue tests in vacuum.

CONCLUSION

The study of the influence of temperature on the low cycle fatigue behaviour of the cast alloy Ti- 48Al- 2Cr- 2Nb has revealed the followings mains observations:

* The cyclic stress-strain behaviour at room temperature is different from that at 750°C. Indeed at room temperature the alloy continuously hardens during cycling until failure and this hardening is more important for the highest total-strain-amplitude and at the beginning of cycling (first ten cycles). On the contrary, at 750°C the alloy presents a rapid stabilisation of the stress-amplitude and for the lowest total-strain-amplitude this stress-amplitude falls down just until failure. This fall is caused by crack propagation. Tests show that the stress-amplitude is lower at elevated temperature than at room temperature.

* A Bauschinger effect is noticed at room temperature like at elevated temperature.

* The fracture surface feature is not influenced by temperature. The kind of rupture is the same at room temperature and at 750°C, i.e. brittle with a translamellar and interlamellar propagation.

* On going work will aim to evaluate the contribution of oxidation, environmental attack on fatigue lives and identify the fundamental deformation mechanisms acting at low and high temperature.

ACKNOWLEDGEMENTS

The authors would like to acknowledge the material supply from SNECMA MOTEURS. The present investigation is conducted within the framework of a national project (CPR "TiAl intermetallic") in collaboration with LTPCM (Grenoble), LSGMM (Nancy), GMP (Rouen), CEMES (Toulouse), LMS (Palaiseau), LMPM (Poitiers), the societies SNECMA MOTEURS and TURBOMECA, with the support of CNRS and DGA.

REFERENCES

1. Davidson D.L., Campbell J.B. (1993). *Metallurgical and Materials Transaction* **24A**, 1555-1574
2. Bowen P., Chave R.A., James A.W.(1995). *Materials Science and Engineering A-Structural Material* 193, 443-456
3. Chan K.S., Shih D.S. (1997). *Metallurgical and Materials Transactions* **A28**, 79-90
4. Pippan R., Hageneder P., Knabl W., Hebesberger T., Tabernig B. (2001). *Intermetallics*, 9, 89-96
5. Campbell J.P., Rao T.V., Ritchie R.O. (1997). *Materials Science and Engineering A-Structural Material* 240,722-728
6. Rosenberger A.H., Worth B.D., Larsen J.M., *Structural intermetallics*, Nathal M.V. et al. eds, (1997). *The minerals metals and materials society*,555-561.
7. Mabru C., Bertheau D., Pautrot S., Petit J., Hénaff G. (1999). *Engineering Fracture Mechanics*, 64, 23-47
8. Recina V., Karlson B. (2000). *Scripta materialia*, 43, 609-615

9. Recina V., Karlson B. (1999). *Materials Science and Engineering A*, **262**,70-81
10. Recina V., Karlson B. (1997). *Structural Intermetallics*, 479.
11. Jones P.E. and D.Eylon (1998). *Fatigue Behavior of Titanium Alloys*, 163.
12. Umakoshi Y., Yasuda H.Y., Nakano T. (1995). *Materials science and Engineering* **A192/193**,511-517.
13. Srivastan T.S., Soboyejo W.O., Stranwood M. (1995). *Engineering fracture Mechanics* **52**, N° 1, 107.
14. Malakondaiah G., Nicholas T. (1996). *Metallurgical and materials transaction* **27A**, 2239.
15. Hardy M.C. (1995). *Titanium'95: science and technology*, 256.
16. Chan K.S. (1997). *Journal of metals*,53-58
17. Umakoshi Y., Yasuda H.Y., Nakano T. (1996). *Philosophical MagazineA*, Vol.73, N°4, 1053-1067
18. Yasuda H.Y., Nakano T., Umakoshi Y. (1995). *Materials Science and Engineering* **A192/193**, 511-517
19. Mercer C., Lou J., Soboyejo W.O. (1999). *Materials Science and Engineering* **A284**, 235-245.
20. Zghal S., Menand A., Couret A. (1998). *Acta Metallurgica* **46**, N°16, 5899-5905.
21. Sastry S.M.L, Lipsitt H.A. (1977). *Metallurgical Transacions A*, vol.8A, 299-308.

Damage under Thermal-Mechanical Loading

Temperature-Fatigue Interaction
L. Rémy and J. Petit (Eds.)

LIFETIME, CYCLIC DEFORMATION AND DAMAGE BEHAVIOUR OF MAR-M-247 CC UNDER IN-PHASE, OUT-OF-PHASE AND PHASE-SHIFT TMF-LOADINGS

T. BECK, R. RATCHEV, M. MOALLA, K.-H. LANG, D. LÖHE
Institut für Werkstoffkunde I, Universität Karlsruhe (TH), Kaiserstr. 12, D-76131 Karlsruhe

ABSTRACT

In-Phase-, Out-of-Phase- and Phase-shift-TMF tests were carried out on MAR-M-247 CC. A minimum temperature of 400°C, maximum temperatures from 800°C up to 1050°C, total mechanical strain amplitudes between 70% and 100% of the thermal strain amplitude and dwell times at maximum temperature from 0s up to 1800s were applied.
Out-of-Phase-TMF leads to higher tensile mean stresses than Phase-Shift-TMF. In-Phase-TMF loadings induce compressive mean stresses. The amounts of the mean stress, the stress amplitude and the plastic strain amplitude generally increase with increasing maximum temperature and dwell time. Out-of-Phase- and Phase-Shift-TMF lead to higher numbers of cycles to failure than In-Phase-TMF. Under Out-of-Phase-TMF a transgranular fracture path is observed even at high maximum temperatures and dwell times. This loading condition leads to compressive stresses at high temperatures and tensile stresses at lower temperatures. Due to that, intergranular creep damage is suppressed. Under In-Phase-TMF the tensile stresses occurring at high temperatures lead to creep induced intergranular crack propagation even at loadings with a dwell time of zero and relatively low maximum temperatures. Under Phase-Shift-TMF loadings transgranular crack propagation occurs at low maximum temperatures, whereas higher maximum temperatures and dwell times lead to intergranular damage. Because the damage mechanisms depend on the phase relationship and the dwell time, the damage parameter P_{OST} which accounts for the plastic strain amplitude and the maximum stress as damage relevant quantities is not able to give a common description of the lifetime behaviour for all considered TMF cycle types.

KEYWORDS
Thermal-mechanical fatigue, Ni-base superalloy, Cyclic deformation, Lifetime behaviour, Damage mechanisms

INTRODUCTION

Thermal-mechanical fatigue (TMF-) loadings are induced by inhomogenous and instationary temperature fields during start-stop-cycles in cooled gas turbine blades and vanes. These loadings are often responsible for the initiation of damage relevant cracks which can propagate until the occurrance of cathastrophic failure of the component and therefore determine the duration of the service life. The phase relationship between temperature and mechanical loading, which has a decisive influence on the predominant damage mechanisms, strongly depends on the considered location within the blade. For example, at the outer side of blades in gas turbines for power generation, the mechanical loading is approximately 180° phase shifted with respect to the temperature-time-course (Out-of-Phase-TMF), whereas on the cooled inner surface, there is nearly no phase shift between temperature and mechanical loading (In-Phase-TMF). At the leading edge of flight turbines, a phase angle of about 135° between temperature and mechanical loading (Phase-Shift-TMF) occurs.

Because data taken from isothermal fatigue tests are principially not suited to study the influence of phase relationships and other quantities on the material's behaviour under thermally induced fatigue as occurring in turbine blades, only results of TMF tests can provide a reliable database for the dimensioning of such components [1, 2]. The present study gives an overview of the behaviour of the cast Ni-base superalloy MAR-M-247 CC under the above mentioned types of TMF loadings. The TMF-tests are accompanied by microstructural investigations in order to achieve a deeper understanding of the complex relationships between the cyclic deformation-, lifetime- and damage behaviour of the material.

TEST MATERIAL

All tests were carried out on the Ni-base superalloy MAR-M-247 CC which is broadly applied for blades and vanes in the first stages of aero gas turbines. The chemical composition is given in Table 1. The Cr- and Al-content result in a good corrosion resistance even at high temperatures. Al- and Ti make the material hardenable by the precipitation of finely dispersed coherent γ'-$Ni_3(Al, Ti)$ particles with axes parallel to the <100> directions of the γ-matrix lattice. The W-, Mo-, Nb-, and Hf- contents stabilise the γ'-precipitates, especially at high temperatures, and result in an additional solid solution hardening of the γ-matrix crystals.

Table 1: Chemical composition of MAR-M 247 CC (Wt.-%)

Ni	Cr	Co	W	Al	Ti	Mo	Hf	Nb	C
bal	8.15	9.32	9.51	5.62	0.7	0.47	1.41	3.2	0.075

After the casting process, the material was solution annealed at 1220°C for 2h, quenched and finally aged at 870°C for 24 hours. Figure 1 shows two optical micrographs of the microstructure in the as-supplied state. In the left picture, the dendritic grain structure of the material can clearly be seen. The higher magnification of the picture on the right hand side reveals finely dispersed coherent γ'-$Ni_3(Al,Ti)$ precipitations inside the grains of the γ-matrix. At the grain boundaries a pronounced coarsening of the γ'-particles as well as some primary segregations can be seen. The volume content of the γ'-phase is about 70%, which leads to a high strength of the material up to temperatures above 1000°C where solution of the γ'-precipitates sets in. The test material was provided as plates, from which solid, cylindrical test pieces were machined with a gauge length of 10mm and a diameter of 6mm within the gauge length [3].

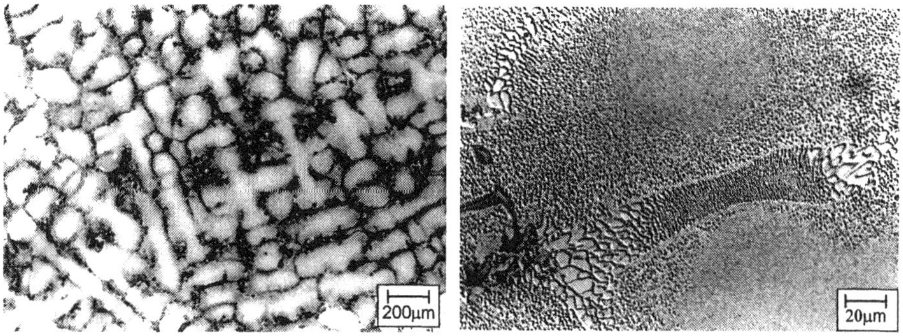

Fig.1: Microstructure of MAR-M-247 CC in the as supplied state

EXPERIMENTAL DETAILS

The TMF experiments were carried out on a servohydraulic fatigue testing machine with 100kN nominal force. The specimens were heated by an inductive system with a maximum power of 5kW. The cooling was achieved by the heat flux into the water cooled, hydraulic specimen grips and, if necessary, additionally by a controlled air jet. The temperature was measured by spot welded thermocouples above and below the upper and the lower end of the gauge length, respectively. For the strain measurement, a capacitive high temperature extensometer was attached using Al_2O_3 bars.

A schematic sketch of the test procedure is given exemplarily for In-Phase- (IP-) TMF in Fig. 2. At the beginning of each test, the specimen is heated up to the minimum temperature at zero stress. Then, still at zero stress, five „reference cycles" are carried out in order to determine the thermal strain ε^{th} as a function of the temperature. After reaching T_{min} again, the machine is switched to total strain control. The subsequent temperature-time courses are the same as during the reference cycles. The thermal strain ε^{th} as a function of the temperature which was determined before and the well known equation $\varepsilon_t^{me}=\varepsilon_t-\varepsilon^{th}$ were used to derive total strain-time courses that result in total mechanical strain- (ε_t^{me}-) time paths with a phase shift of zero against the temperature and a given total mechanical strain amplitude. These cycles are applied to the specimens in close loop control of ε_t. The other types of TMF tests were carried out according to the same procedure with a phase angle between the temperature and the mechanical strain of 180° for Out-of-Phase- (OP-) and of 135° for Phase-Shift (PS-) TMF, respectively. All considered TMF cycles were applied with total mechanical strain amplitudes $\varepsilon_{a,t}^{me}$=70% up to 100% of the respective thermal strain amplitude ε_a^{th}. Dwell times at the maximum temperature T_{max} of 0s, 300s and 600s were applied. T_{max} was varied from 850°C up to 1050°C. The minimum temperature generally was 400°C. A heating as well as a cooling rate of 10°C/s was applied in all tests and the ultimate number of cycles was $N_U=10^4$.

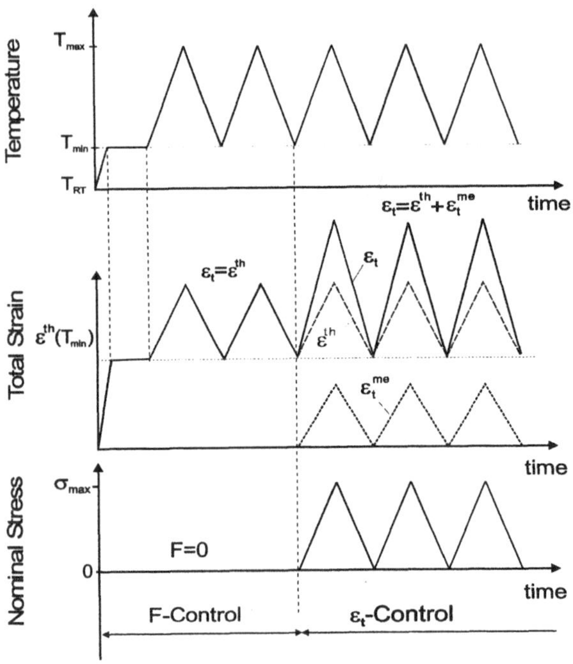

Fig.2: In-Phase TMF test procedure (schematic sketch)

RESULTS AND DISCUSSION

Cyclic Deformation behaviour

In Fig. 3 hysteresis loops of the nominal stress σ_n versus the total mechanical strain ε_t^{me} at N=1 and after one half of the number of cycles to fracture are plotted for OP-, IP- and PS-TMF with a maximum temperature of 950°C and a dwell time of zero. During heating up to T_{max} within the first cycle, compressive stresses occur for OP- and PS-loadings, whereas tensile stresses take place under IP-conditions. The maximum absolute value of σ_n during heating up ist reached at a temperature of about 860°C in the IP- and the OP-cycle. After having reached the maximum (minimum) stress under IP- (OP-) TMF the absolute value of σ_n decreases with increasing T in spite of further increasing (decreasing) absolute values of the total mechanical strain. This is firstly caused by a decreasing yield strength of the material with increasing temperature and secondly by an increasing amount of relaxation processes which overcompensate the effect of work hardening. Under PS-conditions the minimum stress occurs at the minimum value of ε_t^{me} and a temperature of 820°C. Accordingly, no significant relaxation is observed up to that point. However, relaxation during the subsequent unloading leads to a steeper stress-strain path than observed for the other cycle types until T_{max} is reached at $\sigma_n \approx$-400MPa. The further stress-strain course down to ε_t^{me}=0 is parallel to the curves observed under OP- and IP-loading.

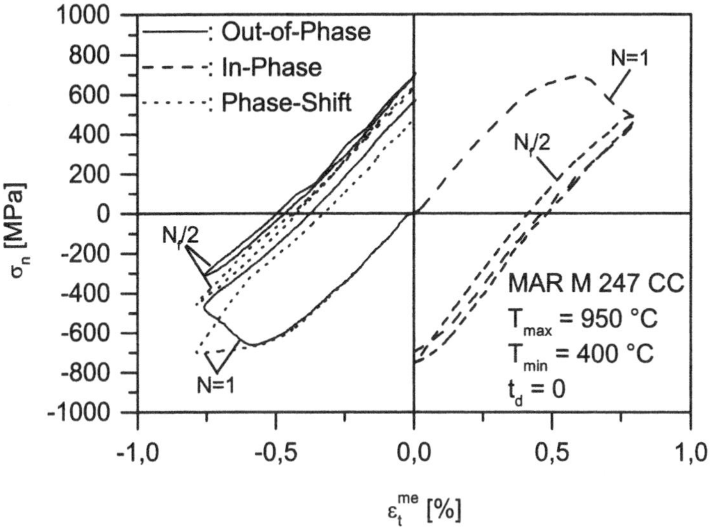

Fig. 3: σ_n-ε_t^{me} hystereses for Out-of-Phase, In-Phase and Phase-Shift TMF without dwell time

Under IP loadings tensile stresses occur at higher temperatures than compressive stresses. This leads to a more pronounced plastic deformation in tension than in compression. For that reason, the hysteresis loop is continuously shifted towards compressive mean stresses until the compressive plastic deformation is equal to the plastic strain in tension as it can be seen for $N=N_f/2$ in Fig. 3. In the OP- and PS- cycles compressive stresses are acting at high temperatures. This leads to tensile mean stresses due to the same mechanisms which are just working in the opposite direction compared to IP-loadings. The mean stresses are higher for the OP- than for the PS-cycle, because in the latter case the unloading starts before the maximum temperature is reached. This leads to a lower amount of relaxation in compression than in the OP-cycle. Generally, at $N=N_f/2$ distinctly lower plastic strain amplitudes are observed than in the first cycle. This is on the one hand due to cyclic hardening, especially within the first cycles where large plastic deformations occur. On the other hand, the mean stresses developing as discussed above cause the maximum absolute stress value generally to occur at the minimum temperature which leads to a further decrease of the plastic deformation.

The history of the hysteresis loops of these tests is evaluated in Fig. 4 as cyclic deformation curves representing the maximum, the mean and the minimum stress (upper diagram) and the plastic strain amplitude (lower diagram) versus the number of cycles N.

The development of the mean stresses occurring under the investigated types of TMF-loadings was discussed above. Nearly constant stress amplitudes are observed after the first cycles. The plastic strain amplitude is distinctly higher for IP- than for OP- and PS-TMF. This is because of extensive creep induced damage which occurs under IP-loadings especially at high maximum temperatures as considered in Fig. 4 because only in the IP-cycle high tensile stresses are acting in the range of T_{max} (also see Figs. 8 and 9 with the respective discussion). That leads to a decrease of the effective cross section of the gauge length due to the formation of pores and intergranular cracks, and therefore causes an increase of the true stress within the material.

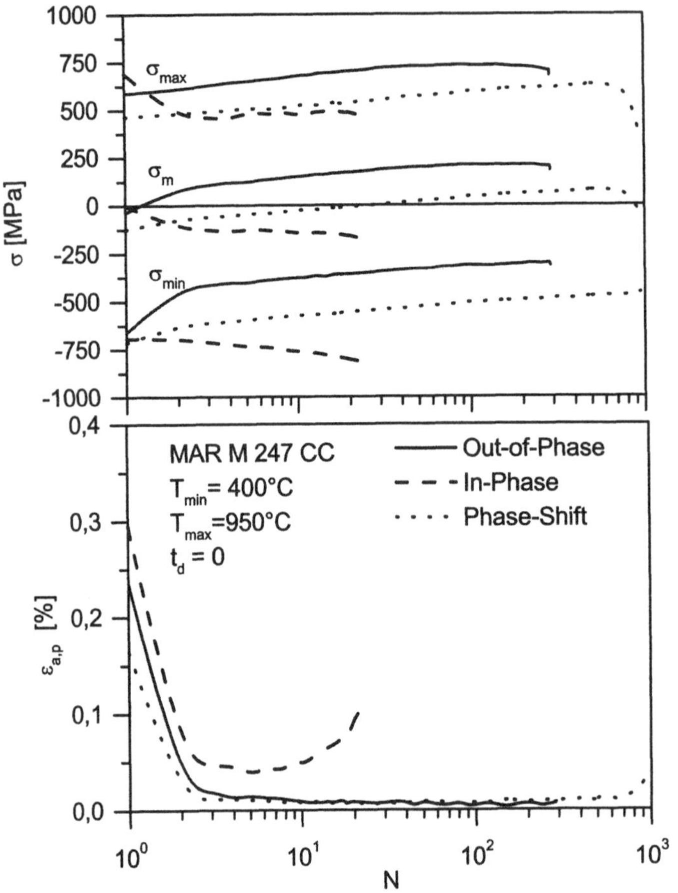

Fig. 4: Maximum-, mean and minimum stress (above) and plastic strain amplitude (below), both versus N for In-Phase, Out-of-Phase and Phase-Shift TMF loadings

During IP-, OP- and PS-TMF dwell times t_d up to 1800s at the maximum temperature were applied. As an example, Fig. 5 shows the influence of this test parameter on the cyclic deformation behaviour under IP-TMF conditions.

As can be seen in the upper graph, the development of compressive mean stresses is enhanced by increasing dwell times. This is due to additional relaxation processes which occur during t_d and lead to an additional tensile plastic deformation with respect to the tests without dwell time. This effect is the higher the longer the dwell time is, because the stress relaxation at T_{max} is a time-dependent process. According to that, the plastic strain amplitude also increases with the duration of the dwell time.

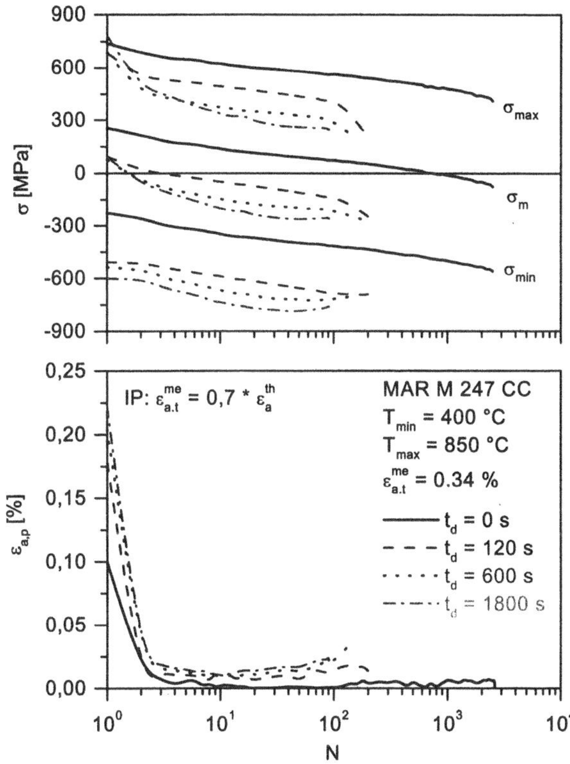

Fig. 5: Maximum-, mean and minimum stress (above) and plastic strain amplitude (below), both versus N for In-Phase TMF loadings with dwell times at T_{max}

For the same reasons increasing dwell times also result in increasing plastic strain amplitudes at OP- and PS-TMF loadings, but tensile mean stresses are built up due to the mechanisms as already discussed with respect to Fig. 3. For the OP-cycles, a similar increase of the absolute value of the mean stresses with increasing dwell time is observed as for IP-TMF.

The influence of dwell times on the cyclic deformation behaviour under PS-conditions is qualitatively the same as for OP-loadings. However, the increase of the plastic strain amplitude as well as of the mean stresses under PS-conditions is less pronounced than under OP-conditions because the relaxation processes during the dwell time at T_{max} which are decisive for that effect take place at lower compressive stresses.

Lifetime behaviour

Figure 6 shows the lifetime behaviour for all considered TMF-cycles with and without dwell times in the representation of the total mechanical strain amplitude versus the number of cycles to failure. The OP-results are represented by open, the IP-tests by solid and the PS-experiments by center crossed data points. The shape of the symbols represents the dwell time as indicated in the figure. The results obtained from OP-TMF tests show a relatively large scatter. However, the data points obtained at t_d=0 can be described consistently by a fit curve according to $\varepsilon_{a,t}^{me}=A\cdot N_f^B$. Dwell times of 300s and 600s reduce the lifetime at OP-TMF by a factor of 2 up to 3. PS- and OP-TMF with t_d=0 reveal nearly equal numbers of cycles to failure. The impact of t_d on the lifetime under PS- is nearly the same as for OP-loading. Except from the data obained at the lowest strain amplitudes investigated, IP-TMF even without any dwell time leads to distinctly lower N_f than OP- and PS-TMF. However, the influence of t_d on the IP-TMF lifetimes seems to be less less pronounced than for the OP- and PS-cycles.

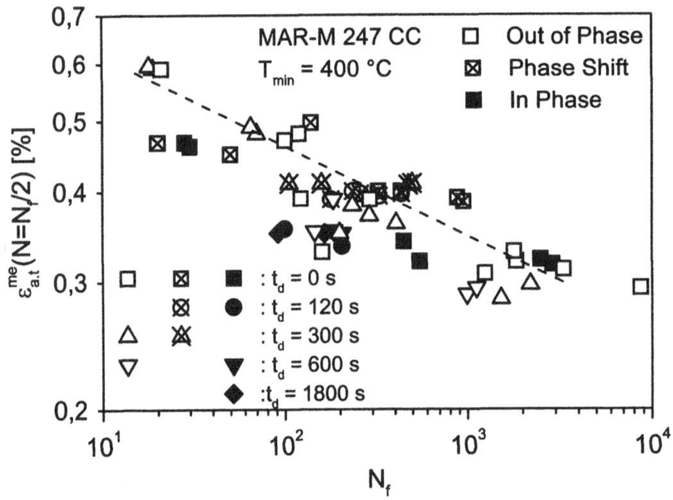

Fig. 6: Lifetime behaviour at In-Phase, Out-of-Phase and Phase-Shift TMF in the representation $\varepsilon_{a,t}^{me}$ versus N_f

For a more detailed discussion of the lifetime behaviour, Fig. 7 shows the damage parameter of Ostergren versus N_f for the same experiments as in Fig. 6. Additionally, separate fit curves were plotted through the data obtained from OP- PS- and IP-tests with t_d=0. Neither the total mechanical strain amplitude nor the Ostergren parameter are able to give a common description of the lifetimes under IP-, OP- and PS-TMF. At a given value of P_{ost}, the highest N_f are generally observed for the OP-, the lowest for the IP-cycles. The values of the PS-tests lie between these results. The relatively large scatter of the data points for all load cycles is due to the relatively coarse grain size of the material (see Fig. 1) which leads to a high scatter of the thermal expansion coefficient and the elastic modulus of the specimens and therefore to deviations of the mechanical loadings, even if identical temperature and total strain cycles are applied.

In spite of the relatively large scatter of the OP-TMF results it becomes obvious that P_{Ost}, which accounts for the decisive influence of both the plastic strain amplitude and the maximum stress on the lifetime behaviour, is able to fairly correlate the results obtained from OP-TMF tests with and without dwell times. This means, that the lifetime reduction with increasing t_d is mainly a result of the additional damage due to the higher maximum stresses and plastic strain amplitudes that arise in this case. For equal P_{Ost} tendentiously lower lifetimes are observed in PS-tests with $t_d>0$ than for $t_d=0$. This leads to the conclusion that the lifetimes under PS-TMF are influenced by damage mechanisms which are not taken into account by $P_{Ost.}$ Surprisingly, under IP-loading the influence of t_d appears even more pronounced in Fig. 7 than in Fig. 6. This is because N_f decreases with increasing t_d in spite of the pronounced reduction of σ_{max} due to relaxation processes that occur during t_d (see Fig. 5). The resulting decrease of P_{Ost} is not fully compensated by the increase of $\varepsilon_{a,p}$ that also arises with increasing dwell time. From these considerations it is concluded that the lifetime under IP-loading is mainly determined by the fact that the tensile stresses acting during dwell times at T_{max} result in a strong increase of creep induced damage, whereas for OP- and PS-loadings, the compressive stresses that are applied during t_d do not result in additional creep damages that are sufficiently high to have a significant influence on the lifetime behaviour.

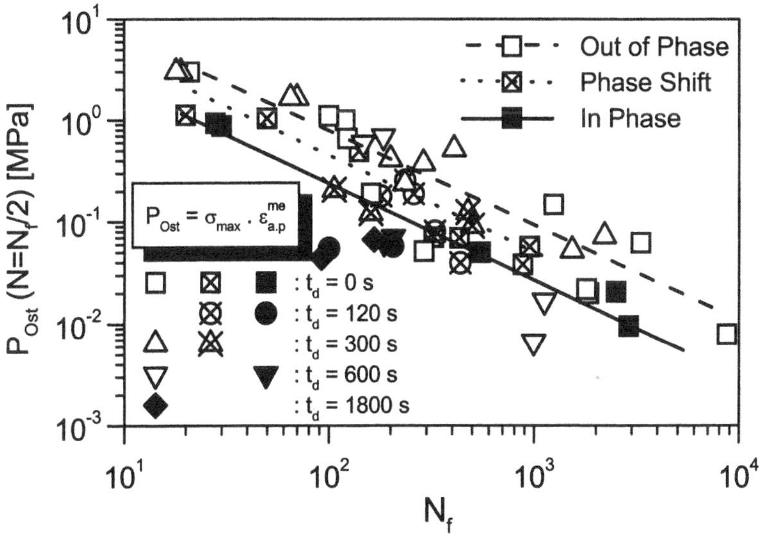

Fig.7: P_{Ost} versus N_f for Out-of-Phase, In-Phase- and Phase-Shift TMF with dwell times

Damage Behaviour

Figure 8 shows optical micrographs of the microstructure along the crack path obtained after OP- (left), IP- (middle) and PS- (right) TMF loading with $t_d=0$. Figure 9 shows the respective micrographs for a dwell time of 300s for OP- and PS-TMF and of 120s for IP-TMF.

For $t_d=0$, OP- and PS-loadings result in a solely transcrystalline crack propagation whereas IP-TMF leads to intergranular fracture even at T_{max}=850°C due to an enhanced creep damage because of the tensile stresses occurring in the range of T_{max}. If dwell times at T_{max} are introduced, the cracks remain transgranular for OP-loadings and intergranular under IP-conditions, but for the PS-cycle a change to intercrystalline crack propagation is observed with increasing t_d. Additionally, under IP-loadings with dwell times, an increasing amount of intercrystalline damage inside the specimen is observed.

These observations correlate perfectly with the lifetime and cyclic deformation behaviour as described above: Because the damage mechanism is independent from the dwell time, all results obtained under OP-TMF can be described well by the Ostergren parameter. Even though intergranular damage is generally observed under IP-TMF, the influence of t_d on the lifetime behaviour cannot be taken into account by this parameter because dwell times cause drastically increasing creep-induced damage inside the gauge length (Fig. 9), which leads to an additional decrease of N_f. Finally, under PS loadings the observed change of the predominant crack propagation mechanism with increasing t_d leads to the tendentially lower N_f observed in the experiments with dwell times.

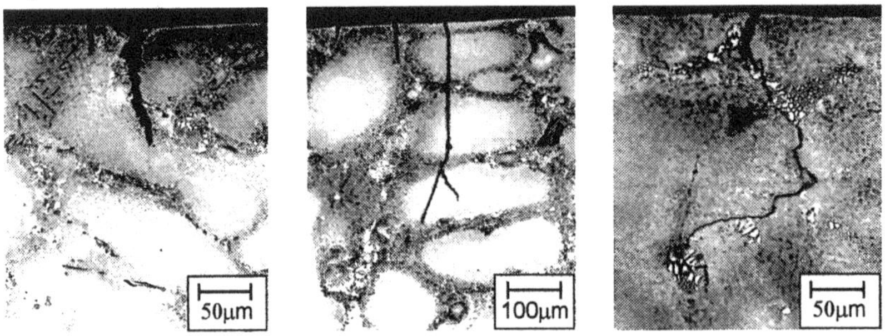

OP: T_{max}=950°C *PS: T_{max}=950°C* *IP: T_{max}=850°C*

Fig. 8: Crack propagation behaviour at TMF cycles with t_d=0

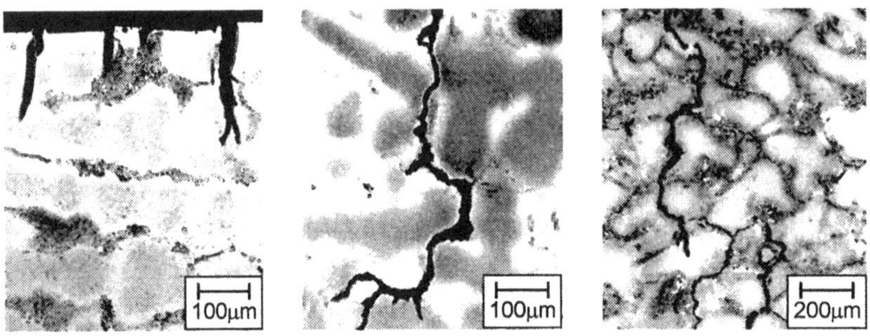

OP: T_{max}=950°C *PS: T_{max}=950°C* *IP: T_{max}=850°C*

Fig. 9: Crack propagation behaviour for t_d=300s (OP and PS) and 120s (IP)

LITERATURE

[1] T. Beck, G. Pitz, K.-H. Lang, D. Löhe, Mat. Sci. Eng. A 234-236 (1997), 719-722

[2] G. Pitz, T. Beck, K.-H. Lang, D. Löhe, Mat.-wiss. u. Werkstofftech. 28 (1997), 142-148

[3] G. Pitz, PhD-thesis, Universität Karlsruhe, 1997

ACKNOWLEDGEMENT

The financial support of the Forschungsvereinigung Verbrennungskraftmaschinen, Frankfurt/Main, Germany during the Project No. 66870 is gratefully acknowledged.

Temperature-Fatigue Interaction
L. Rémy and J. Petit (Eds.)

DAMAGE MECHANISMS UNDER THERMAL-MECHANICAL FATIGUE IN A UNIDIRECTIONALLY REINFORCED SIC-TITANIUM METAL MATRIX COMPOSITE FOR ADVANCED JET ENGINE COMPONENTS

S.HERTZ-CLEMENS*, C. AUMONT** and L. REMY*
**Centre des Matériaux, Ecole des Mines de Paris, CNRS UMR 7633, BP 87, 91003 Evry Cedex, France*
***SNECMA Moteurs, Materials and Process Department, 77550 Moissy-Cramayel, France*

ABSTRACT

The thermal-mechanical fatigue behaviour of a titanium composite reinforced by continuous SiC fibres, the SM 1140+/Ti-6242, was investigated for uniaxial loading, under conditions which simulate service loading in a compressor disc during a flight. The thermal-mechanical tests (100-500°C) give rise to a reduction of fatigue lives compared to isothermal tests (550°C). The fatigue crack growth behaviour for isothermal and non-isothermal conditions was explored. For both types of loading, a steady crack growth rate was observed which was due to fibre bridging. The interfacial damage was defined as the average length of worn carbon-coating interface and was found to be strongly dependent of the loading cycle.

KEYWORDS

Metal matrix composite, thermal-mechanical fatigue, fatigue crack growth, damage mechanisms, fibre bridging

INTRODUCTION

Titanium based metal matrix composites reinforced by unidirectional continuous SiC fibres, are attractive for use in aircraft-engine components, such as compressor discs, because of their higher specific strength and stiffness at medium temperature compared to monolithic materials such as superalloys. But a good understanding of the fatigue behaviour of these materials is necessary for future industrial developments. For that reason, a great number of former studies has concentrated on the longitudinal fatigue behaviour of metal matrix composites in isothermal conditions [1-3]. Some authors have studied the behaviour in non-isothermal conditions [4-5] but always using conventional out-of-phase or in-phase cycles. Little work has dealt with realistic loading of the metal matrix composite in a compressor disc. Furthermore, unidirectionally reinforced metal matrix composites are known to exhibit a good resistance to fatigue crack growth under high frequency cyclic loading due to the fibre bridging effect, but no information is available for low cycle fatigue loading, which is relevant to the present application. The objective of this study was to investigate the thermal-mechanical fatigue behaviour and the fatigue crack growth behaviour in isothermal and non-isothermal conditions of a titanium composite reinforced by continuous SiC fibres, under fairly realistic loading conditions.

MATERIAL AND EXPERIMENTAL DETAILS

The composite material used for this study was made of a Ti-6Al-2Sn-4Zr-2Mo-0.1Si (wt%, so called Ti-6242) matrix reinforced with 12 plies of unidirectional SiC fibres. Ti-6242 is a near-alpha titanium alloy with a grain size about 10 μm. SM 1140+ silicon carbide fibres, made by DERA, are 100 μm in diameter and have an outer carbon-coating of 4-5 μm. The materials were manufactured by SNECMA Moteurs, using a foil-fibre-foil processing route : the fibre alignment was maintained with the use of a fugitive blinder. Unidirectional materials were obtained with measured fibre volume fraction of approximately 0.3. Specimens were cut parallel to the fibre direction from the panel using electrical discharge machining. The specimens used have a rectangular cross-section of 2x8 mm. Titanium tabs were found to be necessary to closely match the requirements of the precisely aligned grips and therefore to minimize possible bending induced by the grips. Tabs were pressed isostatically during processing. The four specimen surfaces were then mechanically polished to remove any surface damage induced from machining ending with 3 μm diamond paste.
Axial strain measurements were made with a high temperature extensometer having a 10 mm gauge length. All tests were performed on specimens with fibres oriented at 0° to the loading axis under load control.

The thermal-mechanical fatigue tests were conducted using servo-hydraulic or electromechanical machines and a computer which generates two synchronous temperature and stress cycles [6]. This computer is used in this way to record stress, mechanical strain and temperature. In the persent work, a single thermal-mechanical fatigue cycle, depicted in Fig. 1, was used. This cycle was derived from calculations of SNECMA Moteurs calculations and simulated the loading of the composite part in the disc during a flight. The tests, conducted under a stress ratio R_σ ($R_\sigma=\sigma_{min}/\sigma_{max}$) of 0, were performed between 100 and 500°C at a frequency of 3.45 10^{-3} Hz (290s cycle). Cooling and heating rates were both 5°C.s^{-1} and a tensile hold time of 90 s was used. Failure is defined as complete separation of the specimen into two pieces.

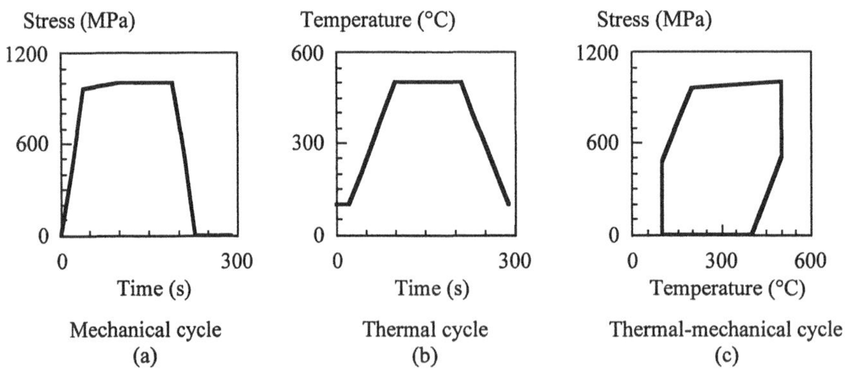

Fig. 1. The thermal-mechanical fatigue cycle for SM 1140+/Ti-6242 : (a) stress versus time ; (b) temperature versus time and (c) stress versus time diagrams

As illustrated in Fig. 2, each specimen used for the fatigue crack growth test was notched at mid-height using electrical discharge machining. The growth of the crack was monitored using the direct current potential drop (pd) technique [7]. Calibration of the voltage measurements had been made with periodic observations of the crack size on both sides. It is important to note that this technique only measures crack length and does not reflect any cracking of the relatively non-conducting fibres. Furthermore, only the total crack length is used(i.e. the notch and the fatigue crack). All tests were performed without exceeding a crack length/width ratio (a/W) of 0.3, excepted for the 37-6 specimen which was unfortunately broken during the last loading. All specimens were pre-cracked at room temperature under a frequency of 10 Hz. The fatigue crack growth behaviour of the SM 1140+/Ti-6242 was examined in isothermal and non-isothermal conditions.

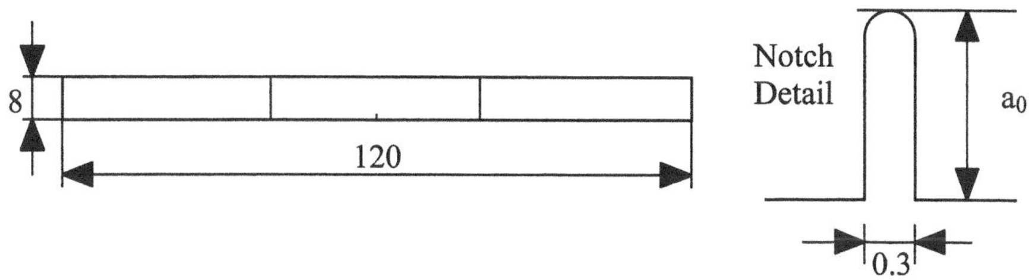

Fig. 2. : Single edge notch specimen geometry

Isothermal tests were performed at 500°C with a stress ratio of 0.1 under two types of solicitation. The first one was a conventional triangular waveform at a frequency of 1 Hz. The second one was chosen to pick up the main characteristics of the thermal-mechanical fatigue cycle : a rapid loading and a slower unloading which lead to matrix relaxation as in the hold time and the unloading at high temperature for the thermal-mechanical fatigue cycle. This test, called the 1-20 test, consists in a loading in 1 s and an unloading in 20 s.

The fatigue crack growth tests in non-isothermal conditions were performed under the same cycle that is used in the fatigue tests for the un-notched specimens. As illustrated in Table 1, a wide range of stress conditions have been successively applied on two specimens : 37-6 and 37-9. In these cases, when the crack propagation was significant the load had been increased by a factor 2.5.

After testing, the fatigue crack growth specimens were polished, without cutting them, to observe the propagation path using optical and scanning electron microscopy (SEM).

Table 1. Characteristics of the fatigue crack growth tests

Specimen	Type	Temperature (°C)	$R\sigma$	Max. Stress (MPa)	a_o (mm)
37-4	TMF	100-500	0	100	1.462
37-5	TMF	100-500	0	250	0.677
37-6	TMF	100-500	0	250 -> 625	0.603
37-9	Isothermal	500	0.1	250 -> 625	0.306

EXPERIMENTALS RESULTS AND DISCUSSION

Thermal-mechanical fatigue results

Figure 3. shows the typical stress-mechanical strain loops of SM 1140+/Ti-6242 loaded in the longitudinal direction between 100 and 500°C under a given stress range for a fatigue cycle number 1 and 260. The mechanical strain is defined as : $\varepsilon_m = \varepsilon_t - \varepsilon_{th}$ with ε_t total strain measured by the extensometer and ε_{th} thermal strain measured during stabilisation cycle without loading before testing. An important plastic strain of the SM 1140+/Ti-6242 is observed at the first cycle which is associated with the plasticity of the matrix and the fracture of the weakest fibres. In the following cycles, the plastic strain shows little variation. The composite behaviour is linear between 100 and 200°C, in fact the Young's moduli of the fibres and the matrix are nearly constant in this range of temperature. From 200°C, the mechanical strain progressively increases whereas the applied stress is nearly constant. During hold time at 500°C, the matrix stress is relaxed and thus, by the mechanisms of load transfer through the fibre/matrix interface, the fibre stress increases. The composite strain is dependent of the fibre strain, so the increase of the fibre stress leads to increase the composite strain.

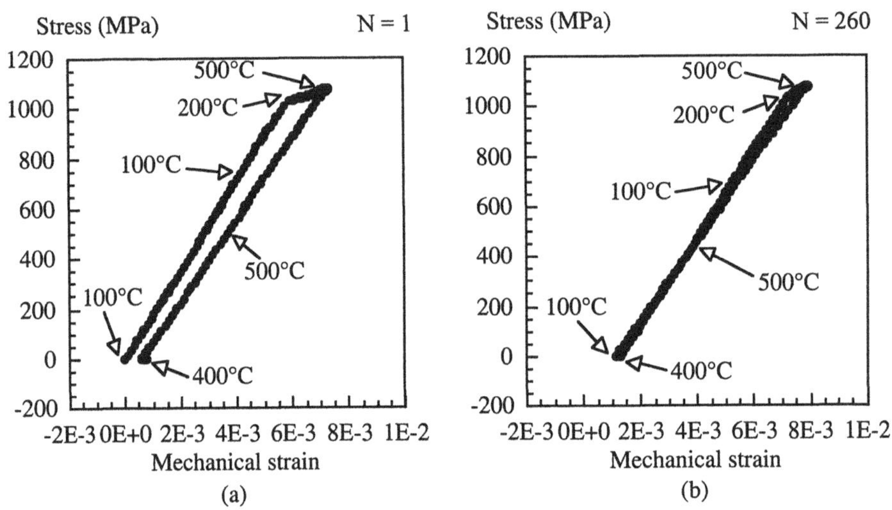

Fig. 3. : Typical stress-mechanical strain response of the SM 1140+/Ti-6242 composite under TMF loading – 1075 MPa : (a) cycle 1st and (b) cycle 260th

Figure 4. displays thermal-mechanical fatigue lives of SM 1140+/Ti-6242 on a stress basis and compares them with results obtained on the same composite under isothermal tests at 550°C in strain control with a R_ε ($R_\varepsilon = \varepsilon_{min}/\varepsilon_{max}$) of 0 [8]. The stress at mid-life is used as usual to plot isothermal test results in Fig. 4.. As can be seen, the thermal-mechanical fatigue conditions give a life slightly shorter than that of isothermal tests. Similar trends were observed in a previous work for SiC/Ti alloys [9] under out-of-phase and in-phase tests. The main reasons of the difference in fatigue lives between isothermal and non-isothermal conditions are the environmental degradation, the effect of mismatch in thermal expansion coefficients between matrix and fibres and the thermo-mechanical loading cycle shape which enhances matrix visco-plastic flow and then overloads the fibres.

Rosenberger and Nicholas [10] have investigated the effect of environment on the fatigue lives of the titanium matrix composites, in isothermal and non-isothermal conditions. They had showed that tests performed in air reduce the fatigue life compared to inert environment. The thermal-mechanical cycle used here enhances the environmental effects by the hold time at high temperature. The effect of oxidation is probably enhanced compared to isothermal conditions.

The composite was processed at high temperature (above 900°C), after cooling a residual stress state arises from the coefficient of thermal expansion mismatch between the matrix and the fibre. The coefficient of thermal expansion of the matrix is more than twice that of fibres. So in the as-fabricated state, the matrix is in a tensile state and the fibre in a compressive state. The mismatch of coefficient thermal expansion leads to a complex stress state near the fibre/matrix interface, which can promote initiation of damage under thermal transient conditions [11].

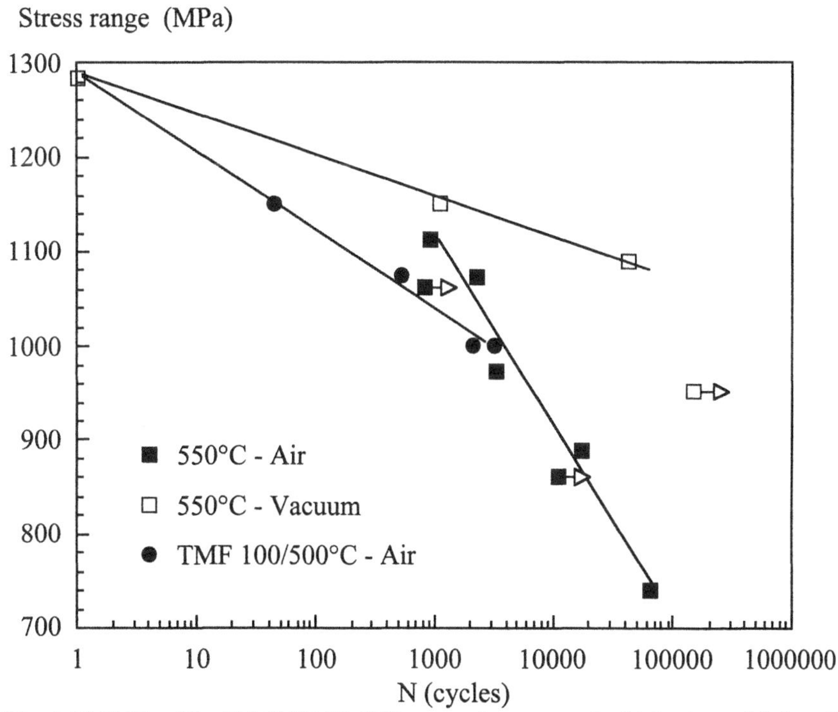

Fig. 4. TMF life of the SM 1140+/Ti-6242 composite compared with isothermal fatigue life

Fatigue crack growth results

Fig. 5. shows the fatigue crack propagation rates measured in SM 1140+/Ti-6242 under a variety of testing conditions. Crack growth rate (da/dN) is plotted versus the total crack length (a), rather than the nominal range of stress intensity factor ΔK used for monolithic materials. The use of conventional ΔK is no longer relevant for the fatigue crack behaviour of metal matrix composites.

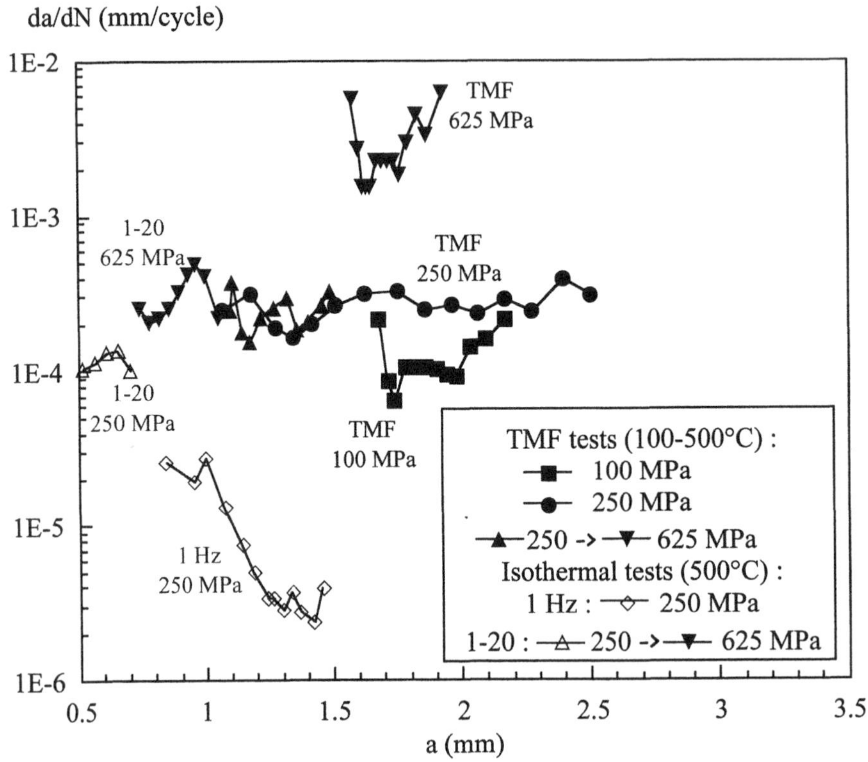

Fig. 5. Fatigue crack growth rates measured in the SM 1140+/Ti-6242 composite in isothermal and non-isothermal conditions

Under all the conditions employed in the present study, there is no crack arrest, the load range is constant in all sequences and the driving force is sufficient to maintain crack propagation. In several works [12-13], the authors have showed that the propagation rate decreases as the crack grows. In this study, only the isothermal test at a frequency of 1 Hz (37-9 test) leads to a decreasing propagation rate. But, the observations of the polished section of the specimen have showed a bifurcation of the fatigue crack. The propagation rate decreases is probably associated with a larger energy dispersion to ensure crack advance. All the other test observations have showed that there is a single fatigue crack growing in the direction of mode I opening.
The 37-6 test shows an increasing propagation rate, which is associated with the catastrophic failure of the specimen in the last sequence. The others conditions lead to a steady propagation state: the average fatigue propagation rates are crack length independent.

In all tests performed, the propagation rate increases with the applied stress, this implies that increasing stress for the two types of loading leads to different fatigue crack growth rates. The thermal-mechanical cycle seems to be more damaging than the isothermal one. For the same applied stress, the fatigue crack growth of the thermal-mechanical tests is larger than the isothermal one. This behaviour is consistent with the comparison of fatigue lives under isothermal and non-isothermal conditions.

Ritchie [14] has showed that a steady propagation state in the metal matrix composite reinforced with unidirectional fibres is often associated with fibre bridging. Fig. 6. shows a typical fatigue crack SEM image in a specimen tested in this study. All fibres along the crack path in the section are unbroken even those far from the crack tip. The crack propagated in SM 1140+/Ti-6242 without breaking any fibre. The fatigue crack growth conditions for all the tests performed are in the bridging fibre regime.
As can be seen in Fig. 6., the fibres which bridges the crack (a) has not the same appearance as the others (b). All the fibres in the crack wake have lost their carbon coating during fatigue crack growth while the others are intact. The lengths where the carbon coating has been consumed, called here worn lengths, have been measured in SEM observations.

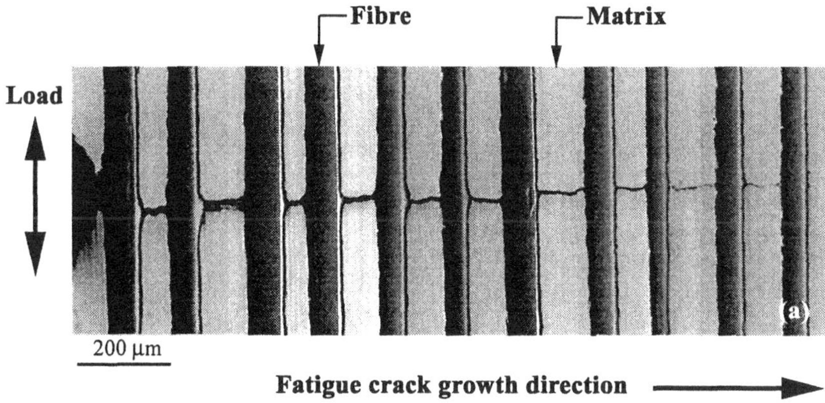

Figure 6. Scanning electron micrograph of a polished surface showing the propagation path of the crack for the thermal-mechanical fatigue crack growth under 250 MPa : (a) and (b) fibre without and with carbon-coating respectively.

Figure 7. displays the worn lengths measured by SEM observations in polished sections of the specimen tested in thermal–mechanical conditions at 250 MPa. The fibres in the notch have their carbon coating consumed, which suggests the wear mechanism is probably thermally activated.
The worn lengths increase with distance from the crack tip, which suggests that the carbon coating consumption rate is associated with crack growth rate. Furthermore a fibre which is not in the crack wake has not consumed its carbon coating. This suggests that the consumption starts when the crack reaches the fibres.
From the worn length defined in each fibre in the crack wake, it is possible to calculate a wear rate of the carbon-coating in conditions defined in this study. For that, it has been supposed that there is no wear during the pre-cracking, the consumption starts when the crack reaches the fibre and continues until test end.
Figure 8. compares the average wear rates (dl/dN) and the average fatigue crack propagation rates (da/dN) under the different testing conditions used in this study. The average rates are calculated for the complete test. The straight line where the two rates are equal divides the figure into two parts: a domain where wear is the predominant phenomenon and an other one where fatigue crack growth is faster than wear. The thermal-mechanical tests seem to be clearly in the domain where the carbon coating is consumed faster than the crack grows. The isothermal crack growth tests are characterised by faster propagation rates.

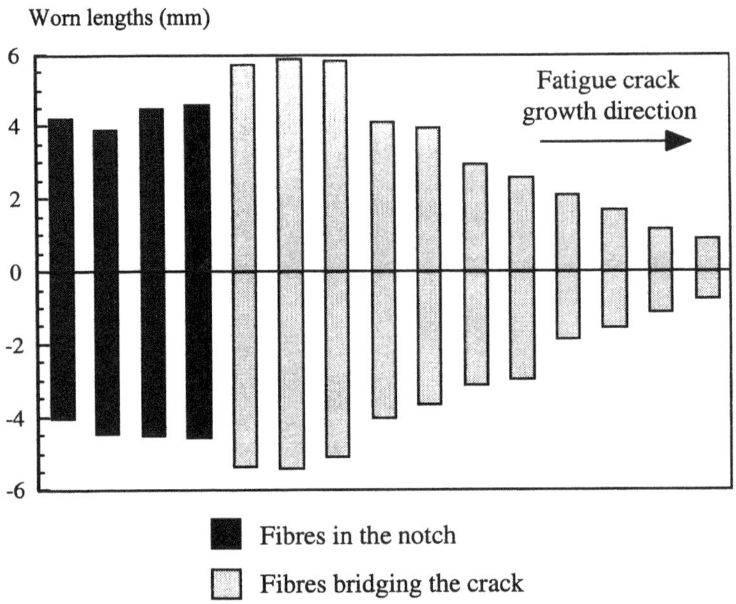

Fig. 7. Worn lengths measured in the SM 1140+/Ti-6242 composite after a thermal-mechanical fatigue crack growth test under 250MPa

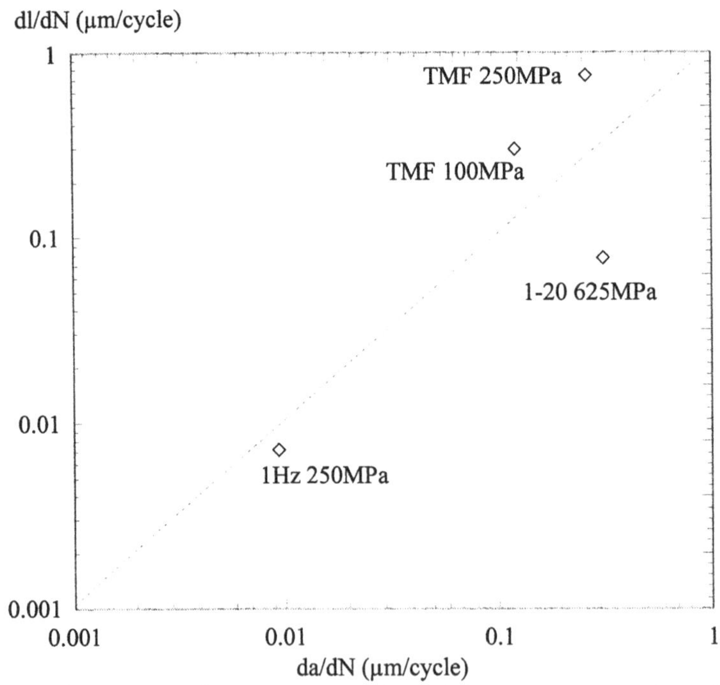

Fig. 8. Comparison between the average wear rates and the average propagation rates for the fatigue crack growth tests

CONCLUSION

The thermal-mechanical fatigue (100-500°C), using a cycle representative of compressor disc loading, appears to reduce the lives of the SM 1140+/Ti-6242 compared to isothermal tests (550°C).
The fatigue crack growth of the SM 1140+/Ti-6242 has been investigated for isothermal and non-isothermal conditions. It has been showed that the non-isothermal cycle leads to higher propagation rates than isothermal tests, at maximum temperature for the same stress range.
The interfacial damage in fatigue crack growth has been measured in fibre bridging regime. It consisted in length where the carbon coating of the fibre has disappeared. The non-isothermal fatigue crack growth tests seem to promote interfacial damage compared to isothermal tests.

ACKNOWLEDGMENTS

The authors would like to acknowledge SNECMA Moteurs for the financial support of this study and provision of material.

REFERENCES

1. Lerch, B., Halford, G. (1995) Materials Science and Engineering A200, pp. 47-54.
2. Guo, S.Q., Kagawa, Y., Bobet, J.L., Masuda, C. (1996) Materials Science and Engineering A220, pp. 57-68.
3. Majumdar, B.S., Newaz, G.M. (1993) In : *Mechanisms and Mechanics of Composites Fracture*, pp. 367-376, Proc ASM Materials congress, ASM International.
4. Gabb, T.P., Gayda, J., MacKay, R.A. (1990). Journal of Composite Materials, 24, pp. 667-686.
5. Jeng, S.M., Yang, J.M., Aksoy, S. (1992). Materials Science and Engineering A156, pp. 117-124.
6. Koster, A., Fleury, E., Vasseur, E., Rémy, L. (1994). In : *Automation in Fatigue and Fracture :Testing and Analysis*, pp.563-580, Amzallag, C., (Eds). Philadelphia, USA.
7. Baudin, G. and Policella, H., (1978). *La Recherche Aérospatiale*, n°4 (Juillet-Août), pp. 195-203.
8. Legrand, N., Rémy, L., Dambrine, B., Molliex, L., (1996). In : *Comptes-rendus des dixièmes Journées Nationales sur les Composites*, pp 1349-1360, Baptiste, D. and Vautrin, A. (Eds). AMAC, Courtaboeuf, France.
9. Castelli, M.G., Gayda, J. (1993) In : *Reliability, Stress Analysis, and Failure Prevention* DE-Vol. 55, pp. 213-221. ASME New-York.
10. Rosenberger, A.H., Nicholas, T. (1997). In : *Composite Materials : Fatigue and Fracture (Sixth Volume)* ASTM STP 1285 E. A. Armonios. Ed. American Society for Testing and Materials pp. 394-408
11. Revelos, W.C., Jones, J.W., Dolley, E.J. (1995). Metallurgical and Materials Transactions A **26A**, pp. 1167-1181.
12. Barney, C., Cardona, D.C., Bowen, P. (1998) International Journal of Fatigue Vol 20 pp. 279-289
13. Cotterill, P.J., Bowen, P. (1993) Composites Vol 24 Number 3 pp. 214-221
14. Ritchie, R.O., (1996). In : *Mechanical behaviour of materials at high temperature*, pp. 461-494, Moura-Branco, C., Ritchie, R.O. (Eds). Kluwer, Dordrecht.

Temperature-Fatigue Interaction
L. Rémy and J. Petit (Eds.)

THERMAL FATIGUE OF A 304 L TYPE STEEL

V. MAILLOT [1], A. FISSOLO [2], G. DEGALLAIX [1], S. DEGALLAIX [1], B. MARINI [2], M. AKAMATSU[3]

[1] Ecole Centrale de Lille - L.M.L.
(URA CNRS 1441) – BP48, 59651 Villeneuve d'Ascq Cedex, France
[2] CEA-CEREM
Commissariat à l'Energie Atomique – Saclay, 91191 Gif sur Yvette Cedex, France
[3] EDF-EMA
Electricité de France - Centre des Renardières - BP N°1 Ecuelles,
77818 Moret sur Loing Cedex, France

ABSTRACT

Various components of nuclear reactors are submitted to very sharp thermo-mechanical loadings. Thermal fatigue cracking has been clearly detected in auxiliary loops of the primary cooling circuit of Pressurized Water Reactors (PWRs). The study presented here is focused on the 304 L type stainless steel used in PWRs. The thermal fatigue behaviour of this steel has been investigated using a specific thermal fatigue test equipment called SPLASH. This test equipment allows the reproduction of multiple cracking networks similar to those detected during inspections. The present study deals with two points :
i) the experimental determination of crack initiation conditions and the morphological description of the growing crack networks;
ii) the multiple crack growth numerical simulation, using a Skelton model, and a generalized Paris law. This modelling, in spite of simplified assumptions, gives predictions in good agreement with observations, as far as the evolution of the mean and deepest cracks during cycling are concerned.

KEYWORDS

Thermal fatigue equipment, 304 L and 316 L(N) type austenitic stainless steel, crack network, simulation of multiple crack growth, numerical modelling

INTRODUCTION

Thermal fatigue causes damage in in-service components, such as moulds or rolling mill cylinders. Some examples are given by Brassers and Rémy [1] or Spera and Mowbray [2]. A similar phenomenon occurs in different types of nuclear reactors. Cyclic temperature variations of the cooling fluid can submit the piping to thermal fatigue. In Liquid Metal Fast Breeder Reactors (LMFBRs) for instance, strong thermal fluctuations are generated by the mixing of two sodium flows at different temperatures : the thermal striping phenomenon causes crack networks, sometimes called crazing. In Pressurized Water Reactors (PWRs), crack networks appear in auxiliary loops, close to a cold water injection site, in spite of relatively small temperature fluctuations [3]. The present study focus on PWR condition, its results are compared with those of a previous study concerning LMFBR condition [4].
Crack networks are obtained in LMFBR or PWR condition with the SPLASH equipment as described in the first paragraph. In most studied networks, the crack depth remains small. The subsequent network parameters are described in the second paragraph. Nevertheless, it was observed in in-service conditions that one or more cracks could grow deep enough for a leak to occur. The propagation mechanisms and the shielding effect of the cracks in the network are therefore investigated using a crack propagation simulation as described in the third paragraph.

THE SPLASH EQUIPMENT

For in-service components, thermo-mechanical loadings usually come from temperature gradients in thickness. Such gradients are produced in SPLASH specimens thanks to the SPLASH equipment. Figures 1 and 2 present respectively the SPLASH testing facility and the SPLASH specimen. A SPLASH specimen is a 240x30x20 mm^3 parallelepiped, continuously heated by an electrical DC current (Joule effect), and submitted to cyclic thermal down-shock (corresponding to a cooling rate of about 500 to 1000°C/s) when water is sprayed on two opposites faces.

Fig. 1. SPLASH facility

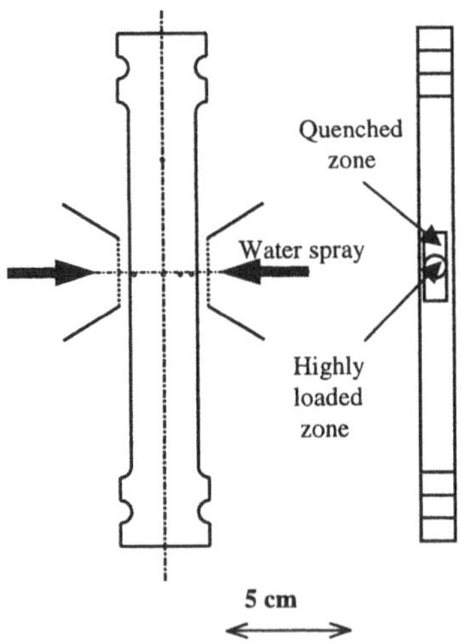

Fig. 2. : SPLASH specimen

Two types of specimens are used : calibration specimens and test specimens. Both are equipped with K-type thermocouples brazed in depth, at 3 and 7 mm from the left and right surfaces, and in the center. Calibration specimens also support K thermocouples brazed at the surface, and are used to determine the parameters of the water spray necessary to obtain the selected temperature range ΔT at the surface. Those K thermocouples are not used on the test specimens, for they could induce premature crack initiation.

PWR and LMFBR thermal conditions are given in table 1.

Conditions	Steel type	Maximum temperature	Temperature range
LMFBR	316 L(N)	550 °C	Between 100 and 300°C
PWR	304 L	320°C	150 or 200 °C

Table 1. Loading conditions

The 316 L(N) and 304 L type steels studied here are two austenitic stainless steels. They differ principally by their molybdenum content. The chemical compositions of those steels are given in table 2.

type	C	Mn	Si	Cr	Ni	Mo	S	P	Cu	Al	B	N	Fe
304 L	0.031	1.48	0.55	19.4	8.6	0.23	0.003	0.028	0.17	0.025	0.0015	0.058	Bal
316 L(N)	0.024	1.82	0.46	17.44	12.33	2.30	0.001	0.027	0.20	-	0.0008	0.06	Bal

Table 2. Chemical compositions in wt%

The number of cycles to initiation N_i is determined from regular observations of the quenched surfaces by optical microscopy, after the removal of the thin oxide layer that forms during cycling. It is considered that initiation occurs when at least one of 50 to 150 μm long crack is observed at the surface. Usually, multiple initiation is observed.
After initiation, the growth and coalescence of the cracks, then the formation of a network are observed during the test. After the network formation, the test is stopped at a chosen number of cycles. After the end of the test, the 3D character of the crack networks is examined, using a step by step removal of thin layers.

NETWORK PARAMETERS

Initiation

Figure 3 gives the evolution of the temperature range as a function of the number of cycles to initiation, for different tests under LMFBR or PWR condition. Some of the tests were performed with a holding time of 20 to 30 s (HT) at the maximum temperature.

Under LMFBR condition, regardless of the holding time, the temperature range strongly affects N_i. When ΔT is between 100 and 150°C, no crack is detected, even after 10^6 cycles, whereas with a ΔT of 300°C, initiation occurs after only 6 000 cycles. Thus, a thermo-mechanical endurance limit is observed. A significant holding time at the maximum temperature increases the number of cycles to initiation.
In PWR condition, multiple cracking is observed around 70 000 - 80 000 cycles for a ΔT = 150°C.
So the initiation has occurred earlier than under LMFBR condition.

A mixed condition test was performed in order to study the influence of the material (304 L or 316 L(N)) on N_i : a 316 L(N) type steel specimen was tested under Tmax = 320°C and ΔT = 150°C. The initiation occurred at 90 000 cycles on both sides of the specimen. It can be concluded from that test that under PWR condition the material (316 L(N) or 304 L) has a relatively small influence on N_i.

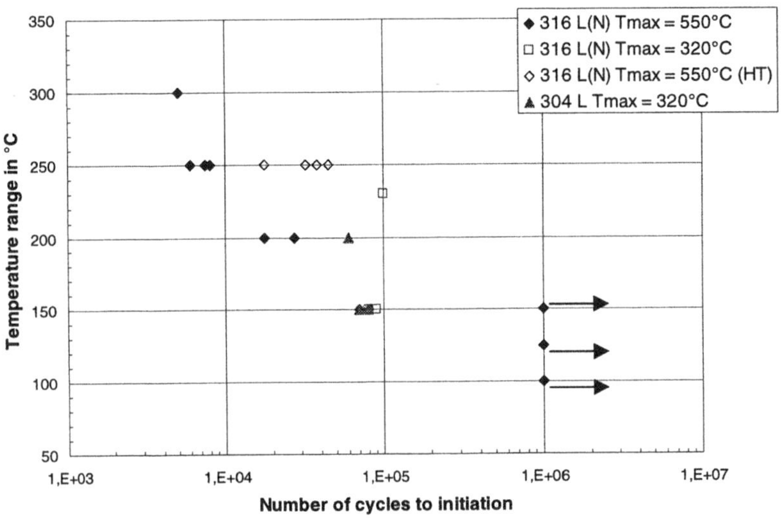

Fig. 3. Temperature range versus number of cycles to initiation

Network morphology

Another important difference between the LMFBR and PWR conditions is the density of cracks observed at the end of the test. As shown in Figures 4 and 5, the networks obtained in LMFBR condition are much denser than those obtained in PWR condition. The average distance between two cracks, on the surface, is of about 50 µm in LMFBR condition and of at least 400 µm in PWR condition.

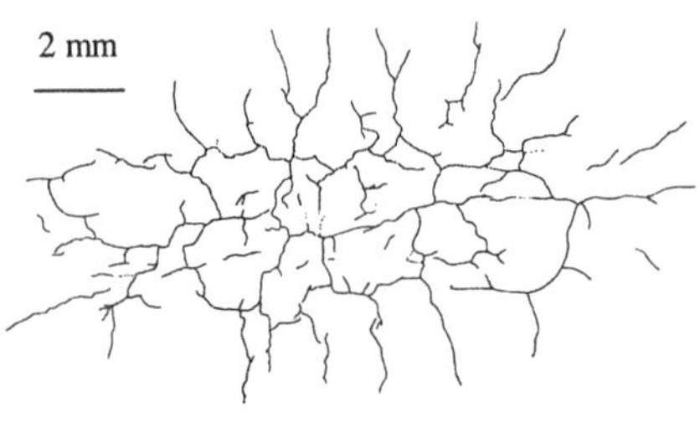

Fig. 4. PWR condition, ΔT = 150°C, Network observed at the surface, after 300 000 cycles

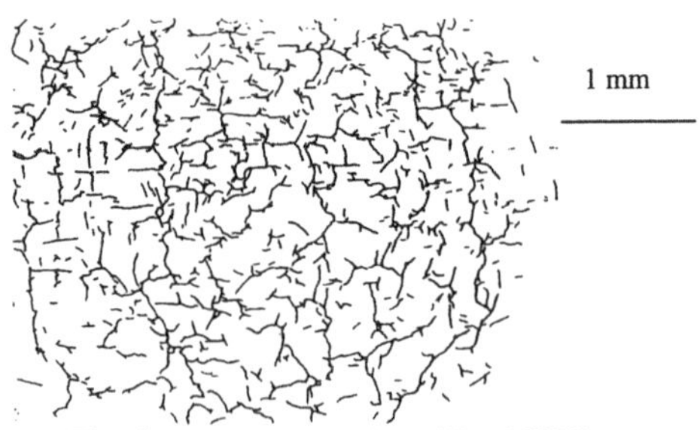

Fig. 5. LMFBR condition, ΔT = 250°C, Network observed at the surface, after 20 000 cycles

In LMFBR condition, the maximum crack depth observed after 300 000 cycles, for $\Delta T = 250°C$, is 1.6 mm [4], whereas it is 2,4 mm in PWR condition, after 300 000 cycles, for $\Delta T = 150°C$. In the following figures (6 and 7), the same network is presented at different depths, respectively at the surface, at 360 µm and at 1 mm. Beneath the surface, the network is far less dense, and only a few cracks reach a 1 mm depth.

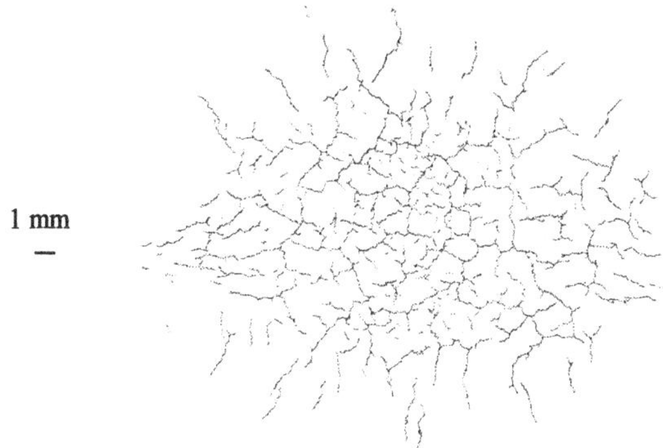

Fig. 6. PWR condition, $\Delta T = 200°C$, network observed at the surface, after 150 000 cycles (at the end of the test)

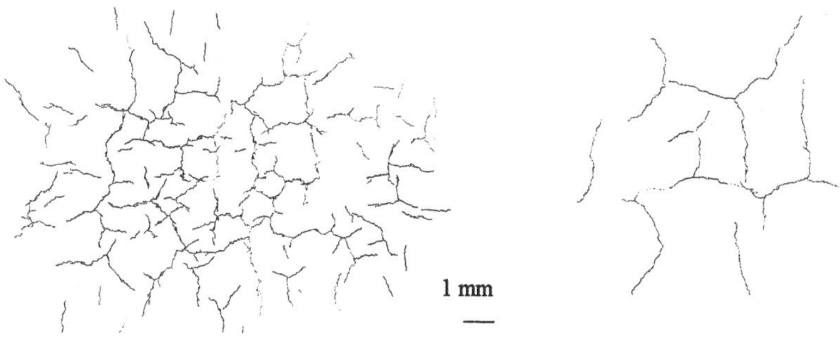

(a) *(b)*

Fig. 7. The same network, observed in depth : 360 µm (a) and 1 mm (b)

It is concluded that LMFBR condition leads to an early initiation of a very dense network of shallow cracks whereas PWR condition leads to a later initiation of deeper cracks, but not as dense. That is why PWR condition seems to be more damaging than LMFBR condition : a deep crack is more dangerous for the integrity of the structure than denser but shallower cracks.

CRACK GROWTH SIMULATION

The 2D modelling was first developped to account for the propagation in depth, of multiple parallel cracks in LMFBR condition. The propagation of an isolated crack was investigated, then that of two cracks, and finally, that of up to ten cracks [5]. The crack initiation is simulated first, and then the propagation.

Initiation

The simulation of crack initiation is performed in two steps. In the first step, a Monte-Carlo randomization method is used to determine the size of the surface grains, according to a distribution law of the experimentally measured grain sizes. The obtained grains are supposed to be square shaped. In the second step, another Monte Carlo randomization gives to each grain a position relative to the surface. The squares then change their shapes to rectangles according to their respective positions as presented in figure 8.

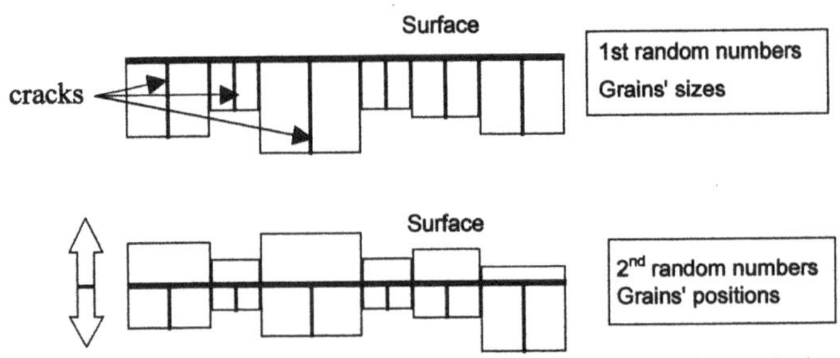

Fig. 8. Illustration of the 2 step Monte Carlo randomization method

In LMFBR condition, the network is dense enough for almost every grain to be cracked. In the modelling, it is assumed that right from the beginning, every grain is completely cracked in its middle, as shown in Figure 8. In PWR condition, this assumption is an oversimplification: only one out of four or five grains is cracked but the method remains the same.
After this simulation of initiation, the simulation of crack growth can be performed.

Propagation

The thermo-mechanical loadings are first calculated. Near the surface, the stresses are supposed to be equibiaxial. A Skelton law [6] is used to define an equivalent stress intensity factor. According to equation (1), the strain loading can be reduced to a stress loading which obeys to the Linear Elastic Fracture Mechanics laws.

$$\Delta\sigma_{eff} = q\Delta\sigma + \Delta\sigma_{pseudo} \text{ and } \Delta\sigma_{pseudo} = \frac{E\Delta\varepsilon_p}{(1-\nu)} \tag{1}$$

$q\Delta\sigma$ represents the part of the cycle when the crack is open, whereas $\Delta\sigma_{pseudo}$ takes into account the plasticity at the crack tip. For the present purpose, q was estimated to be close to 0.6.
$\Delta\sigma_{eff}$ is fitted with a third degree polynomial, for a depth between the surface and 3 mm, as presented in equation (2).

$$\Delta\sigma_{eff}(x) = \sum_{i=0}^{i=3} A_i x^i \tag{2}$$

ΔK_{eff} is then calculated using the superposition method of Buchalet and Bamford [7], as presented in equation (3) for a crack of depth a.

$$\Delta K_{eff}(a) = \sqrt{\pi a}\left(A_0F_1 + \frac{2a}{\pi}A_1F_2 + \frac{a^2}{2}A_2F_3 + \frac{4a^3}{3\pi}A_3F_4 \right) \quad (3)$$

where F_1, F_2, F_3 and F_4 depend on the geometry, and particularly of the other cracks (depths and distances from the studied crack). The values of the F_i are tabulated, in order to make the calculations faster. The growth of each crack then follows a generalized Paris law, as presented in equation (4), for a given number of cycles increment (usually, dN = 500).

$$\frac{da}{dN} = C\left(\Delta K_{eff}\right)^m \quad (4)$$

In order to simplify the calculation, an auto adaptative meshing is used (Figure 9). In this way, only the crack tips are refined, so that the calculations are quicker than they would be if all the elements had the same size.

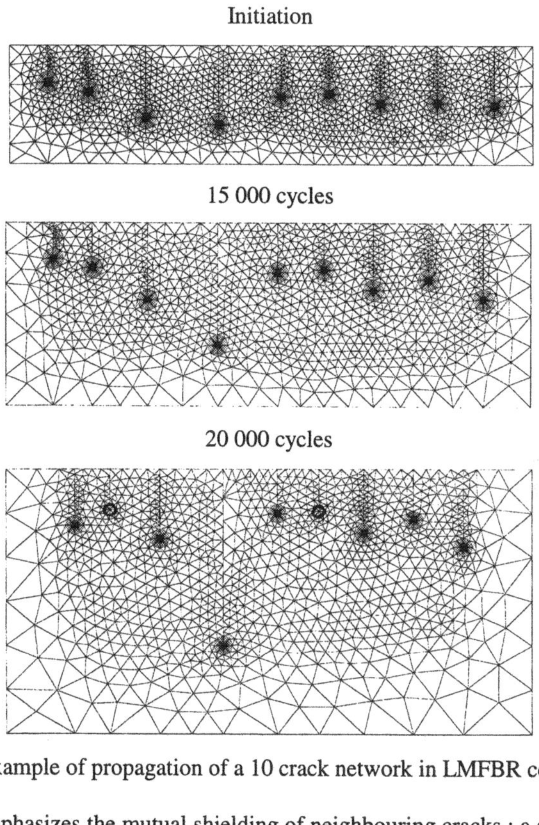

Fig. 9. Example of propagation of a 10 crack network in LMFBR conditions

This simulation emphasizes the mutual shielding of neighbouring cracks : a smaller crack close to a deeper one will slow the growth of the deeper one, whereas its own growth will be slowed even more. Eventually, the smaller crack stops growing altogether.

This effect has been experimentally observed : most of the cracks are no deeper than 100 μm, only a few reach 500 μm, and usually only one is deeper than 1 mm. Mean and maximum crack depths obtained by modelling are close to those observed experimentally in LMFBR condition.

In PWR condition, a few modifications must be brought to the model. First, as the cracks are experimentally deeper, $\Delta\sigma_{eff}$ must be fitted between the surface and 4 or 5 mm at least. In addition, as the number of cracks is smaller, a 2D model, accounting for only a few cracks might not be sufficient to reproduce the experimentally observed shielding effect.

CONCLUSION

The thermal fatigue equipment SPLASH can reproduce multiple cracking networks with different characteristics under PWR or LMFBR condition. The initiation occurs sooner under LMFBR condition and the networks are denser, whereas the deepest crack is not as deep as that observed under PWR condition. This could be explained by a mutual shielding effect that is less important on the less dense networks obtained under PWR condition.

The 2D simulation of crack growth in depth gives good results as far as the shielding effect, the maximum and the average crack depth in LMFBR condition are concerned. Provided a few modifications, still in development, are enforced, to account for the small density of the networks, it should be possible to use this model in PWR condition as well.

REFERENCES

1. Bressers, J. and Rémy, L. (1996) Fatigue under Thermal and Mechanical Loading, *Proceedings of Petten Symposium May 1995*, Kluwer Acad. Pub.
2. Spera, D.A. and Mowbray, D.F. (1976) Thermal Fatigue of Materials and Components *ASTM STP 612*
3. de Keroulas, F. and Thomeret, B. (1990) *Société Française d'Energie Nucléaire* Vol. 1, pp. 107-117.
4. Fissolo, A., Marini, B., Nais, G. and P . Wident, (1996) Thermal Fatigue Behaviour for a 316 L Type Steel, *Journal of Nuclear Materials,* 233-237, pp.156-161.
5. Fissolo, A., Robertson, C., Maillot, V. and Marini, B. (2000) Prediction of Cracking under Thermal Fatigue, *Proceedings of ECF 13, San Sebastian, Spain, 6-9 September 2000.*
6. Skelton, R.P. (1983) Crack Initiation and Growth in Simple Metal Components during Thermal Cycling, *Fatigue at High Temperature,* pp. 1-61, Applied Science Publications.
7. Buchalet, C.B. and Bamford, W.H. (1976) Mechanics of Crack Growth, *ASTM STP 590,* pp. 385-402.

Temperature-Fatigue Interaction
L. Rémy and J. Petit (Eds.)

ACOUSTIC EMISSION ANALYSIS OF OUT-OF-PHASE THERMO-MECHANICAL FATIGUE OF COATED Ni-BASE SUPERALLOYS*

Y. VOUGIOUKLAKIS, P. HÄHNER, V. STAMOS, S. PETEVES, J. BRESSERS
Institute for Advanced Materials, DG-JRC, European Commission,
1755 ZG Petten, The Netherlands

ABSTRACT

The acoustic emission (AE) activity was monitored during the thermo-mechanical fatigue (TMF) testing of Ni-base superalloys (CM186), in both single-crystalline and directionally solidified form, coated by a modified aluminide. The TMF tests were carried out with 180° phase difference between mechanical strain and temperature (out of phase) at two different mechanical strain ranges ($\Delta\varepsilon_m$=0.8 and 1.0%, $R=-\infty$) and with temperature cycled between 350° and 950°C. For reference and AE source identification, TMF with AE monitoring was also performed for the uncoated bulk materials. In-situ monitoring of crack initiation and propagation on the surface of the specimens by means of a fully automated video imaging system was also implemented. Both AE monitoring and video imaging were synchronised with the TMF cycle. The AE data were analyzed statistically with respect to various parameters (such as amplitude, number of counts, duration, rise time, total energy, etc., and the TMF parameters of their occurrence, stress, temperature, cycle number). Within the eleven-dimensional parameter space, lower-dimensional descriptor subspaces were identified by correlation hierarchy that characterize different groups of AE events. From the analysis of the AE signals various types of information were obtained, such as: (1) the effects of the coating on the AE activity, (2) early surface *vs.* bulk damage and the subsequent growth of fatigue cracks, respectively.

KEYWORDS

Acoustic emission, thermo-mechanical fatigue, Ni-superalloy, protective coating

INTRODUCTION

The life of high-pressure gas turbine blades is primarily determined by the interaction of mechanical and thermal loads and environmental contributions. Various protective measures are taken to guarantee efficient operation of the blades up to service temperature and different coatings have been developed to protect the blades against environmental degradation. The failure modes, which must be taken into account for lifetime assessment of such coated systems, include oxidation, corrosion, erosion and microstructural discontinuities caused by interdiffusion between the substrate and the coating. However, the major cause of failure is thermo-mechanical fatigue due to mechanically applied and thermally induced alternate stresses, resulting from the

* work carried out within the Commission R&D programme.

thermal strains generated by the strong temperature gradients experienced during heating and cooling [1].
AE is a non-destructive method that allows the in-situ monitoring of damage during the thermal and mechanical solicitation of a material. As compared to other non-destructive techniques, which provide information about the damage accumulated at a certain stage of life, AE gives access to damage processes inside a material in terms of the rate of damage accumulation, and it is a highly sensitive means for the early detection of impending failure. Hence, the implementation of AE monitoring in a laboratory environment can be instrumental in interpreting the damage mechanisms, as these are triggered and progressing during TMF tests.
In this work results from AE data analysis of TMF tests on single crystal and directional solidified CM186 blade alloys in the uncoated and Sermalloy 1515-coated conditions are reported. Goal of this study is to establish the validity of the information obtained by AE monitoring during TMF and to correlate the AE data with specific damage processes.

EXPERIMENTAL

Specimens were prepared from single crystal (SC) and columnar grained (CG) (directionally solidified) Ni-based superalloy CM186 bars with the crystallographic orientation of the long axis of the samples in the [001] direction.
Threaded end TMF specimens with a solid rectangular cross-section of 12 x 3 mm^2 and a parallel length of 9 mm were machined. The rectangular cross section was adopted to enable observation of the flat surface by means of a computer controlled video imaging system during testing, as discussed later. Some of the SC and CG CM186 samples were coated with a modified aluminide coating, Sermalloy 1515, via an Al/Si slurry diffusion process, with an approximate thickness of 80μm.

Mechanical set-up

The strain-controlled TMF tests were carried out in air on a computer controlled electro-mechanical closed loop testing machine of 100 kN capacity (Schenck Trebel). Asymmetrical triangular out-of-phase cycles with mechanical strain ranges $\Delta\varepsilon_m$ of 0.8% and 1.0%, respectively, and a temperature range ΔT=600°C with temperature cycled between 350°C and 950°C were chosen (minimum to maximum mechanical strain ratio $R=-\infty$, cycle period Δt=200s). So temperature and mechanical strain both varied linearly with time, but with a phase shift of 180°. This cycle imposes maximum mechanical strain/stress at minimum temperature and minimum strain/stress at maximum temperature. Control of the testing machine, the high frequency induction heating system and data acquisition was performed by means of a dedicated computer system using LabVIEW 5.1. Thermal strain compensation was achieved by recording the thermal expansion under zero load as a function of time prior to test initiation using the same temperature cycle as in the subsequent TMF test. Prior to the actual TMF tests, the temperature dependence of the Young's modulus was also determined by static measurements (in thermal equilibrium) in the TMF temperature range.
The tests were started at the minimum temperature and at zero mechanical strain, and conducted until failure, or stopped when the cyclic stress range had dropped below 50% of the previously stabilized value. TMF life is then defined as the corresponding cycle number.
During TMF testing, images of the front face of the specimen were taken at pre-selected cycle numbers by means of a computer controlled video imaging system,

allowing contact-less, in-situ and fully automated monitoring of the evolution of surface damage. To enhance the detection of crack initiation (with a surface length ≥ 15μm), the CCD camera with a flash unit was synchronized with the TMF cycle and images were taken when the imposed stress on the specimen exceeded 80% of the maximum tensile stress of the previous cycle. The images taken covered the whole specimen width and the gauge length. All the images were digitised and stored for post-processing and analysis. The TMF set-up has also been described in Ref. [1].

Acoustic emission set-up

Two broadband piezoelectric transducers were used as sensors to monitor the acoustic emission during the tests. The sensors were attached at the grips of the machine just outside the two cooling plates attached to protect the gripping system and to stabilise the temperature control. The AE signals were pre-amplified and fed into a two-channel acoustic emission digital signal processor. The acquisition of the AE signals was performed by a computer system using MISTRAS 2001. The parameters set in the initialisation file of the software were optimised with respect to the external noise (ambient or induced by the induction heating system), the experimental set-up, the specimen geometry and volume of interest (the bulk inside the gauge length). For the ease of comparability and the analysis of the received signals the same AE settings were used for the whole series of TMF tests. The AE signals were in time-order and seven different AE features were monitored each time: rise time, counts, energy, amplitude, duration, average frequency, counts to peak, as well as the current stress value and the cycle number as input parameters. The evaluation and analysis of the received signals was performed using NOESIS and SPSS software packages.

THERMO-MECHANICAL FATIGUE BEHAVIOR

The TMF results and post mortem fractographic and microstructural analyses of the samples have been reported and discussed extensively elsewhere [1]. As such a few relevant results will be only presented here, namely the observed pronounced cyclic creep during the first cycles caused by the accumulation of compressive inelastic deformation, and the cracking mode (mode I) of the coating.

The TMF tests were conducted under total strain control where the total strain ε_{tot} is the sum of the thermal strain ε_{th} and the mechanical strain ε_m, which in turn consists of an elastic contribution ε_{el} and an inelastic contribution ε_{in}: $\varepsilon_{tot}=\varepsilon_{el}+\varepsilon_{in}+\varepsilon_{th}$. Knowledge of the temperature dependence of the Young's modulus allows the separation of ε_{el} and ε_{in}.

Fig. 1 shows the imposed temperature and mechanical strain as a function of time, together with the observed response in terms of the stress and the inelastic strain for the first cycle. This example refers to the uncoated CM186 in the directionally solidified variant subjected to $\Delta\varepsilon_m$=1.0%. It is seen that during compression at high temperature in the first cycle, there is already an appreciable compressive inelastic strain ($\varepsilon_{in}\approx-0.3\%$), which causes a tensile stress to develop at the low-temperature end of the cycle.

As TMF cycling continues, the accumulation of inelastic compressive strain leads to a stress ratcheting towards tensile stresses (the minimum stress σ_{min} increases from approximately –800MPa to –500MPa, while the maximum stress σ_{max} increases from 0 to 600MPa during the first 100 cycles with more than 80% of the inelastic strain accumulated already during the first ten cycles). As a consequence the specimen experiences damaging tensile stresses during the low-temperature part of the cycles.

This is of particular relevance to the coated specimens investigated, since the coating shows brittle behaviour in the lower range of imposed temperatures.

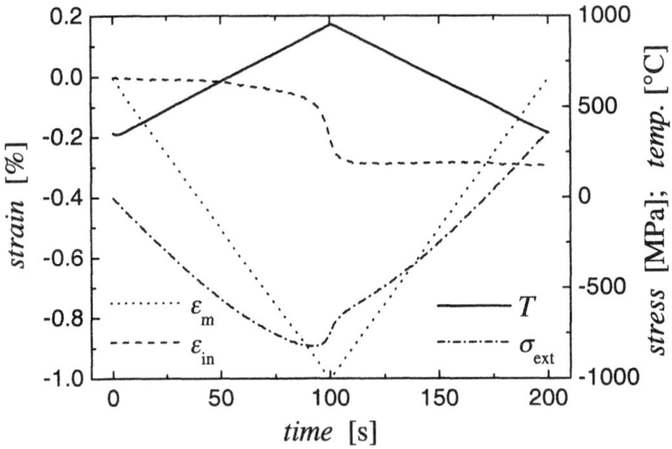

Fig. 1. Thermo-mechanical behaviour of CG CM186 uncoated (Specimen TM7) during the first cycle.

A typical evolution of the maximum and minimum cyclic stresses and the stress range with cycle number *N* is shown in Fig. 2, again for the uncoated CG material at a mechanical strain range of $\Delta\varepsilon_m = 1$ %. The aforementioned gradual shift of the cyclic stress envelopes towards the tensile direction is clearly visible, in particular, during the first 100 to 200 cycles. After a saturation period of virtually constant stress envelopes, the stresses drop abruptly, particularly the maximum stress, towards the end of life due to the propagation of the fatal fatigue crack.

The cyclic stress increase, particularly during the first cycles, is due to the accumulation of inelastic compressive strain, which has to be compensated by elastic tensile strain to comply with the imposed mechanical strain. The cyclic evolution of the inelastic strain is also shown in Fig. 2.

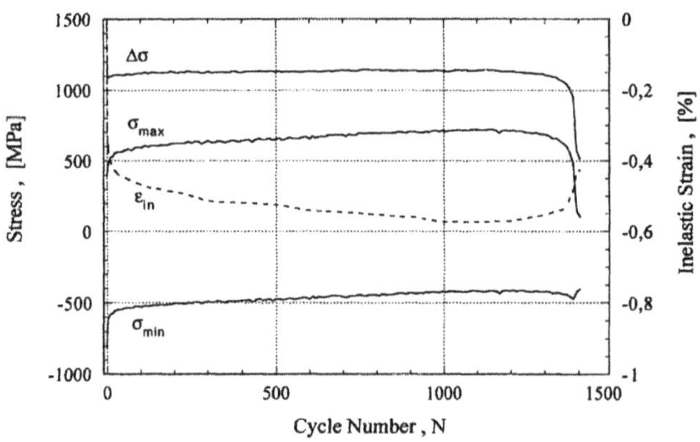

Figure 2: Evolution of the characteristic cyclic stresses and accumulated inelastic strain with cycle number, N, measured on CG CM186 uncoated, at $\Delta\varepsilon_m = 1\%$.

At the same time also the compressive minimum stresses decrease in absolute value, causing the further accumulation of inelastic cyclic compression to decrease.
In the first cycle almost 30 % of the imposed mechanical strain and more than 50 % of the total inelastic strain accumulated during TMF life is transformed into inelastic compression of the sample. The third and fifth cycles already show much smaller increase in inelastic compression (by 0.06 % and 0.02 %, respectively). Consequently, the hysteresis loops stabilize rapidly, after a few cycles, as seen in Fig.3. The pronounced inelastic deformation early in life results in a shakedown of the cyclic minimum stresses to values close to the elastic regime at the maximum temperature, resulting in very narrow and almost closed stress-strain loops during the major part of life as it can be seen in Fig.3.

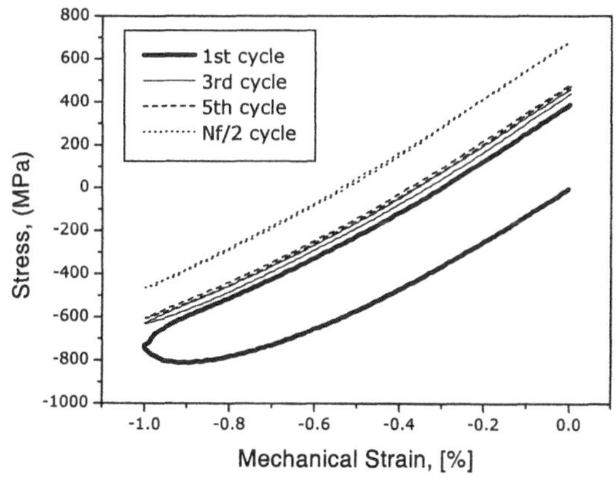

Fig. 3. Stress-mechanical strain hysteresis loops for CG uncoated sample.

The results compiled in Table 1 demonstrate that the effect of the coating on the TMF lives is not unequivocal: it is detrimental in the case of the SC substrate and beneficial in the CG case. At the same time, as far as the SC material is concerned, the presence of the coating has a noticeable effect on the inelastic strain of the first cycle (Fig. 4(b)), whereas inelastic strains are about the same for the coated and uncoated material in the CG case (Fig. 4(a)). Since an excess inelastic compression translates directly into a higher maximum tensile stress (0.1% of strain correspond to 100 MPa, as the Young's modulus is about 100 GPa) it is tempting to attribute the TMF life reduction by the coating in the SC case to the excessive yielding of the coated SC material during the first cycle. In the CG case, on the other hand, the coating may reveal its beneficial effect as an oxidation protection, since there is no significant difference in inelastic behaviour during the first cycle. It should be noted that the video imaging observations showed that the crack initiation life amounts to about 10 – 20 % of the total TMF life (Table 1, cf. also Ref.[1]) without any difference in crack initiation lives between coated SC and CG material.

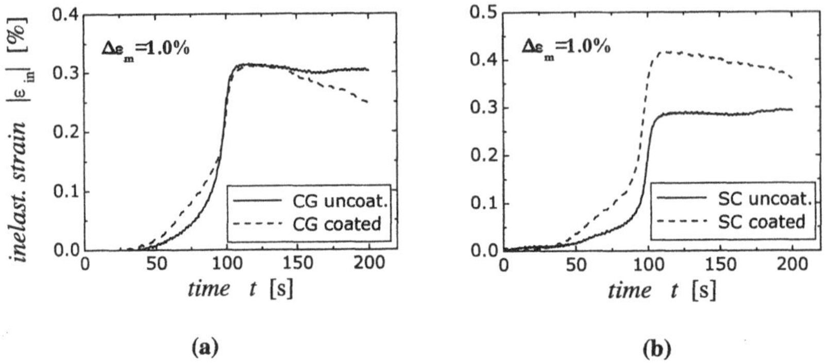

Fig. 4. Comparison of the absolute values of the inelastic strains during the first cycles: (a) Columnar Grained , (b) Single Crystal, both coated and uncoated, respectively.

Microstructural analysis

The coating consists of three layers separated by silicide-rich interlayers (cf. Fig. 5a). In the post-mortem state multiple fracture of the coating is observed. Most of the cracks are confined to the coating in depth (80μm), while they extend over several 100μm or even several mm in length perpendicular to the loading axis. The frequent observation of cracks that had not reached the outer surface (Fig. 5a) suggests that cracks are initiated in the coating at the coating/substrate interface, grow towards the surface and finally propagate into the substrate by Mode I extension, producing a crack front almost parallel to the specimen surface and a crack path perpendicular to the loading direction, see Fig. 5b.

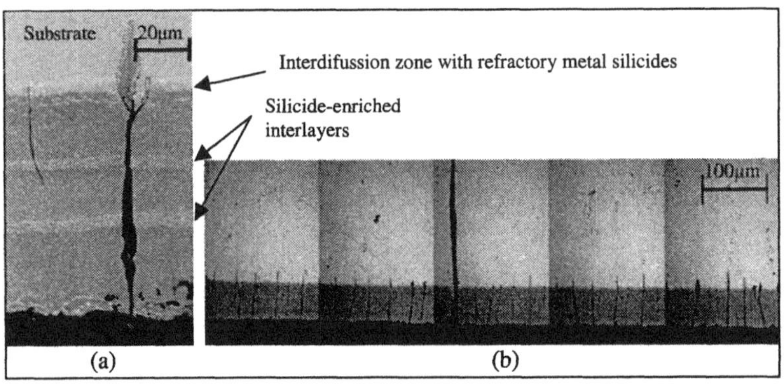

Fig. 5. Post-mortem scanning electron micrographs from a coated CM186 sample (TM9).

Crack nucleation sites appear to relate to refractory metal (W, Ta, Re) silicides adjacent to the coating/substrate interface. The microstructural analysis although could not confirm the above mentioned nucleation sites for all cases, it did prove however, that cracks emerged from the interface and grew outwards while extending along the surface. In a later stage only few of these cracks propagated into the substrate, see Fig.6. The early stage of this crack propagation into the bulk seems to be assisted by oxidation, i.e. to consist in the alternating cyclic formation and fracture of the oxide scale formed during the hot part of the TMF cycle.

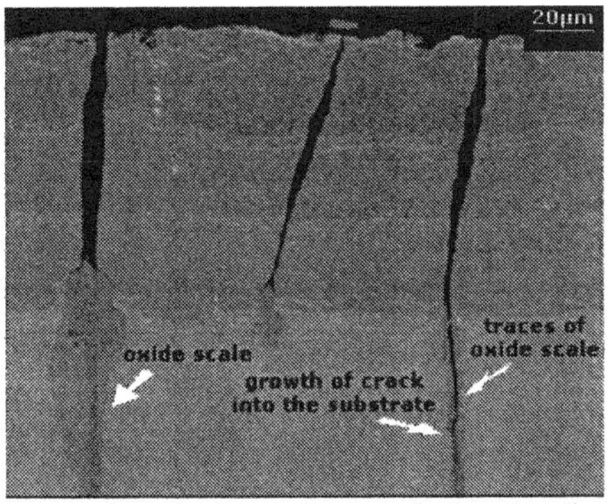

Fig. 6. Post-mortem scanning electron micrograph showing the oxidation-assisted propagation of fatigue cracks into the substrate

AE ANALYSIS

Data filtering

A formula for the reduction of spurious AE data was implemented. This formula was based on AE activity from background sources and from the induction heating system monitored during several "calibration" tests. So, when a certain rule was not fulfilled the signal was excluded from the further steps of the analysis. For the further analysis, the AE signals were separated according to the mechanical load state of the system when they were received: compressive or tensile signals. Table 1 presents the above information for each test.

Table 1. Experimental conditions and AE activity in terms of number of AE events.

SPECIMEN	TM1	TM2	TM3	TM5	TM7	TM9
Material	SC	SC	SC, coated	SC, coated	CG	CG, coated
Mechanical strain range	1%	0.8%	1%	0.8%	1%	1%
TMF lives (# cycles)	1941	9165	1121	5300	1412	2404
AE events in tension	749	1309	2393	7022	742	14693
AE events in compression	2491	4238	4813	5918	850	2629
AE events in Region I	No AE events	No AE events			No AE events	
in tension			1675	6484		6453
in compression			1475	1640		1892
End of Reg. I (cycle #) (% of TMF life)			335 (30%)	2508 (47%)		933 (39%)
First crack at cycle #			200	300		200

Quantification of the AE data

The amount of the AE data presented at Table 1 shows that for the same experimental conditions AE activity in tension is significantly higher for the coated specimens by a factor that ranges from 3 to 20. On the other hand, the AE signals received during the compressive part of the TMF cycles exhibit a less pronounced difference between coated and uncoated specimens.

Furthermore if the overall AE activity during the test is compared between uncoated and coated specimens, two distinct regimes of AE activity are revealed (Region I and Region II in Fig. 7), which may be separated by a quiescent period. In the case of the coated samples both regimes are present, whereas in the case of the uncoated samples only Region II is active during the last 20% of the TMF life (Fig. 7). This observation provides a clear distinction between signals originating exclusively from the coating (Region I) and those involving also the bulk (Region II).

So, signals from Region I relate to damage processes in, or induced by the presence of the coating, while signals from Region II reflect the mechanical response to the TMF loading of the whole material system independently from the presence of the coating.

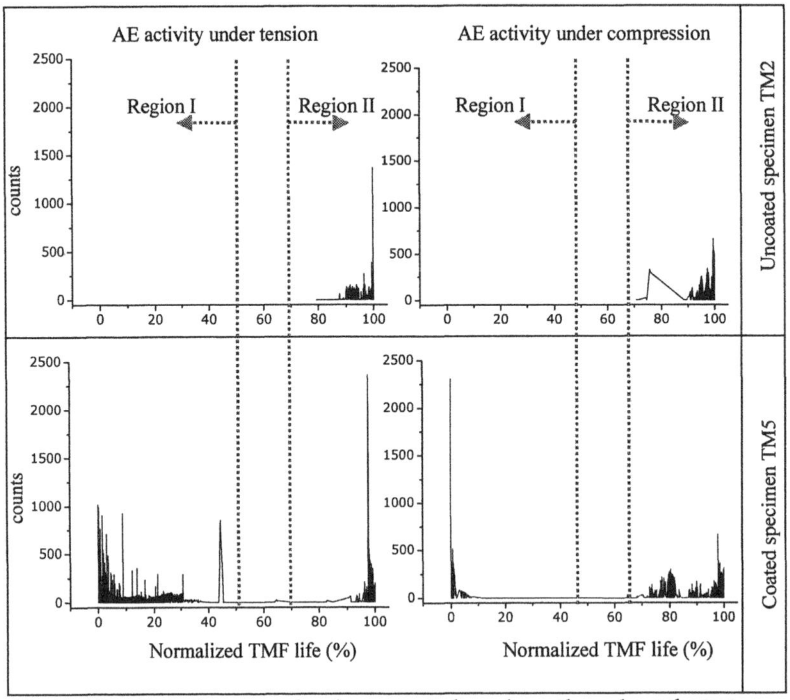

Fig. 7. AE activity for a coated and an uncoated specimen throughout the test.

The focus of interest of this study is to identify and correlate signals of Region I with the onset of damage, the initiation of cracks and with specific micromechanisms at the early stage of TMF life of the coated specimens. This information is to be revealed from a parametric analysis of the AE activity inside Region I, in conjunction with the video imaging observations of the first appearance and the subsequent extension of surface cracks. It should be noted that AE information pertaining to the uncoated specimens will be used only as a reference to identify the beginning of Region II and, thus, to indicate where the damage processes of Region I are exhausted or eventually lead to fatigue crack growth associated with Region II.

For this purpose the AE data from the coated samples have been divided into two data sets and in the following analysis only the first part of the AE data belonging to Region I , as presented in Table 1, is taken into account. Table 1 includes also the number of cycles after which surface cracks became visible, as made available by the in situ video imaging system.

Another observation from the results presented in Table 1 is that although the number of cycles characterizing the end of Region I is quite different for the three coated specimens investigated, its relative duration against the TMF life is comparable for all three tests. This indicates a relation between the extent of Region I (early damage) and the total TMF life. Also the number of signals of Region I received during the compressive parts of the cycles is comparable for the three tests, while their activity fades away well before the end of Region I (Fig. 8). Hence, the AE activity during compression does not seem to have a significant effect on TMF life. What is clearly distinct among the three tests, however, is the AE activity during the tensile parts of the cycles, as seen in Table 1. As such it was decided to use these AE data for the pattern recognition analysis.

Classification of the AE data

The set of seven parameters characterizing the AE events (such as amplitude, energy, duration etc.) was complemented by the set of test parameters of their occurrence, that is temperature, stress and time inside the cycle. Additionally the cycle number was also considered as a parameter because the activation of AE signals shows a most pronounced distribution in the stress-cycle number domain, Fig. 8. Thus, the obtained eleven-dimensional parameter space is referred to henceforth as the extended AE-parameter space.

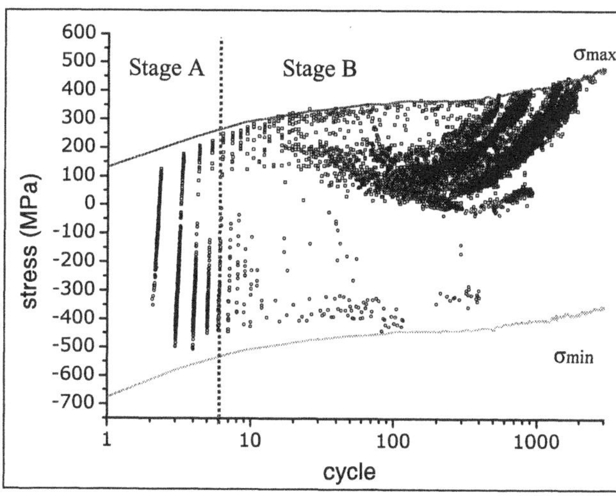

Fig. 8. AE events (symbols) and cyclic maximum and minimum stress envelopes (solid lines) in Region I of a coated sample (TM5).

Before proceeding with the classification analysis of the AE data, some valuable information is already obtained from the graphical representation of the AE activity in the σ_{ext}-N plane, as presented in Fig. 8. Here, any received AE signal enters as a data point irrespective of its AE parameters. This graph provides a clue for the identification of the involved damage mechanisms, their cause of activation and their behaviour during TMF life (see Discussion given below). First, it is apparent that after about ten cycles the distribution of AE activity closely reflects the upper stress envelope of the TMF test, in that AE is received up to the maximum stress. The preceding region is referred to as Stage A in Fig. 8. The total number of Stage A signals is quite large (almost 2000 in TM3 and TM9 and 1000 in TM5) and it is interesting to note that the number received in tension is in inverse proportion to the duration of both the subsequent Stage B and the whole TMF life. This implies that the specimen with the largest number of AE signals under tension during these first few cycles has the shortest TMF life.

Second, when Stage A gives way to Stage B, as noted in Fig. 8, the AE signals are not uniformly distributed between the stress envelopes, but they depend on stress, temperature and cycle number in a characteristic way. They form "branches" of signals, suggesting that the evolution of specific damage mechanisms is being portrayed by each one of these branches of AE events.

The above-mentioned branching behaviour refers to signals received in tension, while compressive AE signals are almost absent in Stage B. It covers the range of TMF lives where surface cracks start to appear, propagate, interact and eventually coalesce as evidenced by the video image monitoring system. As there is a strong interest in interpreting these signals in terms of mechanisms, and in order to substantiate the branching observed in the two-dimensional σ_{ext}-N subspace, a systematic statistical analysis involving the complete extended AE parameter space has been performed. In this analysis only the AE data received in tension have been considered.

The routine used for the classification of the AE data was FORGY©, which is based on the k-means algorithm. It allows customizing the maximum number of iterations used to reach a prescribed convergence criterion. All AE parameters were normalized to lie between zero and one, the Euclidean distances were computed, the initial number of clusters was set to be two, and their centers were initially selected with respect to the cycle number. This pattern recognition methodology is essentially a trial and error procedure meant to identify appropriate parameter combinations for AE cluster description [2,3]. Care was taken to ensure that all the tests were analyzed by using the same settings for cluster analysis.

The final choice of the appropriate parameters used as descriptors for the classification of the data was made with respect to their degree of independence. The AE parameters (i) counts, (ii) amplitude, (iii) duration and (iv) counts to peak were chosen as the appropriate descriptors for the classification, while the temperature and the cycle number were used as two additional descriptors to establish the correlation with the mechanical behaviour. In Table 2, which shows the results of this classification procedure, the clusters are presented in descending order of the radial distance from the origin of their center as defined in the normalized AE descriptor space. Typical classification results obtained by the cluster method are presented in Fig. 9, which confirms the trend of specific clusters to form "branches".

Table 2. The clusters of the AE signals in Region I, Stage B.

Specimen	Number of clusters	Number of AE signals in clusters
TM3	5	69, 107, 126, 325, 65
TM5	6	178, 257, 1072, 2194, 2358,174
TM9	5	485, 443, 3447, 744, 462

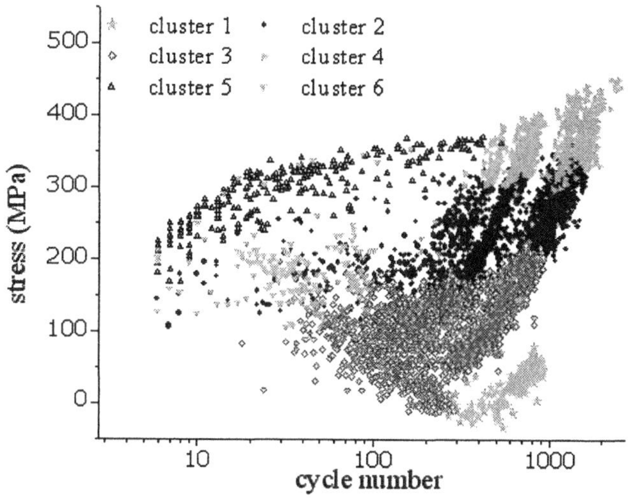

Fig. 9. Cluster analysis result for specimen TM5.

The cluster analysis performed for all three coated specimens revealed similarities among these tests, noting the statistical scatter of the AE features in the clusters of signals. This points to the presence of distinct sequences of damage mechanisms causing specific AE clusters not only in the AE features themselves but also in the stress-temperature-cycle number regime. Moreover, the commonalties in AE clustering demonstrate the usefulness of AE monitoring and analysis in the interpretation of TMF damage evolution and lifing.

Pattern recognition

The results from the cluster analysis revealed in every test at least one cluster (actually two clusters in the case of TM3) which emerged already during the first TMF cycles close to the maximum stresses, i.e. minimum temperature, imposed in the cycles (cf. Cluster 5 in Fig. 9). These clusters correspond to AE signals with high values of all AE features, in particular, amplitude, duration and energy. As far as the time domain with respect to the video imaging observations is concerned, the clusters refer to AE events, which occurred prior to the detection of any surface cracks, continued to be present after the appearance of surface cracks, while their activity was decreasing and finally vanished well before the end of Region I was reached. For reasons to be explained below, the damage mechanism underlying these clusters is referred to as an *athermal process*.

A second distinct group of clusters is the one that exhibits the aforementioned branching behaviour (Clusters 2,3 and 4 for TM5, cf. Fig.9). The initiation of these clusters is closely related to the appearance of surface cracks. The behaviour in the AE parameter domain was found to change gradually during the cyclic evolution of the branches. While the values of duration and amplitude were almost constant, the values of counts to peak and energy started from medium values at the beginning of each branch went through a minimum and finally reached the highest values at the end of the branches. The evolution of the characteristic stresses (Fig. 9) points to a

thermally activated process as the underlying damage mechanism that gives rise to these AE branches.
Those two types of clusters differ not only with respect to their AE features and their stress-temperature-cycle number domain of occurrence, but they also relate to different damage mechanisms, which take place successively and which exhibit distinct temperature dependences reflecting athermal and thermally activated processes, respectively.

Discussion

For an interpretation of those AE clusters in terms of damage processes, it is necessary to consider (i) the characteristic stresses at which AE activity is observed and (ii) whether or not the AE activity is exhausted at a certain stage of TMF life. While the first criterion reflects a possible contribution of thermal activation in the damage process (note that a characteristic stress reflects to a characteristic temperature according to the imposed TMF cycle), the second criterion gives information on the possibly limited number of sources that may cause the AE events.

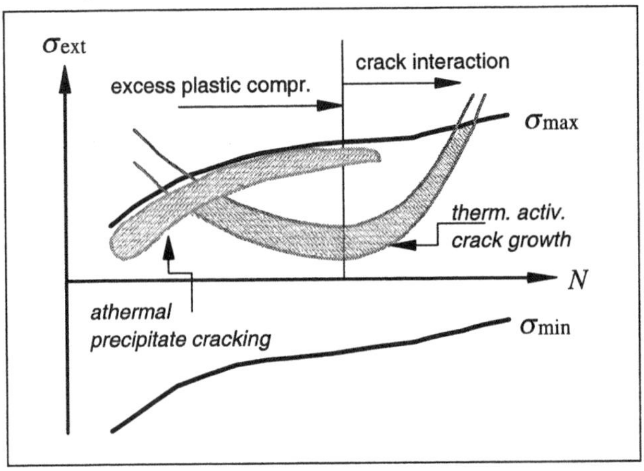

Fig. 10. Schematic of the athermal and the thermally activated AE clusters (hatched).

In view of these criteria the first type of AE clusters can be associated with an athermal damage process, since the corresponding AE activity is maximal at the highest tensile stresses (lowest temperature) available in the TMF cycles, without showing any characteristic temperature dependence. Furthermore, the gradual decrease in activity with increasing cycle number indicates that, given the stress range, only a limited number of `sites´ is available to contribute to this type of AE activity (Fig. 10). Together with the observation that the AE activity commences already in the tensile part of the first cycle and, hence, precedes any other kind of activity in tension, this damage process can be attributed to brittle crack nucleation. Presumably this is associated to precipitate (silicide) cracking at the interface, which commences with the biggest precipitates and subsequently involves smaller and smaller precipitates as the maximum tensile stresses increase with increasing cycle number. The exhaustion of the corresponding AE activity can then be explained by assuming that eventually the available stress does not suffice for further cracking of small precipitates/precipitate fragments.

As opposed to that athermal process, the second type of AE clusters, which exhibit the branching behaviour, is characterized by narrow stress intervals of occurrence (Fig. 10) that carry over to characteristic temperatures if the cyclic evolution of internal stresses is also taken into account. Given the athermal crack nucleation at the interface, which precedes those branches, it is natural to associate the branches with the lateral growth of coating cracks parallel to the surface. To obtain a deeper understanding of the characteristic branch-like appearance of those clusters, the rate of AE activation is written as that of a thermally activated process in terms of the

$$\nu = a\sigma_{\text{eff}}^{n} \exp\left[-\frac{Q}{kT}\right] \tag{1}$$

Arrhenius law

where a is a constant pre-exponential factor, $\sigma_{\text{eff}}=\sigma_{\text{ext}}+\sigma_{\text{int}}$ is the effective stress acting at the sites where damage is likely to occur, Q is an activation energy, k is Boltzmann's constant, and $n>1$ is a stress exponent. Note that both the applied stress $\sigma_{\text{ext}}(t)$ and temperature $T(t)$ are rapidly varying with time t according to the specific TMF cycle applied (Fig. 1), while the internal stress $\sigma_{\text{int}}(N)$ is subject to a slow cyclic evolution which is expressed in terms of a parametric dependence on the cycle number N. Upon differentiating Eq. (1) with respect to t, in order to obtain the maximum activation rate, and eliminating temperature by using the out-of-phase relationship between thermal and mechanical loading, one obtains the characteristic applied stress of maximum activation probability as a function of the cyclic evolution of the internal stress σ_{int}:

$$\frac{\sigma_{\text{ext}}}{E\Delta\varepsilon_{\text{m}}} = \frac{T_{\min}}{\Delta T}\left(b - \sqrt{b^2 - c + \left(\frac{T_{\text{ref}}\Delta T}{T_{\min}^2}\frac{\sigma_{\text{int}}}{E\Delta\varepsilon_{\text{m}}}\right)}\right) \tag{2}$$

where

$$b = 2 + 2\frac{\Delta T}{T_{\min}}\frac{|\varepsilon_{in}|}{\Delta\varepsilon_{\text{m}}} + \frac{T_{\text{ref}}}{T_{\min}} \quad \text{and} \quad c = \left(1 + \frac{\Delta T}{T_{\min}}\frac{|\varepsilon_{in}|}{\Delta\varepsilon_{\text{m}}}\right)^2 \tag{3}$$

and the reference temperature is defined as $T_{\text{ref}}=Q/(nk)$, and E denotes the Young's modulus of the substrate (assumed T independent for simplicity). Moreover, the maximum and minimum temperature and the temperature range $\Delta T=T_{\max}-T_{\min}$ is introduced, as well as the mechanical strain range $\Delta\varepsilon_{\text{m}}$ of the out-of-phase TMF cycle. This result may be used to express the characteristic cycle number dependence of the AE branch of a thermally activated damage process in terms of the cyclic evolution of the internal stress $\sigma_{\text{int}}(N)$ acting at the damage sites, and of the compressive inelastic strain, $\varepsilon_{\text{in}}(N)<0$, that accumulates in the bulk material during the high-temperature compressive phases of the cycle.

In view of Eqs (2,3) the specific branching behaviour observed in Figs. 8 and 9 and schematically summarised in Fig. 10 can be explained as follows. The descending part of the branch is due to considerable plastic compression of the coating in excess to the bulk compression, which occurs at high T. This leads to the development of tensile internal stresses at low T, which will outweigh the residual compressive stress (−200±100 MPa) in the coating, as measured in the as-received state, after a few cycles and then yield appreciable net tensile stresses of up to several 100MPa after say 10 to 100 cycles. According to Eq. (2), these increasing tensile internal stresses will assist the thermally activated propagation of coating cracks so that it is observed at decreasing applied stress levels (downward branching).

However, at some point the downward branching is balanced by the counteracting relief of tensile stresses due to multiple coating cracking. As the post-mortem fractographic analyses have revealed (Fig. 8), dense arrays of coating cracks develop due to the lateral growth of cracks along the surface (average crack spacing ≈60μm, which is to be compared to the average crack depth ≈80μm corresponding roughly to the thickness of the coating). This will tend to shield a considerable part of the applied tensile stress and the internal tensile stresses developed by excess plastic deformation. This kind of crack interaction is the reason why the branch is turned upward and continues until the characteristic stress σ_{ext} exceeds the maximum stress σ_{max} reached in the TMF cycle (as governed by the bulk mechanical behaviour). Then the stress state available is no longer sufficient to maintain the thermally activated surface cracking (Fig. 10).

It should be noted that the discussion given so far refers to a single branch, whereas in Figs. 8 and 9 several of them can be discerned which evolve almost parallel to each other in the σ_{ext}-N plane. It is natural to assume that this multiple branching relates to different crack growth configurations which are associated with the triple-layer structure of the coating (Fig. 5) and which involve slightly different internal stresses assisting/obstructing the propagation of the crack configurations.

CONCLUDING REMARKS

As compared to other non-destructive techniques, which provide information about the damage accumulated at a certain stage of life, acoustic emission allows for monitoring the damage processes inside a material while these occur. This has proven useful for quantifying the damage throughout the life of the specimens, identifying various damage mechanisms and interpreting the way they interact, in particular, in multi-material systems such as the coated system investigated in the present work.

AE monitoring provides complementary and specific information about the interaction of surface, subsurface, and bulk damage processes, which take place during TMF. In combination with the mechanical response and the surface damage analysis by video imaging, it enables for non-destructive damage assessment and it offers possibilities for the TMF lifing of coated material systems.

REFERENCES

1. Peteves S.D., De Haan F., Timm J., Bressers J, Hughes P.M., Moss S.J., Johnson P., Hendersen M. (2000). Proc. of the 9th Int. Symp. on Superalloys (Superalloys 2000), Sept. 17-21, 2000, pp. 655-665.
2. Pappas Y.Z, Markopoulos Y. P., Kostopoulos V., Paipetis S. A (1997). 1st Hellenic Conference on Composite Materials and Structures, Ksanthi 1997, pp.192-206. .
3. Anastassopoulos A. A., Nikolaidis V. N., Philippidis T. P. (1999) Neural Computing & Applications, Vol. 8, pp53-61.

THERMAL FATIGUE OF THE NICKEL BASE ALLOY IN 625 AND THE 2¼ CR-1MO STEEL

Ryuichiro Ebara
Dept. of Advanced Materials Science, Kagawa University
2217-20, Hayashi-cho,Takamatsu,761-0396, Japan
Tamotsu Yamada
Hiroshima R & D Center, Mitsubishi Heavy Industries, Ltd.,
4-6-22, Kan-on-Shin-Machi, Hiroshima,733-8553, Japan

ABSTRACT

Thermal fatigue tests were conducted for Inconel 625 and 2¼Cr-1Mo steel by use of a laboratory made thermal fatigue testing apparatus. The thermal fatigue crack initiation resistance of Inconel 625 is superior to that of 2¼Cr-1Mo steel at 823K. While thermal fatigue crack propagation rate of Inconel 625 is faster than that of 2¼Cr-1Mo steel at 723K and 823K. The lower crack propagation rate of 2¼Cr-1Mo steel can be explained by role of oxide produced at the crack surface during the thermal fatigue crack propagation. The thermal fatigue crack propagated predominantly with a mode of transgranular for both tested materials. The striation was predominantly observed on fracture surface of Inconel 625. The fracture surface of 2¼Cr-1Mo steel was heavily covered by oxide film and the striation like pattern was also predominant on the rugged fracture surface after removing oxide film.

KEYWORDS

Thermal fatigue, crack initiation, crack propagation, Inconel 625, 2¼Cr-1Mo steel, oxide, crack branching, striation, striation like pattern

INTRODUCTION

Inconel 625 and 2¼Cr-1Mo steel have been applied for the heat recovery plant. Authors reported on corrosion resistant properties of the both materials in a molten salt environment of (50% KNO_3 and 50% $NaNO_3$) at temperature of 723K and 823K. Inconel 625 showed high corrosion resistance in the molten salt environment. While, the corrosion resistance of 2¼Cr-1mo steel was strongly dependent on the temperature and Cl^- content of the molten salt [1] .The thermal fatigue might be anticipated due to the frequent start and stop operation of this plant. In this paper it is reported on the thermal fatigue crack initiation and propagation behavior of the both tested materials.

EXPERIMENTAL PROCEDURE

MATERIALS AND SPECIMEN

Chemical compositions and mechanical properties of tested materials are shown in Table1. The shapes and the size of thermal fatigue test specimen is shown in Fig.1. The plate specimens with a fatigue pre-crack were cut off from CT specimens after introducing the fatigue crack at the bottom of the notch. This fatigue pre-crack was introduced with a stress ratio ($\sigma_{min}/\sigma_{max}$) of 0.1 and testing speed of 20 to 30Hz.

Table 1. Chemical compositions and mechanical properties of tested materials

Material	Chemical compositions (mass %)												Mechanical properties			
	C	Si	Mn	P	S	Ni	Cr	Mo	Nb+Ta	Aℓ	Ti	Fe	$\sigma_{0.2}$ (MPa)	σ_B (MPa)	δ (%)	ψ (%)
INCONEL 625*	0.022	0.34	0.33	0.008	0.005	61.15	21.34	9.52	3.73	0.12	0.18	Bal.	508.0	939.5	48.4	60.2
2 1/4Cr—1Mo**	0.13	0.19	0.55	0.007	0.004	—	2.45	1.06	—	—	—	Bal.	466.8	611.9	29.0	76.0

Note : Heat Treatment

* 1,213K × 1.4h WQ

** Normalizing 1,203K × 1.5h, Tempering 1,013K × 0.5h, Annealing 983K × 1.5h

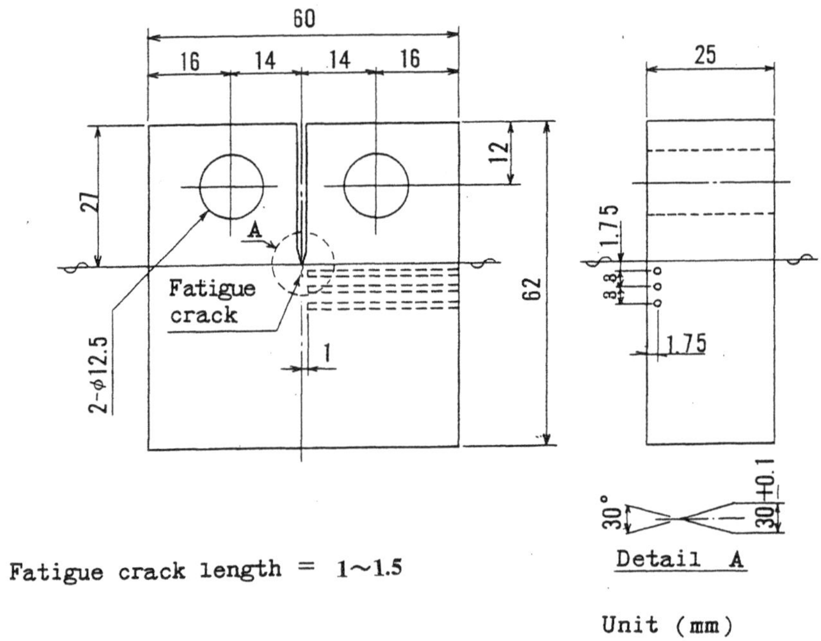

Fig.1 Thermal fatigue test specimen

THERMAL FATIGUE TEST

A laboratory made thermal fatigue testing apparatus was used (Fig.2). This apparatus consisted principally of a heating device using oxygen and LPG gas, a temperature control device for the heated zone , and a rapid cooling device for the specimen. During the thermal fatigue tests the heating and cooling cycles were repeatedly loaded on the specimen lightly clamped on the specimen holder.City water was used as the cooling medium. The heating temperatures were 723K and 823K , and cooling water was sprayed onto the specimen surface through a nozzle. As it was difficult to measure the surface temperature of the specimen, small holes for a thermocouple were prepared to measure the temperature of the notch during the crack propagation tests (the dotted line in Fig.2). Thus the measured temperatures were used as the testing temperatures. The thermal fatigue crack initiation tests and thermal fatigue crack propagation tests were conducted up to 200 cycles at 723K and 823K.The thermal fatigue crack length was measured at every one cycle until 40 cycles, then at each ten cycles up to 200 cycles by use of a viewing microscope with magnification of 200 after interrupting the thermal fatigue tests. The thermal fatigue cracks were examined by an optical microscope and thermal fracture surfaces were examined by a JEOL scanning electron microscope (JXA-73).

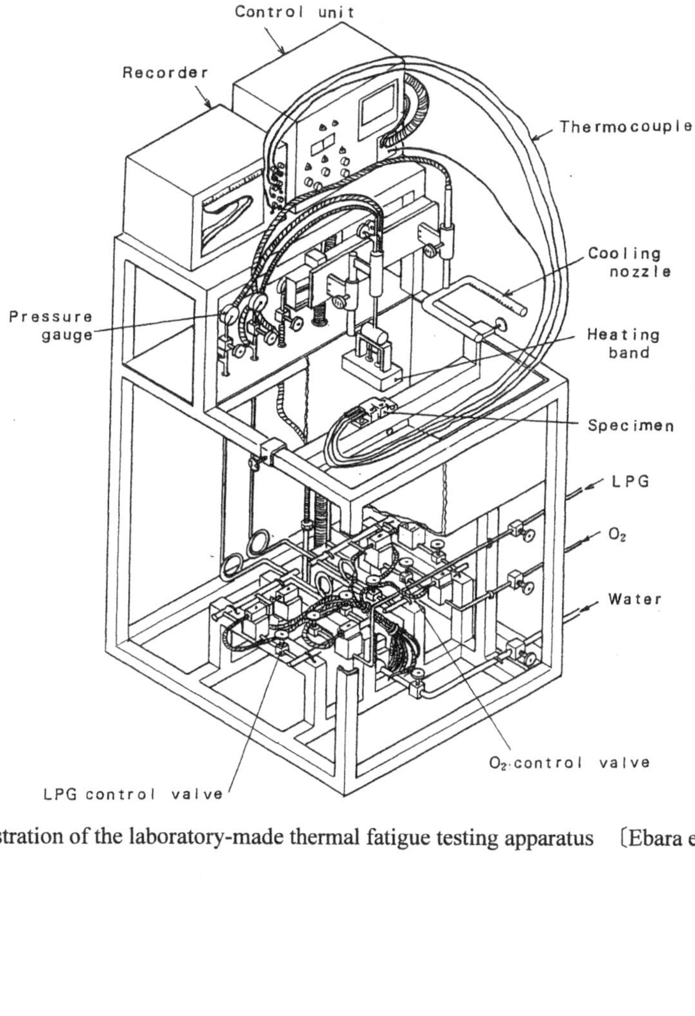

Fig.2 Illustration of the laboratory-made thermal fatigue testing apparatus 〔Ebara et al.(2)〕

RESULTS AND DISCUSSION

The thermal fatigue crack initiated and propagated from the fatigue pre-crack. The number of cycles for fatigue crack initiation of Inconel 625 were 15 at 723K and 10 at 823K. While the number of cycles for thermal fatigue crack initiation of 2¼Cr-1Mo steel was 15 at 723K and 1 at

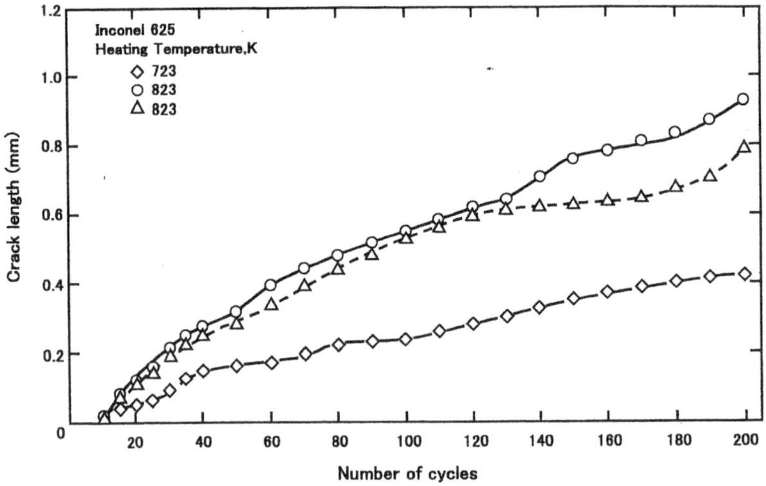

Fig.3 Thermal fatigue crack propagation curves of Inconel 625.

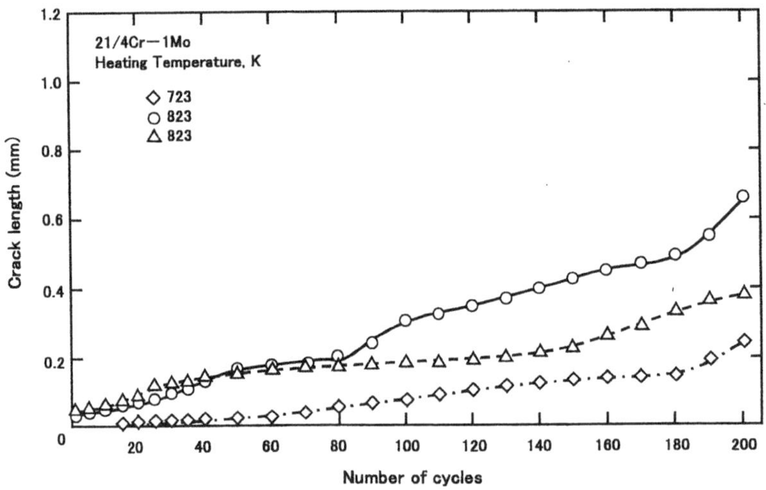

Fig.4 Thermal fatigue crack propagation curves of 2¼Cr-1Mo steel

823K. Thus it is apparent that the thermal fatigue crack initiation resistance of Inconel 625 is superior to that of 2¼Cr-1Mo steel at 823K. The difference of the thermal fatigue crack initiation resistance cannot be fully explained , but it seems to be deeply related to the difference of the properties of the matrix. In general austenitic material such as Inconel 625 with precipitated γ' phase have a higher high temperature strength as compared with the low alloyed feritic steel such as 2¼Cr-1Mo steel. It is also apparent that the lower the temperature difference between heating and cooling, the shorter was the number of cycles for thermal fatigue crack initiation. Fig.3 and Fig.4 show the thermal fatigue crack propagation curves of Inconel 625 and 2¼Cr-1Mo steel, respectively. The crack length of Inconel 625 at 200 cycles were 0.42mm at 723K and 0.86mm at 823K . While the crack length of 2¼Cr-1Mo steel at 200 cycles were 0.25mm at 723K and

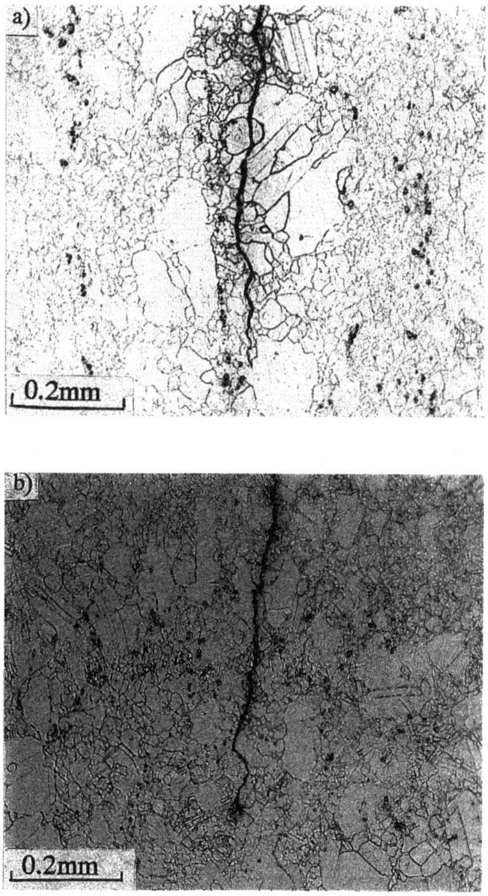

Fig.5 Thermal fatigue crack of Inconel 625, 200 cycles
a) 723K b) 823K

0.52mm at 823K. Thus the crack propagation rate of Inconel 625 was faster than that of 2¼Cr-1Mo steel. The thermal fatigue crack of Inconel 625 showed an inclination to propagate at a constant speed. While for 2¼Cr-1Mo steel ,the crack propagated up to 60 cycles, was arrested between 60 and 140 cycles at 823K. The crack arresting was also observed for 2¼Cr-1Mo steel at 723K. The cause of this phenomena seems to be deeply related to the oxide produced in the crack surface which was observed on hot forging die steel, SKD62 [2]

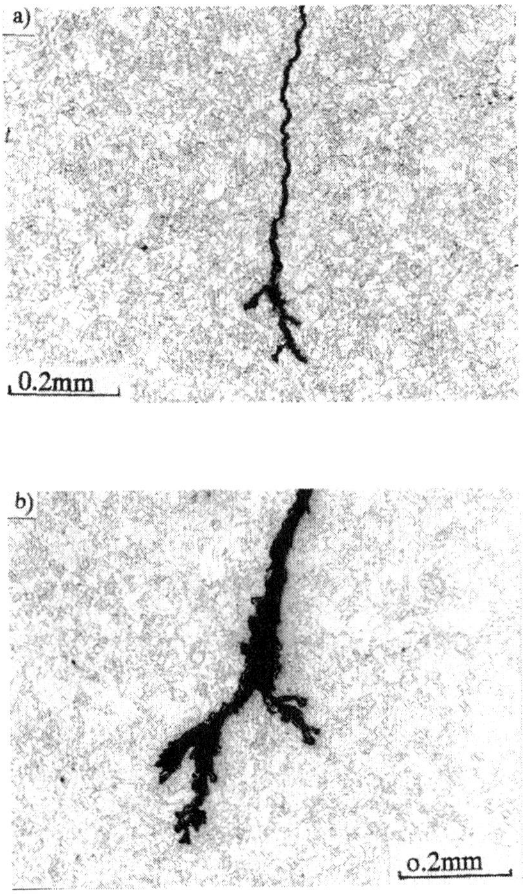

Fig.6 Thermal fatigue crack of 2¼Cr-1Mo steel, 200cycles
a) 723K b) 823K

The thermal fatigue crack for Inconel 625 propagated predominatly with a mode of trasgranular. The tip of the crack was sharp at 723K and was branched at 823K. [Fig.5] . The same crack propagation mode was observed for 2¼Cr-1Mo steel, however the crack tip was rounded and branched [Fig.6] .The typical examples of thermal fatigue fracture surfaces at 823K for Inconel 625 and for 2¼Cr-1Mo steel are shown in Fig.7 and Fig.8 ,respectively.
SKD62 [2] .

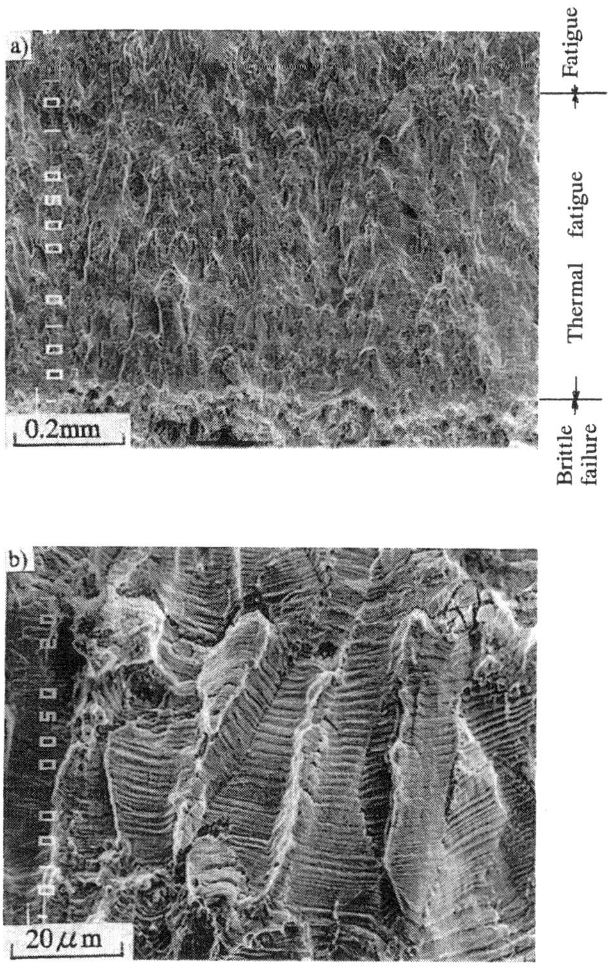

Fig.7 Thermal fatigue fracture surface of Inconel 625, 823K,200 cycles
a) General view b) Thermal fatigue fracture surface

The thermal fatigue fracture surfaces of Inconel 625 were relatively flat. Striation was predominant for Inconel 625. While the thermal fatigue fracture surfaces for 2¼ Cr-1Mo steel were rugged and were covered with oxide. These rugged fracture surfaces can be formed in the results of crack branching during thermal fatigue crack propagation. The striation like pattern was predominantly observed on the rugged fracture surfaces after removing the oxide film.

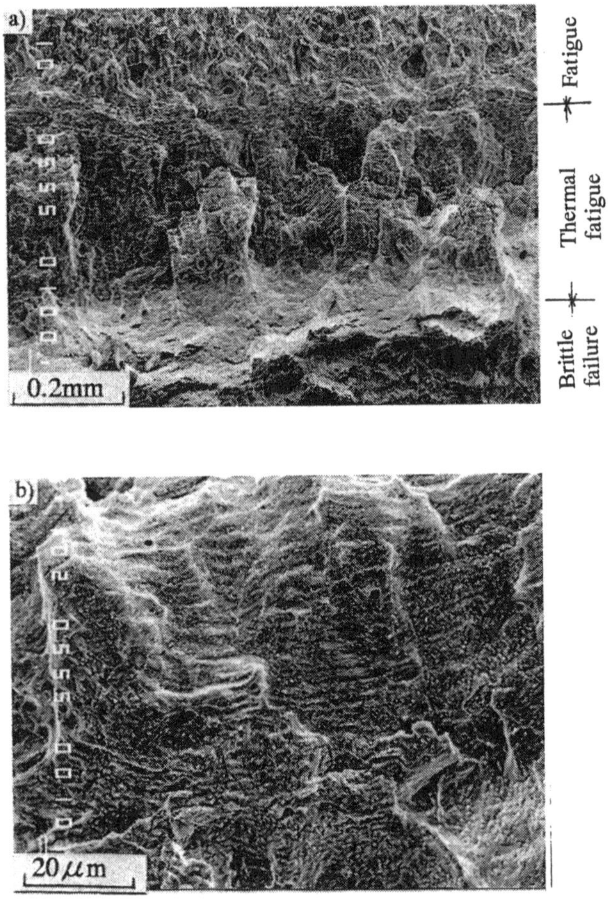

Fig.8 Thermal fatigue fracture surface of 2¼Cr-1Mo steel, 823K,200 cycles
a) General view b) Thermal fatigue fracture surface

CONCLUSIONS

1) Thermal fatigue crack initiation resistance of Inconel 625 is superior to that of 2¼Cr-1Mo steel at 823K.
2) Thermal fatigue crack propagation rate of Inconel625 is faster than that of 2¼Cr-1Mo steel at 723K and 823K. The crack arresting was observed on 2¼Cr-1Mo steel at 723K and 823K.
3) The crack propagated predominantly with a mode of transgranular for Inconel 625 and 2¼Cr-1Mo steel. The crack tip was sharp at 723K and was branched at 823K for Inconel 625. The crack tip was rounded and branched at 723K and 823K for 2¼Cr-1Mo steel.
4) Striation and striation like pattern was predominant on fracture surfaces of Inconel 625 and 2 ¼Cr-1Mo steel ,respectively.

REFERENCES

1. Ebara,R., Nakajima, H.Shouzen ,D.and Yamada,T.(1988) *J.Japan Inst.Metals* **52,**508.
2. Ebara,R.,Yamada,Y.Yamada,T.and Kubota,K.(1987) *J.Materials Science* , Japan **36,**513

Temperature-Fatigue Interaction
L. Rémy and J. Petit (Eds.)

DAMAGE MECHANISMS AND THERMOMECHANICAL LOADING OF BRAKE DISCS

P. DUFRÉNOY[(1)], G. BODOVILLÉ[(2)] and G. DEGALLAIX[(2)]
Laboratoire de Mécanique de Lille, URA CNRS 1441
[(1)] EUDIL, Cité Scientifique, 59655 Villeneuve d'Ascq cedex, France
[(2)] Ecole Centrale de Lille, BP 48, 59651 Villeneuve d'Ascq cedex, France

ABSTRACT

This paper aims at the damage mechanisms of railway disc brakes leading to macroscopic crack occurrence on the friction surface. An analysis of the friction surface of brake discs damaged in service is first carried out to identify two types of cracks. In parallel to this analysis, a numerical simulation is performed in order to determine the thermomechanical loading due to successive brakings, giving additional indications about the damage mechanisms. Results show that thermal fatigue occurs with superposition of friction effects. Both damage surface analysis and numerical calculations give valuable information about the failure mechanisms, and will lead to an improved design of the disc brakes in order to enhance their performances.

KEYWORDS

Brake disc, crack initiation, cracking network, thermomechanical modeling, hot spot, thermal fatigue

INTRODUCTION

For several years, the increase of railway commercial speeds and capacities requires the improvement of the braking performances. Even if dynamic braking systems are often largely used in normal service braking, their performances are not sufficient to ensure an emergency braking at high speed. Then, friction braking systems are important security systems, which have to match severe criteria dictated by the security rules, in terms of stopping distance associated to a maximum average deceleration, under all environmental conditions. As an example, in the case of an emergency braking at 300 km.h^{-1} of the Thalys TGV, the maximum stopping distance is 3500 m with an average deceleration of 1 m.s^{-1} and a braking time of 80 s, corresponding to a dissipated energy of 14 MJ per braking disc. More generally, the growth of dissipated energy in railway braking systems has pushed the disc brakes more and more to their limits. One consequence is the frequent occurrence of cracks [1,2] on the friction surfaces of the discs leading to their early replacement.

Disc brake behaviour is difficult to study due to interactions of thermal, mechanical, metallurgical and tribological phenomena. Many papers, which were moreoften devoted to the thermal or the wear problems, show this difficulty. So, it is of primary importance : i) to have a better understanding of the physical mechanisms activated in the contact, which have a severe detrimental effect on the disc integrity; ii) to develop an efficient modelling, able to provide the designer with satisfactory life prediction. Comparison with experimental results are of course necessary, these tests being either at full scale, or at a reduced scale – provided that similarity rules are respected [3]. The present paper aims to follow this approach. The first part of this paper presents an analysis of the damage observed on out-of-order discs. In the second part, thanks to thermal surface measurements, an observed classification of the thermal gradients is given. In the third part, a numerical thermomechanical model of the disc is presented and the obtained results are discussed.

Braking system and materials

The trailer bogies of the Thalys TGV include two axles, equipped with four disc braking systems. Each system is constituted of one disc and two pairs of pads as shown in figure 1. The disc, with an outer diameter of 640 mm and a thickness of 45 mm, is made of 28CrMoV5-08 steel, manufactured by a forging process. Its chemical compositions are given in Table 1. The heat treatment is an austenitisation at 975°C during 5 h then water quenching, followed by a tempering at 635°C during 9 h and air cooling. The obtained tempered-martensitic microstructure has a yield stress of 970 MPa at 20°C and of 600 MPa at 600°C.

The material pad is a sintered Fe-CuSn metal matrix composite reinforced by ceramic particles. The pads are constituted of 9 cylindrical pins, with a diameter of 40 mm and a height of 25 mm.

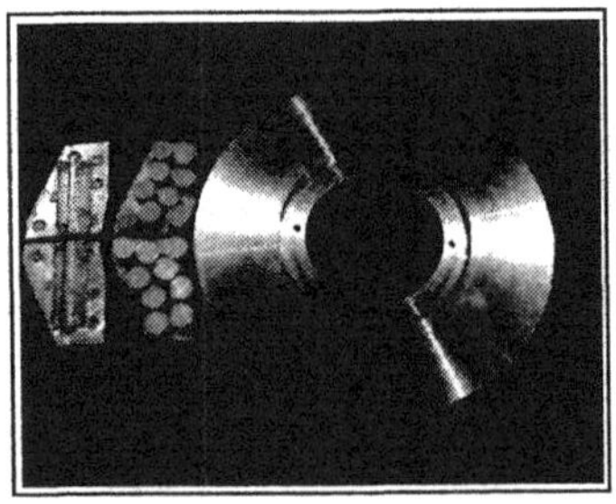

Fig. 1: Disc and pads of a TGV braking system

Table 1: Chemical composition of 28CrMoV5-08 steel (in wt %)

C	Cr	Mo	V	Mn	Si	Ni	S	P
0.24/0.31	1.20/1.60	0.60/0.90	0.20/0.40	0.50/0.90	0.40/0.70	< 0.40	< 0.007	< 0.015

Damage analysis

The friction surfaces of several out-of-order discs were observed. Brake discs are of course security parts and are carefully and regularly controlled by the operator. When a disc presents a crack with a conventional length in surface, a new one immediately replaces it. As a thin oxide layer covered the friction surfaces of the "used" dics, a gradual polishing was first performed in order to eliminate it. A low magnification view is shown in Figure 2. Polishing is more intense from the left to the right. In this and the following figures, the arrow indicates the direction of sliding of the pad.

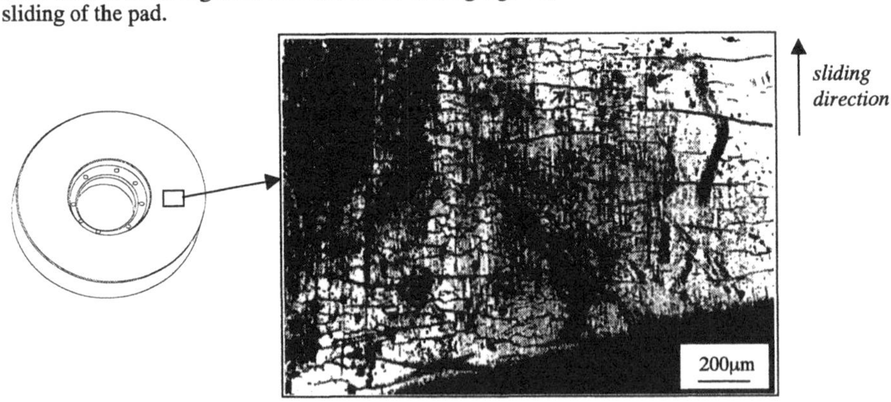

Fig. 2: Friction surface gradually polished

Figure 2 reveals that the friction surface is covered by a relatively dense cracking network, characteristic of thermal fatigue loading. A careful observation of this picture indicates the presence of quasi-closed crack cells on the left. On the contrary, on the right, cracks perpendicular to the sliding direction, i.e. along a radial orientation, are only observed. This means that the depth reached by circumferentially-oriented cracks is not so important when compared with the depth reached by radially-oriented cracks.

A controlled and progressive polishing was performed in order to determine more precisely the characteristics of the different observed families of cracks constituting these networks:

- 1st family: circumferentially-oriented short cracks, with a relatively short length in surface (less than 50 μm) and a low depth (less than 20 μm). They are quasi-systematically tied at least on one side to radially-oriented cracks,
- 2nd family: radially-oriented cracks, with a length in surface which can reach 200 μm and a depth up to about 50 μm. They are probably obtained by coalescences of shorter cracks,
- 3rd family, also constituted by radially-oriented cracks, but with a long length in surface more than 1 mm and a depth greater than 150 μm. These cracks are visible to the eye.

Figure 3 presents views of these different types of cracks observed at higher magnifications. It is observed in particular that the 3rd-family long cracks are more or less situated every 100/150 μm. Contrarily to the others, they appear to be more opened, and often they are filled with oxides and also of particles coming from the disc-pad interfacial layer (the "third body", [4]).

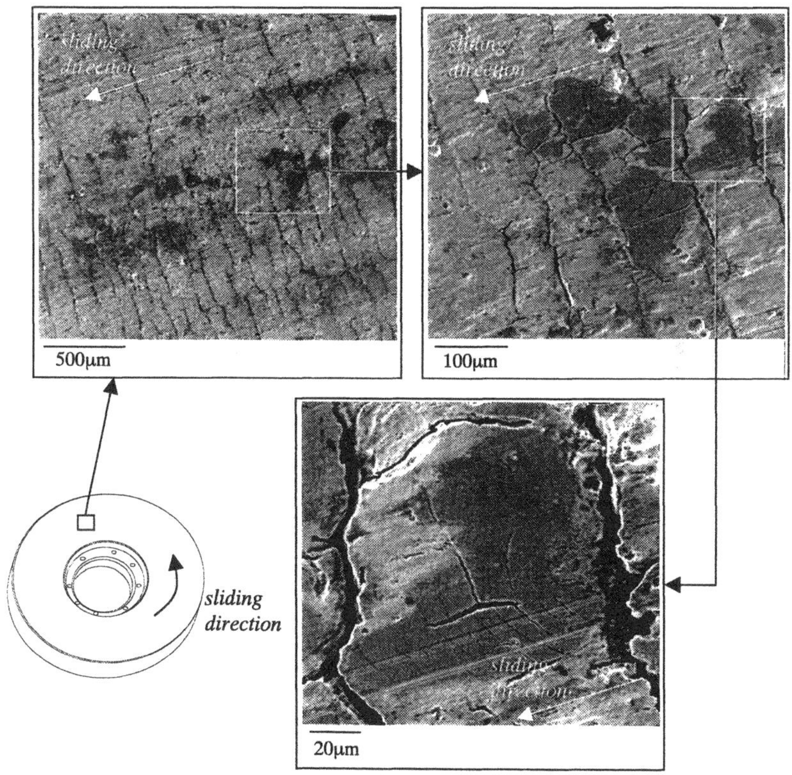

Fig. 3: Microcracks on the friction surface

It has to be noted that the density of the cracking network varies from one zone to another. Some zones of the friction surfaces seem even to be free of cracking network.
If such crazing, classically associated to thermal fatigue loading, is generally not considered as very detrimental for the integrity of the discs, it is not the same case for the long cracks, which are sometimes observed to the eye. It should be emphasised that it is this kind of macroscopic crack whose propagation is carefully controlled.
Figure 4a presents an example of such a macroscopic crack with 66 mm length. Figure 4b shows the in-depth growth of another macrocrack, after machining and opening the crack at low temperature. The measured length in surface was 61 mm, while its depth has reached 15 mm. The irregular shape of the crack front has to be related to the "fire ring" evolution as described in the following. A secondary crack is also visible on the left of Figure 4b: coalescence of this crack with the main crack would probably have happened if the disc had still worked in-service.

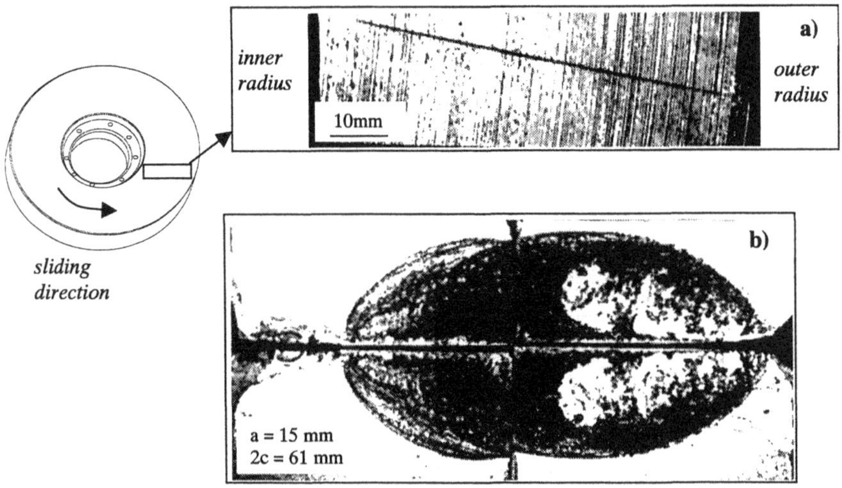

Fig. 4: Macroscopic crack: a) view on the friction surface; b) in-depth growth

Macroscopic cracks were found on the friction surface only along the radial direction, extending from the inner to the outer radius of the disc but without crossing over the whole radial length (figure 4a). They appear rectilinear. Occurrence of such cracks is of course not systematic. It is interesting to highlight that no significant thermal crazing was observed in their close vicinity.
This phenomena, as well as the coexistence of the three families of microcracks, can be explained in terms of crack shielding effects, but also in terms of wear effects due to the pad friction.
It can be thought that there exists a kind of competition between microcrack network development and macrocrack propagation, depending on the local loading on the disc, the severity of the brakings, the pad stiffness, the pad material nature and the physical mechanisms of friction and wear, etc.

THERMAL GRADIENT CLASSIFICATION

Analysis of the temperature distribution on the disc surface is of primary importance for the understanding of the thermomechanical loading. Surface temperature measurements require complex instrumentation. It was experimentally observed that extreme variations of the temperature distribution on the disc surface occur from one braking to another (even under similar conditions) or from one type of pad to another [1,2,5]. By using an infrared camera coupled with an acquisition system, experimental investigations on a full scale test bench with a

TGV disc brake lead to a classification of the main observed thermal gradients [6-8], described in figure 5.

Figure 5a corresponds to the ideal case: the contact pressure is almost uniform and low thermal gradients occur.
Figure 5b is a more general case; the contact pressure is non uniform, high thermal gradients are present and narrow rings of high temperatures, called "fire rings", are observed. The fire rings move along the radial direction during braking. This phenomenon has been described using a thermomechanical numerical model, taking into account contact pressure and heat transfer between disc and pad, thermal distortions of the various components, wear and thermomechanical behaviour of the materials [6,8].
Figure 5c is the case presenting the highest thermal gradients: macroscopic "hot spots" appear on the friction surface. This phenomenon, which appears as a buckling mode of the disc, reduces drastically the contact surface area with very high local temperatures. Understanding of the occurence mechanisms of the hot spots is still under discussion, even if several propositions have already been made [7,9,10].

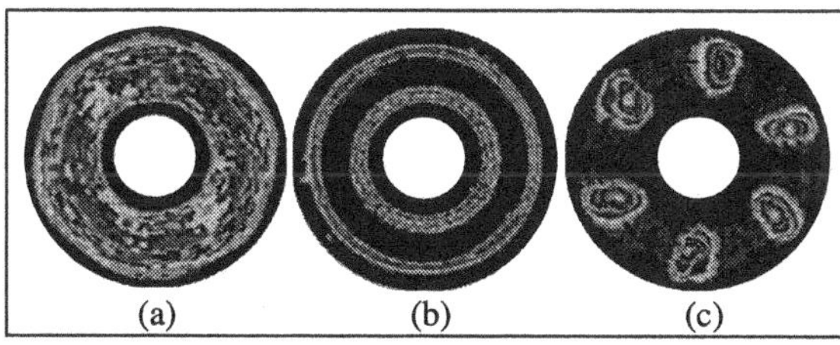

Fig. 5: Thermal gradient classification, from experimental thermographs (see the text)

THERMOMECHANICAL MODELLING

Calculation assumptions

In the friction analysis, it is frequent to consider successively the pad to disc mechanical loading contact and the thermomechanical loading due to the thermal gradients. In the railway application, the normal contact force is low and, on the assumption of a uniform contact pressure, calculation shows that the contact mechanical response is negligible in comparison with the thermomechanical stresses. Therefore, the model presented in this paper considers only the loading due to the thermal gradients, even if contact pressure can be locally much higher and may contribute to the damage enhancement.
Following the above classification and depending on the considered thermal gradient, different models were developed using ANSYS™ 5.5.3 finite element software
Uniform distribution of temperature (Fig. 5a) may be modeled with assumption of uniform contact pressure. Thermomechanical calculations require to introduce the stress-strain behaviour at different temperatures.
For the study of "fire ring" occurrence (Fig. 5b), the thermomechanical simulation requires to consider in addition the contact surface analysis and its variation during braking, together with a wear model.
For the simulation of the macroscopic hot spots (Fig. 5c), it is not necessary to introduce any contact variation during braking, because the hot spots are uniformly distributed on the friction surface and are assumed to be stationary during braking time.

Thermomechanical analysis with assumption of uniform contact pressure

Uniform contact pressure distribution was assumed, leading to an almost uniform heat flux distribution. The thermomechanical analysis was performed using firstly a cyclic viscoplastic model and secondly a linear kinematic model, describing the disc material behaviour. The calculation was carried out for a series of seven successive stop brakings, with the following condition: the speed varies from 300 to 0 km/h during a braking time of 80 s, the time interval between two successive brakings being 1200 s.

Viscoplasticity with disc softening material behaviour. Samrout et al. [11] have identified the thermomechanical behaviour of the 28CrMoV5-08 steel up to 600°C, using a complex visco-plastic model.

Commonly, the kinematic hardening corresponds to the translation of the elastic domain in the stress space and describes the cyclic hardening of a material. The isotropic hardening is associated with the evolution of the elastic domain size as a function of the cumulative plastic strain and describes the cyclic softening of the material. In the present case, these authors have introduced a nonlinear kinematic and isotropic hardening, together with the plastic strain memory effect which induces dependence between the saturated value of the isotropic hardening and the plastic strain amplitude. Temperature influence is taken into account by means of the temperature dependence of three material characteristics and of the thirteen coefficients of the constitutive law.

The model has been implemented in ANSYS™ code. Resolution was then performed using an incremental linearization of the constitutive law with an explicit algorithm, in accordance with the explicit nature of the law. Reasonably good numerical stability was observed.

Figure 6a shows the evolution of the maximum surface temperature during one braking. Temperature rises quickly due to the maximal heat flux at the beginning of braking. Maximal temperature is obtained at almost 2/3 of the time duration, then decreases until the end of braking due to the linear decrease of the heat flux until 0.

Figure 6b gives the temperature distribution in the disc at the time of maximum surface temperature (i.e. t = 48 s) and shows the refined meshing near the contact surface. Nine elements were considered on the half thickness of the disc, with a progressive size (the smallest at the friction surface having a 0.255 mm height).

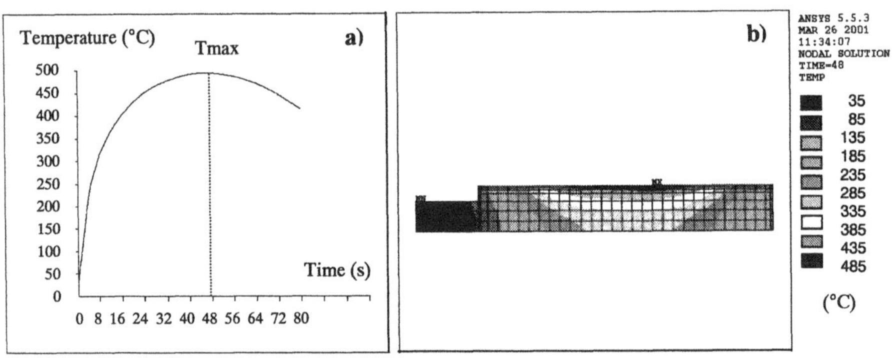

Fig. 6: Temperatures in the disc: a) on the surface during one braking; b) at t = 48 s

Figures 7a and 7b present the stress and plastic strain fields in the circumferential direction at t = 48 s, which corresponds to the maximum values. Similar results are obtained in the radial direction but with lower amplitudes.

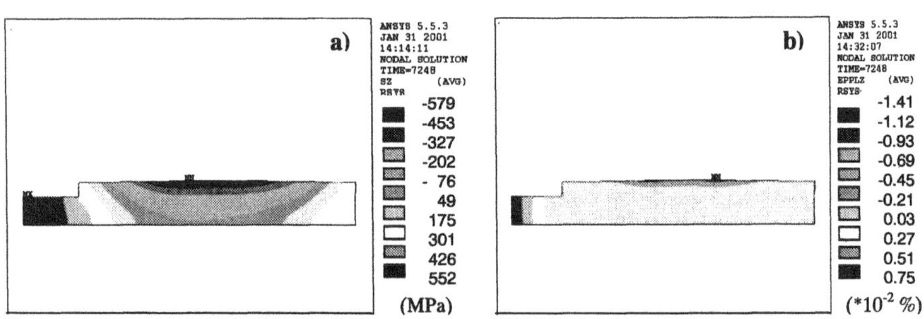

Fig. 7: Results in the disc at t = 48s of the 7th braking:
a) circumferential stresses; b) circumferential plastic strains

Compression occurs due to clamping at the inner and outer radius, induced by cooling as described in figure 7b.
Figures 8a and 8b show the hysteresis circumferential stress-strain loops, respectively in terms of mechanical total (i.d. elastic + plastic) and plastic strains, calculated when the temperature is maximum on the surface of the disc. It appears that residual tensile stresses occur after cooling but with low amplitude. The ratchetting strain decreases braking after braking, that will certainly lead to elastic shakedown a few braking later.

Plasticity with linear kinematic hardening for disc material behaviour. In order to reduce computing time, a simpler constitutive law was used in a second numerical simulation of the same series of seven successive brakings. This law was identified from tension tests performed up to 1100°C and is characterized with a multisurface linear kinematic hardening [8]. Its implementation was done using an implicit integration scheme.
Figures 8c and 8d show the calculated circumferential stress-strain hysteresis loops, respectively in terms of mechanical total and plastic strains, when the temperature is maximum on the surface of the disc. Even if the general trends in stress-strain response are similar to that calculated using the viscoplastic model, the multisurface linear kinematic hardening model leads, in the present case, to higher stress values, lower plastic strain values, and achieves quickly to elastic shakedown as expected. This can be explained by the intrinsic properties of the model, which does not take into account the cyclic softening of the material.
Whatever the used constitutive law, results show that the calculated values of stress and strain are higher in the circumferential direction, than in the radial direction. This is in good agreement with the previous observations of the cracking network, and in particular the predominancy of the circumferentially-oriented cracks on the radially-oriented cracks. Local friction effects increase this trend.

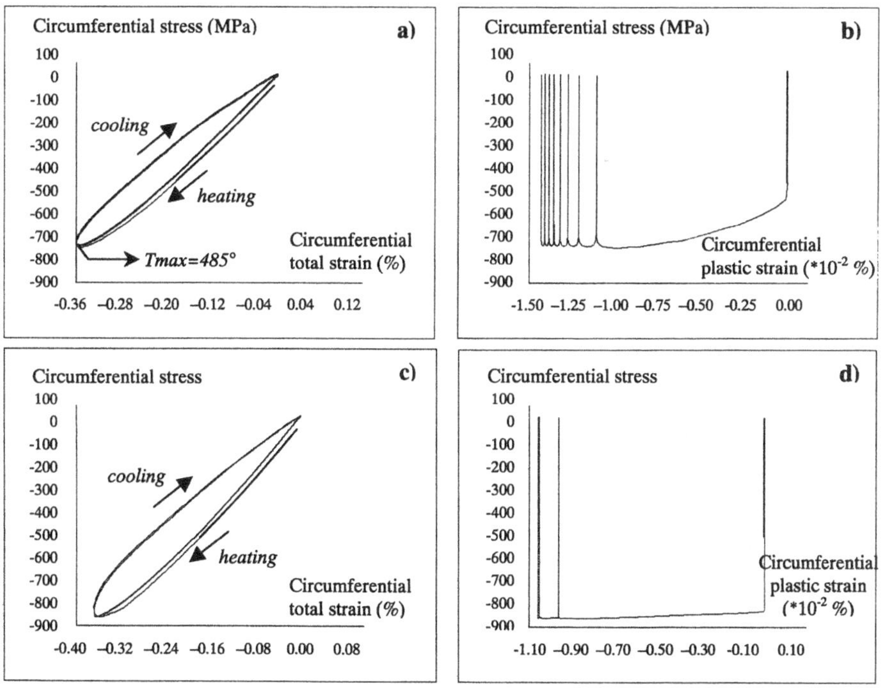

Fig. 8: Stress-strain hysteresis loops during the 7 braking series
Viscoplasticy with softening: a) mechanical total strain; b) plastic strain
Plasticity with linear kinematic hardening: c) mechanical total strain; d) plastic strain

Thermomechanical simulation of macroscopic hot spots
A second numerical simulation has been done in order to describe the occurrence of six macroscopic hot spots.
A new series of seven stop brakings was considered; each braking is from 300 to 0 km/h, during 213 s with 13.9 MJ per disc to be dissipated.
In this approach, thermal and mechanical analyses are uncoupled. While FEM analysis is performed considering the complete geometry of the disc, due to the axial symmetry of the disc, the regular distribution of the hot spots on each face, and their anti-symmetric position on the two faces, it is possible to model only 1/12 of the disc.

Thermal analysis. Figure 9a presents the thermal distribution obtained with the present simulation at t = 72 s, time at which the maximum temperature is reached in these conditions. For each braking, the heat flux is maximum at the beginning of the braking and decreases progressively until zero.
This numerical result can be compared with the infrared cartography obtained at the same instant for the same braking conditions (Fig. 9b). Experimentally, the hot spot angular distribution is very regular, confirming the choice of a reduced angular numerical model. Considering the error of such technique to measure the temperature, in particular the error due to the non-uniformity of the emissivity, calculated and experimental temperatures are of the same order. Moreover, the maximum temperature is reached at the same time.

Thermomechanical analysis. For this calculation with a 3D model, the multisurface linear kinematic hardening model was used, in order to reduce computing time and to permit the calculation above 600°C.
Fig. 9c and 9d present the distribution of the stresses and plastic strains in the circumferential direction at the time of maximum temperature (t = 72 s).
These figures show that thermal distortions induce high compressive state around the hot spot. As the Young modulus decreases as temperature increases, stress amplitude is not maximum on the hot spot itself, where the temperature is maximum. Plastic flow occurs after a time of 8 s, with high compression on the hot spot.

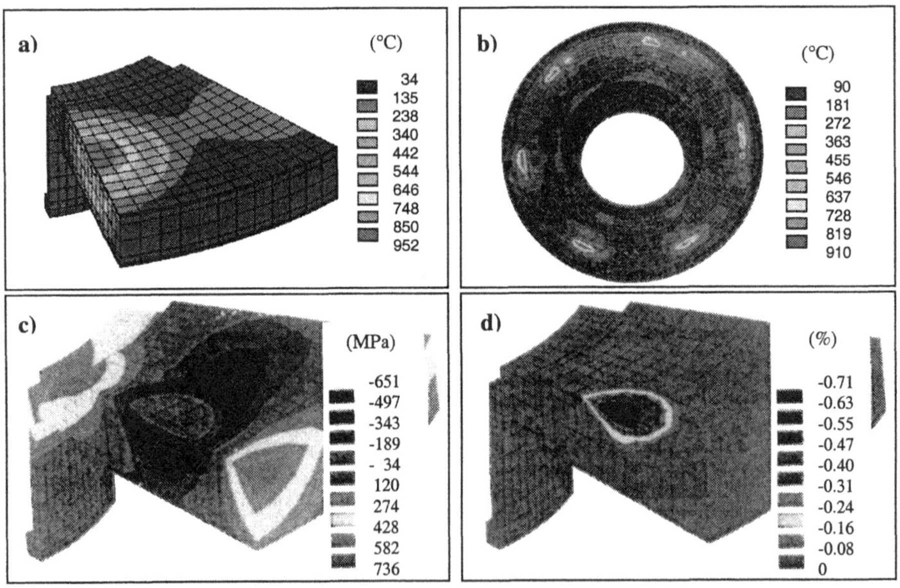

Fig. 9: Results at t = 72 s: a) calculated temperature; b) experimental thermograph; c) circumferential stresses; d) circumferential plastic strains

After cooling, due to plastic flow, residual tensile stresses occur (up to 790 MPa near the center of the hot spot) with a low attenuation of the plastic strains (-0.54 %). Inversion of the stress sign appear near the inner and the outer radius. This is coherent with the experimental observations showing that macrocracks do not pass through the whole radial length.
Results corresponding to the radial direction give similar conclusions, but with lower amplitudes (minimum of –550 MPa during braking and maximum of 692 MPa after cooling).
The mechanical loading amplitude due to friction is more than one hundred times less than the thermomechanical value with the assumption of a uniform contact pressure, but it can of course be locally much higher. It is neglected in the present paper.
Figure 10a presents the evolution of the hysteresis loops corresponding to the series of 7 consecutive stop brakings. The stress and the total mechanical strain are calculated, in the circumferential direction, on the top of the hot spot. As observed experimentally [7,8], it is assumed that the hot spots do not move from one braking to the next one.
Stabilization of the loops is obtained after 6 brakings. Figure 10b gives the stabilised loop, in terms of stress versus plastic strain. The complex shape of the loop is explained by the high variations of the thermal expansion coefficient, particularly at 715°C and 800°C.
With the assumptions adopted for these numerical investigations, results show that a stabilized tension-compression stress-strain loop with a significant plastic flow is obtained. It may be expected that thermal fatigue damage models could give indications on crack initiation.

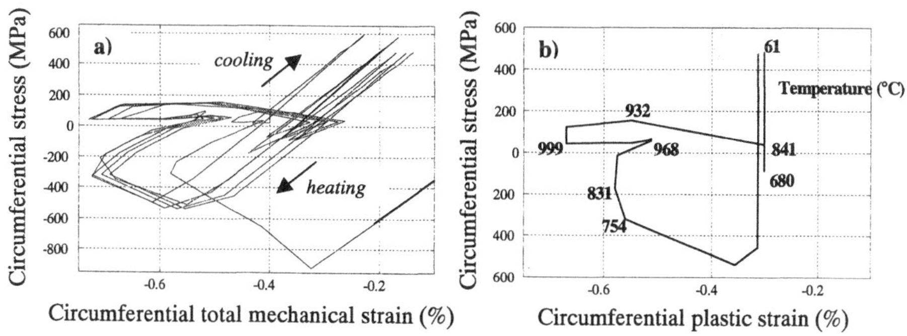

Fig. 10: Hysteresis stress - strain loops:
a) stress - mechanical strains for the 7 brakings; b) stabilized stress - plastic strain loop

CONCLUSION

Railway brake discs are subjected to severe thermomechanical loadings, which may give crack occurrence on the friction surface, leading to their early replacement. Damage observation of several out-of-order discs was performed, showing a quasi-general thermal fatigue crazing and the presence of a few macroscopic radial cracks. A numerical thermomechanical model was developed. Two series of seven consecutive stop brakings have been simulated, in the case of a uniform pressure distribution, and in the case of the presence of hot spots. In both cases, numerical results are in accordance with the experimental observations. The calculated stress-strain loops will be applied furtherly in some thermal fatigue damage models for disc life prediction.

ACKNOWLEDGEMENTS

The authors acknowledge the contribution made to this work by the SNCF railway company.

REFERENCES

1. Anderson, A.E. and Knapp, R.A. (1989) *Int. Conf. on Wear of Materials* **2**, pp. 673-680.
2. Day, A.J. (1990). In *2nd Brakes Workshop*, University of Bradford.
3. Desplanques, Y., Degallaix, G., Copin, R. and Berthier, Y. (2000). In *27th Leeds-Lyon Symposium on Tribology*, Lyon, France. D. Dowson and al (Eds.). Elsevier, under press.
4. Copin, R., Degallaix, G., Desplanques, Y. and Berthier, Y. (2000). In *4th Eur. Solid Mechanics Conf. EUROMECH ESMC4*, Metz, France.
5. Lee, K. and Barber, J.R. (1994) *ASME Journal of Tribology* **116**, pp. 409-414.
6. Dufrénoy, P. and Weichert, D. (1995) *Proc. Instn. Mech. Engrs* Part F **209**, pp. 67-76.
7. Dufrénoy, P., Panier, S. and Weichert, D. (1998). In *JEF98: Journées européennes du freinage*, Lille, France, pp. 245-257.
8. Dufrénoy, P. (1995) PhD thesis, University of Lille, France.
9. Du, S., Zadrodzki, P., Barber, J.R. and Hulbert, G.M. (1997) *J. Thermal Stresses* **20**, pp. 185-201.
10. Fan, X. and Lippmann, H. (1996). *In Asia-Pacific Symposium on Advances in Engineering Plasticity and its Applications AEPA'96*, Tokyo, Pergamon.
11. Samrout, H. and El Abdi, R. (1998) *Int. J. of Fatigue* **20** (8), pp. 555-563.

Temperature-Fatigue Interaction
L. Rémy and J. Petit (Eds.)

LOW CYCLE AND THERMOMECHANICAL FATIGUE OF NICKEL BASE SUPERALLOYS FOR GAS TURBINE APPLICATION

M. MARCHIONNI
CNR-TeMPE Via Cozzi 53 - 20125 Milano, Italy

ABSTRACT

Isothermal low cycle (LCF) fatigue and thermomechanical (TMF) fatigue of some nickel base superalloys have been analysed comparing their fatigue life for different temperatures and strain history.
Fatigue behaviour of CMSX4+Y single crystal alloy for turbine blades can be described either with TMF or with isothermal LCF testing. TMF results on MA6000 and MA760 ODS alloys are strongly different from those obtained by LCF tests, and consequently the selection of testing programme is strictly subjected to material service conditions, from thermal transients, from physical and mechanical material properties at different temperatures. In addition the experimental results on MA6000 alloy are affected by specimen geometry.
The TMF results for different materials show the importance of such experiments in function of the alloy studied. The use of LCF parameters can affect the design procedure especially for the alloys that change strongly their mechanical properties with temperature. In this case the extension of LCF data with TMF parameters is largely recommended for optimising the design procedure.

KEYWORDS

Low cycle fatigue, thermomechanical fatigue, nickel base superalloys, single crystal, ODS alloys.

INTRODUCTION

The study of new alloys, such as nickel base superalloys produced by directional solidification (DS) or single crystal (SX) processes, requires sophisticated evaluation that is capable of characterising the materials under near service conditions and of producing the most accurate parameters to be used in component design.
Since the seventies, materials subjected to cyclic stresses and temperatures have been studied by isothermal low cycle fatigue testing (LCF) and the results have been produced by considering as reference temperature that corresponding to the maximum value of the thermal cycling [1, 2]. After the introduction of thermomechanical fatigue (TMF) as a diagnostic device for the material study [3], the comparison of LCF and TMF testing has determined opposite conclusions. Some scientists have found that LCF and TMF produce comparable results [4, 5], and consequently the application of TMF testing is not considered useful for the evaluation of material property. Besides other scientists have found that the stress history of a thermomechanical cycle can determine results completely different from those produced by isothermal testing [6, 7, 8]. Consequently the use of TMF testing for the characterisation of materials fatigue properties is largely increasing, even if the cost of TMF tests is greatly higher than the isothermal LCF tests. The TMF procedure is particularly recommended for those new materials whose high temperature properties are not known. The more expensive cost of TMF tests can be ascribed either to the complexity of

experimental equipment or to the longer endurance with respect to LCF testing. In the present work a short description of the experimental technique and of the TMF test procedure is shown. In addition some results of TMF testing on new gas turbine alloys for aeronautics application and their comparison with LCF testing are also described.

EXPERIMENTAL TECHNIQUE

When some components of a plant or equipment are subjected to stress and temperature cycling, as in consequence to start and stop of the equipment, they are operating in the thermomechanical fatigue regime. In order to repeat TMF experiments in laboratory scale it is necessary to have some equipments as a servo-hydraulic testing machine, an induction furnace (or a lamp furnace) for a rapid variation of testing temperature, and a micro-computer that can control load or strain and temperature variations according to a programme previously defined. A schematic of the main devices necessary for the experiments is shown in Fig. 1.

TMF tests have been performed by a 250 KN servo-hydraulic testing system. A water-cooled copper induction coil and powered by a 17.5 kV radiofrequency furnace has heated the specimens with a cylindrical gauge length (Fig. 2). Test temperature is controlled by Chromel-Alumel thermocouple spot-welded outside the gauge length in order to avoid the crack initiation in correspondence of the thermocouple welding. Before a series of experiments on a selected material, one specimen is used to optimise the temperature profile in the gauge length.

Taking into account that a variation of temperature during the TMF cycling determines an elongation or a contraction of the specimen, it is necessary to perform a thermal strain calibration in the temperature range selected before each TMF test. A thermal calibration curve is obtained by performing a thermal cycle, repeated at least three times, having the same shape of the TMF cycle. The mean of the cycles recorded forms the material thermal strain curve that will be used during the real TMF test. In practice the total strain supplied by the hydraulic actuator is the sum of the mechanical strain and the thermal strain previously determined. In such way the specimen thermal strain is compensated during all the thermomechanical cycle.

During the TMF experiments, the test temperature, the strain imposed (both the mechanical and total strain), the stress response and the stress-strain cycles are recorded at intervals. In addition an automatic device for the image acquisition from two cameras allows to evaluate the time of crack initiation and the damage evolution.

At the end of the test the computer software can elaborate the TMF results.

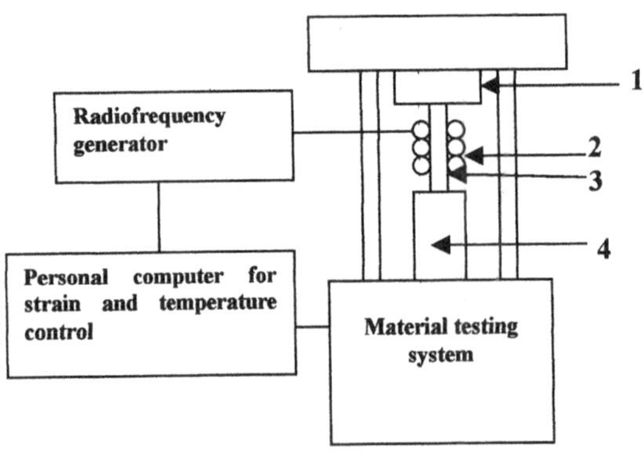

Fig. 1 - Schematic of the equipment for TMF testing. 1) Load cell, 2) Copper induction coil, 3) Specimen, 4) Hydraulic actuator.

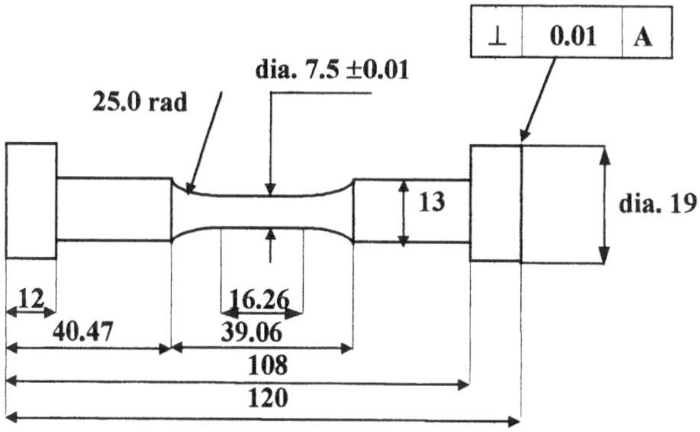

Fig. 2 - Specimen for TMF testing.

TMF TEST DESCRIPTION AND MATERIALS STUDIED

As previously described, the experimental conditions for TMF testing simulate those of the components in service. In the present work the components are vanes or blades of gas turbines for aero engines or power plants for energy production. Generally the shape of the thermomechanical cycle is a compromise between the service conditions determined by finite element analysis and those consistent with the laboratory equipments.

The materials tested are nickel base superalloys produced by directional solidification in order to obtain polycrystal or single crystal microstructure. Among the polycrystal alloys those directionally solidified and oxide dispersion strengthened (currently Yttrium oxide Y_2O_3) are used in vanes for land and jet turbines. Fig. 3 shows an example of TMF diamond cycle that reproduces the variation of the temperature and strain that occur in uncooled vanes for land based gas turbines at plant start and stop. The IP and OP cycles indicates the condition of the temperature respectively in phase and out of phase with the strain.

Single crystal nickel base superalloys, owing to their resistance to higher temperatures and the difficulty of producing components of large size, are mainly used for blades of aero engine turbines (but the recent trend is a process improving for increasing the component size and extending the application to other turbines). The D2 cycle shown in Fig. 4 refers to the conditions of a blade single crystal nickel base superalloy. The cycle has a diamond shape, but temperatures and mechanical strains are different from D1 vane cycle.

Table I - Materials and main parameters of TMF testing.

Material	Property	Use in gas turbine	Cycle shape	T_{max}, (°C)	T_{min}, (°C)
MA6000	ODS	Vanes	IP, OP, D1	1050	550
CMSX4+Y	Single crystal	Blades	D1, D2	1100	600
MA760	ODS	Vanes	D1	1050	550

Legend: IP = in phase, OP = out of phase, D1 = diamond as in figure 3,
D2 = diamond as in figure 4.

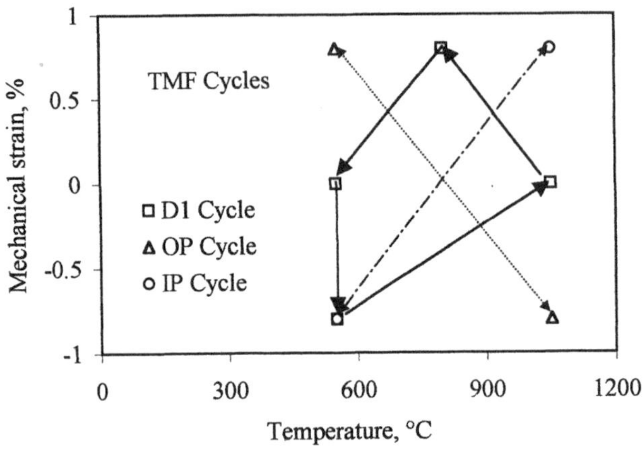

Fig. 3 - TMF diamond cycle (D1), in phase (IP) and out of phase (OP) TMF cycles.

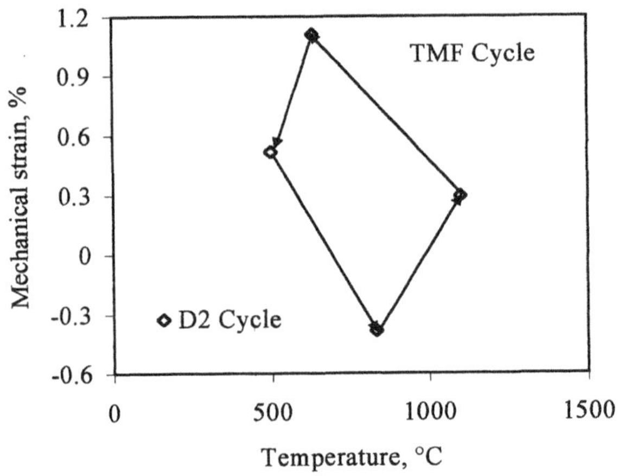

Fig. 4 - TMF diamond cycle (D2) derived from single crystal blades.

Table I presents studied materials, their use, the TMF cycle selected and the test temperature range. The MA6000 nickel base superalloy has been produced by directionally solidification and strengthened by Yttrium oxide dispersion. This alloy is one of the most investigated in our institute either for its application exploitability or for the possibility of comparison with LCF results already obtained within several international projects [9]. Among the other alloys mentioned in the table I the single crystal CMSX4+Y has been used for blades of turbojet and MA760 ODS alloy is studied in order to increase the high temperature mechanical properties in respect to MA6000 alloy.

TMF TESTING RESULTS

MA6000 alloy.

TMF tests have been performed on solid and hollow cylindrical specimens with different wave shapes as described in table I. Hollow specimen geometry is the same of Fig. 2 with a longitudinal central hole of 4 mm diameter. Fig. 5 shows the results and the comparison with LCF tests [9]. TMF tests on solid specimens and cycle D1 exhibit fatigue life comparable to LCF tests at 1050°C for 100 – 200 cycles to failure. At longer endurance LCF fatigue life is sensibly shorter. When an IP and OP cycle is applied or an hollow specimen is tested, TMF life is strongly reduced in respect to the cycle D1, save for IP cycle at low strain and endurance higher than 2000 cycles to failure. The results previously described indicate that it is important to select a TMF laboratory cycle as close as possible to the real strain and temperature of the component in service. In addition they confirm that LCF tests are too conservative in respect to TMF testing for a correct description of thermomechanical fatigue property.

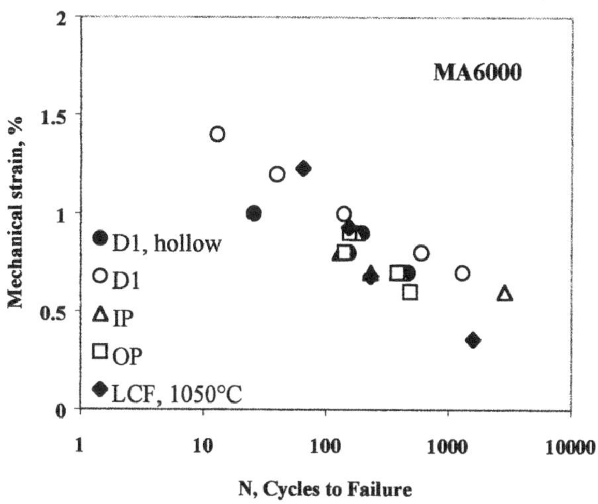

Fig. 5 - LCF and TMF fatigue of ODS MA6000 alloy for different experimental conditions.

CMSX4+Y alloy.

This alloy is a single crystal nickel base superalloy modified by addition of Yttrium oxide particles in order to improve its oxidation resistance [10]. CMSX4+Y alloy is used for blades of gas turbine in aerospace application. Besides the conventional creep and fatigue property the study has been extended to thermomechanical fatigue [11]. The TMF results and their comparison with LCF are reported in Fig. 6. We can observe that the fatigue life is dependent from TMF cycle shape (D2 is

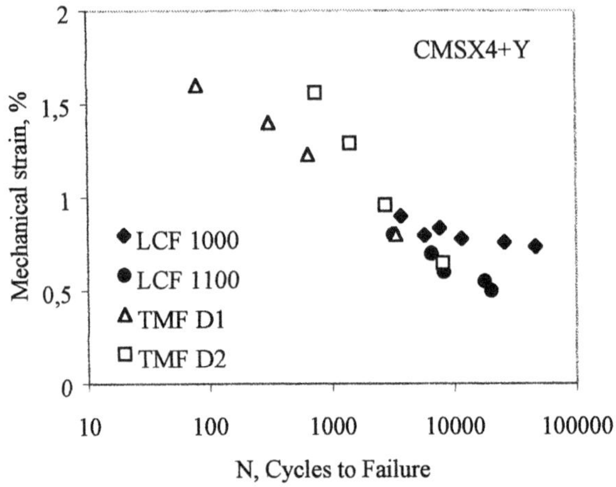

Fig. 6 - LCF and TMF fatigue life of a single crystal alloy.

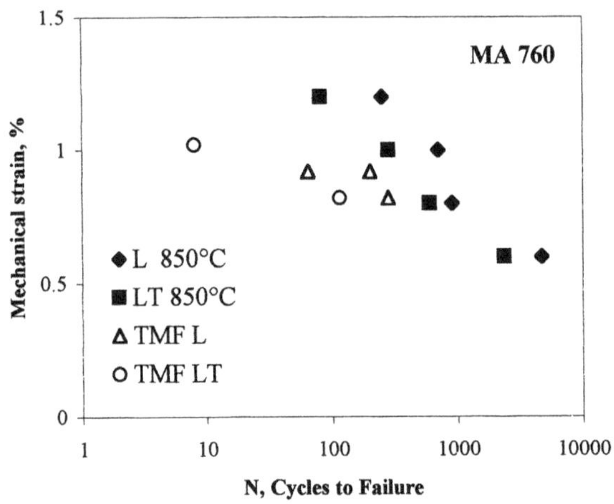

Fig. 7 - Influence of L and LT grain orientation on LCF and TMF fatigue life.

less damaging than D1 mainly at higher strain), while in LCF regime the temperature increasing gives a strong reduction of fatigue life. TMF is less damaging than LCF at 1100°C when strain is higher than 0.8% and comparable for strains lower than 0.8 %. Taking into account that the component strains in service are about 0.5% or lower, LCF at 1100°C can describe the material behaviour during the thermal transients with satisfactory accuracy. However TMF tests confirm the good behaviour of the material for blades even in severe conditions of temperature and strain.

MA760 alloy.
This alloy, as MA6000, belongs to the ODS class, and was produced to obtain an improvement of the high temperature mechanical properties and consequently to increase the design temperature of the component [12]. Due to the anisotropy of the material, LCF and TMF tests have been performed on specimens cut from the bars in two directions in respect to the grains orientation (L longitudinal and LT longitudinal- transverse). Fig. 7 describes the results and the comparison between TMF and LCF. At strain higher than 0.6% TMF life is sensibly lower than LCF at 850°C (this temperature is currently used for component design). The life difference is progressively reduced when total strain decreases and it disappears for strains below about 0.5%, if the trend of TMF curve is considered. The alloy exhibits a strong anisotropic behaviour that gives a large fatigue life decreasing in LT direction.
Therefore the TMF life prediction in both L and LT directions is strongly recommended for a correct use of high temperature material property in component design.

DISCUSSION

The results previously described show that TMF tests are very important for a deep knowledge of new material behaviour at elevated temperature and their use in component design. The reason of the different results of TMF and LCF tests can be ascribed to the materials properties in the temperature range and to the stress variation during service. Therefore several new materials showing best high temperature mechanical properties, at lower temperatures exhibit similar or worse mechanical properties than those of the alloys previously used in the components. In the TMF regime the thermal variations due to the transients in service, give arise variable stresses that the material oppose with different property in function of the temperature reached. It is not the same for LCF regime as temperature is constant. Such behaviour is more apparent when the physical and the mechanical properties of the material change strongly with temperature. With reference to the materials previously described, MA6000 and MA760 alloys, that can be used up to 1100°C, exhibit a low ductility and a marked fatigue crack initiation sensitivity, particularly for temperatures below 900°C. Therefore TMF behaviour is strongly affected by the part of thermomechanical cycle at low temperature and the results of TMF testing are different from those of LCF testing.

CONCLUSIONS

The TMF apparatus and the testing procedure have been described.
The results produced on some nickel base superalloys have been compared with those obtained by LCF isothermal testing.
Fatigue properties of CMSX4+Y single crystal alloy can be described either with TMF or with LCF showing a similar accuracy.
TMF results on MA6000 and MA760 ODS alloys are strongly different from those obtained by LCF tests and consequently the selection of testing programme is strictly subjected to material service conditions, from thermal transients and from physical and mechanical material properties at different temperatures. In addition the experimental results on MA6000 alloy are affected by specimen geometry, and those for MA760 are dependent of the grain orientation.

ACKNOWLEDGMENTS

Most of this research activity has been performed in a European concerted action named COST 501 round II and round III.

REFERENCES

1. Coffin L.F. Jr., Fatigue at Elevated Temperature, ASTM STP 520, Carden, McEvily and Wells Editors, ASTM (1973), 5 –34.
2. Taira S., Fatigue at Elevated Temperature, ASTM STP 520, Carden, McEvily and Wells Editors, ASTM (1973), 80 – 101.
3. Hopkins S.W., Low Cycle Thermal Fatigue of Materials and Components, ASTM STP 612, Spera and Mowbray Editors, ASTM (1976), 157 – 169
4. Malpertu J.L., and Rémy L., Low Cycle Fatigue, ASTM STP 942, Solomon et al. Editors, ASTM (1988), 657 – 671.
5. Shi H.J., Robin C., and Plevinage G., Advances in Fatigue Lifetime Predictive Techniques, Vol. II, ASTM STP 1211, Mitchell and Landgraf editors, ASTM (1993), 105 – 116.
6. Koster A. & alii, Proceedings of "Fatigue under Thermal and Mechanical Loading: Mechanisms, Mechanics and Modelling", Petten (NL), Bressers and Remy Editors, Kluwer Academic Publishers (1995), 25 - 35
7. Sehitoglu H., Fatigue Lifetime Predictive Techniques, ASTM STP 1122, Mitchell and Landgraf Editors, ASTM (1992), 47 - 76
8. Bernstain H.L. & alii, Prediction of Thermal – Mechanical Fatigue Life for Gas Turbines in Electric Power Generation, ASTM STP 1186, Sehitoglu Editor, ASTM (1993), 212 – 238.
9. Marchionni M., Ranucci D. and Picco E., Proceedings of "Fatigue under Thermal and Mechanical Loading: Mechanisms, Mechanics and Modelling", Petten (NL), Bressers and Remy Editors, Kluwer Academic Publishers (1995), 169 – 178.
10. Meyer-Olbersleben & alii, Proceedings of Low Cycle Fatigue and Elasto-Plastic Behaviour of Materials-3 International Conference, K.-T. Rie Editor, Elsevier Applied Science, (1992), 1 – 6.
11. Marchionni M & alii, Proceedings of Materials for Advanced Power Engineering International Conference, Coutsouradis & alii editors, Kluwer Academic Publishers, Liege (B), Vol. II (1994), 989 – 998.
12. Marchionni M., Goldschmidt D. and Maldini M., Journal of Materials Engineering and Performances, Volume 2 (4) (1993), 497 – 503.

HEAT-CHECKING OF HOT WORK TOOL STEELS

B. MIQUEL, S.JEAN, S. LE ROUX, P. LAMESLE and F. RÉZAÏ-ARIA
Tool Surface Assessment Unit, Research Centre on Forming Tools, Materials and Processes
Ecole des Mines d'Albi-Carmaux, F-81000 Albi, France

ABSTRACT

Thermal fatigue (TF) is one of the life-limiting factors of the surface of the hot work tool steels. Bi-axial thermal strains and stresses are the main driving forces for the bi-axial cracking of tools well known as the heat checking. A TF rig using tubular specimens, induction heating and pressure air-cooling is developed. Two tempered martensitic steels 55NiCrMoV7 and X38CrMoV5 are investigated (47 HRC hardness) under a TF between 50°-650°C. The effect of the specimen thickness on the softening (X38CrMoV5) is revealed by the post mortem room temperature microhardness measurements, X-ray residual stress and width broadening evaluations. The initial compressive residual stresses (due to the machining) become tensile very early under TF cycling. The X-ray width decreases with the number of thermal cycle. Cyclic inelastic straining and cyclic temperature tempering explain this softening. Cracking seems start when the oxide scale achieves a «critical thickness». Depending on the thermomechanical stress-state, the damage feature changes from a parallel multi-cracking near the extremities of the specimens (uni-axial loading) to a "cell-type" (or "square-type") cracking at the centre of the specimens (bi-axial loading). Uni-axial cracks can be very much extended. Under TF cycling, secondary cracks initiate progressively perpendicular to major axis of the uni-axial cracks. By increasing the longitudinal over the hoop stress ratio from the extremities ($R=\sigma zz/\sigma\theta\theta=0$) to the centre ($R\approx1$), the density of the secondary cracks enhances. The 55NiCrMoV7 steel presents a lower heat checking resistance. In X38CrMoV5 the localised heat checking and the oxide-scale spalling are observed while in 55NiCrMoV7 the heat checking covers well the whole surface of the TF specimen. Oxidation-TF interactions play an important role in crack initiation and propagation.

KEYWORDS

Thermal fatigue, thermomechanical fatigue, steel, crack initiation, heat-checking, softening.

INTRODUCTION

Thermal fatigue (TF) is one of the life-limiting factors of the surface of the hot work tool steels (HWTS). During hot metal forming the surface of tool in contact with the hot work-piece (in forging) or with the molten alloys (Al or brass in pressure die-casting), is heated in a very short time [1-2]. When the part is formed and during its ejection from a die, the surface of the tools is rapidly cooled down in particular when a cooling or lubricating liquid is employed. Free thermal expansion/contraction of the surface is self-constrained by the bulk of the tools, which stands respectively at lower/higher temperatures in each heating-up/cooling-down operation. The surface is therefore alternatively compressive loaded while temperature increases (compressive thermal chock) and tensile loaded when temperature decreases (tensile thermal chock), Figure 1 [1]. This is also the basic mechanism of TF damage of many industrial components such as turbine blades [3-5], or nuclear parts [6] for example. Bi-axial thermomechanical loading is the main driving force for the bi-axial cracking (heat checking) of any materials. The heat checking is one of the general features of thermal fatigue (TF) damage of HWTS [7-12].

This contribution reports some aspects of TF behaviour and damage observed on quenched and tempered HWTS. A new TF rig is developed. Different tubular TF specimens are employed to achieve various thermomechanical loadings. Softening and heat checking are investigated. These investigations were undertaken in the frame of the French Research Action-II on Forging.

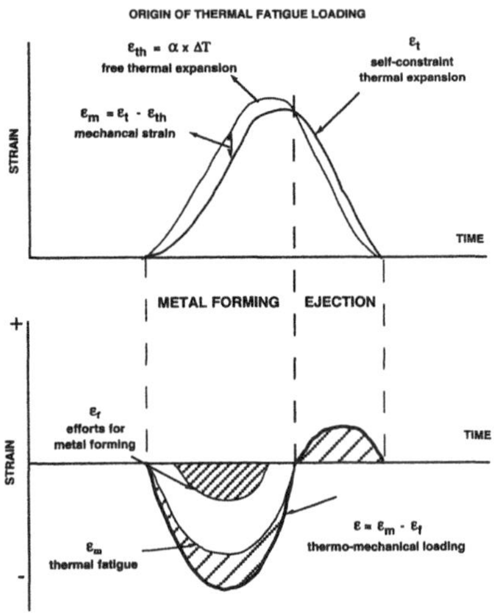

Fig. 1: Schematic presentation of the origin of the thermomechanical loading of HWTS [1].

EXPERIMENTAL PROCEDURE

Steels

Two steels, 55NiCrMoV7 (Thyssen) and X38CrMoV5 (AISI H11, Aubert & Duval) were investigated. They were provided free of charge in the frame of the French Concentrated Action on Forge program. The chemical composition of steels is reported in table 1.

The steels were heat-treated (austenitisation, quenching and tempering) to achieve a martensitic microstructure with a hardness of 472 HV (p=200g). X38CrMoV5 is widely used for dies and matrix in forging or casting. 55NiCrMoV7 has a higher toughness and is generally employed in applications requiring a high resistance die to the mechanical shocks.

Table 1. Chemical composition of steels (major elements in weight %)

Steel	C	Cr	Mn	V	Ni	Mo	V	Si	Fe
55NiCrMoV7 *(55NCDV7)*	0.56	1.10	0.50	0.47	1.70	0.50	0.10	0.20	bal.
X38CrMoV5 *(Z38CDV5)*	0.38	5.05	0.49	0.47	0.20	1.25	0.47	0.92	bal.

Thermal fatigue rig

A TF rig using high frequency induction heating is developed. Tubular specimens with various central cylindrical chambers (wall thickness 5, 7, 10 mm) were designed, Figure 2. The specimens are continuously internally water cooled while the external surfaces are alternatively heated and cooled by compressed air. By modifying the wall thickness, various thermal gradients and therefore different thermomechanical loadings are generated. The external surface of the specimens is mechanically polished down to 1 μm diamond paste.

A 25 kW (100 to 400 kHz) high frequency induction heating system from Celes is used. A cooper coil was constructed such as the thermal stain can be measured during thermal cycling, as it was earlier developed on the single wedged TF specimens [5], Figure 3. The heating and cooling periods are about 5-7s to 15-20s respectively, depending on the specimen thickness and obviously the minimum and maximum temperatures of the thermal cycle at the external surface, Figure 4.

The temperature-time profile is monitored by a spot welded thermocouple type-K (in general with a 0.1 mm diameter). In the first step, the axial and circumferencial thermal gradients on the external surfaces of three dummy specimens were measured by several spot-welded type-K thermocouples (Figure 3). In the present configuration, a thermal gradient less than 15°C is obtained in the central zone (20 mm) of the specimens at the highest temperature of the thermal cycle, 650°C. The minimum temperature of the thermal cycle is 50°C.

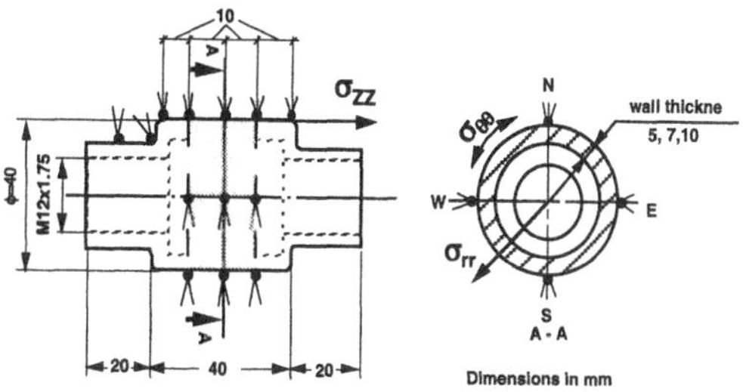

Fig. 2: TF specimens with different thickness and the location of spot-welded thermocouples.

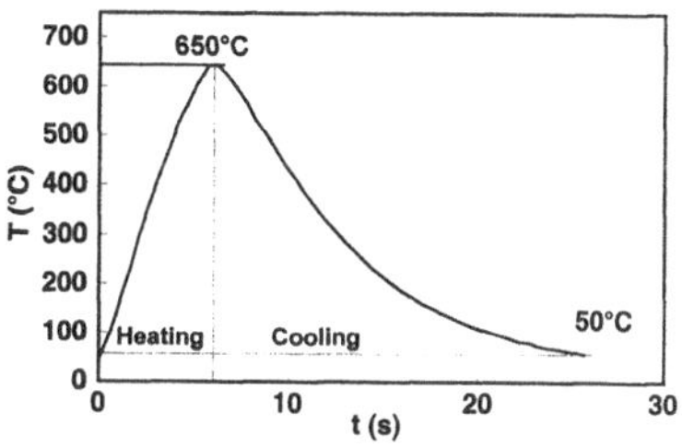

Fig. 3: TF rig (induction coil, pressure air cooling, and specimen) and a typical temperature-time cycle at the centre of the specimen.

Experiments were regularly interrupted to assess by SEM the evolution of the external surface damage. At each interruption, the axial (σzz) and hoop (σθθ) residual stresses were measured by X-ray diffraction at the external surface. Several tests were run from 150 to 6500 or higher thermal cycles. At the end of each test, the residual stresses were in addition measured through the wall thickness by successive electropolishing method [12]. TF specimens were then cut for post-mortem microhardness measurements (200 g, Vickers) along wall thickness [12].

RESULTS AND DISCUSSION

FEM Analysis

Details of thermo-elasto-plastic Finite Element analysis by ABAQUS are reported elsewhere [11-12]. Due to the specimen symmetry, only 1/4 of the specimens was meshed. The constitutive equations parameters were identified at different temperatures using isothermal uni-axial tensile tests. The measured temperature-time cycles were imposed to the nodes of the external elements as boundary conditions. These analyses have revealed that, any point in the specimens, is 3D thermomechanically loaded. The

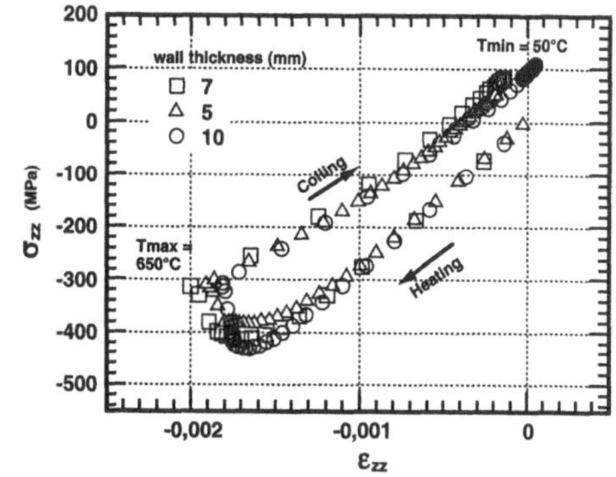

Fig. 4: Calculated hysteresis loops for an element at the centre of the external surface of TF specimens (X38CrMoV5).

radial stress (σrr) is however very small as compared to the σzz and $\sigma\theta\theta$ (at least for the critical elements on the surface). Figure 4 shows an example of the axial stress (σzz)-strain (εzz) hysteresis loops for an element on the centre of the three TF specimens (X38CrMoV5) steel. As can be seen, during heating and cooling the specimen is respectively under compressive and tensile loading. When a thermo-elasto-visco-plastic constitutive equation is used [13-14], the hysteresis loops are shifted to the higher tensile stresses because of the stress relaxation and the strain-induced softening of the steel [15]. Thermomechanical investigations have shown that during an accommodation period the maximum and the minimum stresses increase and then the steel soften continuously [15]. The variations of the first reversal plastic strain as a function of the distance from the external surface of the TF specimens are reported in [11].

Softening

Figure 5 presents an example of the effect of the number TF cycles on the variation of the hardness from the external surface of X38CrMoV5 (7 mm wall thickness specimen). Softening is observed beneath the surface, named "thermomechanically affected zone" (TMAZ). It is found that 55NiCrMoV7, has a lower TF resistance at within the TMAZ, since its hardness reduction after 6500 cycles is more pronounced than X38CrMoV5 after 9500 cycles.

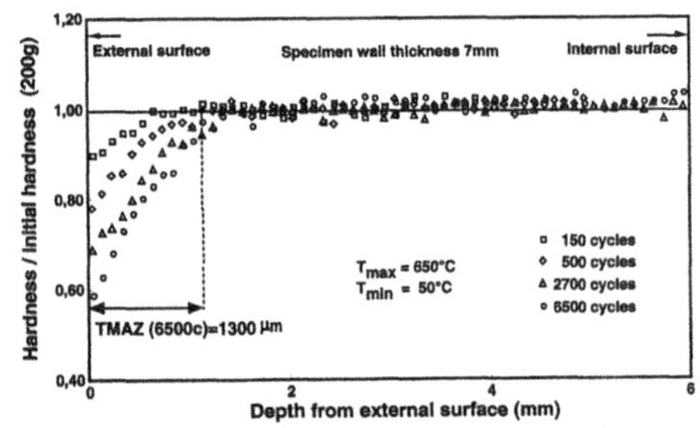

Fig. 5: Variation of the hardness/initial hardness as a function of the distance from external surface of TF-specimens.

Detailed investigations have revealed that hardness reduction can be caused by two coupled cyclic thermally activated processes: "temperature-time" dependant tempering and frequency dependant "plastic strain-cycle" softening. It is known that steels are very prone to the softening under cyclic isothermal [16-17] and non-isothermal low cycle fatigue [15, 18]. The room temperature microhardness measurements of isothermal and cyclic temperature treatments and the isothermal LCF as well as TMF have both shown this softening [11-12, 15]. However, it is not trivial to emphasis the role of each mechanism when the time, temperature and strain are coupled. Based on isothermal and cyclic temperature treatments, an equivalent time-temperature relationship was defined [12] to estimate the hardness reduction of X38CrMoV5 when subjected to temperature cycling. Using this relationship and considering the temperature-time profiles of each element in TF specimens, the hardness reductions due to cyclic tempering was estimated.

An example of such estimation is reported in Figure 6. As can be observed an equivalent time-temperature approach underestimate TF hardness lost. Investigations have shown that in the first hundreds TF cycles the thermomechanical straining is the dominant mechanism of the TF-softening of the quenched and tempered martensitic 5% Cr steels and then progressively the cyclic temperature-time tempering become the prevailing mechanism [121. It is assumed that dislocation rearrangements and annihilations control the softening in early TF cycling and then, carbide coarsening becomes the dominant mechanism.

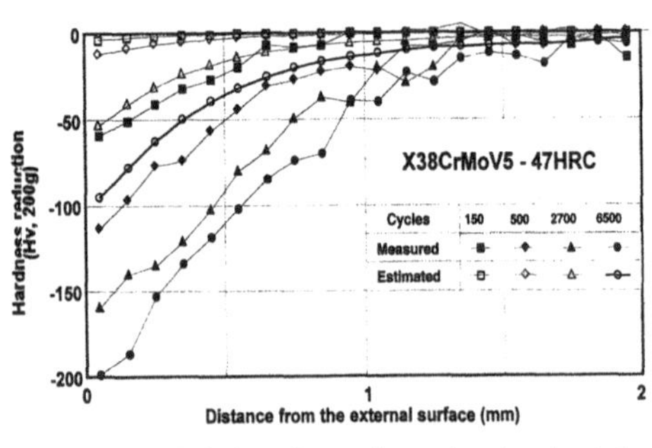

Fig. 6: Variation of experimental and estimated hardness lost in TMAZ as a function of the number of TF cycles (7mm thickness specimen).

On going TEM observations show that under isothermal LCF the initial high dislocation density in tempered martensitic microstructure decreases [19]. In many industrial applications, the "over-heating" of the dies in function is addressed to explain the softening of HWTS and the "plastic strain" induced softening is in general ignored [20]. It should be emphasised that no softening is obviously observed when annealed steel is examined [11-12].

Room temperature X-ray measurements have revealed, Figure 7, that the initial compressive residual stress at the external surface (generated by mechanically polishing) became tensile during the first few hundreds TF-cycles. The variation of the residual stress vs. TF-cycles of X38CrMoV5 steel presents in general first a stabilisation and then decreases with more TF cycle [21]. The measurements have revealed that the half-value breadth of the X-ray diffraction profile decreases also with TF-cycles confirming the same trends [12]. It seems that, because of the inter cell-boundary crack propagation to the base metal, the residual stresses relax. As the hardness measurements, the post-mortem X-ray measurements through-TMAZ by successive electropolishing technique [12], have confirmed the softening of the steel. Based on X-ray assessments one can assume that 55NiCrMoV7 steel has a lower TF-resistance.

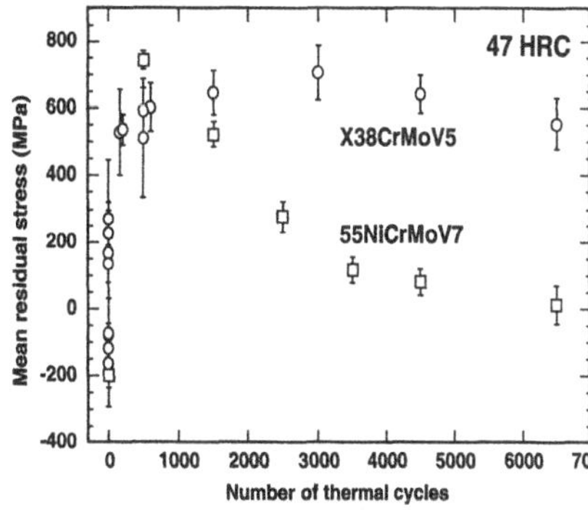

Fig. 7: Variation of mean residual stresses (X-ray) as a function

Damage mechanism

During thermal cycling, first the external surface was oxidised rather homogeneously. The surface colour changes during TF cycling. Post-mortem SEM observations on the specimen cross sections have shown that two oxide-scale layers are formed on the external surface of the X38CrMoV5. The qualitative EDS investigations have revealed different chemical compositions of these oxide-scales. The oxide-scale close to the substrate is rich in chromium. Only one oxide layer is formed on 55NiCrMoV7. Static isothermal oxidation has shown, Figure 8, also the same mechanism, [22]. A higher oxidation thickness was observed in 55NiCrMoV7.

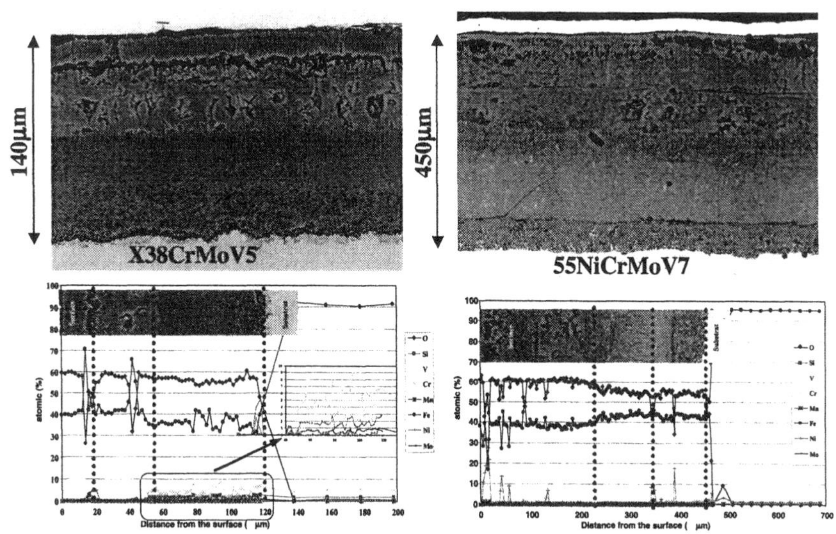

Fig 8: Oxide layers and thickness of two steels at 700°C for 150h (top) and corresponding chemical composition by microprobe analysis (down).

SEM back-scattering observations have shown that by successive TF cycling, tiny hair-like protuberances appear on the outer oxide-scale layers. These protuberances are "superficial" in the beginning and under more cycling and TF-oxidation interactions, they run into V-type inter heat checking boundary short cracks. It was observed that the outer oxide layer cracking (even in the case of very short depth cracks) perturbs the oxide/metal interface and provokes a localised advance of the oxidation inward the base metal. In fact, oxide-scale protuberances and subsequent V-type cracking change the general oxidation of the surface into the localised preferential oxidation beyond the metal/oxide interface. In X38CrMoV5, some V-type cracks are stopped at interior of the first oxide-scale layer, or at the interface of the two oxide-scale layers. Some others have propagated beyond the metal/oxide-scale interface. That creates the sites for preferential wedged-type crater-like cracking under TF solicitations. Only, a certain number of cracks propagate into the base metal, Figure 9. The oxide-scale V-type cracking is also observed in 55NiCrMoV7 steel. It is assumed that the stress-assisted diffusion may alter the local elements distribution ahead of the protuberances and the wedged-type crater-like cracks. The propagation under multi-axial loading can form an interconnected network of cracks resulting in some cases in extraction of the base metal, Figure 9.

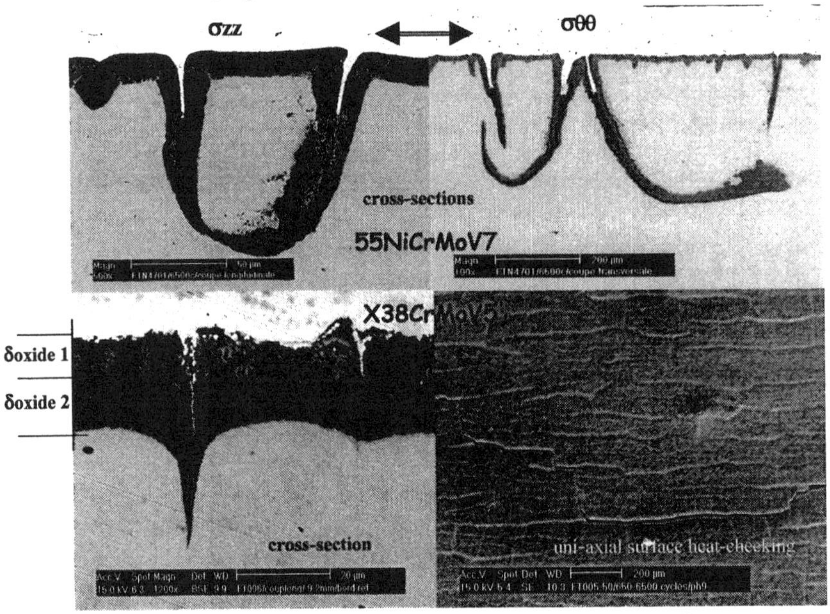

Fig. 9: Different aspects of the crack initiation and short crack propagation in both steels.

Heat-checking morphology

The well known "cell-type" (or "square-type") pattern of the heat checking in the centre of the specimen turns into "rectangular-type" pattern by approaching the extremities, Figure 9. Depending on the steel and the test conditions, very long parallel cracks are initiated along the z-axis (cracking perpendicular to $\sigma_{\theta\theta}$ stress) near the extremities of TF-specimens, Figure 9. FEM analysis and X-ray measurements have shown that the stress ratio $R(=\sigma_{\theta\theta}/\sigma_{zz})$ changes along z-axis from about one (perfect bi-axial stress state) in the centre to about zero (quasi uni-axial stress state) at the extremities of the specimens. As a result, the heat checking pattern changes and density of the secondary cracks (perpendicular to σ_{zz}) decreases, Figures 9 and 10 [22]. Under further TF-cycling, heat checking patterns subdivide and more "cells-type" patterns emerge at the extremity, Figure 10. The cyclic dependence of change of the heat-checking pattern seems indicate that the heat checking is a progressive and therefore a time and frequency dependent process. Using the variation of the σ_{zz} along z-direction, one can in addition determine a "threshold stress" (or more strictly a stress interval) for the transition of the uni-axial to bi-axial heat checking [22]. Crack propagation in mainly transgranular.

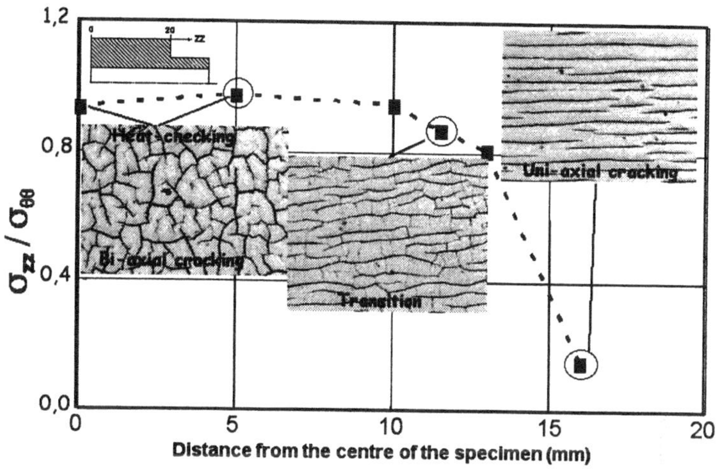

Fig. 10: Variation of the stress ratio (longitudinal/hoop) and the corresponding evolution of the heat checking pattern along the z-direction from the centre (bi-axial loading) to the extremity (uni-axial loading). A 7 mm wall thickness specimen.

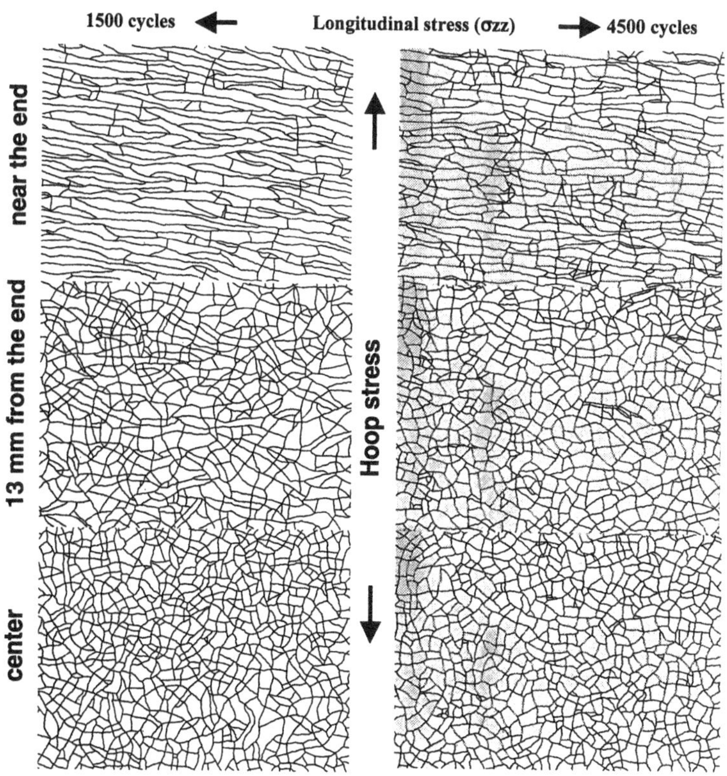

Fig. 11: Image analysis of the heat checking pattern after the specimen surface SEM micrographes and the evolution of the pattern as a function of the number of TF cycles (7mm wall thickness specimen).

SUMMARY

Thermal fatigue behaviour and damage of X38CrMoV5 (AISI H11) and 55NiCrMoV7 have been investigated for 47 HRC. A new thermal fatigue rig was developed. Tubular specimens with different wall thickness were employed. Thermal fatigue softening was shown by the microhardness and X-ray measurements. The coupled cyclic "temperature-time" tempering and « plastic strain-number of cycle » softening govern the TF softening. Different heat checking patterns (cell-type and rectangular-type) are observed depending on the local longitudinal/hoop stress ratio ($R=\sigma_{\theta\theta}/\sigma_{zz}$) and the number of thermal cycles.

ACKNOWLEDGEMENTS

Authors gratefully acknowledge the French Research Action-II on Forging for its support. Contribution of Sabine Caudron for the isothermal oxidation is very much appreciated. Technical assistances of Serge Tovar and Fabrice Rossi are highly acknowledged.

REFERENCES

1. Rézaï-Aria, F. (2000). In : *Voie de progrès dans l'Industrie de la Forge à Chaud, de la Forge par Extrusion et de la Frappe à Froid*, Tome XVII, pp. I-II-1- I-II-5, Cercle d'Etudes des Métaux (Eds.).
2. Bournicon, C. (1991). *Traitement Thermique*, **246**, 70.
3. Glenny, E. (1967). In: *Thermal and High Strain Fatigue*, pp. 346, The Institute of Metals and the Iron and Steel Institute (Eds.), London.
4. Rézaï-Aria, F., François, M. and Rémy, L. (1998). *Fatigue Fract. Engng. Mater. Struct.* **11**, No. 5, 277.
5. Meyer-Olbersleben, F., Goldschmidt D. and Rézaï-Aria F. (1992). In: *The Seventh International Symposium on Superalloys*, pp. 785-794, S. D. Antolovich, R. W. Stusrud, R. A. MacKay, D. L. Anton, T. Khan, R. D Kissinger, D. L. Klarstrom (Eds).
6. Fissolo, A., Marini, B., Berrada, A., Nais, G. and Wident, P. (1995). In: *International Symposium on: Fatigue under Thermal and Mechanical Loading - Mechanisms, Mechanics and Modelling*, pp. 67-77, J. Bressers and L. Rémy (Eds.), Petten, The Netherlands.
7. Northcott, L., and Baron, H. G. (1956). *Journal of the Iron and steel institute*, December, 385.
8. Rousseau, D., Riegert, J. P., Seraphin, L. and Tricot, R. (1977). In: *Colloque sur "Les Aciers à Outils pour travail à chaud"*, pp. 293-321, Cercle des Métaux (Eds).
9. Seux, M. Saint-Ignan, J. C. and Leveque, R. (1988). *Mechanical working and steel processing*, 47.
10. Skelton, R. P. (1983). *Fatigue at High Temperatures*, pp. 1–62, Skelton, R. P. (Eds), Elsevier Science Publishers Ltd., New York.
11. Jean, S., Arcens, J.P., Tovar, S. and Rézaï-Aria, F. (1999). *Matériaux & Techniques*, 23.
12. Jean. S. (1999). PhD Thesis, Ecole des Mines d'Albi-Carmaux, France.
13. Samrout, H. and Abdi, R. El. (1998). *Int. Fatigue* **Vol. 20**, No 8., 555.
14. Kircher, D. (1999). PhD Thesis, Université de Reims Champagne-Ardenne, France.
15. Oudin, A. (2001). On going PhD Thesis work, Ecole des Mines d'Albi-Carmaux, France.
16. Delagnes, D., Rézaï-Aria, F., Levaillant, C. and Grellier, A. (1999). *Matériaux & Techniques*, **N°1-2**, 39.
17. Chai, H. F. and Laird, C. (1993). *Materials Science and Engineering*, 159.
18. Neu, R. W. and Sehitoglu, H. (1998). *Metallurgical Transactions A,* **Vol. 20A**, *Sept., 1755.*

19. Mebarki, N., Delagnes, D., Lamesle, P. and Levaillant, C. (2001). Unpublished results, Ecoles des Mines d'Albie-Carmaux, Farnce.
20. Doege, E., Andreis, G., Dohmann, J. and Walter, S. (1999). In: *Symposium on Neuere Entwicklunggen in der Massivumformung*, pp. 65-88, Siegert, K. (Eds.).
21. Iwanaga, S. (1997). In : *Transactions of the 19th International Die Casting Congress & Exposition, Minneapolis*, pp. 213-219, North American Die Casting Association (Eds).
22. Lamesle, P., Oudin, A., Le Roux, S. and Rézaï-Aria, F. (2001). Unpublished results, Ecole des Mines d'Abli-Carmaux, France.

Temperature-Fatigue Interaction
L. Rémy and J. Petit (Eds.)

THERMOMECHANICAL FATIGUE BEHAVIOUR AND LIFE ASSESSMENT OF HOT WORK TOOL STEELS

A. OUDIN, P. LAMESLE, L. PENAZZI, S. LE ROUX, and F. RÉZAÏ-ARIA
Tool Surface Assessment Unit, Research Centre on Forming Tools, Materials and Processes
Ecole des Mines d'Albi-Carmaux, F-81000 Albi, France

ABSTRACT

The surface of the hot work tools is damaged under coupled non-isothermal fatigue, wear or erosion and reactive environment (oxidation or corrosion). A thermomechanical fatigue (TMF) experiment using tubular specimens is developed. Tests are carried out under out-of-phase TMF cycle with strain ratio R = $-\infty$. The behaviour, the damage and life of two tempered martensitic steels X38CrMoV5 and 55NiCrMoV7 (47 HRC) are assessed. The effect of the maximum temperature (T_{max}) of thermal cycle is examined. Softening is observed for both steels. For a given mechanical strain amplitude ($\Delta\varepsilon_m$), a drastic dependence on T_{max} is observed. X38CrMoV5 has a better TMF life. The oxide-scale cracking and spalling are observed. TMF enhances the surface oxide-scales thicknesses. A phenomenological power law model, $\Delta\varepsilon_m = K(\mathbf{T}_{max}) \cdot N_f^{\alpha}$, is proposed to predict the life under non-isothermal fatigue solicitations. The variation of constant K with T_{max} is expressed by an Arrhenius type equation and the exponent α is taken constant.

KEYWORDS

Thermomechanical fatigue, thermal fatigue, crack initiation, crack propagation, fatigue-oxidation interaction, steel, softening, life prediction,

INTRODUCTION

The surface of hot working tools (HWTS) is damaged under coupled thermomechanically and thermochemically solicitations [1]. The loading and damage are complex combinations of non-isothermal fatigue [1-5], wear [6] (in solid-solid contact, or erosion in solid-liquid contact) coupled with a reactive environment, such as oxidation or corrosion (when HWTS is in intimate contact with a liquid), Figure 1. The well known heat checking (multi-axial cracking) [2-5], is the basic surface damage feature under multi-axial TMF. The severity of loading and damage of HWTS depends on forming techniques (forging, die or gravity casting, rolling, etc.) as well as forming parameters (lower and upper temperatures of thermal cycle, heating and cooling rates, working efforts, etc). In addition, the thermomechanical solicitations differ from one point to another in a HWTS. As hot working tools are "thermal systems" [7], following an early transient situation a thermomechanical "quasi steady-state regime" is established.

Alike many components working at high temperatures, the behaviour and damage as well as the life of HWTS may be assessed through isothermal fatigue [8] or non-isothermal fatigue experiments (thermal fatigue (TF) [9] and TMF).

This contribution deals with an investigation on TMF behaviour and damage of tempered martensitic steels. A TMF experiment is developed using tubular specimens, induction heating. The effect of the upper temperature of thermal cycle is reported. A phenomenological TMF life prediction model is proposed. These investigations have been carried-out with industrial support and so data are qualitatively presented hereafter in order to meet the confidentiality requirements.

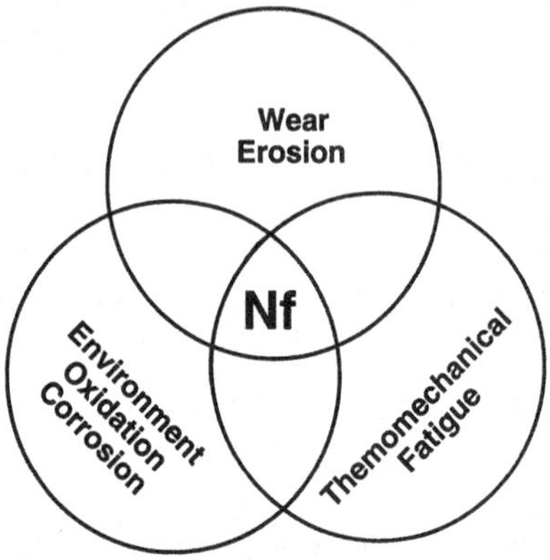

Fig. 1: Coupled life limiting thermomechanical and thermochemical solicitations of the HWTS [1].

EXPERIMENTAL PROCEDURES

Steels

Both X38CrMoV5 (AISI H11) and 55NiCrMoV7 are investigated. Chemical compositions of both steels are given in Table 1. Steels have been annealed, quenched and heat-treated to achieve the tempered martensitic microstructures with an initial hardness of 47 HRC [9].

Table 1. Chemical composition of steels (major elements in weight %)

Steel	C	Cr	Mn	V	Ni	Mo	V	Si	Fe
55NiCrMoV7 *(55NCDV7)*	0.56	1.10	0.50	0.47	1.70	0.50	0.10	0.20	bal.
X38CrMoV5 *(Z38CDV5)*	0.38	5.05	0.49	0.47	0.20	1.25	0.47	0.92	bal.

Specimens are tubular with internal and external diameters of 9 mm and 11 mm respectively, Figure 2. This specimen introduced by L. Rémy et al. is used for TMF assessments of nickel based superalloys [11-12]. The specimen geometry, with a different gripping length, is used in our laboratory for TMF investigations on 9%Cr steels [13] and HWTS [13-14]. Specimens are pre-machined and then heat-treated to achieve the final microstructure as well as the desired hardness before to be re-machined to the final dimensions. Both external and internal surfaces are polished parallel to the loading axis down to Ra=0.015 μm roughness.

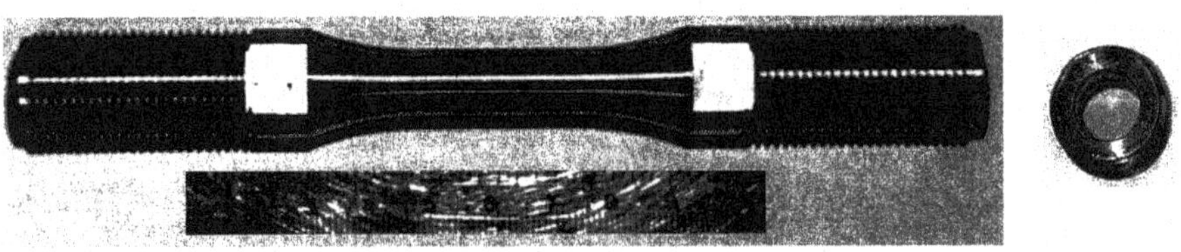

Fig. 2: TMF tubular specimen.

Specimens are induction heated and naturally cooled. The temperature rate is about 4°C/s. Tests are carried-out on a conventional push/pull hydraulic fatigue machine. First, the modulus of elasticity is measured at different temperatures. Then few thermal cycles are performed for thermal stabilisation of specimen, fixtures and extensometer rods. Mechanical strain at T_{min} is nil ($\varepsilon_{m(max)} = 0$) and it is minimum, $\varepsilon_{m(min)}$, at T_{max}, Figure 3. The strain ratio ($R_\varepsilon = \varepsilon_{m(min)}/\varepsilon_{m(max)}$) is $-\infty$. Following thermal stabilisation, the fatigue machine simultaneously imposes two independent temperature-time and mechanical strain-time profiles to the specimen via a software.

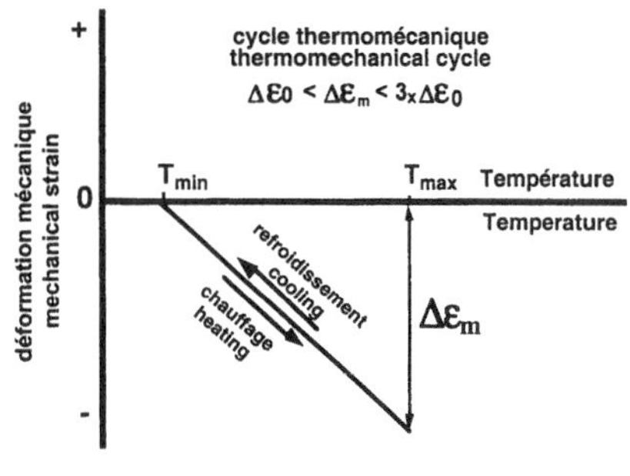

Fig 3: A typical out-of-phase TMF cycle showing the variation of the mechanical strain (ε_m) vs. temperature. $\Delta\varepsilon_0$ is the minimum strain range investigated.

RESULTS AND DISCUSSIONS

Th effect of T_{max} and T_{min} [13-14] was assessed (X38CrMoV5). T_{max} was maintained constant, while T_{min} was changed. Figure 4 shows the drastic effect of T_{max} on TMF life. This effect is a consequence of the inelastic strain enhancement, Figure 4, and the oxidation-TMF interactions. As can be seen, for T_{max} lower than the Primary Tempering Temperature (PTT, about 550°C) a quasi-linear stress relaxation is observed. For T_{max} higher than PTT, the stress amplitude relaxation may not achieve a "stationary" regime. In fact, as the specimen's wall thickness is 1mm and the crack growth rate is high enough, the specimen can break prior to achieve a stationary regime. Such behaviour is not so observed when superalloys are investigated.

Detailed investigation has shown that under conditions reported here, there is an accommodation period where the minimum and maximum stresses as well as the mean stresses change continuously [14]. It was observed that T_{min} [13-14] has less effect on the stress amplitude relaxation (softening) as compared to T_{max}.

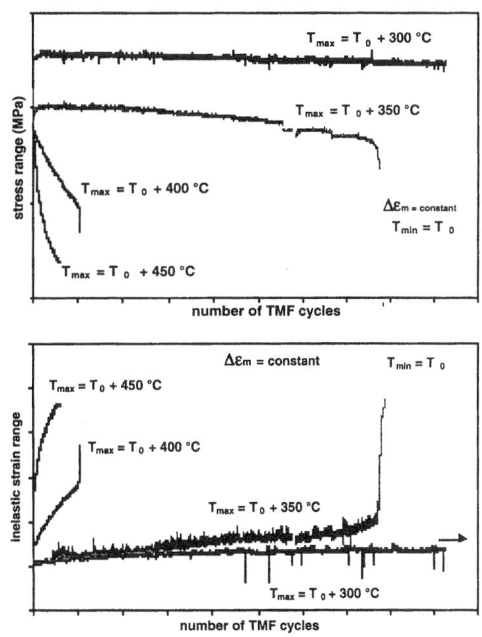

Fig. 4: Effect of Tmax on the variation of the stress range and the inelastic strain range vs. number of TMF. T_0 is the minimum temperature of the thermal cycle.

In both steels, a large inelastic strain is obtained in the first TMF reversal [13-14] as what can occur on the surface of HWTS in the beginning of the hot metal forming operations. In fact, during heating-up, the yield stress of steels decreases, and when $\Delta\varepsilon_m$ is large enough, the inelastic strain can occur. The amplitude of this inelastic strain depends on T_{max} and the amplitude of TMF mechanical strain.

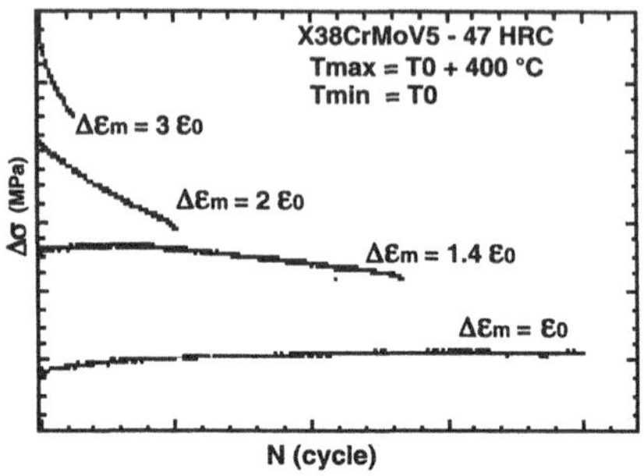

Fig. 5: Effect of mechanical strain amplitude on evolution of the stress amplitude vs. N ($T_{max}=T_{min}+400°C$).

The effect of the mechanical strain amplitude on stress amplitude relaxation is given in Figure 5. It seems that regardless the test conditions examined ($\Delta\varepsilon_m$, T_{max} and T_{min}), each steel reaches more or less rapidly an asymptotic tensile mean stress [14]. This drastic dependence on T_{max} can be explained by the fact that at high temperatures, inelastic deformation and microstructure evolutions (dislocation motion, annihilation and rearrangement, carbide coarsening, etc.) are thermomechanically activated.

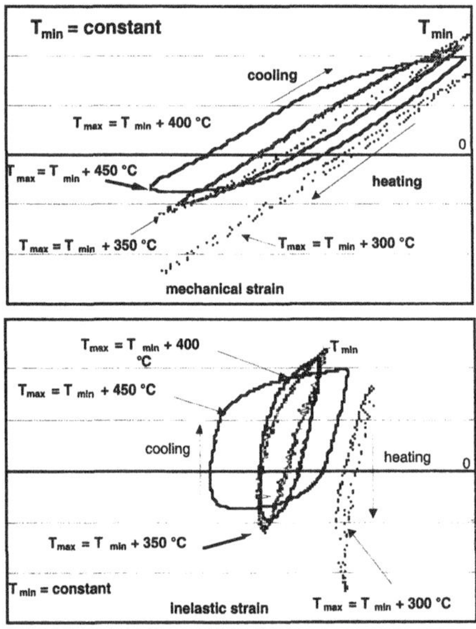

Fig. 6: Stress-mechanical strain and stress-inelastic strain hysteresis loops (X38CrMoV5). $\Delta\varepsilon_m$ and T_{min} are constant.

The effect of T_{max} on the stress-mechanical strain (σ–ε_m) and the stress-inelastic strain (σ–ε_{in}) hysteresis loops at half-life ($N_f/2$) is respectively reported in Figure 6 for X38CrMoV5. For T_{max} higher than the Second Tempering Temperature (STT, about 605°C to 620°C), the inelastic strain increases drastically.

This inelastic amplitude enhancement is important for HWTS. In fact, in many applications the temperature at working surface of HWTS could be higher than STT and that results in inelastic enhancement. In addition the hysteresis loops at $N_f/2$ should be compared with care, since for certain TMF condition the half-life corresponds in fact to the half of the transient regime cycles, while under other condition, it constitutes the half of the stationary regime.

TMF life, N_f, is reported as a function of the mechanical strain amplitude ($\Delta\varepsilon = \varepsilon_{m(max)} - \varepsilon_{m(min)}$) in Figure 7, for various TMF conditions. As can be observed, TMF life depends strongly on T_{max}. Other criteria (inelastic strain range, dissipated energy per cycle, etc.) are reported in [14].

TMF life of both steels is compared in Figure 7. X38CrMoV5 presents a slightly better TMF resistance. However, one can not claim at present whether the higher resistance of X38CrMoV5 is due to higher crack propagation resistance or better crack initiation resistance [14].

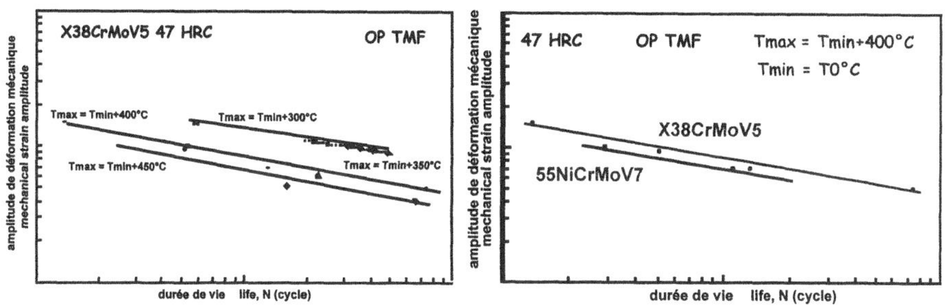

Fig. 7: TMF life as a function of the mechanical strain amplitude (left) and the comparison between two steels (right). T_0 is the minimum temperature in all TMF cycles examined.

Damage mechanisms

SEM observations of the external surface of specimens, Figure 8, have revealed that even under macroscopic uni-axial testing, the surface is damaged by a microscopic complex and "multi-axial" loading (Figure 8c). The oxide-scale cracking perpendicular and parallel as well as making a certain angle with respect to the main TMF loading axis are observed [14]. The oxide-scale cracking perpendicular to the TMF axis is the general feature and the principal cause of the external crack initiation. The oxide-scale spalling is also observed. The spalling contributes to general degradation of steels by successive re-oxidation of the surface. Gradually, the general oxidation transforms to a V-type oxidation, which progresses inward the sub-surface and forms crack initiation sites. This mechanism is comparable to the thermal fatigue crack initiation [9].

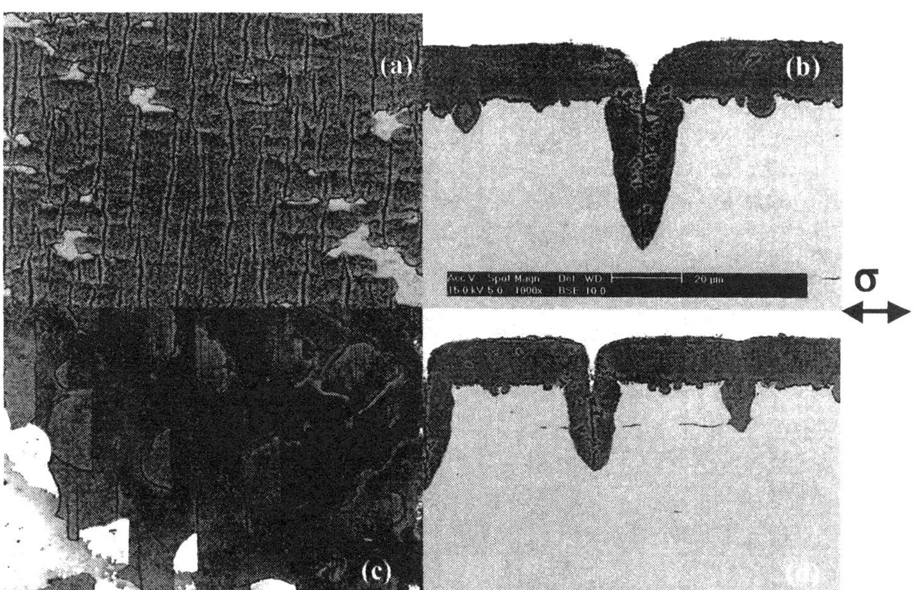

Fig. 8: General features of the external crack initiation, oxide-scale spalling and cracking in depth. Note in (a) and (c) the cracking along the loading axis.

In-situ and continuous macroscopic observations on the external surface of some X38CrMoV5 specimens have revealed the formation of slip bands when the strain amplitude is large enough [14]. These bands may also contribute to the localisation of the oxidation and V-type crack initiation. Observations on longitudinal section of X38CrMoV5 specimens have revealed that two oxide-scale layers are formed. One oxide-scale layer is rich in Cr (in intimate contact with the base steel), and the other is a lower Cr-content oxide-scale (in contact with air). Cr-rich oxide-scale layer is not formed on 55NiCrMoV7 specimens as can be expected from its lower Cr-content. The surface and the sub-surface of TMF specimens or tools can be seen as a multi-layer composite, consisting of one or two oxide-scale layers, and one layer of the substrate which continuously changes its mechanical properties and strength. The oxide-scale thickness measurements have shown that TF and TMF accelerate the oxidation [15].

SEM investigations have revealed a ductile crack propagation aspect in X38CrMoV5, since the fatigue striations are present on the fracture surface of TMF specimens. The fracture surface of 55NiCrMoV7 specimens is covered by an oxide-layer, making difficult to reveal the fatigue striations. A strain based intensity factor was used to rationalise the crack propagation rate by accounting the mean fatigue striation distance per cycle [14].

Life prediction

Based on TMF life results and taking into account that the actual commercial forming simulation softwares, like Forge 2™, consider basically the tools as thermoelastic or thermoelastoplastic, a phenomenological power law predictive TMF life model based on $\Delta\varepsilon_m$ and T_{max} was proposed [13]:

$$\Delta\varepsilon_m = K(\mathbf{T}_{max}) \cdot N_f^{\alpha} \qquad (1)$$

It was observed that α, can be considered as constant for all test conditions examined while $K(T_{max})$ is temperature dependent [14]. As it was observed that TMF life is very much dependant on T_{max}, the variation of this constant with T_{max} was set with an Arrhenius type equation. In addition, transient regimes which occur during early "heating-up" of tools are not taken into the consideration.

The model was used to predict in one hand the TF life of some of specimens tested in our laboratory and on the other hand, to predict the critical regions of an industrial tool in terms of the number of TMF cycles [13]. Forge 2™, software was used to simulate the forging operations. The tool was considered to behave as thermoelastic. T_{max} and mechanical strain amplitudes ($\Delta\varepsilon_m$) of each element were extracted from the forging simulation files and they were then used to predict the fatigue life by equation (1) [13]. For TF specimen, themoelastoplastic analysis was carried-out, using ABAQUS [2, 9]. Figure 9 shows the experimental and the predicted life of TMF as well as TF specimens. Taking into account the

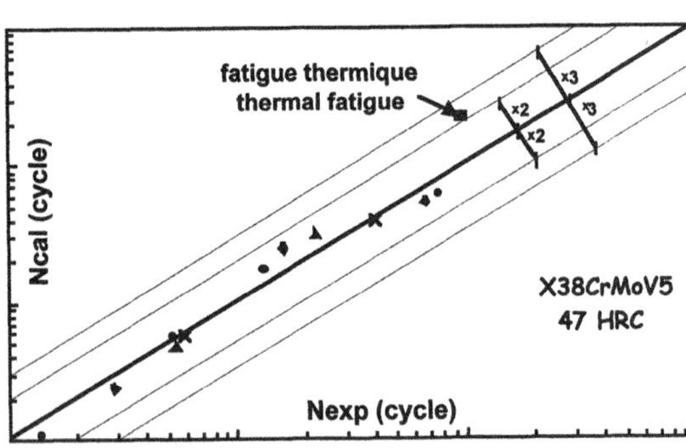

Fig. 9: Comparison between calculated and experimental TMF life.

limited conditions examined, and considering different approximations made, a good estimation of TMF life is achieved (factor 2 to 3).

SUMMARY

A thermomechanical fatigue experiment using tubular specimens is developed. Two tempered martensitic steels, X38CrMoV5 (AISI H11) and 55NiCrMoV7 are assessed under out-of-phase TMF. Softening is observed in both steels. Mean tensile stresses are developed in each steel. SEM observations reveal that the external cracks initiate under cyclic TMF-oxidation interactions. Based on TMF life curves, a phenomenological life predictive model is proposed. The model could predict some laboratory thermal fatigue life results within a factor two to three.

ACKNOWLEDGEMENTS

Authors gratefully acknowledge the French Research Action-II on Forging for its support. Technical assistance of Serge Tovar is acknowledged. Brice Miquel is kindly thanked for performing thermal fatigue experiences.

REFERENCES

1. Rézaï-Aria, F. (2000). In: *Voie de progrès dans l'Industrie de la Forge à Chaud, de la Forge par Extrusion et de la Frappe à Froid*, Tome XVII, pp. I-II-1- I-II-5, Cercle d'Etudes des Métaux (Eds.).
2. Jean, S., Arcens, J.P., Tovar, S. and Rézaï-Aria, F. (1999). *Matériaux & Techniques*, 23.
3. Bournicon, C. (1991). *Traitement Thermique*, **246**, 70.
4. Lévêque, M, (1989). *Traitement thermique*, **no. 231**, 47.
5. Rousseau, D., Riegert, J. P., Seraphin, L. and Tricot, R. (1977). In: *Colloque sur "Les Aciers à Outils pour travail à chaud"*, pp. 293-321, Cercle des Métaux (Eds).
6. Felder, E. (1984). *Revue de métallurgie*, 931.
7. Levaillant, Ch. (1998). In: Colloque sur *"Les Aciers pour Moules et Outils"*, pp. 1.2-1.927-28, Cercle des Mtéaux (Eds.).
8. Delagnes, D., Rézaï-Aria, F., Levaillant, C. and Grellier, A. (1999). *Matériaux & Techniques*, **N°1-2**, 39.
9. Miquel, B., Jean, S., Lamesle, P., LeRoux, S. and Rézaï-Aria, F. (2001). In this conference.
10. Malpertu, J. L. and Rémy, L. (1990). *Metallurgical Transactions A*, **Vol. 21A**, 389.
11. Engler-Pinto, C. C. Jr. and F. Rézaï-Aria, F. (2000). In: *Thermo-mechanical fatigue behavior of materials, third volume, ASTM STP 1371*, pp. 150-164, Sehitoglu, H. and Maier, H. (Eds.), American Society for Testing and Materials, West Conshohocken, PA.
12. Filacchioni, G., Petersen, C., Rézaï-Aria, F. and Timm, J. (2000). In: *Thermo-mecanical fatigue behaviour of materials, ASTM STP 1371*, pp. 239-256, Sehitoglu, H. and Maier, H. (Eds.), American Society for Testing and Materials, West Conshohocken, PA.
13. Oudin, A., Penazzi, L. and Rézaï-Aria, F. (2001). *Matériaux & Techniques*, **N° Hors Série**, 67.
14. Oudin, A. (2001). On going PhD Thesis work, Ecole des Mines d'Albi-Carmaux, France.
15. Lamesle, P., Oudin, A., Le Roux, S. and Rézaï-Aria, F. (2001). Unpublished results, Ecole des Mines d'Abli-Carmaux, France.

A PHYSICAL-BASE MODEL FOR LIFE PREDICTION OF SINGLE CRYSTAL TURBINE BLADES UNDER CREEP-FATIGUE LOADING AND THERMAL TRANSIENT CONDITIONS

A.KOSTER, A.M.ALAM and L.RÉMY

Centre des Matériaux, École des Mines de Paris, CNRS UMR 7633, BP 87, 91003 Évry Cedex, France

ABSTRACT

The damage estimation model developed herein can be used to predict service life under creep-fatigue loading in both isothermal (LCF) and variable temperature (TMF) conditions. This model is based upon a careful identification of basic physical mechanisms taking part in the damage process of a unit microstructural element. The damage is, in fact, considered to be the growth of micro cracks originating from initial material defects. The distribution of these defects has been determined through microscopic observations and their propagation due to fatigue and creep loading is simultaneously calculated in this model. A large near surface pore may develop into a surface crack and its propagation may be accelerated by the weakening of material through oxidation embrittlement. This model has been successful in predicting creep-fatigue life of CMSX4 test specimens at 950°C, and that of tubular specimens under (450-950°C) TMF conditions. It has also been applied to a thermal fatigue (TF), wedge-shaped structural element undergoing thermal transient loading; early crack growth rate has been satisfactorily predicted by this model.

KEYWORDS

Life modelling, Thermomechanical fatigue, Thermal fatigue, Creep, Damage, Single crystal superalloys, CMSX4.

INTRODUCTION

Blades in land based gas turbines used for power generation are now made of single crystal superalloys. These alloys consist of a high volume fraction of the ordered $L1_2$ Ni_3Al phase in the form of regular cuboidal precipitates, γ' cuboids, separated by thin channels of γ matrix. The present work focuses on the situation where life is limited by creep-fatigue loading or thermal mechanical loading (mostly strain-controlled situations). In this case, fracture results from the growth of microcracks from the surface or from internal casting pores.

The prediction of life time is carried out in two steps. First stress-strain hysteresis loops are determined. Secondly, the damage model is used in a post-processing program to evaluate life time. In case of LCF or TMF (volume element) test pieces, these stress-strain loops are simply obtained through test data acquisition.

However, in the case of TF wedge-shaped structural element test specimen, finite element calculations are necessary to determine temperature field and hence the required stress-strain loop at each point of the specimen. The latter requires a constitutive deformation model capable of relating stresses with strains and vice versa. The deformation model used here is a crystallographic model developed by Cailletaud and coworkers [1,2] with phenomenological constitutive equations relating shear strain rate components on every glide system to shear stress components. Two main slip systems are considered in CMSX4 single crystal alloy, i.e. octahedral {111} <110> and cubic {001} <110> slip systems. This deformation model has previously been identified for CMSX4 with the help of cyclic incremental stress strain tests at various temperatures (700, 800, 950 and 1100°C). It was further validated on thermomechanical fatigue tests on tubular volume element specimens.

Prediction of life time under high temperature fatigue, creep-fatigue and thermal transients has been the purpose of numerous life time models [2-4]. In superalloys the importance of oxidation has been recognized [5-7] and has been explicitly taken into account in some models developed for polycrystals [7-9]. Application of these concepts to single crystal superalloys needs numerous adaptations, which are summarized in this work.

EXPERIMENTAL RESULTS

All experiments with which the model predictions shall be compared in the scope of this paper were carried out under the Brite-Euram program # BRPR CT96-0224. They are as follows:

1. Low cycle fatigue tests at a constant temperature of 950°C. Figure 1 presents LCF test results on a graph of applied cyclic strain amplitude versus the number of cycles to failure. We have some high frequency tests without the hold time and a few low frequency tests with hold time ranging from 5 minutes to 30 minutes. One can notice a considerable amount of scatter in these test data.

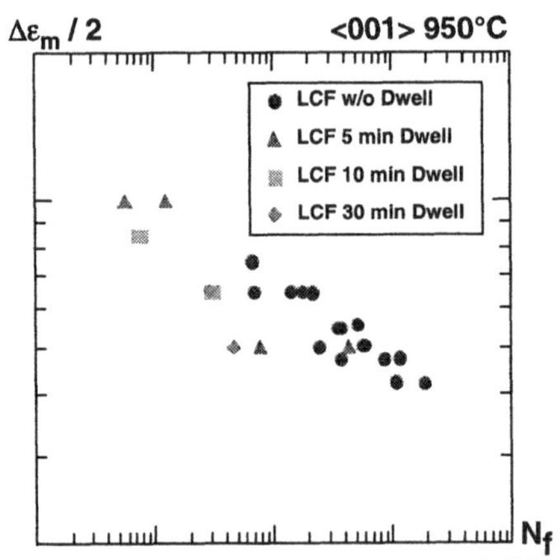

Figure 1. Isothermal low cycle fatigue (LCF) test results for CMSX4 <001> at 950°C.

2. Thermomechanical fatigue tests with 5 minutes hold time were carried out within a temperature range of 450 to 950°C. These tests are of two types: **in-phase** tests, where the strain rises in time while the temperature increases and **out-of-phase** tests in which strain and temperature are varied in opposite order. Figure 2(a) presents the in-phase and out-of-phase cycles used in this study and figure 2(b) shows the test results.

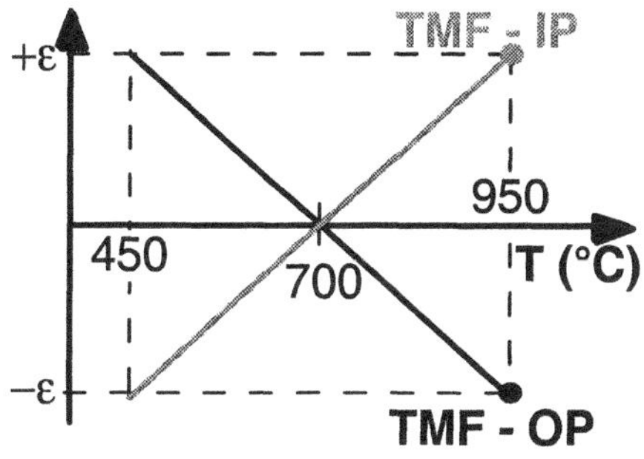

Figure 2(a). Thermomechanical fatigue cycles with a dwell of 5 minutes at the highest temperature of the cycle.

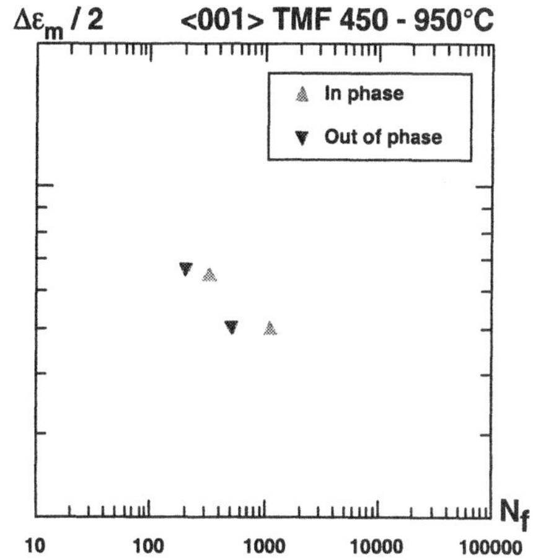

Figure 2(b). Thermomechanical fatigue test results on a graph of mechanical strain amplitude $\Delta\varepsilon_m/2$, as a function of N_f.

3. A thermal fatigue test carried out on a wedge-shaped specimen. For this test, the main crack growth rate has been measured by macroscopic observations of specimen when test was interrupted at regular intervals.

All test specimens considered in this paper are oriented in <001> direction parallel to the loading axis which corresponds to the major thermal-mechanical stress axis in blades during operation.

DAMAGE MODELLING

General Framework

This model considers damage as the growth of microcracks originating from casting pores. The size distribution of these pores is introduced in the model in a semi-probabilistic way and it is based on tedious microscopic observations [3]. A major surface crack, **a,** results from the failure of a microstructural element λ (figure 3). The initiation time for this surface crack depends upon the probability of the presence of a pore near the surface. Its propagation is assisted by oxidation, which is introduced as the weakening of material due to embrittlement. While calculating damage of a microstructural element λ, the volume effect on its critical strength τ_c is introduced with the help of a Weibull type defect distribution. Besides surface crack, other internal cracks propagate under creep-fatigue condition without any interaction with ambient environment. The global damage results from the propagation of all internal as well surface cracks and can be decomposed into fatigue and creep damages.

$$D = D_{fatigue} + D_{creep}$$

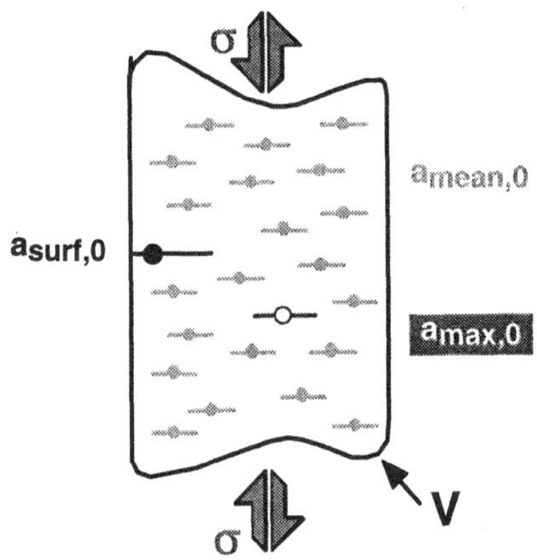

Figures 3 (a). Schematic diagram of the physical model showing a volume element with internal and surface pores

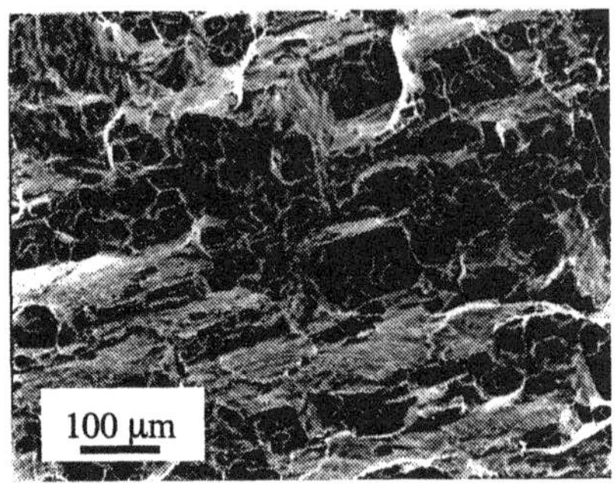

Figure 3 (b). Micrograph of the fracture surface of a specimen, tested in LCF with tension dwell at a constant 950°C temperature.

A fatigue damage law in which we can introduce the interaction with oxidation and creep damage, should describe the failure of microstructural element λ. For the fatigue damage equation we consider either Basquin's relation:

$$\Delta\tilde{\tau} \cdot N^b(\lambda) = 2\tau_c$$

or Tomkins [10] type relation, adapted for single crystal case, for crack growth under extended plasticity conditions:

$$\frac{da}{dN} = \Delta\varepsilon_p \cdot \left(\frac{1}{\cos\frac{\pi}{2}\frac{\Delta\tilde{\tau}}{2\cdot\tau_c}} - 1 \right) \cdot a$$

where: τ_c is the critical shear strength of the microstructural element λ, $\Delta\varepsilon_p$ is the plastic deformation range and $\Delta\tilde{\tau}$ is the effective shear stress range defined as follows:

$$\Delta\tilde{\tau} = \frac{\Delta\tau}{1-D}$$

where D is the damage resulting from crack growth $D = f\left[\frac{a}{w}\right]$. Interaction with creep is introduced in the effective shear stress as:

$$\Delta\tilde{\tau} = \frac{\Delta\tau}{(1-D_{creep})\cdot(1-f\left[\frac{a}{w}\right])}$$

The creep damage is estimated using the classical Rabotnov law [4, 11]:

$$dD_{creep} = (1 - D_{creep})^{-k} \cdot \left(\frac{\tau}{A}\right)^r dt$$

This formulation is defined for isothermal loading conditions and is easily adapted for thermomechanical loading and thermal trasients.

Oxidation Interaction

In the model the interaction between oxidation and creep-fatigue damage is introduced by assigning a lower critical strength in the area embrittled by localized oxidation, figure 4(a). This procedure was previously described in some detail for polycrystalline superalloys [6, 8, 9], in cast or wrought forms. In other words, it's considered that an oxide spike of length l_{ox} induces an embrittled zone of depth l_f in λ which exhibits lower mechanical properties than a non-oxidized material. Embrittled zone is considered to be greater in size than the oxidized zone.

$$l_f = P \cdot l_{ox}$$

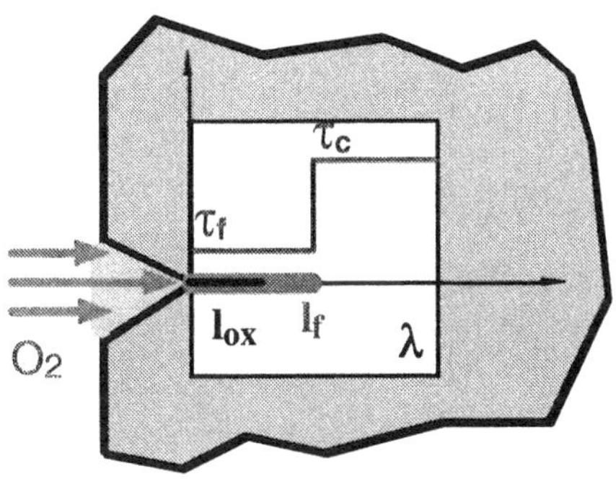

Figure 4(a). Schematic diagram explaining oxidation embrittlement in a unit microstructural element. The embrittled zone is considered to exhibit lower strength as compared to the non-oxidized material zone.

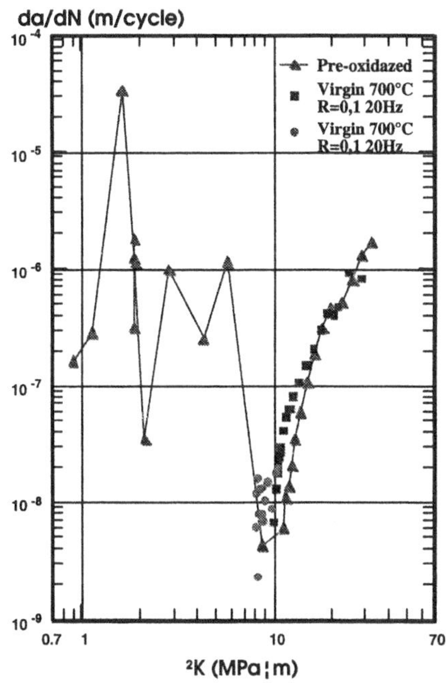

Figure 4(b). Comparison of crack growth rate data da/dN = f(ΔK) between non-oxidized and pre-oxidized CT specimens One of the specimens was pre-oxidized for 300 hours at 950°C before failure at room temperature

Oxidation kinetics and for that matter any oxidation-loading interactions need to be identified through specimen observations and measurements. The growth of an oxide spike is found to follow a power law of the following type:

$$dl_{ox}^{4} = \alpha^{4}(T).dt$$

where: $\alpha(T)$ is called the "oxidation constant" at a given temperature T [6, 7]. Any possible effect of the loading conditions is taken into account by introducing a corrective function dependent on the plastic deformation range:

$$\alpha = \alpha_0(T) \cdot f(\Delta\varepsilon_p)$$

The evolution of the oxidation constant $\alpha_0(T)$ with temperature is assumed to follow an Arrhenius law:

$$\alpha_0^4(T) = \alpha_{0,0}^4 \cdot e^{\left(\frac{-Q}{RT}\right)}$$

where: $\alpha_{0,0}$ and Q are two constants which need to be identified from experimental results.

Material Embrittlement due to Oxidation

In our model we consider that the depth of the embrittled zone is greater than that of an oxide spike. One way to estimate the size of the embrittled area is to perform crack growth tests on non-oxidized CT (compact tension) specimens and obtain a reference curve. Then we can compare this curve with the one obtained with a pre-oxidized specimen. This could be done as shown in the figure 4(b). In this figure we can notice that for a lower level of ΔK the crack growth rate for the oxidized specimen (300 hours at 950°C) is higher than that of a non-oxidized specimen; and after some crack growth both curves become identical. The crack length where crack growth rates are different, gives a direct access to the embrittled zone size. Such observations were carried out for different pre-oxidation times and the value of the factor **P** between l_f and l_{ox} was estimated from these results.

MODEL APPLICATIONS

Isothermal Low Cycle Fatigue with or without Dwell

Figure 5(a) presents a comparison between experimental and calculated results for isothermal low cycle fatigue at 950°C. Experimental and calculated results are in pretty good correlation i.e. less than a factor 2 for most of the points. The model is, therefore, capable of predicting lives for LCF tests with or without the dwell (5, 10 and 30 minutes). A few points are however out of this range; in part due to a considerable amount of scatter in the experimental results (figure 1).

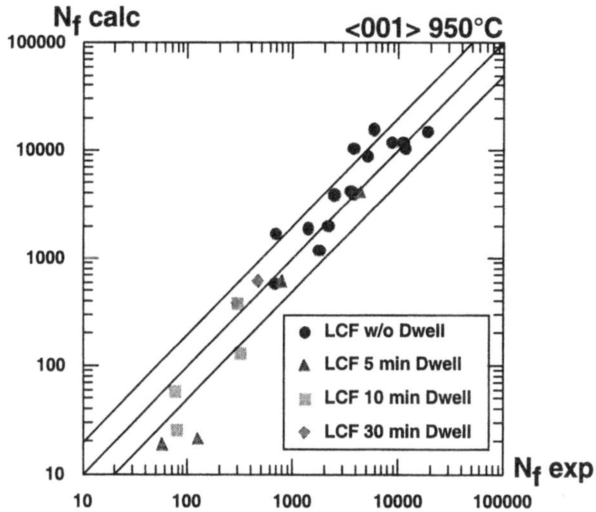

Figure 5 (a). Comparison of the experimental and calculated lives in case of 950°C LCF tests with or without the hold time.

Thermomechanical Fatigue with Dwell

Figure 5(b) presents a comparison between experimental (figure 2) and calculated results for variable temperature thermomechanical fatigue with 5 minutes dwell time.

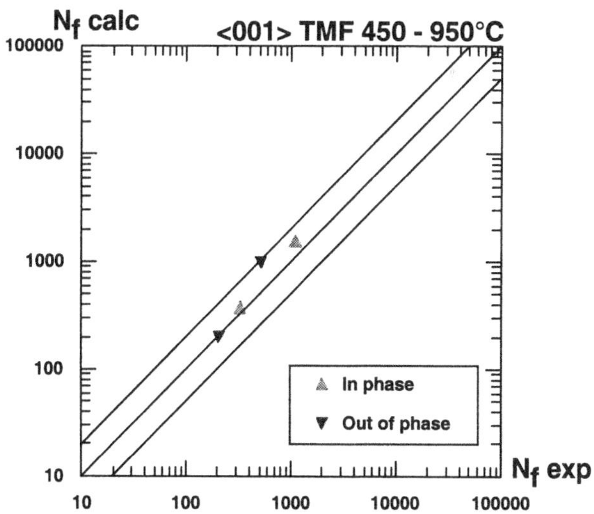

Figure 5 (b). Comparison of the experimental and calculated lives in case of TMF tests with hold time.

There are four tests with in-phase and out-of-phase cycles. We can see that all experimental results can be correctly predicted by these calculations.

Thermal Fatigue with Dwell

Thermal fatigue tests were performed on a wedge-shaped structural element (figure 6a) with its leading edge oriented parallel to <010> direction. The test rig and experimental procedure are presented in [12]. The thermal cycle applied to the leading edge is presented in the figure 6(b).

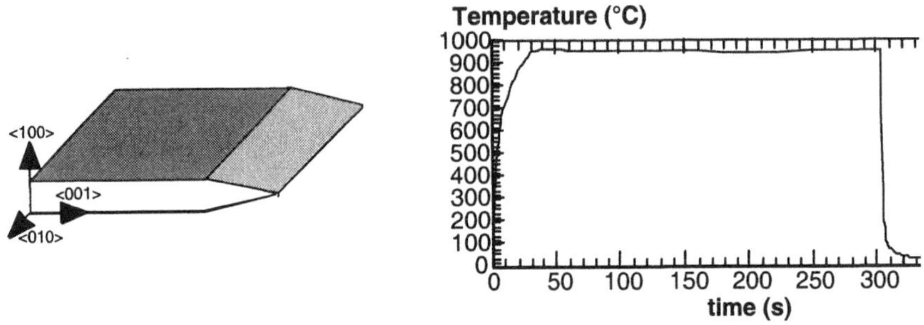

Figure 6. a) The structural element type of test specimen used for thermal fatigue test; b). Temperature cycle applied to the leading edge of the thermal fatigue test specimen.

It is obvious that due to this particular shape of the specimen, the temperature field on the rest of the specimen needs to be calculated. The same is done with the help of finite element calculations of heat transfer between the furnace and the test specimen. Once the temperature field is known, resulting deformations and stresses are calculated. For this purpose we use a crystallographic deformation model developed by Cailletaud and coworkers [1, 2]. The deformation model was, in first place, validated on volume element TMF test specimens (figure 7 a and b), and was found to give satisfactory results. The same model is used here in order to calculate stress and strain cycles on every point of the wedge-shaped specimen.

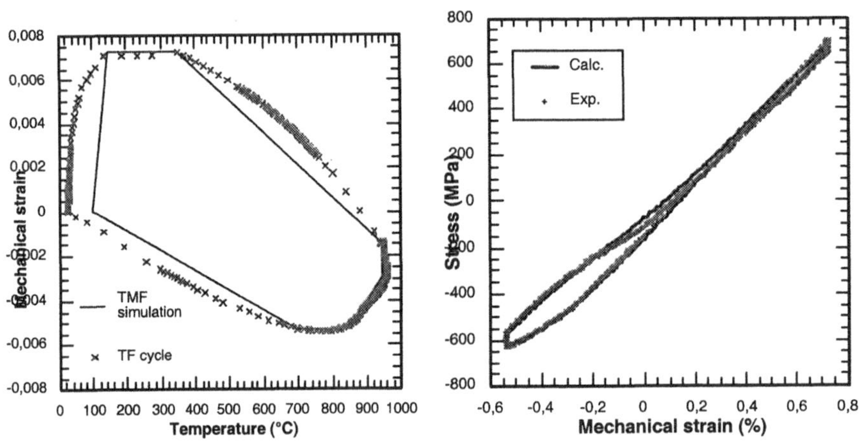

Figure 7. Diagrams showing comparisons of the experimental ε_m-T and σ-ε_m loops with the ones calculated through deformation model used in this study.

The application of the damage model on TF specimen is carried out in order to calculate the early crack growth rate only, since as long as the crack depth is small with respect to the specimen size, the effect of redistribution of stress due to its propagation can be neglected. Appropriate calculations with a propagating long crack would require recalculation of temperature, stress and strain fields at every crack increment. For this purpose we define four zones starting from the leading edge, in which the temperature and other load cycles can be assumed to be uniform. That is how we calculate lives of four different zones each one undergoing a different temperature and stress cycle. The main crack is made to advance with successive failure of these zones. In this way the calculations were carried out for a maximum crack length of 1.4 mm only.

Figure 8 presents, on a graph, the measured and experimental crack length **a** as a function of the number of thermal fatigue cycle **N**. We can notice that we have a fairly good correlation between the measured and experimental crack growth rates up to a crack length of 1 mm. The effect of redistribution of stresses starts to be felt beyond this length, and our stress calculations, made with an intact structural element, no longer remain adequate.

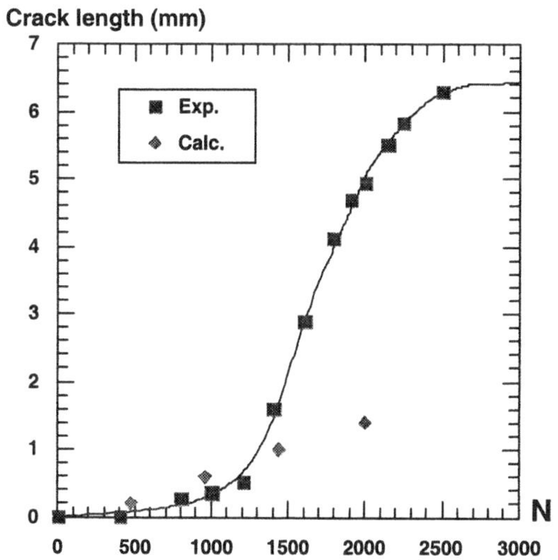

Figure 8. Comparison of the experimental and calculated crack growth results.

CONCLUSIONS

A damage model has been developed for service life predictions of single crystal turbine blades, operating under thermomechanical creep-fatigue conditions.

The damage is defined as the propagation of microcracks originating at casting defects. The distribution of such pores is taken into account in a semi-probabilistic manner.

Weakening of material due to localized oxidation embrittlement has been observed in experiments and the same is introduced in the model considering the oxidation-creep-fatigue interactions.
The model gives satisfactory life predictions for CMSX4 isothermal LCF tests with or without dwell and for variable temperature TMF tests with dwell period.

A convenient adaptation of this damage model has been made for the case of thermal transient test on a wedge-shaped structural element type of specimen. The calculated crack growth is in fairly good correlation with the experimental one, at least up to a crack length of 1 mm.

Acknowledgment

This work was funded by Brite-Euram program # BRPR CT96-0224.

REFERENCES

1. Méric L., Poubanne P., and Cailletaud G. (1991), *J. of Eng. Mat. Technol.*, vol. 113, pp. 162-170 and pp. 171-182.
2. Rémy L., and Skelton R.P., (1992), *High Temperature Structural Design, ESIS 12* (edited by M. Larsson), Mechanical Engineering Publications, London, pp.283-315.
3. Probst-Hein M., Eggeler G., (2001), *Predictive microstructural assessment and micromechanical modelling of deformation and damage accumulation in single crystal gas turbine blading* (MICROMOD-SX) Final report for the Brite-EuRam III, project.
4. Lemaitre J., Chaboche J.L., (1985), *Mécanique des matériaux solides*, Dunod, Paris.
5. Coffin L., (1973) *Fatigue at elevated temperatures*, ASTM STP 520, Philadelphia, p.5—34.
6. Reuchet J., and Rémy L., (1983), *Metall. Trans.* 14A, pp. 141-149.
7. Rémy, L. (1993), *Behaviour of defects at high temperatures*, ESIS 15, R.A. Ainsworth and R.P. Skelton (Eds), Mechanical Engineering Publications, London, pp. 167-187.
8. Rémy L., Koster A., Chataigner E., and Bickard A., (2000), *Thermal-mechanical fatigue and the modelling of materials behaviour under thermal transients*, Third ASTM Symposium on Thermomechanical fatigue behaviour of materials: vol 3, ASTM STP 1371, H. Sehitoglu and H. J. Maier, Eds., American Society for Testing and Materials, West Conshohocken, PA, pp. 223-238.
9. Koster A., Rémy L., (1999) *An oxidation-creep-fatigue damage model for fatigue at high temperature and under thermal transients*, "Fatigue '99", proceedings of the 7th international Fatigue Congress, X. R. Wu and Z. G. Wang, Eds., Beijing, China, June 8-12, Higher Education Press and EMAS, Vol. 4, pp. 2139-2144.
10. Tomkins B., *Phil. Mag.*, (1968), vol. 18, pp. 1041-1066
11. Rabotnov Y. N., (1969), *Creep Problems in Structural Members,* North Holland, Amsterdam.
12. Koster A., Chataigner E., Rémy L., (1996), *Thermal fatigue, a useful tool to assess low cycle fatigue damage in superalloys for components experiencing thermal transients*, 81st AGARD Structures and Materials Panel, "Thermal-mechanical fatigue of aircraft engine materials", Banff, Canada, 2-4 octobre, AGARD-CP-569, pp. 8-1/8-8.

Crack Growth

Temperature-Fatigue Interaction
L. Rémy and J. Petit (Eds.)

HOW FAR HAVE WE COME IN PREDICTING HIGH TEMPERATURE CRACK GROWTH AND THE CHALLENGES THAT REMAIN AHEAD

ASHOK SAXENA
School of Materials Science and Engineering, Georgia Institute of Technology
Atlanta, GA, 30332-0245, USA

ABSTRACT

Extending the operating life of high temperature components beyond their original design life has considerable advantages. Fracture mechanics is used extensively to predict the remaining life and safe inspection intervals as part of maintenance programs for these systems. The presence of creep deformation and time-dependent damage accumulation presents very significant challenges in accurately predicting life of these components. Therefore, the emphasis in this paper is on time-dependent fracture mechanics (TDFM) concepts. A critical assessment of the current state-of-the-art of TDFM concepts, test techniques, and analytical procedures is made to demonstrate the potential of this technology. In addition, future developments that are needed to enhance the application of this technology are also described and limitations of the current approaches are also discussed.

KEYWORDS

Creep, Fatigue, Crack growth, Hold-time, Fracture

INTRODUCTION

Predicting the design life or the remaining life of power-plant components, chemical reactor pressure vessels, and hot-section components of land, sea and aeronautical gas turbines is important for economic and safety reasons. Extending the life of existing equipment is a multi-billion dollar business on one hand. However, on the other, the economic advantages of saving or delaying capital investments must always be weighed against the increased risk of catastrophic failures that can cause total shut downs and lead to loss of human lives. Therefore, use of accurate life prediction models that also include the assessment of flaw tolerance of critical components and combined with state-of-the-art nondestructive inspection methods offer the best defense against risk of fracture.

Fracture mechanics based models are widely used in risk/remaining life applications in several industries that operate mechanical equipment. The focus of this paper is on high temperature components in which failures occur by creep damage and by other high temperature damage mechanisms such as environmental degradation and creep-fatigue-environment interactions. The methodology being presented applies to a broad class of high temperature components.

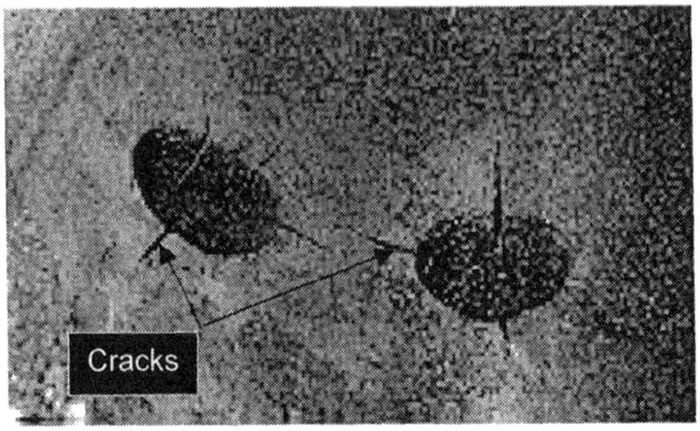

Fig.1- Cracks in a steam header that was in service for 25 years at a temperature of 540 C. The cracks were discovered during a maintenance inspection The hole diameter is approximately 30 mm and can be used as a reference for sizing cracks in the picture.

Figure 1 shows cracks that were found in the interior of a steam header during an inspection after approximately 25 years of operation at a maximum service temperature of 538 C. During a complete operating cycle, the stress-time history of critical locations (locations were cracks were found) in the header include a period in which the stress rises with time, a hold period and a time during which the stress decreases with time. The three segments of the stress-time history correspond to a start-up period, steady-state operation and shut-down. Frequently, due to thermal stresses during cold starts, the load history can experience stress transients that are significantly higher in magnitude than the steady-state stresses and can cause significant damage accumulation. The role of creep and creep-fatigue is very important in the development and propagation of cracks that may be present in these components. Figure 2 shows a schematic of a comprehensive methodology for ensuring structural integrity of elevated temperature components. The importance of the role of crack growth models under creep and creep-fatigue condition is apparent.

Time-dependent fracture mechanics (TDFM) concepts have been developed over the past 20 years to address high temperature crack growth [1-7] under sustained and cyclic loading. Methodologies have been developed for remaining life prediction using these concepts and have been applied to problems such as the header problem mentioned above and others such as steam pipes and turbine casings [8-10].

In this paper, the fracture mechanics concepts and models that are available to support the prediction of crack growth under the loading conditions described above are reviewed, primarily with the intention of discussing limitations of the various models and the challenges ahead to address the needs listed in Fig.2 that are not within the capabilities available today. We will focus on models that are currently available for describing the constitutive behavior of materials that are used in crack growth predictions, and also models for predicting creep crack growth, fatigue crack growth with and without a hold time, with some mention of thermal-mechanical fatigue crack growth. The time-dependent damage mechanisms considered include creep damage, microstructural degradation and damage caused by environmental effects such as oxidation.

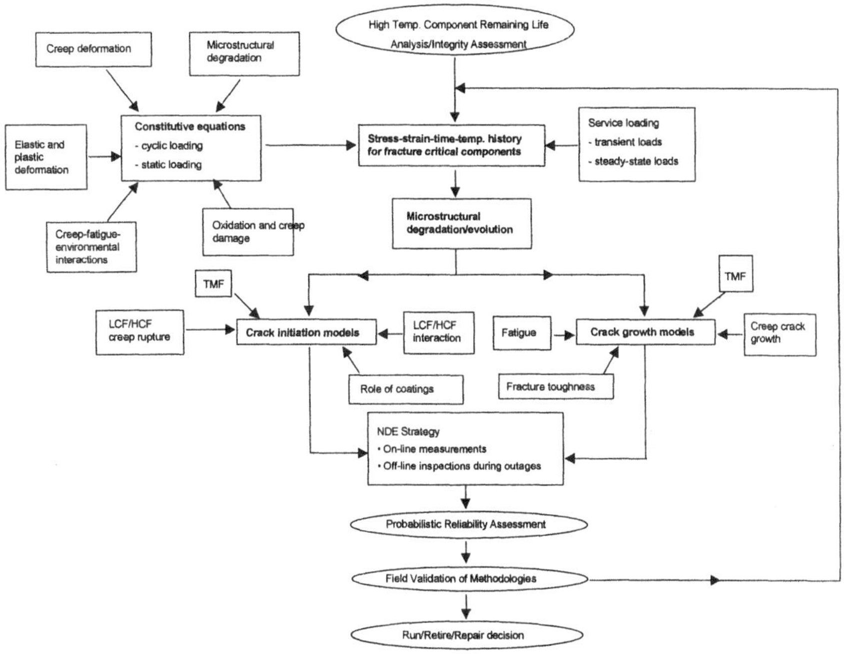

Fig.2- A methodology for assessing integrity of structural components that operate at high temperatures.

CREEP CRACK GROWTH

When a constant load is suddenly applied to a cracked body at elevated temperature, creep deformation accumulates in the crack tip region due to high stresses resulting from the stress concentration. In some materials, that are called creep-ductile materials, considerable creep deformation accumulates prior to crack extension. Thus, the crack extension occurs in the presence of substantial creep strains and the crack tip lags considerably behind the advancing creep zone boundary. In other materials that are known as creep-brittle materials, the crack extends rapidly as the creep strains accumulate and in the steady-state, the creep zone boundary and the crack tip move at equal rates. Thus, to an observer situated at the moving crack tip, it appears that the stress distribution ahead of the crack tip is constant and is uniquely determined by the magnitude of the applied stress intensity parameter, K. A necessary condition for an ideal steady-state to exist is that the size and shape of the creep zone be uniquely determined by K. In practice, certain amount of time and crack extension may be required prior to the achievement of steady-state conditions. During this transient period, the relationship between crack growth rate and K cannot be unique. In the following discussion, the approaches used for characterizing creep crack growth in creep-ductile and creep-brittle materials are discussed. This will then be followed by a discussion of crack growth under creep-fatigue conditions.

Crack Growth in Creep-Ductile Materials

Examples of creep-ductile materials include materials such as Cr-Mo steels, austenitic stainless steels, and Cr-Mo-V steels extensively used in pressure vessels and in rotors of steam turbines. Typically, the creep ductility in these materials exceeds 5% as a rule of thumb. In applying time-dependent fracture mechanics (TDFM) to creep-ductile materials, an assumption is made that the crack tip is essentially stationary. This implies that the elastic stresses due to crack growth in the forward sector of the crack tip are overwhelmed by the creep strains that continue to accumulate due to high stresses in that region. The assumption of a slowly moving crack makes it possible to use stationary crack tip parameters for correlating creep and creep-fatigue crack growth rates. The uniaxial version of the creep constitutive law used for describing the materials is given by the following equations.

$$\dot{\varepsilon} = \dot{\sigma}/E + A_1 \varepsilon^{-p} \sigma^{n_1(1+p)} + A\sigma^n \tag{1a}$$

$$\dot{\varepsilon} = \dot{\sigma}/E + \left[A_1(1+p)\right]^{1/(1+p)} \sigma^{n_1} \Big/ \left[(1+p)t^{p/(1+p)}\right] + A\sigma^n \tag{1b}$$

Equations 1(a) and 1(b) are equivalent forms in which σ = stress, ε = strain, t = time, and the dots indicate derivatives with respect to time, E = elastic modulus, A_1, n_1, and p are regression constants that describe the primary creep behavior and A and n are similar constants that describe the secondary creep behavior of the materials at a constant temperature.

The time rate of crack growth, da/dt, is characterized by the C_t parameter [1] for a wide range of creep conditions that include small-scale creep and extensive creep. C_t can be measured at the load-point in a test specimen if the crack size, applied load, size and geometry of the specimen are known. In addition, it is also necessary to know the K-calibration function and the expression for measuring the C^*-Integral [5,6] for the specimen geometry. This information is readily available for common crack growth specimens. These correlations have been experimentally shown to be valid for primary, secondary and combined primary and secondary creep conditions [2]. Figure 3 shows a correlation between da/dt and C_t obtained for a large compact type (CT) specimens with a width of 254mm [4]. In this specimen, a significant portion of the creep crack extension occurred under both small-scale and extensive creep conditions. The value of C_t in these tests first decreases (see the direction of the arrows in Fig. 3) to a minimum value and then increases after a minimum value is reached. This implies that small-scale creep conditions dominate the initial portions of the test and extensive creep conditions dominate after the minimum C_t value has been reached. The crack growth rate is described by:

$$da/dt = b\,[C_t]^q \qquad (2)$$

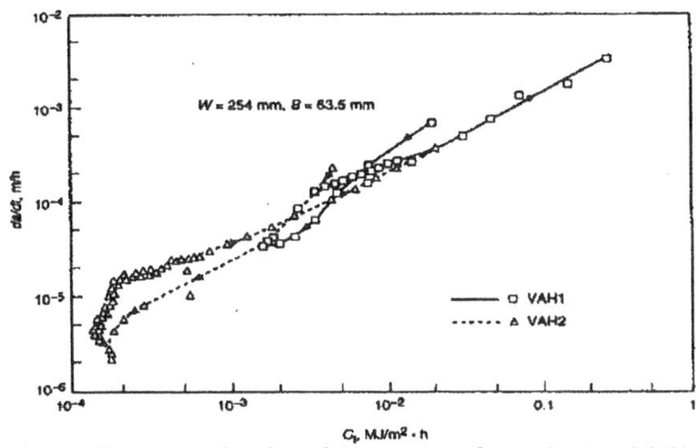

Fig. 3- Creep crack growth rate as a function of C_t parameter for a 1Cr-1Mo-0.25V steel at 538 C obtained using 254 mm wide compact type specimens. The arrows on the trend lines indicate the order in which the crack growth data were collected. Note the initial decrease in da/dt due to small-scale and transition creep conditions and the subsequent increase in da/dt for both specimens after extensive creep conditions are established. The load for VAH1 was higher than the load for VAH 2[4].

Where, b and q are regression constants obtained from the slope and intercept of the best fit straight line through the creep crack growth rate data in Fig. 3.

The methods of estimating C_t in test specimens and in components are reviewed elsewhere [2]. Under extensive creep conditions, C_t becomes identical to the C^*- Integral [1,2] and it characterizes the amplitude of the crack tip stress fields. In the small-scale creep regime, C_t is directly related to the rate of expansion of the creep zone size [3]. Thus, direct relationships have been identified that uniquely relate the globally measured parameter C_t to the local crack tip quantities which are expected to dominate the kinetics of the damage processes and determine the creep crack growth rate.

Crack Growth in Creep-Brittle Materials

Figure 4(a) shows a relationship between creep crack growth rate and the stress intensity parameter for a highly cold-worked C-Mn steel at a temperature of 360 C, which is below what would be considered as the temperature where creep begins to be of concern [11]. The correlation between da/dt and K is apparent and that between da/dt and C_t for the same data is non-existent as shown in Fig. 4(b). Similar results have been shown for other materials and the readers are encouraged to read about it in detail in a special issue of Engineering Fracture Mechanics [12].

During the initial period following application of the load, transient conditions exist in creep-brittle materials. The transients are observed in the form of an incubation period during which time-dependent creep damage accumulates at the crack tip. Some models have been proposed to address the incubation period and are described in reference [13]. A second type of transient could be in the form of crack growth during which the creep zone size and shape has not achieved steady-state conditions. A parameter equivalent to the C_t that involves a combination of K and time has been proposed to characterize the creep crack growth rate under these conditions [14]. This parameter is essentially equal to C_t because it is uniquely related to the rate of expansion of the creep zone size but also considers an additional variable related to the shape of the creep zone which also evolves during this transient period making it distinct from the steady-state condition when the creep zone size is uniquely characterized by K.

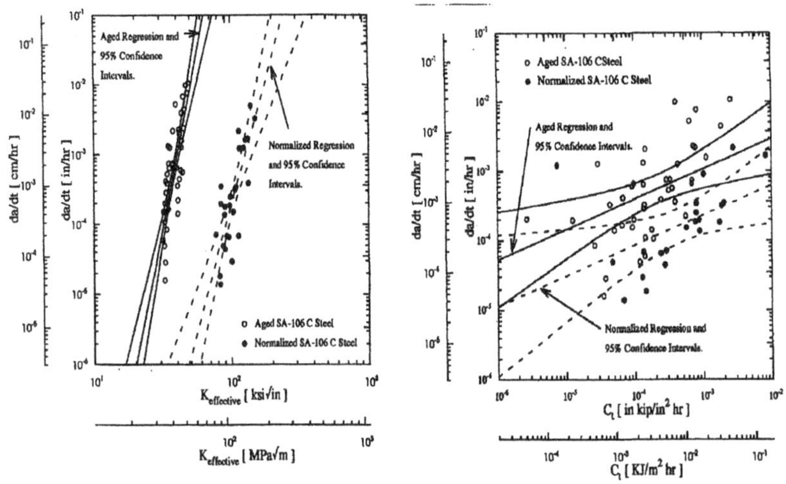

(a) (b)

Fig. 4- (a)Creep crack growth rate as a function of K for a C-Mn steel at 360 C. (b) The same data as in (a) plotted as a function of the C_t parameter showing a lack of correlation [11]

Limitations of the Creep Crack Growth Methodology

The techniques for measuring creep crack growth rates have been standardized in an American Society for Testing and Materials (ASTM) standard [15]. It can be therefore concluded that the for long-term sustained loading conditions, the fracture mechanics methods for characterizing crack growth rates for certain type of materials are reasonably well established. However, considerable limitations also still remain which include the following.

- The current methods of estimating C_t are limited to the creep constitutive relations described by equation 1 and several high temperature materials are not well represented by this equation. This is not a fundamental limitation on C_t but one that applies to the currently available methods for determining it.
- The deformation based global fracture mechanics parameters are only valid when creep cavitation damage is limited to a small region in the vicinity of a crack. If the

damage is widespread, other approaches based on damage mechanics are perhaps more appropriate.

- The CT specimens currently used for determining the creep crack growth behavior represent high crack tip constraint conditions and may not be fully representative of the loading conditions in pressure vessels with considerably lesser constraint.
- In creep-brittle materials, considerable amount of crack extension can occur under transient conditions. The approach for characterizing the crack growth rate under such transient conditions are only preliminary proposals [14] and are not well established.
- Several high temperature materials in gas turbine applications are single crystal of directionally solidified with strong directional characteristics. This problem is being addressed but will need considerable more attention in the future [16].

CREEP CRACK GROWTH IN WELDS

Several high temperature cracking problems originate in welds. The creep deformation rates in the weld metal region can be substantially different from the creep rates in the base metal region. If the interface is very distinct and the crack lies along the interface with loading that is perpendicular to the crack face, the parameters C_t and C^* are valid parameters except that the stress at the interface will be influenced by the mis-match in the creep deformation rates between the base metal and the weld metal. Figure 5 shows the da/dt versus C_t behavior for cracks located in the weld metal region and along the fusion line of a 2.25Cr-1Mo weld. Clearly, the rates along the fusion line were higher than the crack growth rates of cracks in the weld metal region.

There is currently no rigorous creep crack growth theory for predicting cracking in welds that have microstructural gradients which is the case for number of welds where the fusion region cannot be described as a sharp interface; it is rather a region of finite width that consists of the heat-affected zone (HAZ) on the base metal side. The microstructure of the HAZ may consist of a coarse grain region, for example. Also, when the crack location is away from the interface, no rigorous fracture mechanics parameters can be defined to characterize the crack growth rate.

A major concern in evaluating large welds is the variability in the weld metal creep deformation resistance, the geometry of the weld and microstructural variations. For example, variations in trace element content and microstructure can significantly reduce creep ductility, creep deformation resistance, or both. An alloy that is ductile in the ferritic condition can become brittle in the coarse grain bainitic condition. Likewise, base metals and weld metals that look very similar in their microstructure can exhibit quite different creep resistance due to minor variations in the chemical composition. Variations in creep deformation resistance between base metal and weld metal can often cause strain concentrations in the weaker metal and accelerate crack formation and growth rate. It is important to clearly understand the role of chemical composition and microstructure on the creep deformation rates and creep ductility prior to sorting out the effects of these variables on creep crack growth rates.

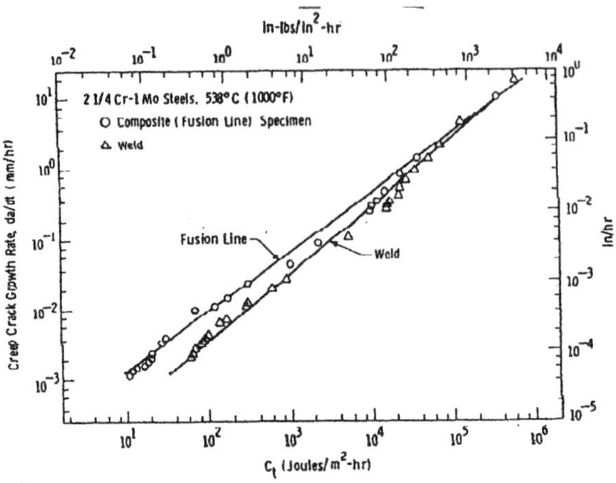

Fig. 5- Comparison between creep crack growth rates for base metal and fusion line regions in a 2.25Cr-1Mo steel [8]

CREEP-FATIGUE CRACK GROWTH

Crack Growth During Continuous Cycling

At uniform, constant elevated temperature, the fatigue crack growth per cycle, da/dN, in a cracked body continues to be characterized by the cyclic stress intensity parameter, ΔK, and the load ratio, R as in the sub-creep temperature range. It is well established that additional variables such as frequency, ν, and the waveform also enter into the equation. Thus, functionally, the fatigue crack growth rate is represented by the following equation.

$$da/dN = f(\Delta K, R, \nu, \text{wave form}) \quad (3)$$

When the load levels are high enough to cause cyclic plasticity, ΔK may be replaced by the cyclic J-integral, ΔJ as defined by Dowling and co-workers [17]. The above relationship is valid only for isothermal conditions. Frequently, cyclic loads in elevated temperature components are caused by the high thermal gradients during the start-up cycle. The resulting high thermal stresses can easily cause significant amounts of plasticity in regions where cracks are present. Since the flow properties of the material are temperature dependent, the above simple equation that applies to isothermal conditions may not be any longer valid. Since the cyclic flow properties of the material are temperature dependent, it can no longer be ensured that the cyclic stress-cyclic strain properties through out the body are uni-valued, a necessary condition for the ΔJ-integral to be path-independent. The latter condition is necessary for the crack tip stress and strain ranges to be uniquely correlated with the magnitude of ΔJ. A question then arises about the limits of applicability of ΔJ under these conditions and what other parameters, if any, are available to characterize fatigue crack growth conditions under thermal-mechanical fatigue conditions.

Crack Growth During Hold Time

The application of C_t has been extended to situations involving time-dependent crack growth rates during hold times between loading and unloading events [18]. Crack extension under such conditions are classified as creep-fatigue crack growth. The average value of da/dt during hold time, $(da/dt)_{avg}$, and the average value of C_t during hold time, $(C_t)_{avg}$, are shown to uniquely correlate with each other for different amounts of hold time. Figure 6 shows an example of $(da/dt)_{avg}$ and $(C_t)_{avg}$ data for Cr-Mo steels for various hold times and it also includes creep crack growth data for the material at that temperature [19]. In this material, creep crack growth rates and creep-fatigue crack growth rates are indistinguishable within the normal scatter band. It can thus be argued that creep-fatigue interaction in this material consists of reinstatement of the stress redistribution after every unloading and loading portion of the overall cycle. In other words, the creep accumulation during the hold time is reversed by plastic loading at the crack tip during unloading. The reversal of this creep can be partial or it can be for practical purposes complete depending on factors such as the applied load level, creep and plastic properties of the material. If the cyclic plastic zone is large in comparison to the creep zone that develops during the hold time, the reversal of creep will be nearly complete. If on the other hand, the creep zone is large in comparison to the cyclic plastic zone, there will be little or no reversal of creep strains during unloading. references [2] and [20-22], deal with the estimation of the creep reversal parameter that can be used to estimate the magnitude of the $(C_t)_{avg}$ as a function of accumulated cycles.

The observation from that the creep and hold-time creep-fatigue crack growth rates in Fig. 6 are indistinguishable, implies that the fatigue and creep mechanisms must be quite distinct. In other words, cyclic loading/unloading bears no fundamental change on the subsequent crack growth mechanism during the sustained loading period. Noting that fatigue damage occurs by transgranular mechanism while creep occurs by grain boundary cavitation in creep-ductile materials, perhaps this result is not unexpected. However, this should not be interpreted as a general result. At the very least it is limited to creep-ductile materials and not expected to automatically apply to creep-brittle materials. There are no experimental data currently available to guide our thinking in this regard.

Crack Growth During Combined Fatigue and Hold Time

The simplest model for combining the effects of cyclic loading and hold time will be to linearly sum the crack growth during the two segments of the cycle. Such a model been referred to the damage summation hypothesis in the literature. This leads to the following equations for the crack growth during one complete cycle,

$$da/dN = (da/dN)_{cycle} + (da/dN)_{time} \qquad (4a)$$

$$da/dN = C(\Delta K)^m + b_1[(C_t)_{avg}]^q\, t_h \qquad (4b)$$

Where, da/dN = total crack growth per cycle including loading, unloading, and hold time, $(da/dN)_{cycle}$ = crack growth rate for the same value of ΔK except with no hold time, and $(da/dN)_{time}$ = crack growth during the hold time, t_h. An alternate way to sum damage is to use the dominant-damage hypothesis according to which da/dN is given by,

$$da/dN = \max [(da/dN)_{cycle} , (da/dN)_{time}] \quad (5)$$

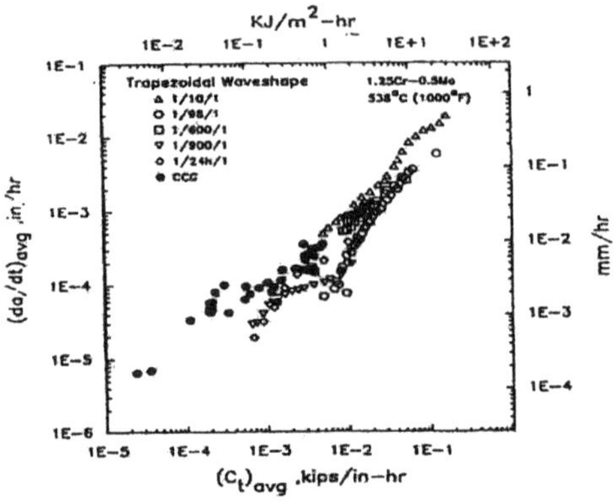

Fig. 6- Crack Growth rates at various hold times ranging from 10 seconds to 24 hours and including creep crack growth rates for 1.25 Cr-0.5 Mo steel [19]

For long hold times, the difference between the two approaches is negligible. For very short hold times approaching the value of zero where the cycle-dependent component dominates, the two approaches are also essentially the same. However, for intermediate hold times, differences between the two are expected and there is not sufficient data to allow one to choose between the two approaches. These equations only consider the influence of cyclic loading on the crack growth behavior during subsequent hold time. We must also consider the influence of creep deformation during the hold time on the crack growth rate during the subsequent cyclic loading. This becomes relevant in the short to intermediate hold times of less than 100 seconds. It can be argued that creep deformation can blunt the crack substantially and decrease the amount of cyclic crack growth. Indeed, studies have shown that at low ΔK values, the da/dN for cycle with a short hold time is less than the da/dN for the corresponding ΔK without the hold time [20]. Therefore, to address this shortcoming in both models, it is necessary to add an interaction term in equations (4) or (5). However, more experimental guidance is necessary to formulate such a term.

DISCUSSION OF FUTURE NEEDS

The potential of time-dependent fracture mechanics (TDFM) in establishing design life of new components, or safe inspection intervals for components in service, or for performing risk assessments is obvious. The technology has come a long way in the past 20 years but still much remains to be done to develop total confidence in the approach. A brief description of these needs is provided in this section.

High temperature components are usually subjected to varying temperatures that can range from temperatures well into the creep range to temperatures where creep damage may either be marginal or not significant. A majority of tests and analyses are performed assuming isothermal conditions in which the influence of environment is not explicitly included. More research in understanding the creep-fatigue-environment interactions is necessary for accurate life predictions.

The cracking in large number of high temperature components is due to transient thermal stresses but life prediction estimates assume isothermal conditions. Considerable research is needed in analytical methods for treating crack growth under thermal-mechanical loading and new test methods are needed that provide crack growth data under temperature gradients. The limitations on parameters such as ΔJ under thermal gradients should be explored.

An area that has not been explored much is that of load interactions during crack growth at elevated temperature. In the presence of transient thermal stresses, it becomes quite important to treat the effects of overload on the crack growth rate during the subsequent hold time.

There are significant opportunities for developing standard methods for creep-fatigue crack growth testing. These tests are highly specialized and require very precise controls and measurements. The data analysis is also complex so that forcing some uniformity on how data are treated will also help the overall goal of developing a well accepted life prediction methodology.

Extension of these methods to directionally solidified alloys, single crystal alloys, and to intermetallics is needed. These materials can exhibit a range of behavior not seen in Cr-Mo ferritic and austenitic stainless steels. For example, depending on the loading conditions and orientation, the same alloy may behave as a creep-ductile or a creep-brittle alloy. Solving this problem will require good numerical simulations that are now well within the capability of the current technology.

Monitoring of service experience is very important in determining which aspects of the problem deserve a priority over others. Service experience is also important to validate the models after they are developed and implemented.

SUMMARY AND CONCLUSIONS

Considerable progress has occurred in the recent years in predicting crack growth behavior in elevated temperature components. Crack tip parameters for characterizing high temperature fatigue crack growth, creep crack growth and crack growth during hold time between fatigue cycles are described in this paper. Similarly, well developed test methods are available for characterizing the crack growth behavior in such materials. Several areas have been identified in which more research is needed to further this technology.

REFERENCES

1. Saxena, A.,(1986) "Creep Crack Growth Under Non-Steady State Conditions", Fracture Mechanics: Seventeenth Conference, ASTM STP 905, 185.

2. Saxena, A. (1998) Nonlinear Fracture Mechanics for Engineers", CRC Press, Boca Raton, Florida.
3. Bassani, J.L., Hawk, D.E. and Saxena, A. (1989) "Evaluation of the C_t Parameter for Characterizing Creep Crack Growth in the Transient Regime", Nonlinear Fracture Mechanics: Time-Dependent Fracture, ASTM STP 995, 7.
4. Saxena, A., Yagi, K. and Tabuchi, M. (1994), "Crack Growth Under Small-Scale and Transient Conditions in Creep-Ductile Materials", Fracture Mechanics: Twenty Fourth Volume, ASTM STP 1207, 481.
5. Landes, J.D. and Begley, J. A. (1976) "A Fracture Mechanics Approach for Creep Crack Growth", ASTM STP 590, 481.
6. Nikbin, K.M., Webster, G.A., and Turner, C.E. (1976) "Relevance of Nonlinear Fracture Mechanics to Creep Crack Growth", ASTM STP 601, 47.
7. Saxena, A (1980) "Evaluation of C^* for Characterizing Creep Crack Growth in 304 Stainless Steel", Fracture Mechanics: Twelfth conference, ASTM STP 700, 131.
8. Liaw, P.K., Saxena, A. and Schaefer, J. (1989) *Engineering Fracture Mechanics,* **32**, 675.
9. Liaw, P.K., Saxena, A., and Schaefer, J. (1989) *Engineering Fracture Mechanics,* **32**, 709.
10. Saxena, A., Liaw, P.K., Logsdon, W.A., and Hulina, V.A. (1986) *Engineerng Fracture Mechanics*, **25**, 289.
11. Gill, Y. (1994), "Creep Crack Growth Characterization of SA-106 C Carbon Steel" *Ph.D. Dissertation,* Georgia Institute of Technology, Atlanta, GA.
12. Saxena, A. and Yokobori, T. editors (1999) Special Issue on Crack Growth in Creep-Brittle Materials, *Engineering Fracture Mechanics,* **62**, No.1.
13. Austin, T.S.P. and Webster, G.A. (1992), *Fatigue and Fracture of Engineering Materials and Structures*, **15**, 1081.
14. Hall, D.E., McDowell, D.L. and Saxena, A. (1998) *Fatigue and Fracture of Engineering Materials and Structures*, **21**, 387.
15. ASTM Standard E1457-2000 (2000), "Standard Test method for characterizing Creep Crack Growth in Metals", Annual Book of ASTM Standards, **03.03**.
16. Gardner, B., Saxena, A. and Qu, J. (2001), " Creep Crack Growth parameters for Directionally Solidified Superalloys", Proceedings of the Eleventh International Conference on Fracture, Hawaii, Dec. 3-7, 2001 (in press).
17. Dowling, N. E. and Begley, J.A. (1976) "Fatigue Crack Growth Under Gross Plasticty and the J-Integral", ASTM STP 590, 82.
18. Saxena, A., and Gieseke, B., "Transients in Elevated Temperature Crack Growth" (1987), Proceedings of the International Symposium on High Temperature Fracture Mechanics and Mechanisms, Dourdan, France, 19
19. Yoon, K.B., Saxena, A., and Liaw, P.K. (1993) *International Journal of Fracture,* **59**, 95.
20. Grover, P.S. and Saxena, A. (1999) *Fatigue and Fracture of Engineering Materials and Structures*, **22**, 111.
21. Adefris, N., Saxena, A. and McDowell, D.L. (1996), *Fatigue and Fracture of Engineering Materials and Structures,* **19**, 387
22. Adefris, N., Saxena, A. and McDowell, D. L. (1996), *Fatigue and Fracture of Engineering Materials and Structures,* **19**, 401.

ENVIRONMENTAL EFFECTS ON NEAR-THRESHOLD FATIGUE CRACK PROPAGATION ON A TI6246 ALLOY AT 500°C

C. SARRAZIN-BAUDOUX AND J. PETIT
Laboratoire de Mécanique et de Physique des Matériaux - UMR CNRS n° 6617, ENSMA - B.P. 109 - Chasseneuil de Poitou - 86960 Futuroscope Cedex-France

ABSTRACT

The cracking behavior of a Ti6246 alloy is studied in the near-threshold fatigue crack propagation regime with a special attention to possible coupled effects of corrosion and creep. Tests were conducted at 500°C in selected environmental conditions (high vacuum, controlled leak low pressure, controlled partial pressure of water vapor in pure argon) and at different frequencies. The near-threshold crack propagation is shown to be highly sensitive to the environment with a predominant detrimental influence of water vapor even under very low partial pressure. A modeling is proposed accounting for the influence of partial pressures of water vapor and oxygen, of test frequency and effective stress intensity factor, and introducing a single adjustable parameter accounting for the nature of the alloy and its sensitivity to environment. Applicability of this model is validated on an IMI834 alloy.

KEYWORDS

Titanium alloys, fatigue, crack propagation, threshold, environment, corrosion, temperature.

INTRODUCTION

Failure of structural materials operating in various environments due to cracking remains a safety and economic problem despite the effort that has been devoted to understand the phenomena of fatigue, stress corrosion and creep. Because of their corrosion resistance, high specific strength and low density, titanium alloys are used in turbine engines where they are subjected to cyclic conditions in aggressive environments such as moist air at elevated temperature. A detailed characterization of these alloys is thus required in order to ensure a good damage tolerance during their operational life. Following an investigation of the influence of environment on the fatigue crack growth behavior of a Ti-6Al-4V titanium alloys at 300°C by Sarrazin-Baudoux et al. [1], and a detailed analysis of the influence of environment on the fatigue crack growth in a Ti6246 alloy tested at 500°C in moist environments (Sarrazin-Baudoux et al. [2]), this paper deals with the modeling of the latter fatigue crack growth behavior at 500°C.

MATERIAL AND EXPERIMENTAL PROCEDURES

The Ti6246 alloy (5.68 Al, 1.98 Sn,3.96 Zr, 6.25Mo) used in this investigation is β-forged at 950°C. The heat treatment consists of 930°C for two hours, followed by water quenching, aged at 900°C for one hour and air cooled, held at 595°C for a total aging time of eight hours and air cooled. The alloy contains 75% of α grains and displays a Widmanstätten structure, consisting of intermeshing colonies of α platelets contained in large prior β grains (300 μm), the size of the actual α grains not exceeding 50 μm. The mechanical properties are given in Table 1. Fatigue crack growth experiments are carried out on Compact Tension C(T) specimens (10 mm thick and 40 mm wide) in accordance with ASTM Test Method for Measurements of Fatigue Crack Growth Rates (E 647-88) using a servo-hydraulic machine equipped with an environmental chamber and a furnace allowing testing in ambient air, high vacuum ($3x10^{-4}$ Pa) and controlled atmospheres such as humidified argon with controlled partial pressure of water vapor, at temperatures ranging up to 500°C. Crack lengths are tracked using a DC (electrical) potential drop technique [2]. The specimens are submitted to sinusoidal loading at frequencies varying from 35 Hz to 10^{-3} Hz with a load ratio (R) of 0.1 or at variable R. Crack closure is detected using a capacitive displacement gauge and determined by means of the offset compliance technique [3,4]. K_{max}-constant tests for near threshold propagation are conducted at increasing steps of K_{min}, the decreasing steps of ΔK being similar as for the constant R tests and K_{min} being at any time higher than the stress intensity level for crack closure and so closure is eliminated in all the explored range.

The environmental effect is studied in various gaseous atmospheres controlled by mean of a mass spectrometer and high performance hygrometers. For the different environmental conditions used, the partial pressures of oxygen and water vapor are given in Table 2.

Table 1 – Mechanical properties of Ti6246

Temperature	σ_y (MPa)	σ_u (MPa)	R (%)	K_{IC}(MPa√m)	E (GPa)
Room T	987	1098	10.2	75	122
500°C	680	800	–	–	102

Table 2–Environmental conditions for propagation tests

Environments	Partial pressure H_2O (Pa)	Partial pressure O_2 (Pa)	Partial pressure N_2 (Pa)	Total pressure(Pa)
Ambient air	1300	$2x10^4$	$8x10^4$	10^5
Humidified argon	3000	< 1	-	10^5
Dry Argon	3	< 1	-	10^5
Medium vacuum	1	-	-	1.33
Rough vacuum	100	-	-	133
High vacuum	$<2x10^{-4}$	$<10^{-6}$	-	$3x10^{-4}$

A conventional approach of the study of the near-threshold fatigue crack propagation (FCP) is based on experiments conducted at R=0.1, and the effective propagation is determined by mean of closure correction. More recently, K_{max} constant tests have been recommended to reach directly the effective behavior without closure correction which can be more or less controversial [5], especially when interaction can occur between closure and environment, for example when there exists a competitive influence of closure induced by oxide wedging and a limitation of the access

of the embrittling species (i.e. water vapor molecules) up to the crack tip by oxide deposits [6], or an activation of environmental embrittlement by rupture of the oxide film induced by closure as observed on Ti alloys at 500°C [7].
Hence, this study considers effective data provided by tests with closure correction when closure-environment interaction are not observed, or by tests performed at sufficiently low K_{max} level to avoid any influence of closure.

EXPERIMENTAL RESULTS ON THE INFLUENCE OF ENVIRONMENT ON FATIGUE CRACK GROWTH.

To illustrate the influence of environment on the fatigue crack propagation behavior at room temperature, da/dN is plotted with respect to ΔK_{eff} in figure 1 for tests performed at constant K_{max} < 24 MPa√m at 500°C in air, high vacuum, dry and humidified Argon which contains a partial pressure of water vapor of 3Pa and 3 kPa respectively (see Table 2). The test frequency is 35 Hz. In all the explored range, the CP rates in ambient air and in humidified Argon are similar and much faster than in high vacuum specially in the low rate range with the presence of the plateau for rates close to 10^{-8} m/cycle typical of an environmental effect as observed in several Ti alloys at 300°C [1] and 500°C [2,8].

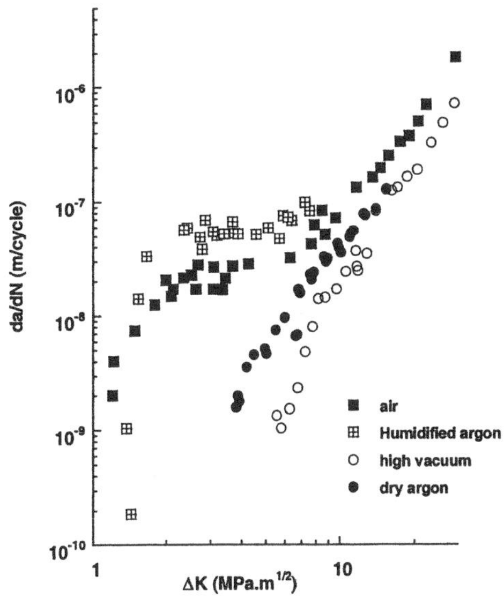

Fig. 1. da/dN vs ΔK_{eff} relation in high vacuum, ambient air and humidified argon at room temperature at 35 Hz.

These observations are in accordance with a great sensitivity of the material to the environment at 500°C and support an environment effect involving water vapor. The test performed in dry argon leading to growth rates slightly higher than in high vacuum further indicates that a very low amount of water vapor is sufficient for an environmental effect but needs more time to induce this effect.

Tests performed in laboratory air at different frequencies ranging from 35 Hz to 0.05 Hz in air (Figure 2) show clearly that decreasing frequency induces accelerated rates up to a saturation of the environmental effect at 0.5 Hz leading to growth rates similar as those at 0.1 Hz. Crack growth data with a triangular loading at 0.05 Hz and trapezoidal loading with dwells of 90s and 180s fall in the scatter band of the saturated regime. This result indicates that the environmentally-induced embrittlement process results more from a corrosion fatigue mechanism than a creep fatigue process. It can be described by a relation in accordance with the following model initially proposed by McClintock [9]:

$$da/dN \propto \Delta CTOD$$

where ΔCTOD is the crack tip opening displacement range.

Taking into account the existence of an effective threshold range for environmentally assisted propagation, this relation can be derived as :

$$da/dN = (B/\sigma E).(\Delta K^2 - \Delta K_{th}^2)$$

where E is the Young modulus, σ a stress parameter and B a dimensionless parameter.

Figure 3 illustrates more precisely the kinetics of the effect of water vapor with tests run under low pressure of water vapor and at very low frequency.

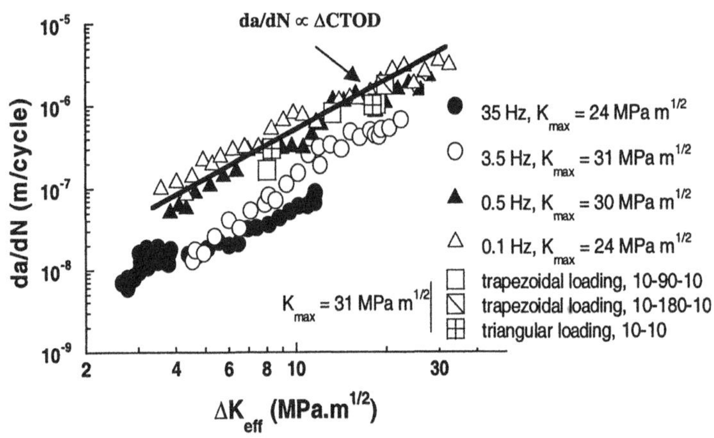

Fig 2. Influence of test frequency and of dwell time on effective crack propagation in lab. air at 500°C.

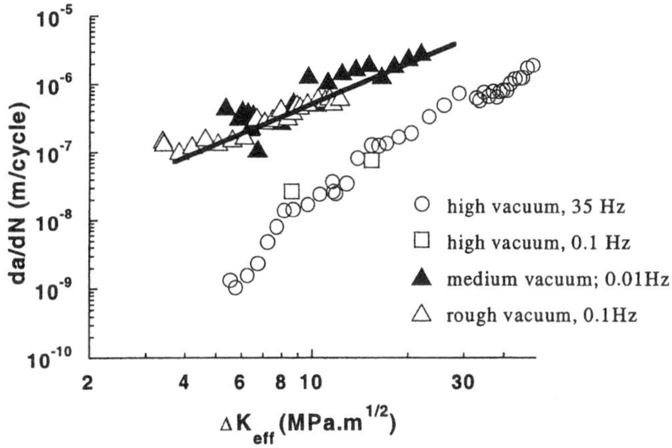

Fig. 3. Influence of low pressure of water vapor and test frequency on effective fatigue crack propagation at 500°C.

A first experiment was performed at 0.01 Hz under a medium vacuum of 1.3 Pa. The partial pressure of oxygen was below the detection threshold of the analyzer while that of water vapor was of 1 Pa (see table 2). The data are similar to those in air at 0.5 Hz and 0.1 Hz and are consistent with the saturated corrosion-fatigue mechanism described above. A second test was performed at 0.1 Hz in a rough vacuum (total pressure of 133 Pa, partial pressure of water vapor of about 100 Pa and oxygen not detectable). The same result was obtained. In contrast, a frequency of 0.1 Hz had no effect on the propagation in high vacuum. The results of these experiments are very convincing of the detrimental effect of water vapor which is shown to be active even under very low partial pressure and in absence of any detectable trace of oxygen.

MODELING OF ENVIRONMENTALLY ASSISTED CRACK PROPAGATION

The effect of water vapor for titanium alloys has been reported earlier at room temperature [10, 11] and 300°C [1]. It has been shown to be controlled by the exposure $P_{H2O}/2f$ (P_{H2O}: partial pressure of water vapor and f the test frequency) as previously defined [10]. The critical exposure can be estimated about 10^{-2} Pa×s and the substantial enhancement of the growth rates of the saturated regime can be attributed to the adsorption of water vapor on freshly created surfaces at the crack tip [12-14] followed by an embrittling mechanism. Effective data of tests performed at 500°C for different environment-frequency couples are plotted in the figure 4.
The interpretation of the curves could appear confusing, but a superposition model has been proposed which can be written as follow :

$$\left(\frac{da}{dN}\right) = (1-\psi)\left(\frac{da}{dN}\right)_0 + \psi\left(\frac{da}{dN}\right)_e \qquad (1)$$

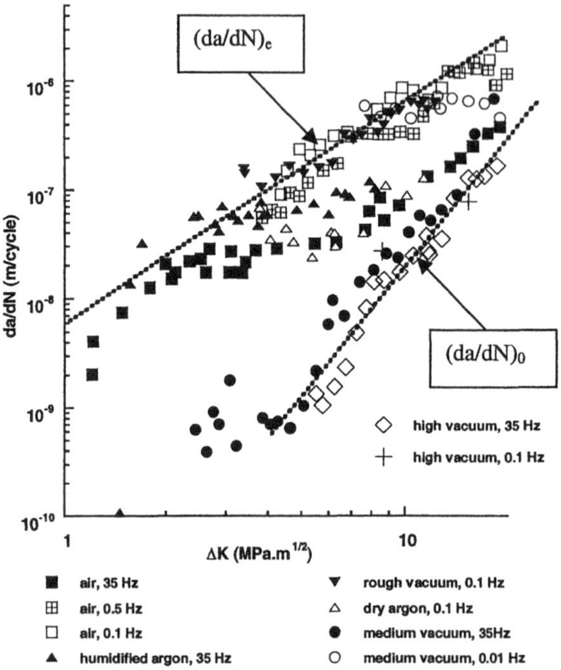

Fig. 4. Illustration of crack propagation curves obtained under different environments at different frequencies.

The expression (1) well accounts for environment effect with propagation curves evolving from the fatigue-corrosion regime for $\Psi = 1$ in the near threshold range with a full effect of environment, to the intrinsic regime $(da/dN)_0$ for $\Psi = 0$ without any environment influence at higher ΔK. The transition is more or less pronounced according to the environment and the test frequency. It can be extracted from the previous relation the expression of Ψ:

$$\Psi = \frac{(da/dN) - (da/dN)_0}{(da/dN)_e - (da/dN)_0}$$

The expression of Ψ is considered as a function of the exposure as defined by Wei [10] which is $P_{H2O}/2f$.. To account for the competitive adsorption of water vapor and oxygen, the ratio of the partial pressure of water on the sum of the partial pressures of oxidizing species present in air, $(P_{H2O}/(P_{H2O}+P_{O2}))^{1/4}$, and the size of the freshly created surfaces at each cycle which is proportional to ΔK^2 is also considered. The experimental measurements of Ψ versus X with $X=(P_{H2O}/2f)*(P_{H2O}/(P_{H2O}+P_{O2}))^{1/4}*\Delta K^{-2}$ are plotted in figure 5 showing a typical S curve.

Taking for Ψ an hyperbolic function, $\Psi = \left[1/2\left\langle th(\log X) + 1\right\rangle\right]^{1/3}$, the final expression becomes after developing and adjustment as:

$$\Psi = \left[\frac{(X/3)^2}{\left[(X/3)^2 + 7\,(X_{CR}/3)^2\right]}\right]^{1/3}$$

gives a good correlation with experimental data as it can be shown in the figure 5.

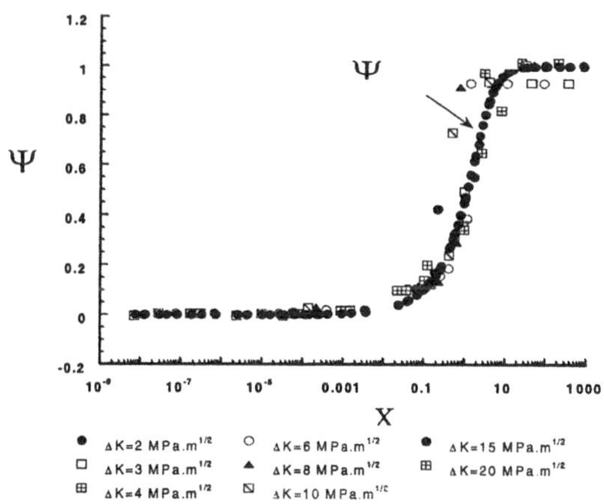

Fig. 5. Evolution of Ψ versus X

Finally for the Ti6246 alloy, the following expression for the crack growth can be written:

$$\left(\frac{da}{dN}\right) = 1.8.10^{-12}.(1-\psi)\Delta K^4 + 6.10^{-9}.\psi\left[\Delta K^{-2} - \Delta K_{th}{}^2\right]$$

with ΔK_{th}=1.2 MPa√m, threshold asymptotic value and X_{CR} = 1.13

Crack growth rates calculated from the model present good agreement with experimental crack growth rates as illustrated in the figure 6 for numerous diagrams corresponding to various amount of water vapor environments in a great range of ΔK and without any adjustable parameter, apart from X_{CR} which is characteristic of the material. In figure 7 a similar comparison is done for some experimental data available on a IMI834 alloy [10] with X_{CR} = 3.6. Here again model and experiments are in accordance. Even if a more complete validation of this model would required experimental data on other alloys tested in various environments and in an extended range of temperature, this comparison suggests that the model can give a good description of the environmentally assisted fatigue crack growth behavior of Ti alloys at high temperature.

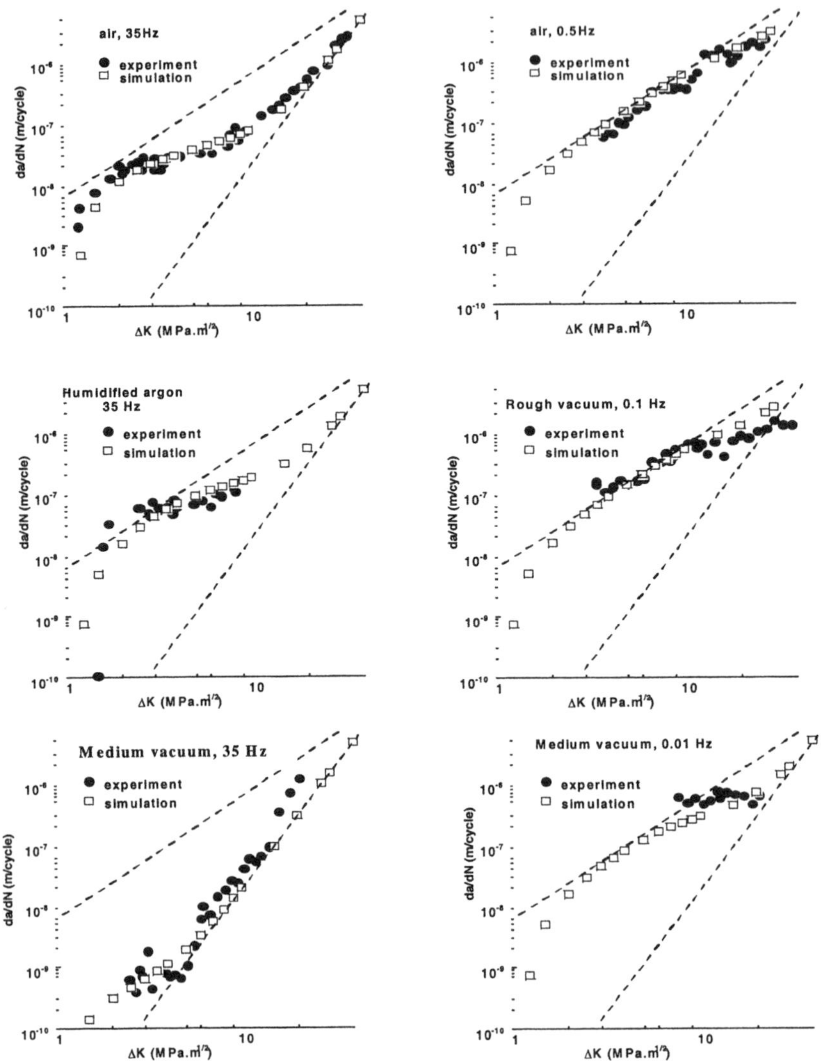

fig. 6 . Comparison of experimental data and modeling for Ti6246

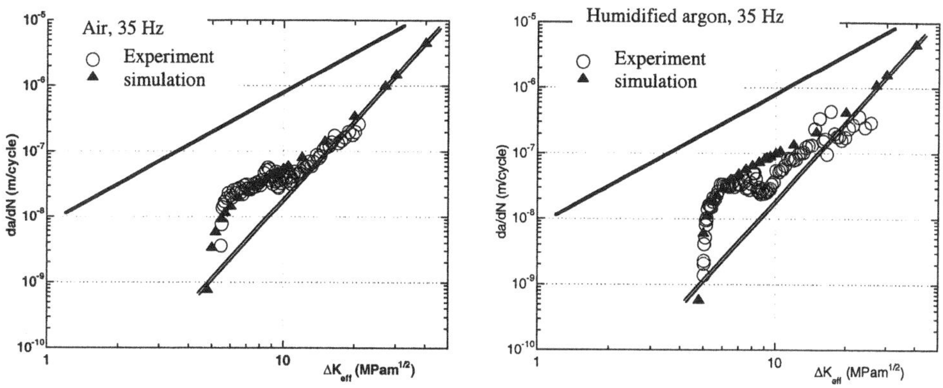

Fig. 7 . Comparison of experimental data and modeling for IMI 834

CONCLUSION

Experiments conducted on a Ti6246 alloys at 500°C in various gaseous environments with controlled contains of water vapor and oxygen have demonstrated the detrimental influence of water vapor even under very low partial pressure.

A superposition model is proposed for environmentally assisted fatigue crack propagation at 500°C accounting for the influence of partial pressure of active gas, test frequency and effective stress intensity factor range. Introducing a parameter accounting for the sensitivity of the alloy to the environment, a good accordance is obtained between experimental data and calculated for two Titanium alloys tested in various gaseous atmospheres.

REFERENCES

1. Sarrazin, C., Lesterlin, S. and Petit, J. (1997) "Atmospheric influence on fatigue crack propagation in titanium alloys at elevated temperature", *Elevated temperature effects on fatigue and fracture, ASTM STP 1267*, Ed. R.S. Piasick, R.P. Gangloff and A. Saxena, American Society for Testing and Materials, Philadelphia , pp. 117-139.
2. Sarrazin-Baudoux, C., Chabanne, Y., and Petit, J. (2000) "Mean stress and Environmental Effects on Near-Threshold Fatigue Crack Propagation on a Ti6246 Alloy", *Fatigue Crack Growth Threshold, Endurance Limits and Design, ASTM STP 1372*, J.C. Newman, Jr. And R.S. Piasick, Eds., American Society for Testing and Materials, Philadelphia, , pp. 341-360
3. Elber, W. (1971) "The Significance of Crack Closure" *Damage Tolerance in Aircraft Structures, ASTM STP 486,* pp.230-242.

4. Kikukawa, M., Jono, M. and Mikami, S. (1982) "Fatigue Crack Propagation and Crack Closure Behavior under Stationary Varying Loading-Test Results on Aluminum Alloy", *Journal of the Society on Materials Science Japan,* **31** , pp.438-487.
5. Newman J.C., Jr. and Piasick R.S. (1998) *Fatigue Crack Growth Threshold, Endurance Limits and Design, ASTM STP 1372,* American Society for Testing and Materials, Philadelphia, pp.
6. Kwon, J.H. (1985) "Influence de l'Environnement sur le Comportement en Fatigue d'un Acier E460 et d'un Alliage léger 7075 près du Seuil de Fissuration", *Doctorate Thesis*, University of Poitiers, France,.
7. Lesterlin., S. (1996) "Influence de l'Environnement et de la Température sur la Fissuration par Fatigue des Alliages de Titane", *Doctorate Thesis*, University of Poitiers, France.
8. Petit, J., Sarrazin-Baudoux, C., Chabanne, Y., Lütjering, G., Gysler, G., and Schauerte, O. (1998), "Environmental Interaction in High Temperature Fatigue Crack Growth of Titanium Alloys", *Fatigue Behavior of Titanium Alloys*, R.R. Boyer, D. Eylon and G. Lütjrring eds., TMS pub., Warrendale, Pennsylvania, USA., pp. 203-210.
9. McClintock, J. (1968) "Plasticity aspects of fracture", *Fracture*, H. Leibewitz Ed., , pp. 48.
10. Gao, S.J., Simmons, G.W., and Wei, R.P. (1984) "Fatigue Crack Growth and Surface Reactions for Titanium Alloys Exposed to Water Vapor", *Mat. Sc. and Engng.*, **62**, , pp. 65-78.
11. Wanhill, R.J.H. (1976) "Environmental Fatigue Crack Propagation in Ti-6Al-4V Sheet", *Metallurgical Transactions*, **7A**, pp.1365-1373.
12. Langmuir, I. (1918) "The Adsorption of Gases on Plane Surfaces of Glass, Mica and Platinum", *Journal of American Chemical Society*, **40**, pp. 1361-1403
13. Rehbinder, P.A. (1971) "Les phénomènes de surface dans la déformation et la fracture des solides", Séminaire de mécanique des surfaces, ISMCM.
14. Piascik, R.S. and Gangloff, R.P. (1991) "Environmental Fatigue of an Al-Li-Cu Alloy: Part I. Intrinsic Crack Propagation in Hydrogenous Environments", *Metallurgical Transactions*, **22A**, pp. 2415-2428.

Temperature-Fatigue Interaction
L. Rémy and J. Petit (Eds.)

GROWTH BEHAVIOUR OF SMALL SURFACE CRACKS IN INCONEL 718 SUPERALLOY

M. GOTO and T. YAMOMOTO
Department of Mechanical Engineering, Oita University, Oita,870-1192, Japan
N. KAWAGOISHI
Department of Mechanical Engineering, Kagoshima University, Kagoshima, Japan
H. NISITANI
Department of Mechanical Engineering, Kyushu-Sangyo University, Fukuoka, Japan

ABSTRACT

In order to study the fatigue damage of the Ni-base superalloy Inconel 718 at room temperature and 500℃, plain specimens were fatigued under constant stress amplitude and in two-step loading. Experimental results showed that the resistance to slip and crack initiation decreased due to the softening of the matrix at elevated temperature. With regard to the effect of the elevated temperature on the crack growth, the growth rate ***dl/dN*** of large cracks was accelerated, whereas the growth of microcracks less than 50μm was suppressed relative to R.T.. There was no effect of the stress change on the crack growth behaviour for low-to-high and high-to-low block loading at room temperature, nor for low-to-high block loading at 500℃. For high-to-low block loading at 500℃, however, the growth behavior of microcracks less than 50μm at the second stress level was affected by the first-stress revel. The results are discussed from the viewpoint of the softening of the matrix and of the oxide films formed at the elevated temperature.

KEYWORDS

Fatigue, elevated temperature, crack initiation, small crack growth, stress change, Inconel 718

INTRODUCTION

The service fatigue life of machines and structures is increasingly evaluated on the basis of the growth behaviour of fatigue cracks, and a considerable database of standard crack growth rates has been accumulated. This is particularly the case for Ni-base superalloys used in gas turbine engines. Extensive growth rate data for long cracks [1,2] has been included in such databases, however, relatively little small crack growth data exists [3-7]. It is well known that the fatigue life of smooth structures is dominated by the growth life of small cracks [8]. Therefore, the estimation of small crack growth behaviour is crucial to increase the safety of machines and structures.

In the present study, fatigue tests of the nickel-base superalloy Inconel 718 were carried out at room temperature and at 500℃ to clarify the growth behaviour of a small surface crack initiated in a plain specimen under constant stress amplitude. The growth behaviour of small cracks was analyzed by means of replicas taken from the surface. In addition to the experiments under constant stress amplitude, fatigue tests under two-step loading (low-to-high and high-to-low block loading) were performed. The effect of the stress change on fatigue damage was investigated through its effect on the growth of a small crack. Finally, the growth behaviour of a small crack is discussed from the viewpoint of the softening of matrix and the oxide films formed at the elevated temperature.

EXPERIMENTAL PROCEDURES

The material was a precipitation-hardened nickel-base superalloy Inconel 718. The chemical composition (wt.%) of the alloy was 0.02C, 0.07Si, 0.09Mn, 0.005P, 0.0008S, 18.64Cr, 3.08Mo, 0.16Co, 0.02Cu, 0.53Al, 0.96Ti, 23.89Fe, 0.0044B and the remainder nickel. The alloy was solution treated for 1 hr at 982℃, water quenched, then aged for 8 hrs at 720℃ and 8 hrs at 621℃ followed by air cooling. The mechanical properties after the heat treatment were 1201MPa 0.2% proof stress, 1444 MPa ultimate tensile strength, 1805 MPa true breaking stress at room temperature, and 1030 MPa 0.2% proof stress, 1240 MPa ultimate tensile strength, 1420 MPa true breaking stress at 500℃.

The cylindrical test specimens, see Fig.1, were machined from the bars after heat treatment. Although the specimens have a shallow transverse notch, the strength reduction factor for this geometry is close to unity, so that the specimens can be considered as plain specimens (the behaviour of cracks in the specimens can be used to simulate that in plain specimens). Before testing, all the specimens were electropolished to remove about 20 μm from the surface layer followed by polishing down to a 0.5μm finish, in order to facilitate the observations of changes in the surface state.

All the tests were carried out using a rotary bending fatigue machine with a capacity of 100 Nm operating at 50 Hz. The specimens were fatigued at room temperature (R.T.) and at 500℃ in

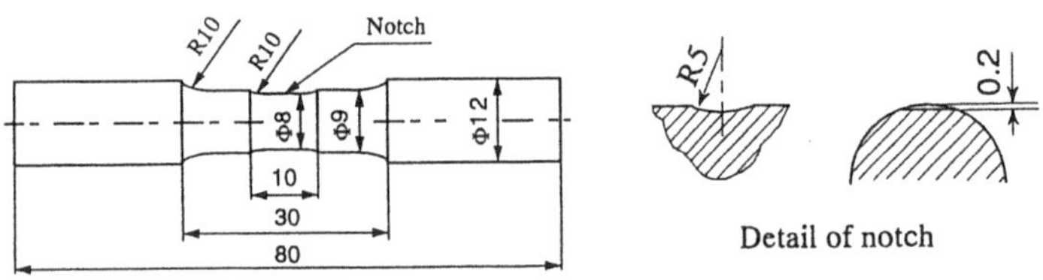

Fig.1. Dimensions (in mm) of the specimen and the transverse notch.

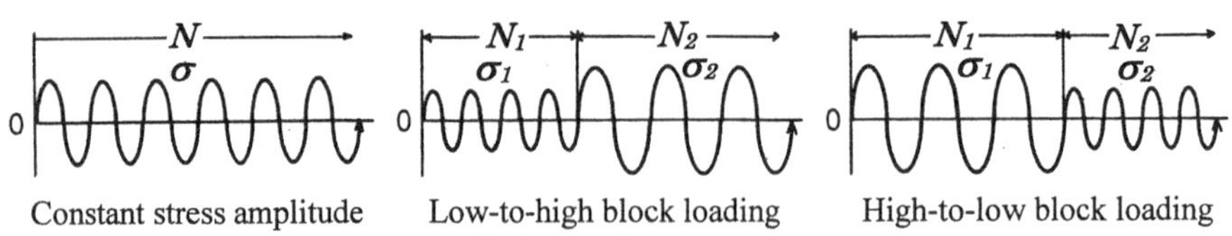

Fig.2. Loading patterns.

air under constant stress amplitude and in two-step loading (low-to-high and high-to-low block loading). Fig.2 shows the loading patterns used in the tests.

The observation of fatigue damage on the surface and the measurements of crack length were made via plastic replicas using an optical microscope at a magnification of $\times 400$. The value of stress, σ_a, means the nominal stress amplitude at the minimum cross section.

EXPERIMENTAL RESULTS AND DISCUSSION

Figure 3 shows the S-N curve at R.T. and at 500℃. Fatigue life at stresses beyond 650MPa is larger at R.T. than at 500℃. The fatigue limit stress at 10^7 cycles, σ_w, increases from σ_w = 450MPa at R.T. to σ_w = 625 MPa at 500℃.

Fatigue behaviour under constant stress amplitude

Figure 4 shows the change in the surface states under σ_a = 750 MPa at R.T. and 500℃. At R.T., slip bands are observed at $N = 4 \times 10^3$ cycles. A major crack, which led to the final fracture of the specimen, is initiated from the slip bands at around $N = 16 \times 10^3$ cycles. After initiation, the crack grows and the specimen was broken at $N_f = 5.51 \times 10^4$ cycles. For the case at 500℃, many slip bands are generated at the early stage of cycling (N < 2000 cycles). The density of slip bands is larger at 500℃ as compared to R.T.. At N = 4000 cycles, a propagated crack of l = 43μm is observed. Thus, the resistance to crack initiation is decreased at the elevated temperature.

Figure 5 shows the growth curves of major cracks (ln$l$ vs N relation) at R.T. and 500℃. The life to initiate a crack of 0.02mm under the same stress amplitude is smaller at 500℃ than at R.T.. With regard to the crack growth behaviour, the slope of the growth curve under the same stress is larger at 500℃ than at R.T. except for the microcracks less than 50μm. For crack lengths l = 20 to 50μm, however, the slope at 500℃ tends to be small. This trend is remarkable for the stresses close to the fatigue limit, i.e. after a crack propagates to 20 μm at σ_a = 650 MPa, the growth rate decreases rapidly followed by a temporary arrest. Beyond the arresting period, the

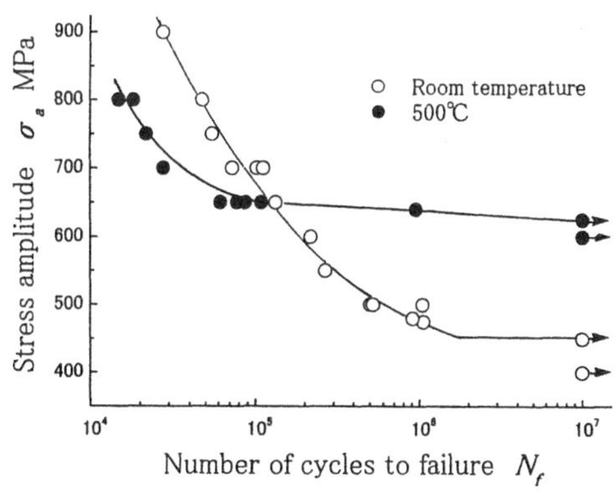

Fig.3. S-N curves at room temperature and 500℃.

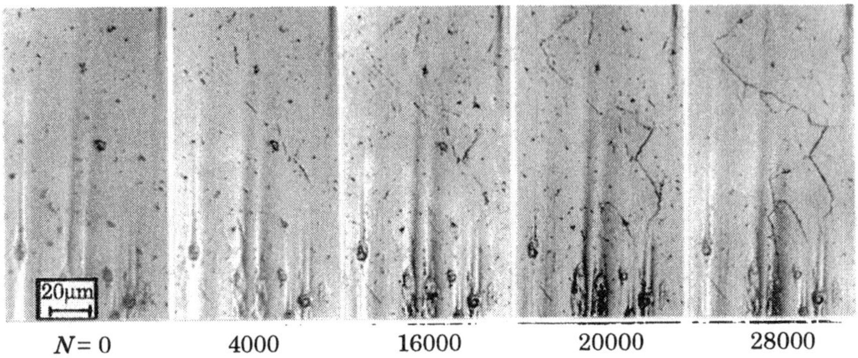

(a) $N_f = 5.51 \times 10^4$, ⇔: Axial direction.

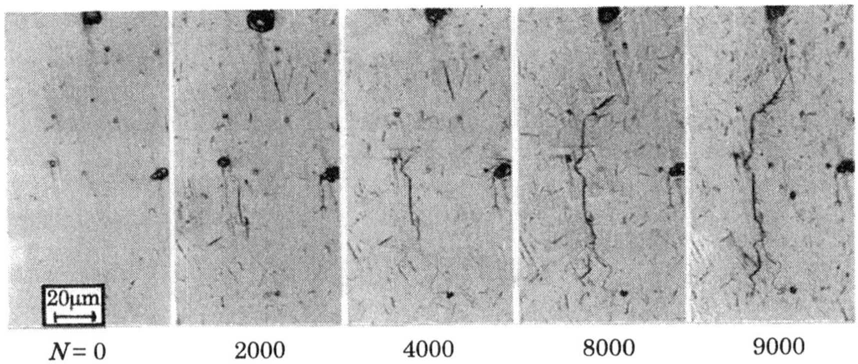

(b) $N_f = 2.19 \times 10^4$, ⇔: Axial direction.

Fig.4. Change in surface states around a major crack at $\sigma_a = 750$ MPa: (a) at R.T.; (b) at 500℃.

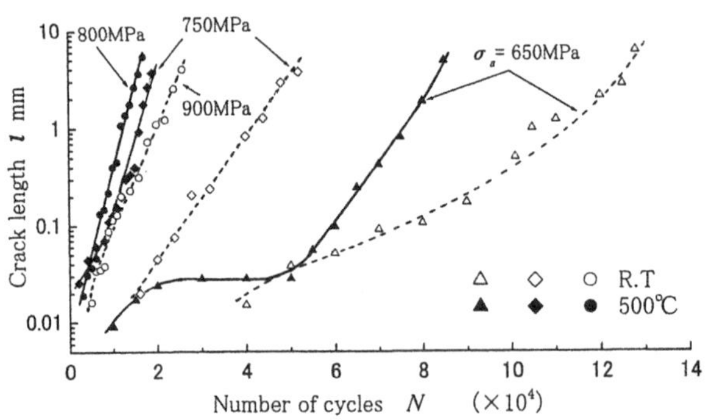

Fig.5. Major crack growth curves at room temperature and 500℃.

crack grows again and faster at 500℃ than at R.T.. Thus, the elevated temperature has the effect of suppression, the growth rate of a microcrack smaller than 50μm, but it increases the growth rate for a crack longer than $l = 50\mu m$. The suppression of the microcrack growth may be main reason for the increased fatigue strength at a low stress.

At the elevated temperature, the matrix is softened and oxide films cover the surface. It may be considered that the softening promotes crack initiation and crack growth but the oxide films formed coherently over the surface suppress crack initiation. The experimental results showed that crack initiation and the large-crack growth rate are enhanced at the elevated temperature, which results from the softening of the matrix. Conversely, the suppression of crack initiation due to oxide films may be negligible. The thickness of oxide films at 500℃ was 0.48μm after 1hr heating and 0.73μm after 6 hrs heating. The influence of oxide films on fatigue behaviour depends on the material type, temperature ranges, etc., see Kanazawa et al. [9] have reported that the fracture origin of a low-alloy steel SCMV2 in high-temperature fatigue was inside the specimen because of the suppression of surface crack initiation due to the presence of oxide films with a thickness over 1μm. On the other hand, the growth of a microcrack in the length range of 20 to 50 μm is enormously suppressed at 500℃, because the formation of oxide films on crack faces and the softening of the matrix, promoting plasticity and oxide-induced crack closures. Thus, the fatigue strength in the long-life region, especially the fatigue limit which is determined by the limiting stress for microcrack growth, is much higher at 500℃ than R.T..

Figure 6 shows the ln***l*** vs N/N_f relation. The relation at R.T. is approximated by a straight line for $\sigma_a > 700$ MPa, and a concave curve for $\sigma_a \leq 650$ MPa. This means that the crack growth characteristics at R.T. depend on the stress amplitude. The relation at 500℃ is approximated by a curved line excepting $\sigma_a = 650$ MPa. For a crack larger than 50μm, especially, the relation can be expressed by a straight line.

Figure 7 shows the ln***l*** vs $(N-N_{0.1})/(N_f-N_{0.1})$ relation at 500℃. $N-N_{0.1}$ is the number of cycles counted after the crack length has reached 0.1mm and $N_f-N_{0.1}$ is the crack growth life from l = 0.1mm to the fracture. The relation is approximated by a straight line independent of the stress amplitude, whereas the ln***l*** vs N/N_f relation at $\sigma_a = 650$ MP is different from other stress amplitudes. This indicates that there is no essential difference in crack growth characteristics at 500℃ between $\sigma_a = 650$ MPa and $\sigma_a > 700$ MPa.

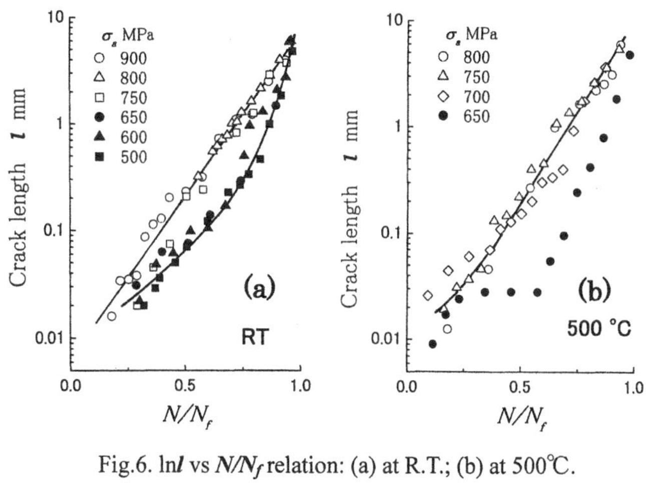

Fig.6. ln***l*** vs N/N_f relation: (a) at R.T.; (b) at 500℃.

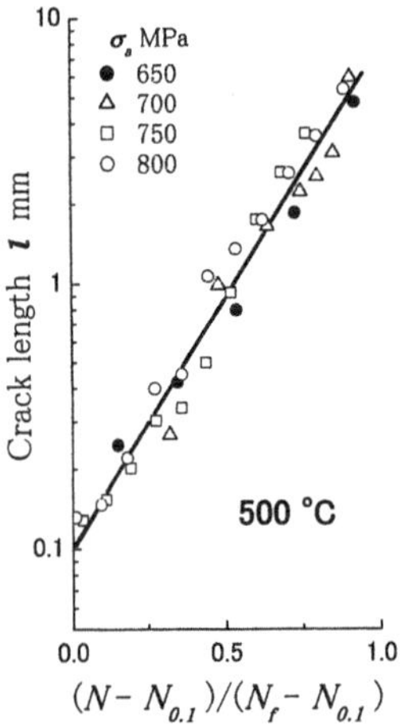

Fig.7. ln$l$ vs $(N-N_{0.1})/(N_f-N_{0.1})$ relation.

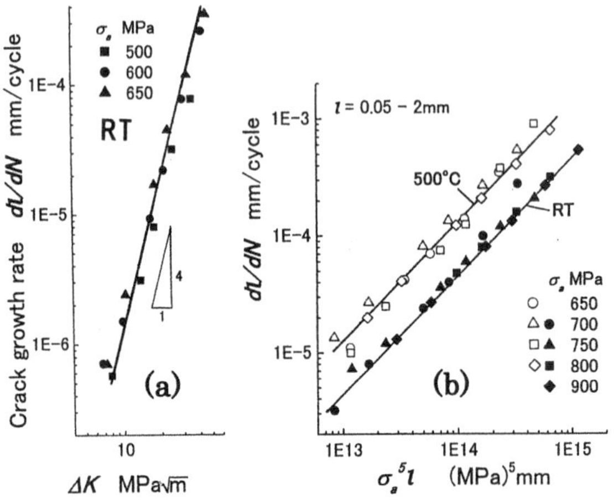

Fig.8. Crack growth data: (a) dl/dN vs ΔK relation; (b) dl/dN vs $\sigma_a{}^n l$ relation.

Figure 8 shows the dl/dN vs ΔK relation and the dl/dN vs $\sigma_a{}^n l$ relation [10-13]. At R.T., dl/dN under a low stress amplitude correlates with ΔK, but with $\sigma_a{}^n l$ under a high stress amplitude. At 500℃, dl/dN correlates with $\sigma_a{}^n l$. The value of n in $\sigma_a{}^n l$ is a material constant and is about 5 at both temperatures. This is directly related to the parallel S-N relation at a high stress amplitude for both temperatures. The stress intensity factor range was calculated as $\Delta K = (2/\pi)\Delta\sigma(\pi l/2)^{1/2}$.

Fatigue behaviour under two-step loading

Experiments at room temperature. The stresses used for the two-step loading at R.T. were σ_a = 500 and 700 MPa. The mean fatigue life under the constant stress amplitude was $N_f = 6.92 \times 10^5$ cycles for σ_a = 500 MPa and $N_f = 9.54 \times 10^4$ cycles for σ_a = 700 MPa. Figure 9 shows the ln$l$ vs N/N_f relation under the constant stresses σ_a= 500 and 700 MPa. The crack growth characteristics

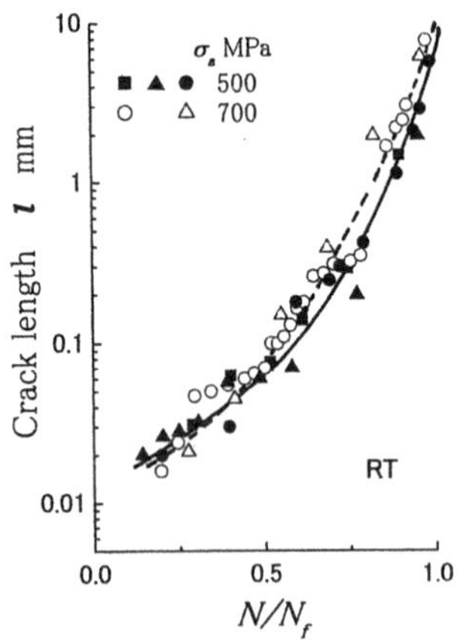

Fig.9. Crack growth characteristics for σ_a = 500 and 700 MPa (at R.T.).

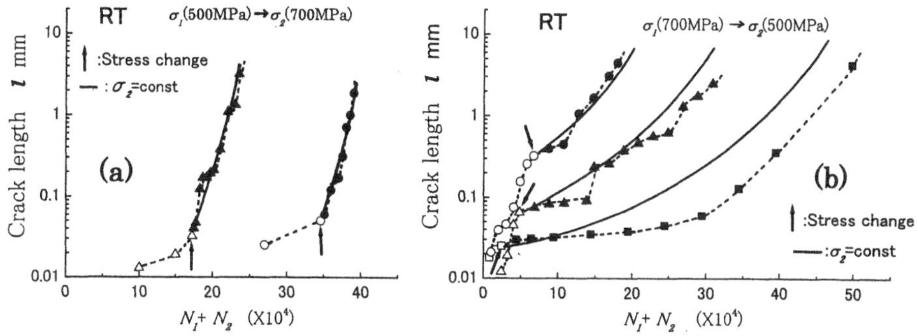

Fig.10. Crack growth curve under the two-step loading at R.T.: (a) low-to-high block loading; (b) high-to-low block loading.

Table 1. The values of cumulative cycle ratio for the two-step loading (at R.T.).

Stress (MPa)	N_1 (cycles)	$\Sigma(N_1/N_{f1})$	N_2 (cycles)	$\Sigma(N_2/N_{f2})$	$\Sigma(N/N_f)$	Symbols in Fig.10
500	173000	0.25	79000	0.83	1.08	▲
⇓	346000	0.50	54000	0.57	1.07	●
700	346000	0.50	44000	0.46	0.96	
	554000	0.80	20500	0.21	1.01	
700	24000	0.25	481700	0.70	0.95	■
⇓	48000	0.50	284000	0.41	0.91	▲
500	67000	0.70	125500	0.18	0.88	●

at σ_a = 700 MPa are different from those at σ_a = 500 MPa, because the shift of crack growth law from ΔK to $\sigma_a^n l$ occurs at around σ_a = 700 MPa (see Fig.6(a)). However, the difference in the relation for both stresses is not large.

Figure 10 shows the crack growth curve for (a) low-to-high and (b) high-to-low block loading. The ↑ symbol indicates the position where the stress is changed from σ_1 to σ_2, σ_1 and σ_2 being the first and second stress levels, respectively. The full lines show the growth curve under the constant stress of σ_2 (no repetitions of σ_1). There is no significant difference between the growth data after the stress change and each full line for both the loading patterns. Thus, it can be concluded that the crack growth under σ_2 is not affected by the history of the first stress cycling.

Table 1 shows the values of cumulative cycle ratio $\Sigma(N/N_f)$. The values of $\Sigma(N/N_f)$ are roughly equal to unity ($0.88 < \Sigma(N/N_f) < 1.08$). Thus, Miner's rule nearly holds in this case. The $\Sigma(N/N_f) \fallingdotseq 1$ results from the lack of an effect of history of σ_1 on the subsequent growth behaviour under σ_2 and the negligible difference in the ln$l$ vs N/N_f relation between σ_a = 500 and 700MPa.

Experiments at the elevated temperature. The stresses used for the two-step loading at 500℃ were σ_a = 650 and 800 MPa. The mean fatigue life under the constant stress amplitude was $N_f = 8.39\times10^4$ cycles for σ_a = 650 MPa and $N_f = 1.50\times10^4$ cycles for σ_a = 800 MPa. Figure 11 shows the ln$l$ vs N/N_f relation at the constant stresses σ_a = 650 and 800 MPa. There are large differences in the relation between σ_a = 650 and 800 MPa, caused by the temporary arrest of microcrack growth at σ_a = 650 MPa.

Figure 12 shows the crack growth curves for (a) low-to-high and (b) high-to-low block loading. For the case of low-to-high block loading, there is no significant difference between the growth data after the stress change and the full lines which represent crack growth curves under the constant stress of σ_2. Thus, it can be concluded that the crack growth behaviour under the second stress is not affected by the history of the first stress cycling. For the case of high-to-low block loading, the effect of σ_1 on the growth behaviour under σ_2 is determined by the length of crack initiated under σ_1. That is, when the stress is changed before crack initiation (● symbol), no effect of the first stress on the growth behaviour under σ_2 is observed. A similar result is obtained for a large crack (l > 50μm) initiated at σ_1 (■ symbol). However, when the length of a crack initiated at σ_1 is about l = 20 to 50μm, the growth behaviour is quite different from that of a crack initiated at σ_2 (▲ symbol). Namely, in the two-step loading, the temporary crack arrest

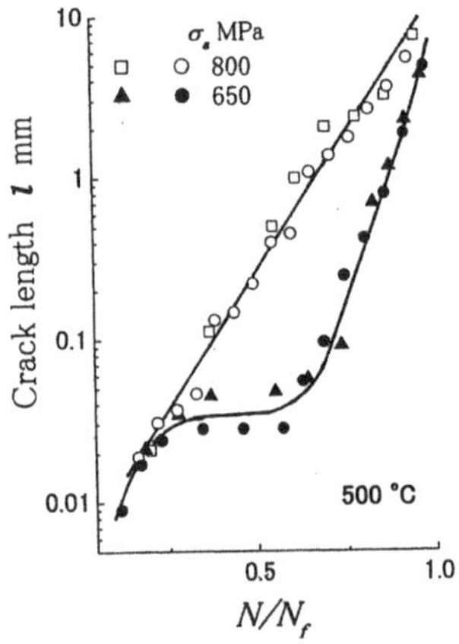

Fig.11. Crack growth characteristics for σ_a = 650 and 800 MPa (at 500℃).

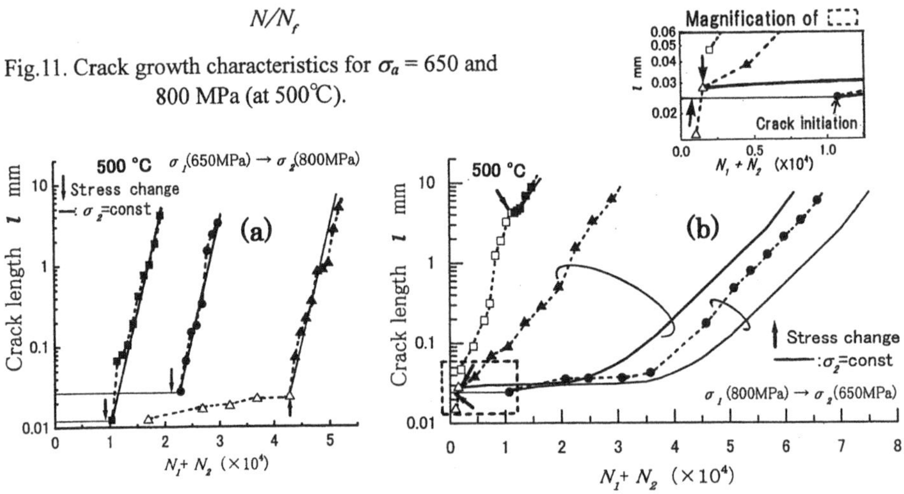

Fig.12. Crack growth curve under the two-step loading at 500℃: (a) low-to-high block loading; (b) high-to-low block loading.

Table 2. The values of cumulative cycle ratio for the two-step loading (at 500℃).

Stress (MPa)	N_1 (cycles)	$\Sigma(N_1/N_{f1})$	N_2 (cycles)	$\Sigma(N_2/N_{f2})$	$\Sigma(N/N_f)$	Symbols in Fig.12
	8300	0.10	12700	0.84	0.94	■
650	20900	0.25	11200	0.74	0.99	●
⇓	41900	0.50	10900	0.72	1.22	▲
800	41900	0.50	8200	0.54	1.04	
	62900	0.75	10900	0.72	1.47	
	71200	0.85	7000	0.46	1.31	
	700	0.05	43900	0.52	0.57	●
	700	0.05	67000	0.80	0.85	
800	1500	0.10	59500	0.71	0.81	
⇓	1500	0.10	29000	0.35	0.45	▲
650	3700	0.25	22700	0.27	0.52	
	7500	0.50	13300	0.16	0.66	
	11300	0.75	4200	0.05	0.80	■

usually observed for a crack initiated at $\sigma_2 = 650$MPa disappears, whereas for a crack larger than l = 50μm, no significant difference in dl/dN between the constant amplitude and two-step loading is observed. Although the mechanism for the disappearing of growth arrested is not clear, it may result from decreased crack closure rather than from the decreased crack growth resistance relating to the change in material properties caused by the loading under σ_1. Such a decreased crack closure may result from the large opening ratio of a crack initiated at a high stress and thin oxide films due to the small time (number of cycles) required for crack initiation. Goto et al. [14] have reported that, for high-to-low block loading of a heat-treated 0.45% C steel at room temperature, the growth rate under low stress is accelerated as the result of cyclic softening caused by the repetitions of high stress.

Table 2 shows the values of cumulative cycle ratio $\Sigma(N/N_f)$. All the values of $\Sigma(N/N_f)$ are within the range $0.45 < \Sigma(N/N_f) < 1.47$. For low-to-high block loading, $\Sigma(N/N_f)$ tends to be larger than unity ($0.94 < \Sigma(N/N_f) < 1.47$) in spite of no influence of the first stress on the subsequent crack growth. This tendency comes from the stress dependency of the ln$l$ vs N/N_f relation (Fig.9). For high-to-low block loading, $\Sigma(N/N_f)$ is less than unity ($0.45 < \Sigma(N/N_f) < 0.85$). The values of $\Sigma(N/N_f) < 1$ can be explained by the stress dependency of the ln$l$ vs N/N_f relation and the acceleration of microcrack growth rate under the second stress. Therefore, when examining the fatigue damage under complex loading it should be taken into account that the value of $\Sigma(N/N_f)$ has a distinct tendency determined by the loading pattern and temperature.

CONCLUSIONS

In order to study the fatigue damage of an Inconel 718 superalloy at room temperature and at 500℃, plain specimens were fatigued under constant stress amplitude and two-step loading. The fatigue damage was discussed based on the behaviour of crack initiation and small crack growth. The main conclusions may be summarized as follows:

1. The resistance to slip and crack initiation decreased due to the softening of matrix at elevated temperature.
2. With regard to the effect of the elevated temperature on the crack growth, the growth rate ***dl/dN*** of large cracks was accelerated as compared to R.T., however the growth of microcracks less than 50μm was suppressed. The suppression of microcrack growth tends to increase with a decrease in stress; especially it was remarkably large when the stress approaches the fatigue limit.
3. The suppression of microcrack growth may be caused by the plasticity and oxide-induced crack closure that results from the softening of the matrix and from the formation of oxide films on crack faces.
4. The ***dl/dN*** of a small crack at room temperature correlates with ΔK for a low stress amplitude and a term $\sigma_a^5 l$ for a high stress amplitude. At 500℃, ***dl/dN*** of a small crack larger than 50μm correlates with $\sigma_a^5 l$ at all stress levels.
5. For the two-step loading at room temperature, the subsequent crack growth under the second stress: σ_2 is not affected by the history of the first stress: σ_1.
6. With regard to the two-step loading at 500℃, for low-to-high block loading, no effect of the first stress on the subsequent crack growth behavior under σ_2 was observed. For high-to-low block loading, however, the growth behavior of cracks less than 50μm under σ_2 was affected by the cycling of σ_1. These results were explained from the viewpoint of the softening of matrix and the oxide films formed at the elevated temperature.
7. The cumulative cycle ratio $\Sigma(N/N_f)$ was calculated for the two-step loading. The values of $\Sigma(N/N_f)$ exhibit a distinct tendency determined by the loading pattern and temperature. This tendency can be explained by the crack growth behaviour.

REFERENCES

1. King, J.E. (1987) *Mater. Science Technology* **3**, 750-764.
2. Bache, M.R., Evans, W.J. and Hardy, M.C. (1999) *Int. J. Fatigue* **21**, S69-S77.
3. Lankford, J. and Cook, T.S. (1981) *Int. J. Fracture* **17**, 143-155.
4. Sheldon, G.P., Cook, T.S., Jones, J.W. and Lankford, J. (1981) *Fatigue Eng. Mater. Strict.* **3**, 219-228.
5. Boyd-Lee, A. and King, J.E. (1994) *Fatigue Fract. Mater. Struct.* **17**, 1-14.
6. Okazaki, M., Ohshima, S. and Nohmi, S. (1994) *J. Soc. Mater. Science Jpn* **43**, 860-866.
7. Kawagoishi, N., Chen Q., Nisitani, H., Goto, M. and Tanaka, H. (1997) *Trans. Jpn Soc. Mech. Eng.* **63**, 2298-2302.
8. Nisitani, H. and Goto, M. (1986) In: *The behaviour of Short Fatigue Cracks*, pp.461-478, Miller, K.J. and de los Rios, E.R. (Eds). Mech. Eng. Publications, London.
9. Kanazawa,K. and Nishijima, S. (1997) *J. Soc. Mater. Science Jpn* **46**, 1396-1401.
10. Nisitani, H. (1981) In: *Mechanics of Fatigue-AMD*, pp.151-166, Mura.T. (Ed). ASME.
11. Nisitani, H., Goto, M. and Kawagoishi, N. (1992) *Eng. Fract. Mech.* **41**, 499-513.
12. Goto, M. and Knowles, D.M. (1998), *Eng. Fract. Mech.* **60**, 1-18.
13. McEvily, A.J. and Ishihara, S. (2001) *Int. J. Fatigue* **23**, 115-120.
14. Goto, M., Nisitani, N., Miyagawa, H. and Yanagawa, Y. (1990) *Jpn Soc. Mech. Eng. International Journal* Ser.I, **33**, 249-255.

Temperature-Fatigue Interaction
L. Rémy and J. Petit (Eds.)

THE EFFECT OF TEMPERATURE ON CRACK BEHAVIOR IN AN 7175 ALUMINUM ALLOY UNDER MODE I + STEADY MODE III

F.S. SILVA[1] and A.C.M. PINHO[2]
Mechanical Engineering Department, Minho University
Campus de Azurém, 4800 Guimarães, Portugal

ABSTRACT

This paper deals with the role of temperature in the behaviour of fatigue cracks in an aluminum alloy (7175). Push-Pull tests ($R = -1$) with an imposed steady torsion were carried out at different temperatures (-100°C; +20°C; and +100°C) in air. The experiments were carried out with round specimens with a pre-crack. A DCPD controlling system was used to measure the crack growth on this study. At each temperature the role of steady torsion on the shape of the $da/dN\text{-}K_{eq}max$ curve was assessed. The crack propagation rate was studied. Roughness influence on crack closure and cyclic plastic properties were studied in order to understand their influence on propagation of fatigue cracks. The effect of temperature in the referred subjects was discussed. The existence of a ductile-brittle fatigue transition temperature, parallel to the monotonic ductile-brittle transition temperature, although these transition temperatures are not the same, was assessed.

KEYWORDS

Multiaxial fatigue; steady torsion; high temperature; low temperature; cyclic properties.

INTRODUCTION

Complex stresses prevail in multiple engineering situations. Steady torsion (*Mode III*) plus alternated tension (*Mode I*) are common in many applications such as rotating parts. The use of aluminum alloys is more and more demanded in order to reduce costs, particularly in automotive and aeronautical industry. Finally, many of the components should be prepared to operate over a wide range of temperatures which may vary from -50°C to 50°C.

Many studies exist in uniaxial fatigue at room temperature and at high temperature. However, very few studies deal with complex stresses at low temperature. Influence of complex stresses, and in particular steady torsion imposed to alternated tension, at different temperatures, is still a relatively new subject.

Some authors [1-10] studied the effect of an imposed steady torsion on an alternated stress, at ambient temperature. Round specimens with a circumferential crack were

[1] Email: Fsamuel@dem.uminho.pt
[2] Email: acmpinho@dem.uminho.pt

used [1-6],with a semi-elliptical crack [7-10]. Most of them [1-9] conclude that an imposed steady *mode III* decrease the alternated *mode I* crack propagation rate and increase the fatigue threshold, K_{th}[6]. This is atributed to crack closure due to the surface roughness, or the so called "factory roof effect" which is a result of the imposed steady torsion. Other studies [11-12], on the effect of this imposed steady *mode III*, made with a titanium alloy, a 7175 aluminum alloy, and two steels (4340 and C45E), with round specimens with a semi-elliptical crack, at room temperature, suggest that in addition to surface roughness cyclic properties appear to have influence on crack growth. However, there aren't many studies on fatigue behavior at low temperatures, mainly because it is accepted that low temperatures provide better fatigue resistance than room temperature [13-14]. When complex stresses are present it is even more difficult to find investigations on fatigue properties of materials at low temperature.
The aim of this work is to understand the influence of an imposed steady stress of torsion in the fatigue behavior of components with alternated tension (R = *-1*) in a temperature range from –100°C to + 100°C.

MATERIALS AND METHODS

Material and Specimens

The material used in this investigation is an Al 7175-T73511 alloy. The chemical composition, obtained with X-Ray fluorescence, in (wt.%), is Si:0,048; Fe:0,18; Cu:1,57; Mn:0,052; Mg:2,54; Cr: 0,22; Ni:0.003; Zn:6,4; Ti:0,025, delivered by Pechiney. The specimens used are round specimens according to ASTM E 606-80 with a pre-crack [11-12](fig. 1b). Mechanical properties are listed in table 1.

Table 1. Mechanical Properties of Al 7175

Temperature °C	$\sigma_{ced\,(0,2\%)}$ *MPa*	σ_r *(MPa)*	*E* *(MPa)*	ε_r (%)
- 50°C	477	561	$0,937*10^5$	10,6
+20°C	461	535	$0,736*10^5$	13,8
+ 100°C	417	493	$0,672*10^5$	13,9

Methods

Fatigue tests were conducted at various temperatures (*-100°C; 20°C and 100°C*) in laboratory air using a controlled sinusoidal loading (R= *-1*) at a frequency of *8 Hz* on a servo-hydraulic testing machine. Cyclic curves were obtained with the multiple step test method, under deformation control.

The steady torsion was introduced using an assembly with dead weight [12] (fig. 1a). At each temperature both the role of steady torsion on the shape of the da/dN-$K_{eq}max$ curve and the *a-N/Nf* curves were studied. A Pulsed DCPD system, operating with a high resolution [15], was used to measure the crack growth. The deepest point of the crack was measured. Fracture surface examinations were made using an optical microscope. The high temperature tests were carried out with a heating furnace with three heating levels and the temperature was controlled within 1 °C . The low temperature tests were carried out in a cooling chamber, with nitrogen, with an accuracy of ± 1 °C.

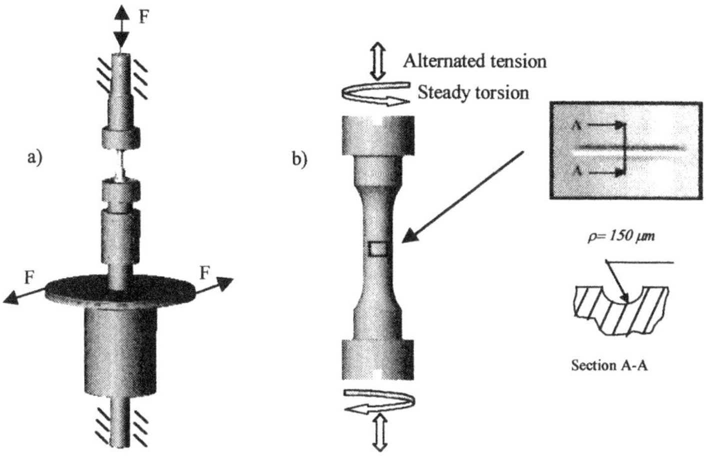

Fig. 1 a) Assembly to introduce steady torsion; b) specimen with pre-crack detail.

da/dN curves were plotted against a maximum equivalent stress intensity factor, $K_{eq}max$, for two reasons.

1. Crack propagation plane changes its direction when an alternated torsion is imposed to an alternated stress. Fig. 2 shows an example of a fatigue crack, growing in a plane perpendicular to the alternated stress (without steady torsion) (fig. 2a), and another crack (with steady torsion)(fig. 2b) growing in a plane which is deflected in relation to the usual one, on alternated stress.

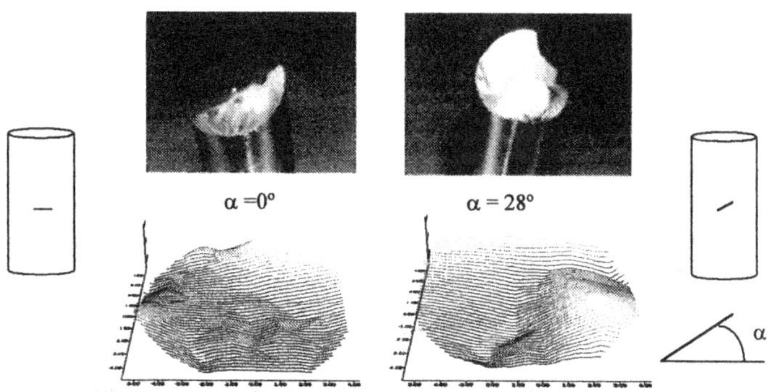

Fig. 2 Example of crack propagation planes at $T=+20°C$: a) without steady torsion, $\Delta\sigma= 256$ MPa; $\tau= 0$ MPa; b) with steady torsion. $\Delta\sigma= 256$ MPa; $\tau= 256$ MPa

Due to the deflection of the crack, when is imposed the steady torsion, the alternated stress will produce mode *I* and mode *III* at point B, the deepest point of the crack.

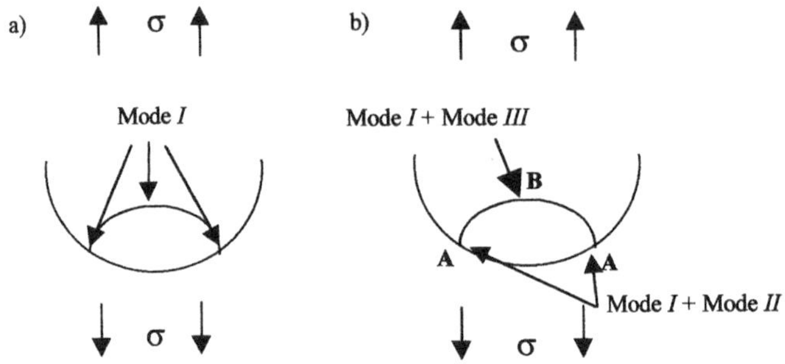

Fig. 3 Different modes present on a crack. a) without steady torsion; b) with steady torsion (inclined crack).

Pook, [17] suggests that K_I and K_{III} can be calculated using the aproximate expressions:

$$K_I = K_A \cos^2 \alpha ; \qquad K_{III} = K_A \cos\alpha \sin\alpha \tag{1}$$

where K_A is K_I for pure mode I ($\alpha = 0°$). K_A was calculated using an expression proposed by Murakami [18] for semi surface cracks in round bars under uniform tension. K_{eq}. was obtained using the expression proposed by Tanaka [19]:

$$\Delta K_{eq} = \left[\Delta K_I^{\ 4} + 8\Delta K_{II}^{\ 2} + \frac{8\Delta K_{III}^{\ 4}}{(1-\upsilon)} \right]^{1/4} \tag{2}$$

2. According to ASTM 647,

$$\Delta K = K_{max} \ for \ R \leq 0.$$

Thus, for $R = -1$, a maximum equivalent stress intensity factor must be used.

RESULTS

Crack growth data is presented in figs. 4, 5, and 6. In these figs. it is shown the effect of steady torsion on crack propagation rates from –100 °C up to +100°C. The related optical fotographs show both the fracture morphology and surface roughness of the fractures at the same temperatures.

In order to understand the effect of steady torsion both on crack initiation and crack propagation fig. 7 shows *da/dN* – $K_{eq}max$ curves with the effect of temperature in propagation rates, with and without steady torsion. Fig. 8 shows the relative lives of the tests. Stresses and lives for the specimens tested in this study are given in table 2.

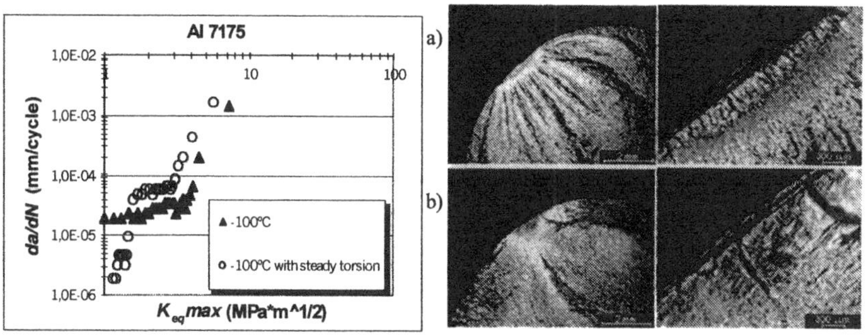

Fig. 4. Fracture morphologies and crack growth data. *T= -100°C*. a) without steady torsion ; b) with steady torsion.

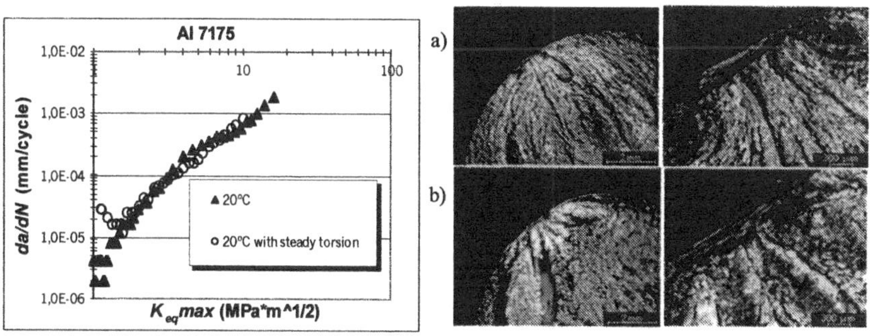

Fig. 5. Fracture morphologies and crack growth data. *T= +20°C*. a) without steady torsion; b) with steady torsion.

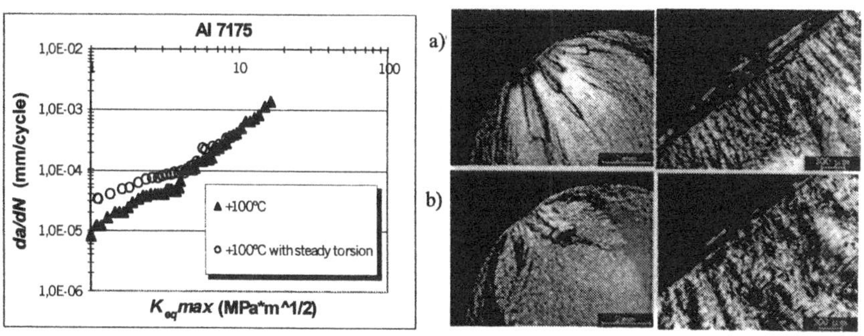

Fig. 6. Fracture morphologies and crack growth data. *T= +100°C*. a) without steady torsion; b) with steady torsion.

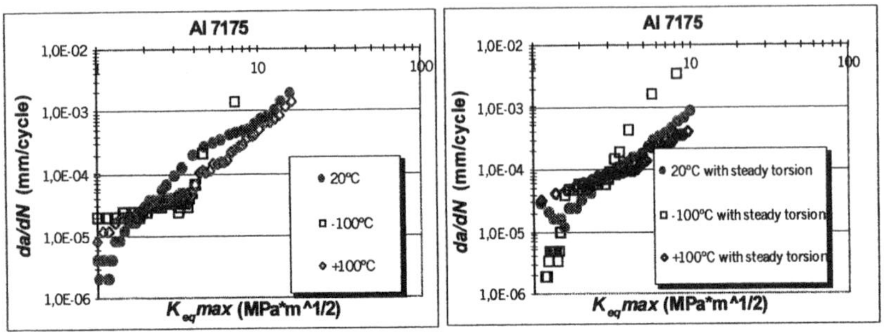

a) b)

Fig. 7. Crack growth data at different temperatures. a) without steady torsion; b) with steady torsion

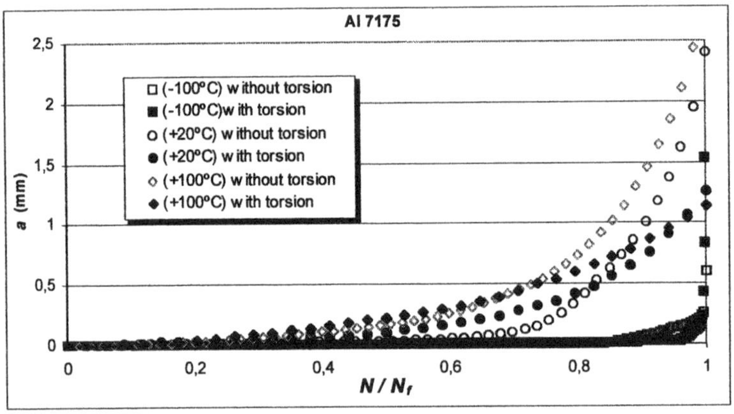

Fig. 8. *a-N/Nf* curves for –100°C, +20°C, and 100°C, with and without steady torsion.

Table 2. Testing Conditions of Al 7175. R= -1; f = 8 Hz.; N_f – Average life until complete fracture;

Loading Conditions	***T* [°C]**	***Δσ [MPa]***	***τ [MPa]***	**N_f [cycles]**
Without torsion	-100	256	0	47.105
With Torsion	*-100*	*256*	*256*	*73.863*
Without torsion	+20	256	0	13.952
With Torsion	*+20*	*256*	*256*	*11.739*
Without torsion	+100	256	0	12.770
With Torsion	*+100*	*256*	*256*	*7.990*

From figs. 4 to 8 and from table 2 the results can be summarized as follows:

1. When a steady torsion is imposed to an alternated tension,

- Crack propagation rate increases for low temperatures, -100°C (fig.4);
- Crack propagation rate doesn't change much at +20°C and +100°C (figs.5-6);
- At +20°C and +100°C, steady torsion reduces the relative life spent on crack initiation. At low temperatures, - 100°C, steady torsion increases the relative life spent on crack initiation (fig. 8).

2. With and without steady torsion,

- Crack propagation rate decreases with temperature. For low temperature crack propagation is much faster than for room and high temperature. (fig.7);

On fig. 9 are the monotonic and cyclic properties at different temperatures.

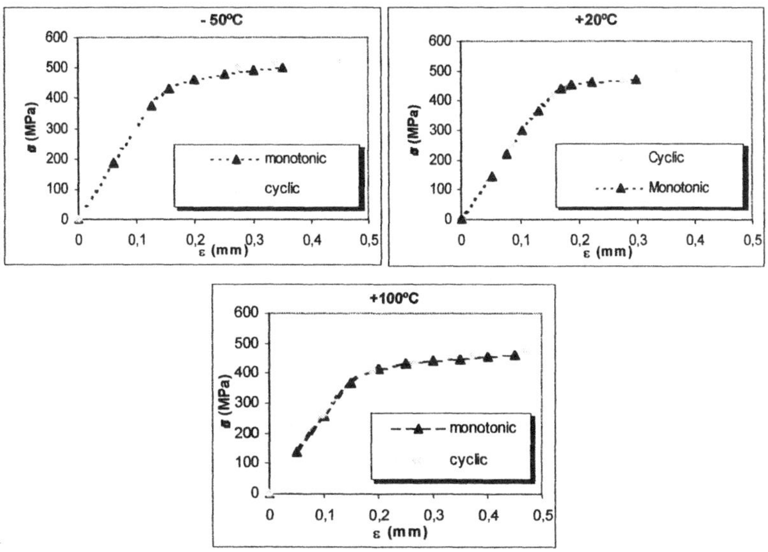

Fig. 9. Monotonic and cyclic curves for different temperatures.

On fig. 9 it is observed that at + 20°C and +100°C the aluminum alloy is almost neutral, while at low temperatures, -50°C, this alloy becomes a cyclic hardening alloy.

DISCUSSION

Tschegg [5,6] considered that two main mechanisms should be assessed when studying alternated *mode I* + steady *mode III*: closure effects caused by the macrofaceted structure; and work hardening at the crack tip. He concluded that strain hardening is not a relevant mechanism, and closure, due to roughness, is responsible for retardation on crack propagation rates.

By the fracture morphologies observed on figs. 4-6 it is clear the effect of steady torsion upon morphology. It is clear the friction between surfaces due to the rotating effect caused by the imposed torsion.

Contrary to other materials, in Al 7175 roughness does not seem to increase with temperature. For Al 7175, the fracture surface is very irregular either at –100°C or at +100°C. In materials such as Ti6Al4V, C45E or 4340 steels [11-12,16] fracture surface is very smooth for low temperatures and roughness increases with temperature. By the observation of figs 4 through 6 one should conclude that for Al 7175 closure effects due to roughness should be important for all temperatures, within the range –100°C to +100°C. Thus, it can also be concluded that, when steady torsion is imposed to alternated tension, retardation effects on crack propagation, should be present at the range of temperatures studied. This is not the case in this study. There is no significant retardation effect at any temperature, and at –100°C the opposite occurs. These observations lead us to conclude that closure effects due to roughness are not an important mechanism when is imposed a steady torsion, at least on Low Cycle Fatigue ($N_f < 10^5$). There were two main differences between tests made by Tschegg and Hourlier [5,6] and tests on this study. Tschegg studied mainly the threshold regime, and used round specimens with a circumferential crack. Hourlier and Tschegg [2-3, 5-6] also studied the effect of stress ratio, *R*. In this work, it is observed the behavior of round specimens, in low cycle fatigue regime, and with semi-elliptical cracks, exclusively for *R*= *-1*. The reason why roughness and 'factory roof effect' are considered the main mechanisms on crack retardation effect, instead of plastic properties, in those studies [2-3, 5-6], may be related with the low stresses and low plastic deformation involved. Another reason may be related with the crack front which is different in a circumferential crack and in a semi-elliptical crack.

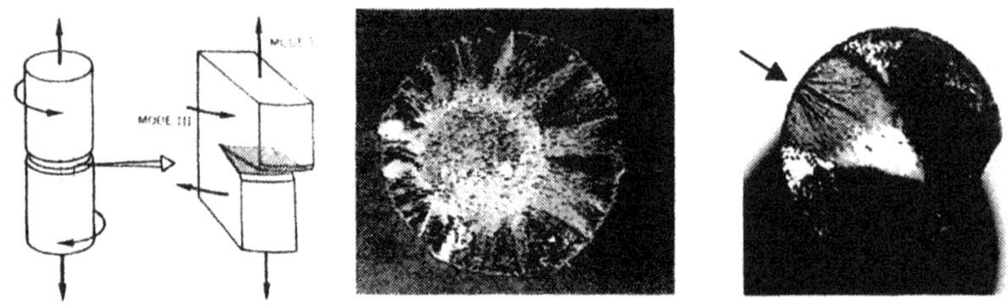

a) b) c)

Fig. 10: a) Distortion of fracture surface in a circumferential crack; b) Fracture morphologie for a circumferential crack; c). Fracture morphologie for a semi-elliptical crack

From the observation of fig. 10b and 10c, it is easy to conclude that, as the crack grows, for the same crack length and specimen size, it is easier for the specimen with circumferential crack to have a distortion, due to steady torsion, (fig. 10a) than for the specimen with semi-elliptical crack, since the remainder resistant area of the specimen is smaller for the specimen with a circumferential crack. The consequence may be a more pronounced influence of roughness and particularly of 'factory roof effect' on crack closure, in circumferential cracks than in semi-elliptical cracks. Thus, with semi-elliptical cracks, and in low cycle fatigue, it is acceptable that roughness and 'factory roof effect' may have less influence on crack propagation than cyclic plastic properties.

A second observation concerns the low temperature. At –100°C, without steady torsion (fig.7a), crack propagation rate is higher than at other temperatures. As the material becomes more resistant at low temperature, it should be expected a slower crack propagation rate, but this does not happen. This result is also reported by Fuchs and Stephens [13] and suggests that it may exist a 'fatigue ductile-brittle transition temperature' similar to the monotonic ductile-brittle transition temperature. This 'fatigue transition temperature' also exists for materials less sensitive to monotonic ductile-brittle transition temperature, as is the case of materials such as aluminum and titanium. Tests with Ti6Al4V [16] show an identical behavior.

From fig. 4 it can be observed that at low temperature, contrary to tests at +20°C and +100°C, steady *mode III* causes a substantial increase in crack propagation rate. Under these circumstances, the material becomes more sensitive to steady torsion. Although roughness effect is clearly seen in fig. 4b, crack propagation becomes faster. This effect may be related to the 'fatigue ductile-brittle transition temperature'.

In relation to total life it is interesting to observe that steady torsion reduces specimens life at +20 and +100°C, but increases the total life at –100°C (table 2). It seems that steady torsion retards the initiation of the crack at low temperatures and accelerates its initiation at positive (>0°C) temperatures.
Because this study concernes the low cycle fatigue regime (N_f< 100.000 cycles) the mechanism responsible for this behavior may be related with cyclic plastic properties of the material and with its variation with temperature. In fact, in fig. 9 it is observed that, while at +20°C and +100°C, Al 7175 is nearly neutral, at low temperatures (-50°C) it is cyclically hardening. Thus, the hardening behavior at the crack tip may be the mechanism responsible for crack retardation, when steady torsion is imposed. Steady torsion increases the cyclically plastic zone size and changes the plastic behavior at the crack tip, mainly when there is a stress concentration, as exists in this work at the pre-crack. Recent studies [11-12] on different materials at room temperature, with semi-elliptical crack, have shown that cyclic behavior is the main mechanism, instead of roughness, when steady torsion is imposed to alternated tension. In these studies [11-12], when steady torsion is imposed to alternated tension, total life decreases in cyclic softening or neutral materials, and when the material is cyclically hardening total life increases.
Thus, it seems that there is a strong relation between cyclic properties and fatigue behavior, mainly in LCF, in specimens with semi-elliptical cracks, and in the presence of notches.

CONCLUSIONS

The main conclusions of this work are:

- With semi-elliptical cracks and $R = -1$, closure effects due to roughness does not seem to be a relevant mechanism on crack propagation rates, in spite of the very clear effect of torsion on fracture morphologies;

- It seems that exists a 'fatigue ductile-brittle transition temperature' even for materials less sensitive to monotonic ductile-brittle transition temperature, such as Al 7175;
- At positive temperatures (T > 0°C) steady torsion reduces slightly total life of specimens and does not have much influence on crack propagation rates;
- Below the 'fatigue ductile-fragile transition temperature' the effect of a steady torsion becomes the opposite, increasing the total life of the specimen but increasing also the propagation rate.
- Cyclic plastic properties may be the mechanism responsible for the fatigue behavior, in Low Cycle Fatigue, and on crack initiation from notches, when steady torsion is imposed to an alternated tension.

REFERENCES

1. Hourlier, F., Pineau, A., (1982) *Fatigue Fract. Engng. Mat. Struct.* 5, pp. 287-302
2. Hourlier, F., Pineau, A., (1979) *Mémoires Scientifiques Révue Metallurgie*, pp. 175-185
3. Hourlier, F., Mclean, D., Pineau, A., (1976), *Metals Technology*, pp.154-158
4. Hourlier, F., Hondt, H., Truchon, M., Pineau, A., (1985), *ASTM STP 853*, pp. 228-247
5. Tschegg, E.K., Stanzl, S.E., Mayer, H., Czegley, M., (1992) *Fatigue Fract. Engng. Mat. Struct.* 16, pp. 71-83
6. Tschegg, E.K., Mayer, H.R., Czegley, M., Stanzl, S.E., (1991) *ESIS 10*, pp. 213-222
7. Pinho, A.C.M., (1996), PhD Thesys,
8. Freitas, M. M, François, D., (1995) *Fatigue Fract. Engng. Mat. Struct.* 2, pp. 171-178
9. Fonte, M.A., Freitas, M.M., François, D., (1994) *4th Int. Conf. On Biaxial/Multiaxial Fatigue*, pp. 159-170
10. Fonte, M.A., Freitas, M.M., (1997) *Fatigue Fract. Engng. Mat. Struct.* 20, pp. 895-906
11. Silva, F.S.; Pinho, A.C.M.; (2001) *10th Int. Conf. Fracture*, Honolulu
12. Silva, F.S.; Pinho, A.C.M.; (2001) *6th Int. Conf. Biaxial Multiaxial Fatigue Fracture*, Lisbon, pp. 663-672
13. Fuchs, H.O., Stephens, R.I., (1980) *Metal Fatigue in Engineering*", John Wiley & Sons, , pp. 176-181
14. Frost, N.E. ; Marsh, K.J.; Pook, L.P.; (1974) *Metal Fatigue*, Clarendon Pres, Oxford, , pp. 90-92
15. Silva, F.S.; Pinho, A.C.M.; Peixinho, N.; Meireles, J., (2000) *Damage and Fracture Mechanics VI*, Wit Press, pp. 353-364.
16. Silva, F.S.; Pinho, A.C.M.; (2001) *CREEP7*, Japan, pp. 347-352
17. Pook, L.P., (1972) *Engng. Fract. Mech.* 8 pp.267-276
18. Murakami, Y., (1987) *Stress Intensity Factors Handbook*, Pergamon Press, Oxford, Vols. 1 e 2, , pp. 661
19. Tanaka, K., (1974) *Engng. Fract. Mech.* 6, pp.493-507

HIGH TEMPERATURE FATIGUE CRACK GROWTH RATE IN INCONEL 718 : DWELL EFFECT ANNIHILATIONS

S. PONNELLE[1-3], B. BRETHES[2], A. PINEAU[1]

(1) Centre des Matériaux, Ecole des Mines de Paris, BP 87, 91003 Evry cedex, France UMR CNRS 7633.

(2) SNECMA Moteurs – YKOM1- Etablissement de Villaroche, 77550 Moissy Cramayel, France.

(3) SNECMA Moteurs– YKOG2- Etablissement de Gennevilliers, 291 av. d'Argenteuil 92234 Gennevilliers Cedex, France.

ABSTRACT

Inconel 718 superalloy is a widely used material for turbine discs working at elevated temperatures (up to 650°C). In this material tested at high temperatures, it has often been reported that fatigue crack growth rates (FCGR) increase when a dwell time is hold at maximum load. This dwell effect is known to be linked to mechanical and environmental interactions.

In the present study, an industrial turbine disc provided by SNECMA Moteurs was used to characterize fatigue crack growth behaviour under loading conditions including 300s dwell times, at 650°C. Partial unloading before dwell application considerably reduced the hold time effect. With a 20% load decrease before dwell time, FCGR are unaffected by the dwell period. Simultaneously, the fracture mode changes from a purely intergranular cracking to a mixed mode involving transgranular fracture. These results are highlighted by FEM calculations, in order to establish a relationship between the stress-strain field at the crack tip and the dwell effect.

Last but not least, a strong influence of the local microstructure in the disc has been found, dwell effect sensitivity depending on relative orientations of crack front and δ-phase alignments. For some specimens, this phenomenon has even led to a dwell-effect annihilation.

KEYWORDS

Fatigue Crack Growth Rate (FCGR), Creep-fatigue, Dwell Time Effect, Ni based Superalloy

INTRODUCTION

Some of the most critical components in aeronautical engines are turbine discs which can be exposed to high stresses (up to 1000 MPa) and to temperatures close to 650°C. Thereby, nickel based superalloys are usually used for these discs because of their good mechanical properties at elevated temperature. Among them, Inconel 718 is a largely used alloy because of its good formability and weldability properties.

At such temperature and for low frequency of cycling, the Fatigue Crack Growth Rate (FCGR) is found to be time-dependent (intergranular fracture and growth rate depending on the time spent at maximum load). A similar behaviour is observed when the material is submitted to a trapezoidal wave shape cycle. Conversely, a higher frequency and/or lower temperature lead to a cycle-dependent regime (transgranular fracture and growth rates depending on the cycle amplitude) [1].

Material variables such as grain size, hardening precipitates morphology, chemical composition and mechanical behaviour can modify crack growth resistance as well as the interaction between the mechanical cycling and the environmental conditions [2]. Oxidation effects are known to be responsible for the increase of the FCGR when a hold time at maximum load is applied during a trapezoidal fatigue cycle. The fracture mode and the FCGR depend on crack tip oxygen partial pressure PO_2 [3]. Below a critical transition value (PO_2~10^{-3} mbar), the oxide formed at the crack tip is a chromium rich protective oxide, the fracture mode being transgranular. For oxygen partial pressure higher than 10^{-3}mbar, a nickel rich oxide is formed, which lets oxygen diffuse into the material leading to a vacancy over-saturation in the material due to nickel atoms consumption for the oxide build up. Such phenomenon is common to other nickel based superalloys such as Alloy N18 [4]. The embrittlement of grain boundaries is believed to be due to the transport of oxygen and vacancies to grain boundaries assisted by deformation (i.e. by dislocations). Hence, under air conditions and for low fatigue frequency at 650°C, applying a hold time of 300s at maximum load (creep-fatigue cycles) always leads to dramatic FCGR increase, by up to two orders of magnitude [2, 5, 6]. For these conditions the fracture mode is purely intergranular.

In these studies, the kind of material used and cycling conditions are not reflecting the reality of component loading conditions. The first difference is that these studies involved laboratory materials such as bars or plates, with microstructures which can be largely different from those of industrial forged discs. This seems to be important since Lynch and al. [7] have shown a possible absence of hold time effect in forged disc of Direct Aged (DA) 718 alloy.

The second difference lies in the real mechanical loading which is not as simple as a trapezoidal cycle. To be closer to reality, one should consider a schematic flying mission with loading ramp to the maximum load, corresponding to the take-off, then a slight unloading to reach the cruise load which is then kept constant for a while before the complete unloading and the engine stop. Weerasooriya and Nicholas [8] and more recently Heuler and al. [9] have observed a strong retardation effect of a peak load applied before the hold time of a creep-fatigue cycle. This kind of behaviour has also been mentioned in other superalloys such as N18 [4], Astroloy [10] or René 95 and Udimet 720 [11].

The present study focuses on these dwell effect annihilations and brings highlights on microstructural features and on mechanical and environmental interactions. Moreover, finite element calculations were performed to investigate the crack tip strain-stress state modifications which lead to these annihilation effects.

MATERIAL AND EXPERIMENTAL PROCEDURE

Material

All the specimens of the study were cut from a forged turbine disc produced by SNECMA Moteurs. The fabrication route starts from a double melted billet (VIM - VAR). The chemical composition is given in table 1. The forming of the disc consists of a sequence of forging and circular rolling and final die stamping at different temperatures. Conventional heat treatment is then applied consisting in a first heat treatment at 955°C for 1 h followed by air cooling, then in an ageing treatment at 720°C for 8 h, followed by furnace cooling down to 620°C, and final ageing at 620°C for 8 h before air cooling. These treatments lead to a fine grained austenitic γ matrix (ASTM 9-10), strengthened by the precipitation of γ' ($Ni_3(Ti\text{-}Al)$, FCC, 20 nm spheres) and γ'' ($Ni_3(Nb\text{-}Ti)$, BCT, 20 nm discs). Volume fraction of γ'' precipitates is about 15%.

Ni	C	Cr	Fe	Mo	Nb	Ta	Ti	Al
Balance	0.03	17.94	18.00	2.97	5.27	0.01	1.01	0.46

table 1 : chemical composition of Inconel 718 disc investigated (wt %)

A third type of precipitate is also present in the alloy, the orthorhombic δ phase Ni_3Nb. Bigger globular δ particles are found both at grain boundaries and inside the grains. These precipitates look like ellipsoid discs, 1 to 4 μm long and around 0.4 μm thick and are present in the material before the last forming operations of the turbine disc. This results in a Forming Induced Arrangement (FIA) of the globular δ phase according to the flaw pattern during forming. One of the characteristics of the FIA is the δ phases alignments which can be compared to sheets of discs reflecting the history of component deformation during forming [12, 13].

Experimental procedures

Conventional CT type specimens 10 mm thick and 40 mm wide (B=10mm, W=40mm) were used.
During the test, crack growth is measured by DC Potential Drop (DCPD) technique. Each specimen is fatigue pre-cracked at room temperature at high frequency (20 to 40Hz) under sinusoidal wave cycles with a positive load ratio of R=0.1. Fatigue pre-cracking is finished at high temperature (650°C) by initiating crack growth under triangular cycles of 20s until 1mm crack growth is reached. The FCGR tests are then conducted at 650°C with a stress ratio of R=0.1, maximum load being constant at each cycle.
Three types of cycles are applied : the two first types are classical fatigue triangular cycles of 20s (called 10-10 cycles) and creep-fatigue trapezoidal cycles with the same rate of loading and unloading and a 300s hold time at maximum load (10-300-10 cycles). The third type has been chosen to be closer to realistic loading conditions and presents a peak load before dwell time. The definition of this cycle is given in figure 1, where D% refers to the unloading amplitude before dwell application.

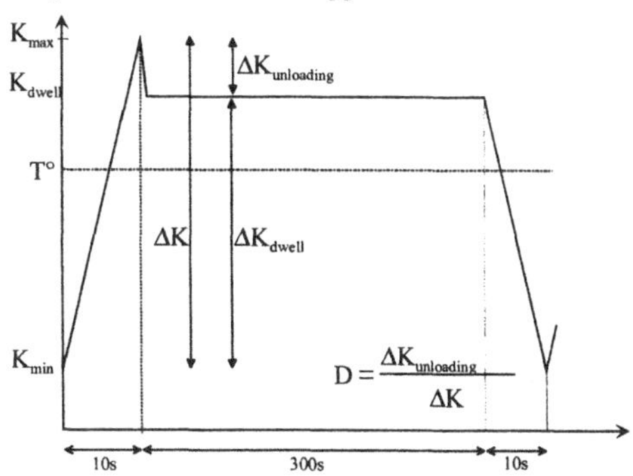

$$\text{Unloading : } D\% = \frac{\Delta K_{unloading}}{\Delta K} = \frac{K_{peak} - K_{dwell}}{K_{peak} - K_{min}}$$

$K_{max} = K_{peak}$
$\Delta K = K_{max} - K_{min}$
$R = K_{min}/K_{max}$
$\Delta K_{dwell} = K_{dwell} - K_{min}$

figure 1 : Creep-fatigue cycle with a peak load before dwell time

On each CT specimen, the small scale of plastic zone size allows to apply several programs of propagation to test several conditions of cycling. For each program, at least 2 mm crack growth is applied to make sure reaching stabilised regime.

EXPERIMENTAL RESULTS

Microstructural effects

These effects have been investigated in detail elsewhere [12, 13]. Here, only the main results are given.

A strong interaction between crack front position and orientation with FIA was observed. Figure 2 reports results obtained on CT specimen for radial crack propagation. In this figure, the sketch in the caption represents a section of the disc in the plane *r-z*, where *z* is the axis of the disc and *r* the radial direction. Black lines refer to as FIA and particularly to the orientations of δ phase alignments. When specimens are extracted from the disc in position A, a strong hold time effect is measured on FCGR and fracture mode is purely intergranular. When extracted in position B, this means when crack front is perpendicular to FIA, no dwell effect is observed. A 5 minutes hold time at maximum load does not produce any increase of growth rate compared to 10-10 triangular fatigue cycles. Fracture mode changes to mixed mode, as observed for 10-10 cycles. Under creep-fatigue conditions, when crack front is perpendicular to FIA, transverse delamination along δ phase alignments occurs, which changes the stress state in the bulk of the specimen from plane strain to a multi-layer material under plane stress conditions. This leads to a reduction of the crack tip driving force. This results in a strong anisotropy of FCGR in the disc which can be seen in varying extraction position and orientation of 3D defects [13].

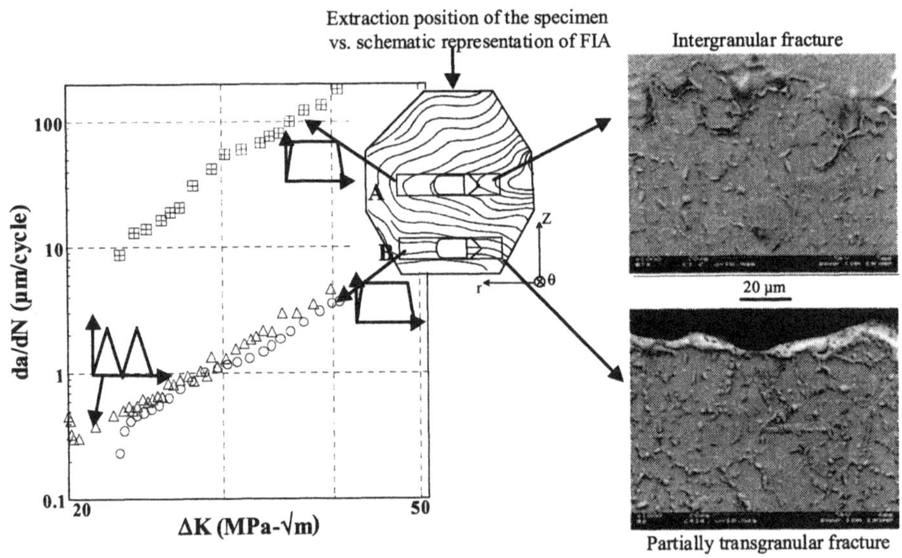

figure 2 : Dwell effect annihilation due to FIA/crack front interaction

Effect of peak load before hold time

The effect of partial unloading before the application of dwell time at each cycle is reported in figure 3a. FCGR da/dN versus ΔK is plotted for unloadings ranging from D=0% (no unloading, creep-fatigue cycle) to D=50% (see figure 2), tests being performed on CT specimen, extracted from the disc where hold time effect is significant. The values reported on *x* axis correspond to ΔK calculated with the maximum load of the cycle. In spite of the scatter observed, it can be seen that as soon as unloading is applied (even with D=5%), a significant reduction of FCGR is observed compared with classical creep-fatigue cycles. For a 20% unloading ramp before dwell time, hold time effect has almost disappeared. For 50% unloading, FCGR is equivalent to triangular fatigue cycle.
The FCGR measured as a function of percentage of unloading and normalised by FCGR with no peak load is plotted in figure 3b at K_{peak} = 30MPa√m. This figure shows the immediate reduction of FCGR with unloading and that for unloading > 20%, dwell time effect does not longer exists. Moreover, fracture surfaces exhibit parts of transgranular fracture as soon as an unloading is applied at each cycle. The part of transgranular fracture is more important with the increase of unloading. For 20% of unloading, the fracture mode is mixed as observed in continuous triangular fatigue. This suggests that the detrimental environmental effect during dwell time does not occur when an unloading higher than 20% is applied.

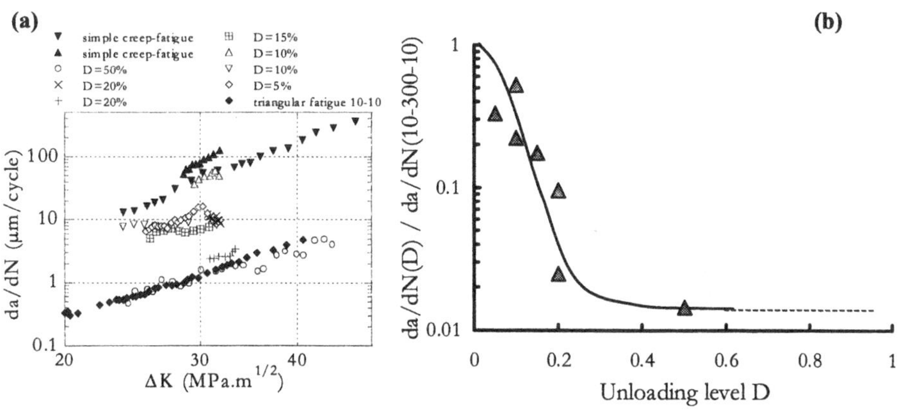

figure 3 : Effect of a unloading before hold time in creep fatigue cycling
(a) : da/dn vs. ΔK at peak load ; (b) FCGR at K_{peak} = 30 MPa√m for an unloading of D% normalised by FCGR at 30 MPa√m for creep-fatigue cycle.

DISCUSSION

True effect of partial unloading ?

Experimental results show a significant decrease of FCGR due to unloading before dwell time. One could suggest that the effect of a D% unload is just the consequence of applying a dwell time at a level D% lower than the reference creep-fatigue cycle. A simple way of accounting for the effect of a hold time D% lower than peak load is to adopt a linear cumulative rule considering FCGR due to the fatigue cycle at K_{peak} and the growth rate due to hold time at K_{dwell}. This kind of model is known to be efficient for trapezoidal creep-fatigue growth rate prediction [8]. The crack growth during the hold time can be assessed by data obtained in creep crack growth experiments (*da/dt×hold time*) or in creep-fatigue experiments which have shown a purely time dependent behaviour for this extraction position [14] :

$$\frac{da}{dN}_{cucmulative} = \left(\frac{da_{fatigue}}{dN}\right)_{Kpeak} + \left(\frac{da_{creep}}{dt}\right)_{Kdwell} \cdot t_{dwell\ time} \quad (1)$$

The comparison between experimental data and this cumulative rule is given in figure 4. The dotted line corresponds to an experimental FCGR equal to the calculated one. This figure shows that the linear cumulative rule overestimates FCGR when unloading is applied before dwell. This means that the effect of the unloading part is not a simple decrease of loading level. Moreover, it cannot either be considered as the result of a simple overload as usually qualified in fatigue since the peak load is present at each cycle.

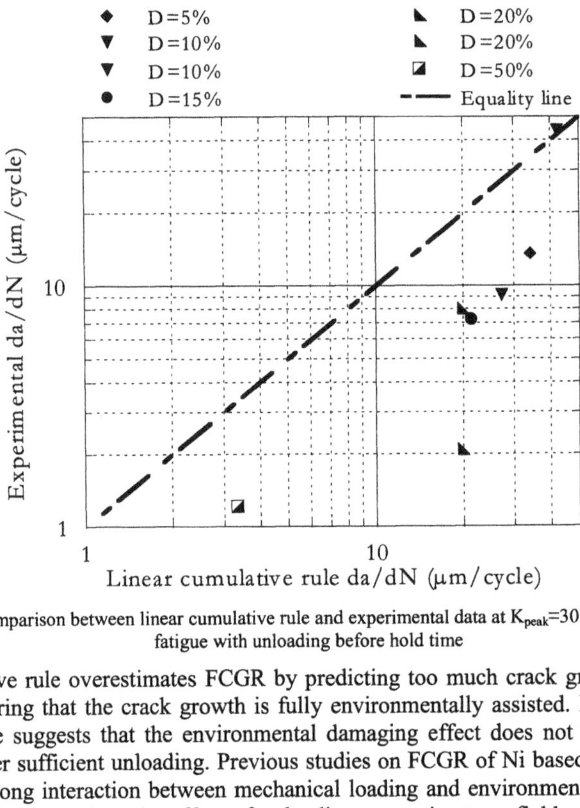

figure 4 : Comparison between linear cumulative rule and experimental data at K_{peak}=30MPa√m for creep-fatigue with unloading before hold time

The cumulative rule overestimates FCGR by predicting too much crack growth during hold time, considering that the crack growth is fully environmentally assisted. But the change in fracture mode suggests that the environmental damaging effect does not take place during hold time after sufficient unloading. Previous studies on FCGR of Ni based superalloys have shown the strong interaction between mechanical loading and environmental damage [3, 4]. Hence, in order to analyse the effect of unloading on strain-stress field at crack tip and its consequences on environmental damaging process, FEM calculations of stationary cracks in a CT specimen have been conducted by simulating several unloading levels before hold time.

Strain and stress field at crack tip due to unloading before hold time

FEM calculation conditions : At 650°C, the material exhibits an elasto-visco-plastic behaviour. Constitutive relations according to Lemaitre and Chaboche model [15] has been identified, including an isotropic hardening, two kinematic hardening relations and a Norton-type viscosity law [14]. The identified parameters give a good representation of stress relaxation during hold time and cyclic softening experimentally observed for the alloy. 2D calculations were performed with the hypothesis of a plane strain state, which is representative of the stress-strain state in the major part of a CT specimen at the simulated loading (30µm wide cyclic plastic zone at first approximation).To describe crack tip field as well as possible in the plastic zone, meshes have been refined to 5 µm square elements near the crack tip. Indeed, 10µm and 20µm square elements have also been tested, which led to the same stress level from the second mesh but were judged too big compared to plastic zone size. Crack length and peak load applied for the simulated cycles are chosen so that K=30 MPa√m at peak load. Although it would have been necessary to simulate a propagating crack to reproduce closure effects [16], these closure effects do not seem to play a major role [17]. Hence the stress-strain field is analysed for the case of a stationary crack, at the first cycle as well as after ten cycles to take into account the softening behaviour of the material.

Stress field : The opening σ_{yy} stress profiles for 0%, 10% 20% and 50% unloading before dwell time are plotted in figure 5 just after the unloading ramp (full lines) and after 300s of hold time (dotted lines).

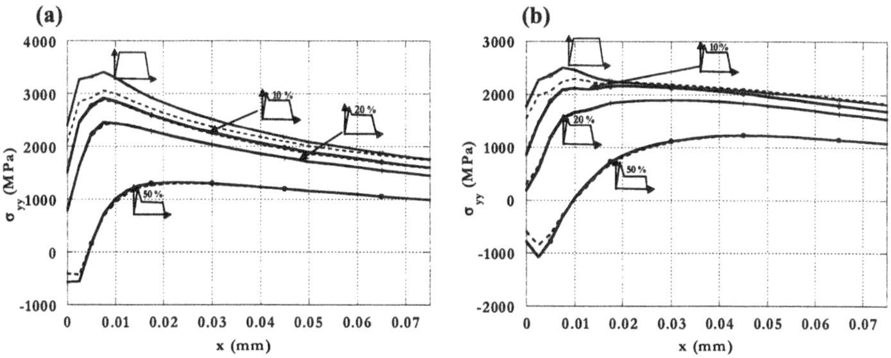

figure 5 : Stress profile in the ligament behind crack tip as a function of unloading level
(a) : at first cycle ; (b) for the 10th cycle.
Full line corresponding to the beginning of the hold time (just after unloading), dotted line corresponding after 300s hold time.

The first observation is that stress triaxiality induces high stress level at crack tip. The second point is that the reduction of stress level at crack tip at the beginning of dwell time is more important than the percentage of unloading applied to the specimen. For 20% of global unloading, stresses are significantly lower at crack tip compared to the cycle with no peak load. The absence of hold time effect for this level of unloading suggests an effect of load level at crack tip to allow oxygen and/or vacancies created by oxidation to move to grain boundaries and embrittle them. For D=50%, the crack tip is under compressive load, which possibly stops diffusion of any species in this zone.

The last observation from figure 5 is that large stress relaxation occurs during the hold time for the creep-fatigue cycle whereas stress levels are almost identical from the beginning to the end of the hold time (full lines and dotted lines surimposed) when 10% or 20% of unloading is applied before the hold time. This stress relaxation for the creep-fatigue cycle is induced by important viscoplastic strain during the hold time. Hence, the level of unloading influences the quantity of viscoplastic strain at crack tip during hold time. This also induces a difference in strain rate.

Strain rate at crack tip : The evolution of calculated cumulative visco-plastic strain rate ($\dot{\varepsilon}_{vp}$) at second gauss point located at 3.75 µm from the crack tip during hold time is reported in figure 6. The unloading ramp creates a rapid decrease of $\dot{\varepsilon}_{vp}$. At the beginning of dwell time, $\dot{\varepsilon}_{vp}$ is about ten times lower for 10% or 20% unloading applied than for conventional creep-fatigue cycle. During hold time, stress relaxation occurs and $\dot{\varepsilon}_{vp}$ continues to decrease for both creep-fatigue cycle and complex cycle, but the difference between the two types of cycles is still present.

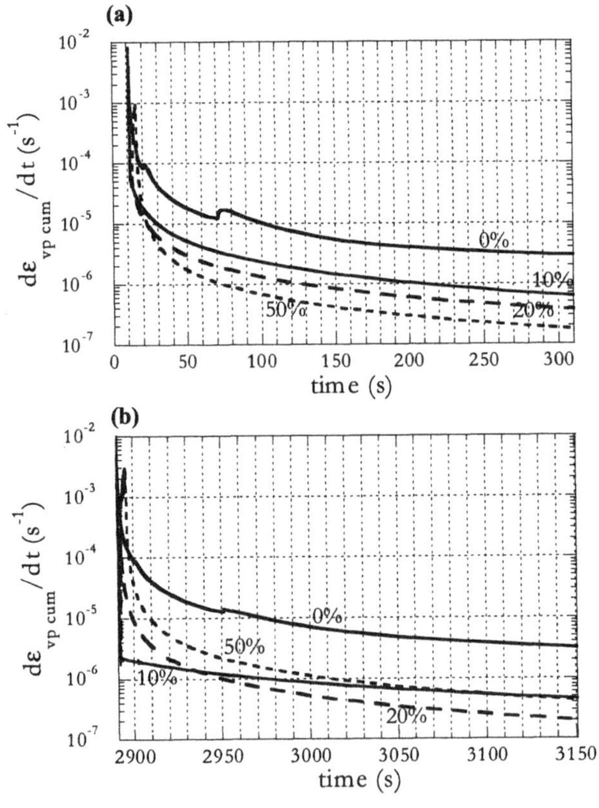

figure 6 : Cumulative visco-platic strain rate $\dot{\varepsilon}_{vp\,cum}$ during dwell time depending on unloading level (a) : First cycle ; (b) 10th cycle.

Effect of unloading on environment assisted crack growth

The results plotted in figures 3 and 6 clearly show that a reduction of the local creep strain rate produces a significant retardation in crack growth rate. Moreover, the observations of fracture surfaces showed changes in fracture mode from intergranular to mixed mode when a sufficient unloading before dwell was applied. This suggests an interaction between the crack tip stress-strain field and the environment.

Here it is worth noting the study performed by Molins et al. [3]. These authors applied a transition in partial pressure of oxygen PO_2 (from low PO_2 to higher PO_2) during the hold time of a creep-fatigue cycle. They showed that it is necessary to wait for about 200s before exposing crack tip to oxidising environment to avoid any damaging effect of environment during hold time [3]. This suggests that after 200s of dwell time, visco-plastic strain rate has become sufficiently low to avoid any transport of damaging species during hold time. This results also in a drastic reduction in crack growth rate and a change in fracture mode.

The material used by Molins et al. [3] for these experiments had a slightly different microstructure (grain size of 6 ASTM compared to 9 ASTM for our case) but the material behaviour can be considered as similar in both studies. Hence, our FEM calculations can give an approximation of the limit of visco-plastic strain rate under which environmental effects are annihilated.

For classical creep-fatigue cycles (figure 6), the value of $\dot{\varepsilon}_{vp}$ after 200s of hold time is 3.9 10^{-6} s^{-1}. For an unloading of 20%, this limit of 3.9 10^{-6} s^{-1} is reached after only 30s dwell time. Hence, the time during which material is exposed to the intergranular transport of damaging species is much lower when unloading is applied before hold time. Moreover, figure 5 also shows that for a classical creep-fatigue test, the opening stress level after 200s of hold time is higher than the stress level at the beginning of hold time when an unloading level as low as 10% is applied. This suggests that the strain rate level plays a major role compared to stress level for activating environmental damaging processes.

CONCLUSIONS

The following conclusions can be drawn from this study concerning the hold time damaging effect in Inconel 718 tested at 650°C. When applied at peak load of the cycle, this hold time is known for increasing crack growth rate due to environmental effects. Nonetheless, the present study indicates that the microstructure and the cycle shape may have a strong influence on this acceleration on FCGR :

1. It has been shown that the FCGR is much lower when the crack front is perpendicular to the alignment of δ phase. This effect is related to delamination along the δ phase alignments which produces a change in stress state.
2. Experimental results confirmed that the dwell time effect is annihilated when a 20% unload is applied before the hold time. This annihilation coincides with a change in fracture mode from purely intergranular to mixed fracture. This suggests the reduction of environmental damaging effect during hold time after unloading.
3. FEM calculations were performed to investigate the crack tip stress-strain state. The first effect of the unloading before dwell time is a reduction of the opening stress level at crack tip which reduces intrinsic crack tip driving force. The second effect is an increase and an acceleration of the stress relaxation at crack tip. This indicates that a rapid reduction of the crack tip strain rate prevents intergranular damage related to environmental effect.

4. The crack growth resistance of Ni based superalloys for elevated temperature creep-fatigue conditions is linked to the capacity of these alloys to produce rapid stress relaxation, leading to an important decrease of strain rate. These results suggest that to improve resistance of aeronautical superalloy to hold time effect is to realize a good compromise between creep resistance and rapid stress relaxation.

REFERENCES

1. Khobaid M., Ashbaugh N. E., Hartman G. A., Weerasooria T., Maxwell D. and Goodman R.C. (1988) *AFWAL-TR-88-4062, Wright-Patterson AFB, OH.*
2. Ghonem H., Nicholas T. and Pineau A. (1993) *Fat. Fract. Engng Mat. Structures*, **16**, n°5 & 6 pp. 565-590.
3. Molins R., Hochstetter G., Chassaigne J.C. and Andrieu E. (1997) *Acta Materialia*, **45**, n°2, pp. 663-674.
4. Andrieu E. and Pineau A. (1998) *Euromech Mecamat 98, 3rd European Mechanic of Material Conference on Mechanics and Multiphysics Processes in Solids : Experiments, Modeling and Applications,* E. BUSSO and G. CAILLETAUD (Eds), EDP Sciences, pp PR9-3 - PR9-11.
5. Pédron J. P. and Pineau A. (1982) *Materials Science and Engineering*, **56**, pp. 143-156.
6. Sadananda K. and Shahinian P. (1980) I*n : Creep-Fatigue Environment Interactions, ed by R.M. Pelloux, N.S. Stoloff, TMS-AIME*, pp. 86-111.
7. Lynch S. P., Radtke T.C., Wicks B.J. and Byrnes R.T. (1994) *Fat. Fract. Engng. Mat. Struct.*, **17**, n°3, pp. 313-325.
8. Weerasooriya T. and Nicholas T. (1988) *In : Fracture Mechanics, 18th symposium.* ASTM STP 945, Reed D.T. and READ R.P. (Eds), pp.181-191.
9. Heuler P., Affeldt E. and Wanhill R. J. H. (2000) In : *The 13th European Conference on Fracture. Fracture Mechanics : Applications and Challenges*. Fuentes et al. (Eds), ESIS-Elsevier.
10. Prigent P. (1993) Phd Thesis, Ecole Nationale des Ponts et Chaussées, France.
11. Telesman J. and Kantos P. (2000), *Private communication.*
12. Ponnelle S., Brethes B. and Pineau A. (2000) In : *The 13th European Conference on Fracture. Fracture Mechanics : Applications and Challenges*. Fuentes et al. (Eds), ESIS-Elsevier.
13. Ponnelle S. Brethes B. and Pineau A. (2001) *To appear in : Superalloys 718, 625, 706 and Various Derivatives*, E.A. Loria (Eds), The Mineral & Material Society.
14. Ponnelle S. (2001) PhD thesis, Ecole des Mines de Paris, France
15. Lemaitre J. and Chaboche J. L. (1985) *Mécanique des matériaux solides*, ed DUNOD
16. Pommier S. and Bompard P. (1999) *Fat. Fract. of Engng. Mat. Struc.* **23**, n°2, pp. 129-139.
17. Pommier S. *private communication, 1997.*

Temperature-Fatigue Interaction
L. Rémy and J. Petit (Eds.)

A CORRELATION OF CREEP AND FATIGUE CRACK GROWTH IN HIGH DENSITY POLY(ETHYLENE) AT VARIOUS TEMPERATURES

G. PINTER, W. BALIKA and R.W. LANG
Institute of Materials Science and Testing of Plastics
University of Leoben, A-8700 Leoben, Austria

ABSTRACT

The creep crack growth (CCG) and fatigue crack growth (FCG) behaviour of two PE-HD pipe grades was studied based on a linear elastic fracture mechanics (LEFM) methodology. The FCG-tests were performed under a sinusoidal load at a frequency of 1 Hz and R-ratios (F_{min}/F_{max}) of 0.1, 0.3 and 0.5; the test temperatures were 23 (only FCG), 60 and 80 °C. The results showed that FCG rates in PE-HD are caused by a combination of cyclic-induced and creep-induced damage, depending on the mean stress level. While for given values of K_{Imax} (FCG tests) and K_I (CCG tests), respectively, at low test temperatures the cyclic component of the applied stress dominates crack growth rates with CCG rates (R = 1) being lower than the FCG rates (R < 1), at high test temperatures the creep component becomes increasingly important in affecting crack growth rates so that CCG rates even exceed FCG rates. The point of inversion from fatigue to creep dominated failure on the temperature scale apparently depends on molecular and morphological characteristics of the PE-HD type and occurs at around 80 °C for PE-HD 1 and around 60 °C for PE-HD 2 in this investigation.

KEYWORDS

poly(ethylene), creep crack growth, fatigue crack growth

INTRODUCTION

Based on empirical findings of the failure behaviour of pressurised plastics pipes in laboratory tests and on experiences with long-term field failures of pipes, linear elastic fracture mechanics (LEFM) concepts are frequently applied to also study the crack growth behaviour of pipe grade poly(ethylenes) (PE). In the past the crack propagation behaviour of various PE types was investigated using crack growth tests under static loads (so called creep crack growth (CCG) tests) [1-4]. More recently cyclic loading conditions (fatigue crack growth (FCG)) were used by some research groups to characterise the crack propagation resistance of PE [5-8].

However, while some studies on the failure behaviour of notched specimens under static and fatigue loading conditions for a given PE type have been reported [9,10], so far hardly any information is available on a correlation between the kinetics of crack growth behaviour under static and cyclic loads using LEFM concepts. Hence, it is the main objective of this paper to investigate the effects of R-ratio (minimum load to maximum load, F_{min}/F_{max}, in a fatigue cycle) on FCG behaviour at various temperatures and to compare the results with the kinetics of crack growth under static loads. To gain some insight into the micromechanisms of crack growth under the various loading conditions, fracture surfaces were studied systematically via scanning electron microscopy (SEM).

GENERAL BACKGROUND

According to the concepts of linear elastic fracture mechanics (LEFM), the growth rate of a sharp crack in linear elastic materials with small scale plasticity is governed only by the applied stress intensity factor, K_I (index I stands for opening mode or pure tensile loading conditions), which describes the local crack tip stress and strain field [11]. The stress intensity factor, K_I, can in general be expressed as:

$$K_I = \sigma \cdot \sqrt{a} \cdot Y \qquad (1)$$

where σ is the applied stress, a the crack length and Y is a non-dimensional correction function accounting for crack and component geometry as well as the type of loading.

Using these concepts CCG rates, da/dt, may then be plotted as a function of K_I and FCG rates, da/dN, may be controlled by the stress intensity factor range at the crack tip, $\Delta K_I = K_{Imax} - K_{Imin}$ [12]. Frequently, over a certain crack growth rate range an extended linear section is revealed on a double logarithmic scale by many plastics, indicating a power law relationship of the form

$$\frac{da}{dt} = A \cdot K_I^m \qquad \text{for CCG} \qquad (2)$$

$$\frac{da}{dN} = A' \cdot \Delta K_I^{m'} \qquad \text{for FCG} \qquad (3)$$

where A, A' and m, m' are constants, which depend on the material as well as on test variables such as temperature, environment, frequency and stress ratio. However, this relationship generally holds true only over an intermediate range of crack growth rates. When investigating a wide range of da/dt (da/dN), deviations from the power law may be observed as illustrated schematically in Fig. 1. That is, crack growth rates in region I decrease rapidly to vanishingly small values as K_I (ΔK_I) approaches the threshold value, K_{Ith} (ΔK_{Ith}), and they increase markedly in region III as K_I (K_{Imax}) approaches the material's fracture toughness, K_{Ic}, and crack propagation becomes unstable.

EXPERIMENTAL

Two commercial grades of high density poly(ethylene) (PE-HD), produced by PCD Polymere Ges.m.b.H. (now Borealis AB, Linz, A) were used in this investigation. Relevant material properties for both material types are listed in Table 1.

For the CCG and FCG tests specimens of the compact type (C(T)) configuration (see Fig. 1) with a specimen width, W, of 40 mm were machined from 10 to 12 mm thick compression moulded plaques. All specimens were kept at a controlled temperature of 23 °C and a relative humidity of 50 % for at least 14 days before testing. Precracks were introduced into the test specimens prior to the fracture mechanical experiments by pressing a fresh commercial razor blade with a nominal thickness of 0.1 mm at room temperature into the V-notch tip.

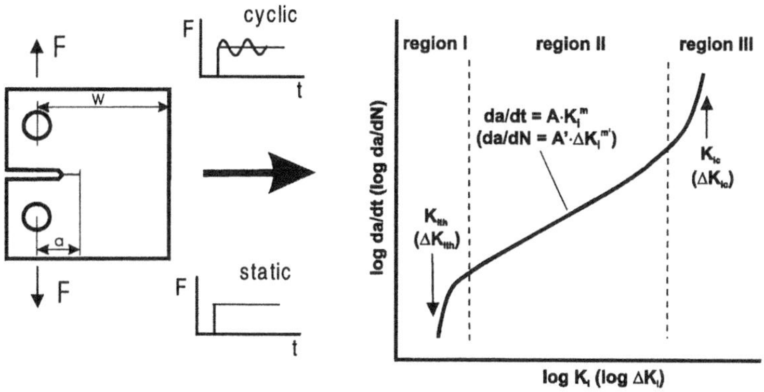

Fig. 1: Schematic crack growth behaviour of polymers under static and fatigue loads (F = force, W = specimen width, t = time)

Table 1: Material characterisation and material properties

Material-Code	ρ (23 °C/50 % r.h.) [g/cm³]	X_C [%]	L_C [nm]	M_n [kg/mol]	M_w [kg/mol]	E (23 °C/50 % r.h.) [N/mm²]	σ_Y (23 °C/50 % r.h.) [N/mm²]
PE-HD 1	0.954	60	13	16	290	950	24
PE-HD 2	0.963	77	21	16	320	1400	30

(ρ = density; X_C = degree of crystallinity; L_C = lamella thickness; M_w = weight average molecular mass, M_n = number average molecular mass; E = elastic modulus; σ_Y = yield stress)

The CCG experiments were performed in a test apparatus, designed and constructed at the Institute of Materials Science and Testing of Plastics (University of Leoben, A). FCG testing was conducted with a servo-hydraulic closed-loop testing machine (MTS Systems GmbH, Berlin, D) under sinusoidal load control at a frequency of 1 Hz (to minimise hysteretic heating effects) and at R-ratios (F_{min}/F_{max}) of 0.1, 0.3 and 0.5. Both, CCG and FCG tests were performed in distilled water at 23, 60 and 80 °C, respectively, to simulate environmental conditions equivalent to hydrostatic stress rupture tests of pipes. Crack lengths values were monitored with the aid of travelling microscope units equipped with linear variable transducers (LVTD) for displacement measurements.

Fractographic investigations of specific fracture surface details were carried out using a scanning electron microscope (SEM; Zeis, Oberkochen, D). Prior to the investigations all specimen were sputter coated with a 15 to 20 nm thick layer of gold. The operating voltage was 10 kV.

RESULTS AND DISCUSSION

Crack Growth Behaviour

In order to verify the applicability of LEFM, constant ΔK_I and constant K_I experiments, respectively, were performed. Typical results are illustrated in Fig. 2 as da/dN and da/dt, respectively, versus the normalized crack length, a/W. The data depicted for both materials show remarkably constant crack growth rates with very little scatter over the entire a/W range, thus providing good support for the applicability of LEFM to these materials.

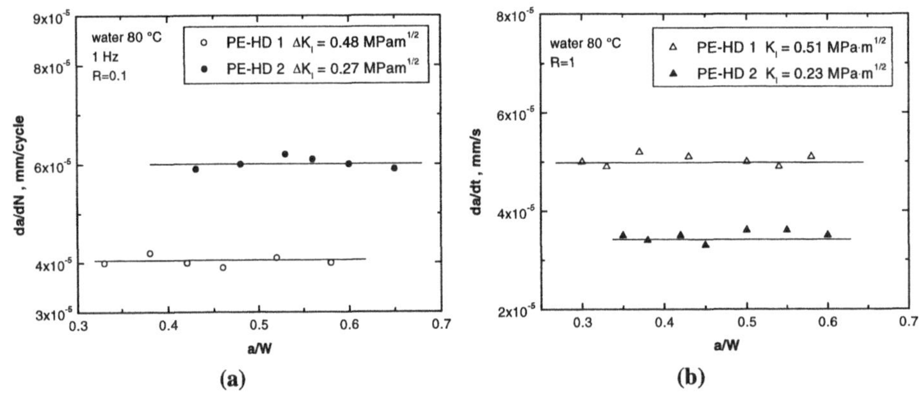

Fig. 2: FCG rates (a) and CCG rates (b) in PE-HD under constant-ΔK_I and constant-K_I conditions, respectively

The FCG behaviour of the two PE-HD types at different temperatures and an R-ratio of 0.1 is compared in Fig. 3. While PE-HD 1 exhibits superior FCG resistance over the entire temperature range, for both materials the FCG curves are shifted towards lower ΔK_I values with increasing test temperature. The improved behaviour of PE-HD 1 is believed to be a result of the higher density of tie-molecules and the lower yield stress [13, 14].

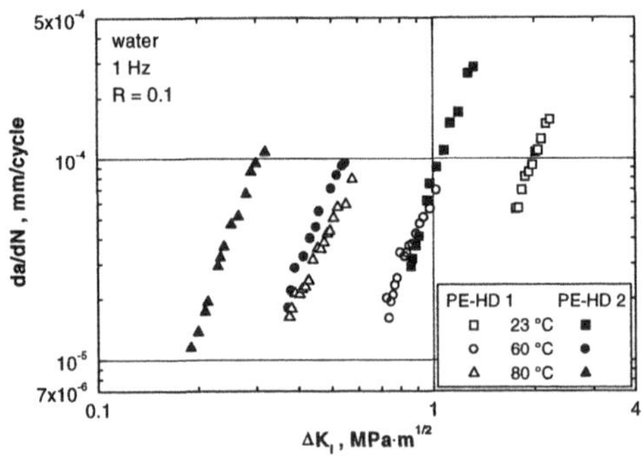

Fig. 3: Influence of test temperature on FCG behaviour in PE-HD

FCG data for the three test temperatures illustrating the effects of variations in R-ratio at a given frequency are shown as a function of the applied stress intensity factor range, ΔK_I, in Fig. 4. Whereas the FCG resistance in terms of ΔK_I is markedly reduced for PE-HD 2 at all temperatures as the R-ratio is increased, PE-HD 1 exhibits this effect only at 80 °C; at 60 °C the FCG curves for R = 0.1 and 0.3 and at 23 °C the curves for all R-ratios coincide.

Apparently mean stress effects on the fatigue response of PE-HD are controlled by conflicting processes. On the one hand there may be a tendency for higher crack growth rates at higher R-ratios as a result of more creep crack extension associated with the higher K_{Imax} and mean stress intensity levels. Alternately, as the maximum plastic zone dimensions are expected to be controlled by K_{Imax}, higher R-ratios will lead to more extended plastic zone dimensions (craze dimensions), which act to blunt the crack and result in an increased tendency for strain energy dissipation, thus acting to reduce crack growth rates [12, 15, 16].

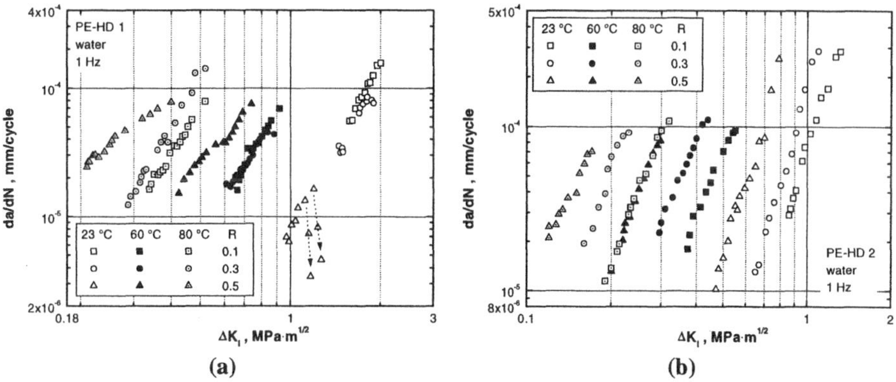

Fig. 4: FCG rates of (a) PE-HD 1 and (b) PE-HD 2 for various R-ratios and temperatures as a function of ΔK_I

The just described phenomena are especially of relevance for PE-HD 1. At higher temperatures FCG rates are enhanced with higher R-ratios, whereas at lower temperatures larger plastic zones (crazes) with increasing R-ratio are responsible for relatively decreasing crack growth rates and even the arresting of cracks (i.e., in the case of R = 0.5, 23 °C). Such crack arrests could only be reinitiated by an increase in load. Another explanation for the coinciding curves at lower temperatures could be a decreasing influence of creep-induced damage.

In order to further investigate the effects of temperature on the significance of creep-induced and fatigue-induced damage, the FCG data of Fig. 4 are plotted in Fig. 5 in terms of K_{Imax} together with data from CCG tests (the latter corresponding to the limiting case of a FCG test with an R-ratio of 1). In terms of K_{Imax} both materials exhibit lower crack growth rates at higher R-ratios at 23 °C due to the reduced ΔK_I-range. At higher temperatures, however, the differences between the crack propagation rates for various R-ratios vanish, so that at 60 °C for PE-HD 2 and at 80 °C for PE-HD 1 the curves for all R-ratios coincide. Apparently, at higher temperatures the decrease in da/dN at higher R-ratios (associated with the decrease in ΔK_I range) is almost balanced in PE-HD by an increase in da/dt (associated with the higher average K_I level), thus providing further evidence that creep-induced damage is more pronounced at higher temperatures.

In other words, while at low test temperatures the cyclic component of the applied stress dominates crack growth rates with CCG rates (R = 1) being lower than the FCG rates (R < 1), at high test temperatures the creep component becomes increasingly important in affecting crack growth rates so that CCG rates even exceed FCG rates at given values of K_{Imax}. The point of inversion from fatigue to creep dominated failure on the temperature scale apparently depends on molecular and morphological characteristics of a given PE-HD type and occurs at around 80 °C for PE-HD 1 and around 60 °C for PE-HD 2.

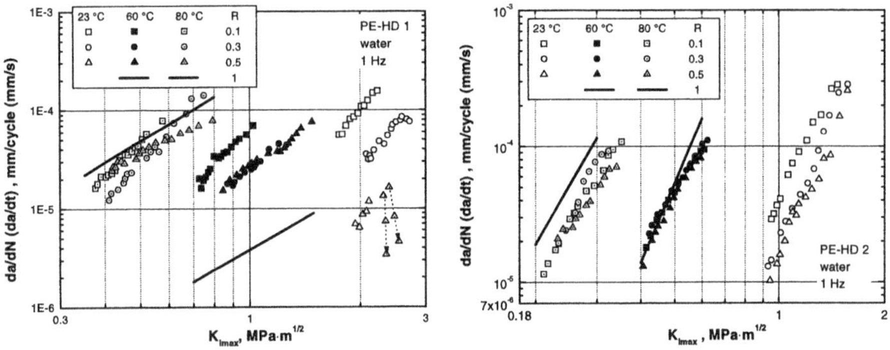

(a) (b)

Fig. 5: FCG rates of (a) PE-HD 1 and (b) PE-HD 2 for various R-ratios and temperatures as a function of K_{Imax} and comparison with CCG data (R = 1)

Fracture Surface Morphology

Generally the fracture surfaces of both materials reveal the remnants of voids and fibrils, the typical attributes of craze formation and breakdown (see Fig. 6). Comparing the fracture surfaces of the two PE-HD types at equivalent ΔK_I values (Fig. 6a,b), it becomes apparent that the fibrils of PE-HD 2 are considerably less drawn than those of PE-HD 1, which on the one hand reflects the differences in the yield stress values of these materials and their effects on crack tip craze development. On the other hand, the higher tie molecule and interlamellar entanglement density of PE-HD 1 acts to stabilise the craze fibrils in the craze extension process prior to craze breakdown, leaving a more tufted structure with remnants of more highly stretched fibrils on the fracture surface. The higher tie molecule and interlamellar entanglement density of PE-HD 1 is of course also the prime reason for the superior CCG and FCG resistance of this PE-HD type [13, 14].

In Fig. 7 fracture surface details of PE-HD 2 tested at 60 °C under cyclic loads with different R-ratios and with K_{Imax} of 0.45 MPa·m$^{1/2}$, and under static load with a K_I value also of 0.45 MPa·m$^{1/2}$ are compared. For all of these test conditions nearly equivalent crack growth rates of approximately $3 \cdot 10^{-5}$ mm/cycle (mm/s) were determined.

Of special relevance to the observations in Fig. 7 it has been pointed out previously [16] that some influence of R-ratio at constant K_{Imax} values on crack tip craze dimensions may be anticipated for viscoelastic materials, since a change in R-ratio also implies a change in the loading history. From the load-time traces illustrated in Fig. 8 it is evident that the loading rate (dF/dt and hence dK/dt) and the load-time integrated area per cycle at a constant value of the maximum load decrease and increase, respectively, as the value of R increases. Both of these factors will have some tendency to increase the crack-tip craze dimensions and the fibril extension with increasing R-ratio by decreasing the craze stress as a result of the smaller local strain rate, and by promoting creep and stress relaxation locally at the crack tip due to the higher average load.

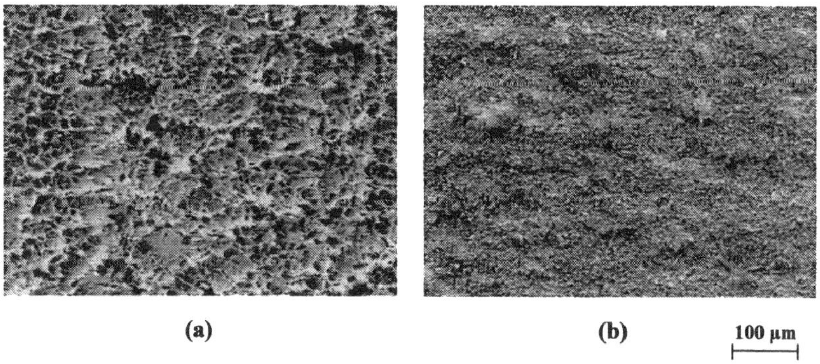

Fig. 6: Comparison of the fracture surface of PE-HD 1 (a) and PE-HD 2 (b) at 80 °C and ΔK_I = 0.32 MPa·m$^{1/2}$

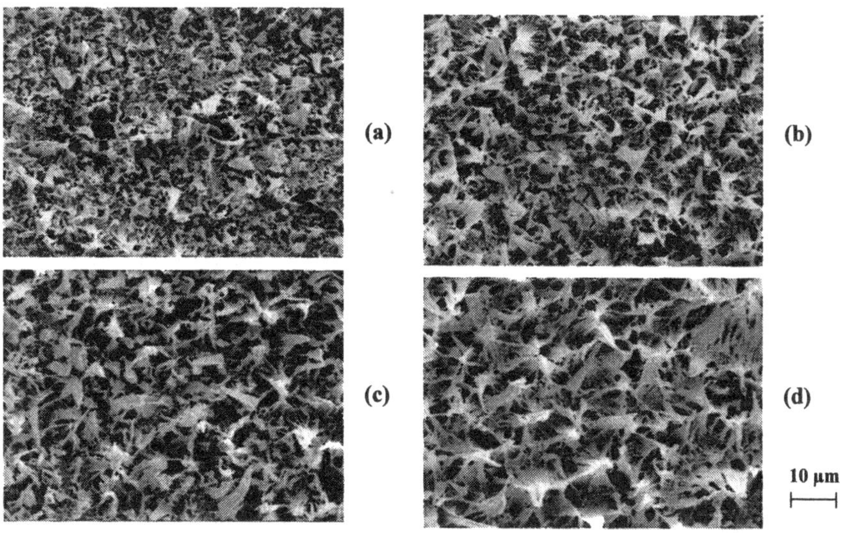

Fig. 7: Comparison of the fracture surface of PE-HD 2 at 60 °C, constant K_{Imax} resp. K_I values of 0.45 MPa·m$^{1/2}$ and equal crack growth rates; (a) fatigue: R = 0.1, (b) fatigue: R = 0.3, (c) fatigue: R = 0.5, (d) static (R = 1)

Indeed, significant differences in crack tip craze zone dimensions were observed during the crack growth experiments, with larger crack tip craze zones being generated at given K_{Imax} values with increasing R-ratio. Hence, the pronounced influence of R-ratio (CCG tests corresponding to R = 1) on the micromorphology of fracture surfaces of PE-HD 2 in Fig. 7, with more highly stretched fibrils as the R-ratio is increased, apparently reflects the corresponding increase in craze zone dimensions.

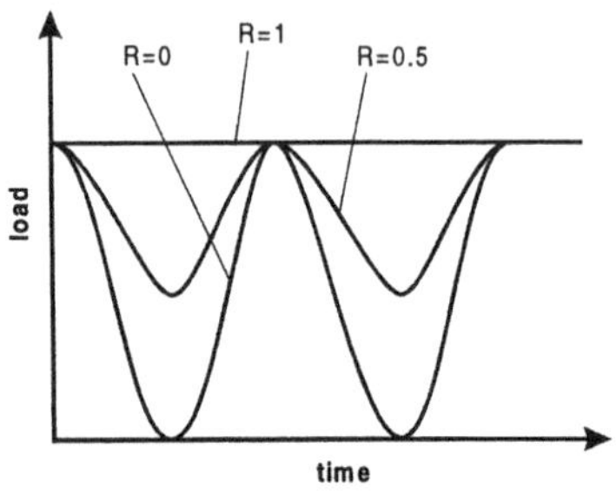

Fig. 8: Comparison of two cyclic loads with a sinusoidal waveform at a constant maximum load but with different load-ratios, R

CONCLUSIONS

Based on FCG experiments with two types of PE-HD at various R-ratios from 0.1 to 0.5 and on CCG experiments (corresponding to an R-ratio of 1) in the temperature range from 23 to 80 °C, it could be shown that FCG rates in PE-HD are caused by a combination of cyclic-induced and creep-induced damage, depending on the mean stress level. While for given values of K_{Imax} (FCG tests) and K_I (CCG tests), respectively, at low test temperatures the cyclic component of the applied stress dominates crack growth rates with CCG rates (R = 1) being lower than the FCG rates (R < 1), at high test temperatures the creep component becomes increasingly important in affecting crack growth rates so that CCG rates even exceed FCG rates. The point of inversion from fatigue to creep dominated failure on the temperature scale apparently depends on molecular and morphological characteristics of a given PE-HD type and occurs at around 80 °C for PE-HD 1 and around 60 °C for PE-HD 2.

The differences in the crack growth behaviour of the two materials were interpreted in terms of molecular and morphological structure (i.e., interlamellar tie molecule and entanglement density, effects of the degree of crystallinity on yield stress) and on the resulting crack tip craze formation and breakdown processes. The mechanisms inferred were corroborated by fracture surface observations.

REFERENCES

[1] Lustiger, A. and Markham, R.L. (1983) *Polymer*, **24**, 1647.
[2] Egan, B.J. and Delatycki, O. (1995) *J. Mater. Sci.*, **39**, 3351.
[3] Brown, N. and Lu, X. (1995) *Polymer*, **36**, 543.
[4] Brown, N., Lu, X., Huang, Y.L., Harrison, I.P. and Ishikawa, N. (1992) *Plastics a. Rubber a. Composites Proces. a. Appl.*, **17**, 255.
[5] Bucknall, C.B. and Dumpleton, P. (1995) *Plastics a. Rubber Proces. a. Appl.* **5**, 343.
[6] Yeh, J.T. and Runt, J. (1991) *J. Polym. Sci.: Part B: Polym. Phys.* **29**, 371.
[7] Strebel, J.J. and Moet, A. (1991) *J. Mat. Sci.* **26**, 5671.
[8] Strebel, J.J. and Moet, A. (1995) *J. Polym. Sci.: Part B: Polym. Phys.* **33**, 1969.

[9] Young, P., Kyu, T., Suehiro, S., Lin, J.S. and Stein, R.S. (1983) *J. Polym. Sci. Polym. Phys. Ed.* **21**, 881.
[10] Reynolds, P.T. and Lawrence, C.C. (1991) *J. Mater. Sci.* **26**, 6197.
[11] Kinloch, A.J. and Young, R.J. (1983). *Fracture Behaviour of Polymers*, Applied Science Publishers Ltd., Barking.
[12] Hertzberg, R.W. and Manson, J.A. (1980). *Fatigue of Engineering Polymers*. Academic Press, New York.
[13] Pinter, G. (1999). Dissertation, University of Leoben, Austria.
[14] Pinter, G. and Lang, R.W. (2001) *Polymer*, in preparation.
[15] Clark, T.R., Hertzberg, R.W. and Manson, A. (1990) *J. Testing a. Evaluation* **18**, 319.
[16] Lang, R.W. (1984). Dissertation, Lehigh University, USA.

Temperature-Fatigue Interaction
L. Rémy and J. Petit (Eds.)

INFLUENCE OF TEMPERATURE ON FATIGUE CRACK PROPAGATION MICROMECHANISMS IN TiAl ALLOYS

G. HÉNAFF, C. MABRU, A. TONNEAU & J. PETIT
LMPM / ENSMA, 1 Avenue C. Ader BP 40109
F - 86961 FUTUROSCOPE CHASSENEUIL FRANCE

ABSTRACT

In view of the introduction of TiAl-based alloys into structural components the fatigue crack propagation behavior of these materials at in-service temperatures must be assessed. With this respect literature reports about an "anomalous" temperature dependence of the fatigue crack growth resistance of TiAl alloys. In such cases the higher resistance is observed at elevated temperature, above the brittle-to-ductile transition, while the lowest resistance is obtained just below this transition and the room-temperature is intermediate between these two situations. However, as for conventional engineering alloys, their fatigue crack propagation resistance results from a complex balance between processes of different nature. These differences would be related to different contribution of intrinsic resistance, crack tip shielding by oxide-induced closure and environmental effects. However up to now no study has systematically investigated the influence of temperature on these different processes. The present study is precisely tackling the issue of identifying the influence of temperature on the various micromechanisms involved in the fatigue crack growth process and their interaction with temperature by conducting series of tests on a quaternary alloy Ti-48Al-2Cr-2Nb at different temperatures and under different environmental conditions.

KEYWORDS

Gamma titanium aluminide; fatigue crack propagation; crack closure; influence of environment.

INTRODUCTION

TiAl-based intermetallic compounds have received considerable interest during the last years since they now appear as potential high temperature structural materials for advanced aerospace and automotive applications in the temperature range of 600-850°C. Hence the determination of their fatigue crack growth (FCG) resistance and the understanding of the crack growth mechanisms operative at these temperatures are key issues before they can be integrated with a sufficient level of confidence in component integrity.

A literature survey indicates that the analysis of the effects of temperature on the fatigue crack growth resistance of γ-based alloys is highly debatable. Balsone et al. [1] reported that for a duplex alloy tested at 1 Hz in air, the crack growth rates for temperatures ranging from 25 to 954°C fall in a very narrow band. Chan and Shih [2] stated a similar behavior for a lamellar alloy that exhibited the same crack growth resistance at 25 and 850°C in air as well as in vacuum. Soboyejo and co-workers [3-5] however observed an improved resistance at 700°C with respect to the room temperature behavior. Finally investigations on fatigue crack propagation behavior of γ-alloys at temperatures around the ductile-brittle transition temperature pointed out another general trend: different alloys (G7 [6], K5 [7], XD^{TM} [8]) have been proved to offer the best fatigue resistance at 800°C and the lowest resistance at 600°C. This kind of behavior has been qualified as "anomalous" [9]. Indeed in such cases it is implicitly assumed that FCG resistance should decrease as temperature is raised. However it should be emphasized that such anomalous temperature dependence is observed on the global fatigue crack growth resistance in ambient air without differentiation between the various mechanisms involved in the growth process. Although differences in chemical composition, processing and heat treatments might of course partly account for such discrepancy in the elevated-temperature behavior, one has also to recognize that this global behavior might actually result from a complex balance between various mechanisms which can be themselves temperature dependent.
Thus as TiAl compounds undergo a ductile-to-brittle transition one might indeed expect some consequences on the intrinsic fatigue crack growth mechanisms. In addition, intermetallic compounds, especially aluminides, are also well known to be prone to moisture embrittlement [10, 11]. This embrittlement might in turn be temperature dependent and thus induce variations in the magnitude of environmental fatigue crack growth enhancement. Besides it is now well established that the near-threshold behavior can be deeply affected by crack tip shielding induced by closure [12]. In particular in an active environment like ambient air the magnitude of oxide-induced closure effects can be important [13]. In addition this phenomenon can obviously be enhanced by temperature. Now most of the investigations did not experimentally determine the crack opening stress intensity factors. The magnitude of closure effects was generally derived from oxide thickness measurements. Therefore a different balance between these effects could also account for the discrepancy evoked here above as temperature varies. Finally it should be noticed that almost no data on the behavior in the temperature range 25°C-600°C are available. Indeed, although such data may not be relevant with respect to in-service temperature conditions, they might be useful to clarify the influence of temperature on the various processes involved in FCG behavior.

The present paper precisely aims to elucidate this issue by identifying and quantifying the influence of temperature on the intrinsic fatigue crack growth resistance and on extrinsic factors affecting the global behavior such as closure or environmentally-assisted fatigue crack propagation mechanisms. In particular the magnitude of shielding induced by closure was in each case investigated using direct measurements by means of a compliance technique or indirect estimations using a variable load ratio, K_{min} increasing test procedure. In addition FCG data at 150°C and 500°C are also included in order to in order to clarify the influence of temperature in this intermediate range.

EXPERIMENTAL PROCEDURE

Material

The material is a quaternary alloy of nominal composition Ti-48Al-2Mn-2Nb, provided by the IRC (Interdisciplinary Research Center in Materials for High Performance Applications / The University of Birmingham) as a piece of ingot produced in a large-scale plasma furnace. The material is tested in the as-cast condition. The microstructure is nearly fully lamellar, consisting of coarse alternating γ and α_2 plates grains with a grain size of approximately 400 µm [14]. Chemical composition is given in Table 1. Tensile properties determined at 20, 750 and 900°C are given in Table 2. It should be added that Young's modulus values used in the following sections have been precisely determined by a resonance technique for temperatures ranging from 20 to 900°C [15].

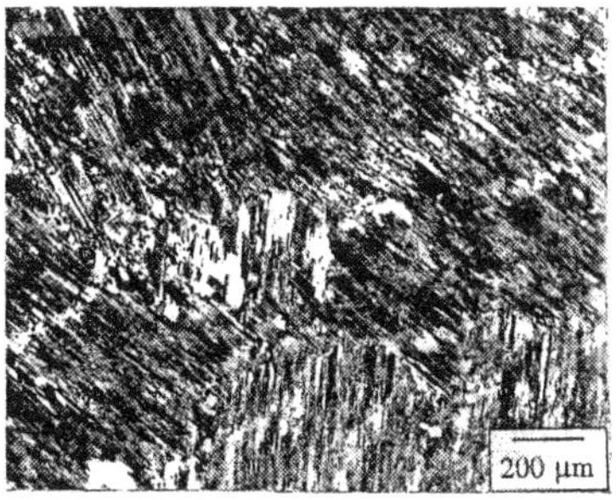

Figure 1: Microstructure of the as-cast Ti-48Al-2Mn-2Nb.

Table 1: Chemical composition of the Ti-48Al-2Mn-2Nb alloy (at. %).

	Ti	Al	Mn	Nb	O
At. %	47.9	48	2	1.9	0.2

Table 2: Tensile properties as a function of temperature with a strain rate: $3.6 \times 10^{-4}\ s^{-1}$ (Y. S. : Yield Stress; U. T. S. : Ultimate Tensile Stress).

Temperature (°C)	0.2% Y. S. (MPa)	U.T.S. (MPa)	Elongation (%)
20	380	465	1.03
750	280	515	4.58
900	282	355	23.2

Testing

Most of the fatigue crack growth experiments were carried out on CT specimens (W=22mm, B=5mm). The fatigue crack propagation tests were performed on servohydraulic machines equipped with an environmental cell and/or a resistance furnace allowing various test conditions. The environmental conditions used are described in [15]. Crack closure measurements were performed at test frequency according to the unloading compliance method using, at room temperature, a back face gauge and, at elevated temperature in air, a sensor measuring the rod displacement more precisely than the LVDT signal of the actuator. The opening load (P_{op}) value was then estimated as the load corresponding to the point of deviation from the linear portion of the load versus differential displacement curve. At high temperature in vacuum, a different specimen geometry (K_{tR}) was used [15]. In addition since it was not possible to perform direct crack closure measurements, several test methods were used to indirectly evaluate the crack closure loads, namely constant load ratio $R=K_{min}/K_{max}$ and selective constant K_{max}-increasing R load ratio monitoring procedures [15].

RESULTS AND DISCUSSION

Temperature effects on fatigue crack propagation in air

Fatigue crack growth rates obtained at various temperatures for R=0.1 are plotted in Figure 2. It is observed that increasing the temperature from room temperature up to 800°C does not significantly modify the fatigue crack growth response of the material, excepted at 500°C where the FCG is lower than in other cases under 10^{-7} m/cycle, inducing a much lower threshold value. One can also only notice a lower resistance at 750°C above 10^{-7} m/cycle. The near-threshold behavior does not seem affected by temperature. Besides no influence of creep accompanied by extensive blunting of the crack tip, as observed by Zhu and co-workers [16], was noticed in the present case. These findings are consistent with results reported by Balsone et al. [1] and Chan and Shih [2]. However, as stressed in the introduction, it can be argued that opposing mechanisms might be responsible for this nearly unchanged behavior. The following sections address this issue by examining the influence of temperature on these mechanisms.

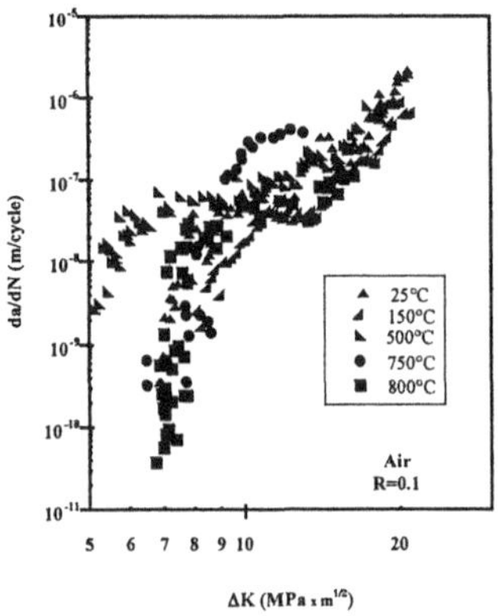

Figure 2: Influence of temperature in ambient air (R=0.1).

Influence of temperature on intrinsic fatigue crack growth

The intrinsic fatigue crack growth behavior has been determined under high vacuum conditions at different temperatures. The results are plotted in Figure 3 with respect to $\Delta K_{eff}/E$ in order to account for elastic modulus variations. In addition similar data obtained in the case of a conventional titanium alloy are also included for comparison purpose. It can be seen that under 500°C the behavior is unaffected by temperature. For temperatures above 500°C the fatigue crack growth resistance is improved mainly in the high crack growth rate regime. This improvement results in da/dN curves with a lower slope. However, it should be noticed that the near-threshold behavior is unaffected when temperature is raised from room temperature up to 850°C. Interestingly it can be noticed that the behavior exhibited by this intermetallic alloy is not so much different from that observed in a conventional titanium alloy, suggesting that the intrinsic mechanisms governing propagation are not so much different either.

Therefore, as the intrinsic near-threshold resistance is not influenced by temperature, possible interactions between temperature and closure and/or environmental effects must be investigated.

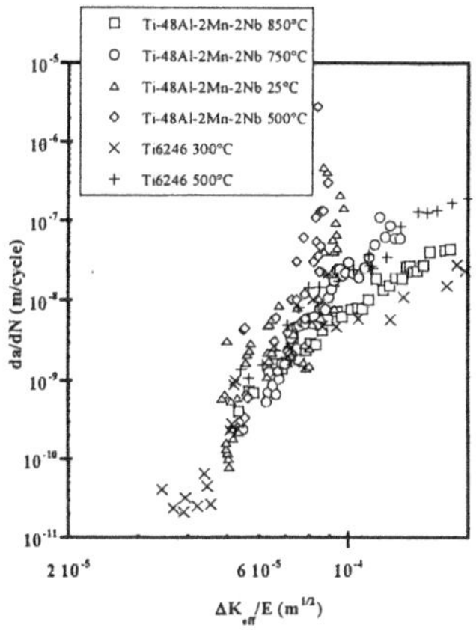

Figure 3: Intrinsic fatigue crack propagation resistance as compared to a conventional Titanium alloy (data from [17, 18]).

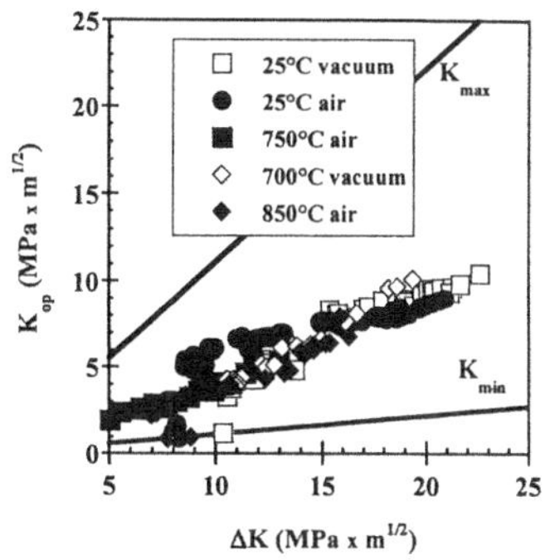

Figure 4: Crack opening stress intensity factors under various environments and temperatures.

Closure effects

The global resistance observed in air under a wide temperature range could be dependent on the variation in the magnitude of crack closure. In particular, as mentioned in introduction, one would expect an enhanced contribution of the oxide-induced mechanism at elevated temperatures.

Figure 4 presents a compendium of crack opening stress intensity factors K_{op} measured as a function of the applied ΔK value under various environmental conditions and for different temperatures. It can be seen that the closure behavior is not dependent on these parameters. Indeed opening loads determined in vacuum are nearly independent of the temperature. In addition K_{op} values obtained at room temperature and at elevated temperature in air (750 and 850°C) are nearly identical. This suggests that oxidation does not promote closure at elevated temperature. These findings are in agreement with the results from Rosenberger et al. [19] who noticed that closure corrections do not modify the relative position of da/dN curves at elevated temperatures. They, however, somewhat contradict the conclusions reached by other authors [4, 9]. These authors did not experimentally determine the opening loads; they derived their values from measurements of the oxide layer thickness using a micromechanical model developed by Suresh [13]. They concluded that oxide-induced closure plays a dominant role at 800°C. as a consequence the poorest crack-closure corrected behavior would be obtained at 800°C while the best resistance would be exhibited at room temperature. Obviously this is not the case in the results presented here, as shown in Figure 5. The effective curves derived from crack closure measurements are shifted to the left but their relative positions are generally almost unchanged. It is further remarkable that the behavior at 500°C does no longer differ from those observed at lower and higher temperatures. Actually this test produced an anomalous closure behavior since closure effects were almost negligible in this case. The reasons for this remain unclear. However it has been previously shown that closure effects in this alloy can be related to the load history [14], which could explain the observed discrepancy. In the present investigations the maximum oxide thickness was estimated from post-mortem observations of the oxide layer on fracture surfaces produced at 800°C to be 0.4 μm. This value is always lower than the cyclic crack tip opening displacement even at low ΔK values [20]. It is then concluded that at this temperature, and consequently at lower temperatures where oxidation is reduced, oxide wedges in the crack wake do not induce significant closure effects. The role of chemical composition on the oxide thickening of cracked surface would need to be investigated because it could explain the differences in oxide layer thickness between the present results and those obtained on different alloys by other authors [4, 9]. Indeed Balsone et al. [1] using an alloy of similar composition (namely Mn and Nb alloying) also found temperature-independent threshold values in ambient air. Anyway, as oxide-induced closure does not appear as the prevailing closure mechanisms at any temperature nor under any environmental condition, it is suggested that the roughness-induced mechanism is responsible for the observed closure behavior. This assumption is supported by in-situ observation of the crack opening and closure kinematics [20].

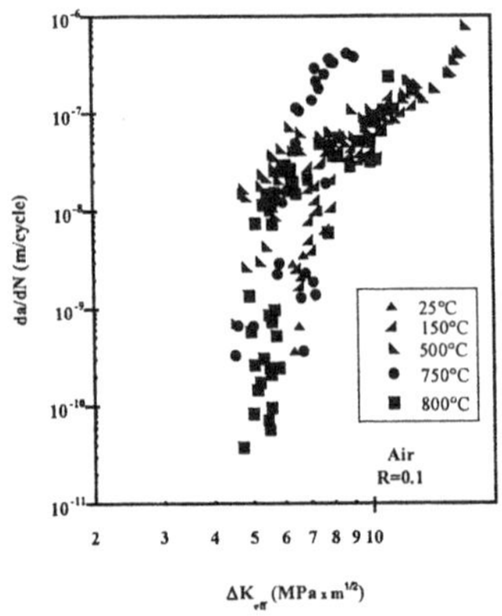

Figure 5: Effective propagation curves at different temperatures in ambient air.

Influence of environment

As a consequence of the nearly temperature-independent global resistance (Figure 2) and the lack of temperature influence on the intrinsic resistance and on the closure behavior observed in the present study, it turns out that the contribution of environment, which is marked, is also temperature independent as shown on Figure 6. This observation is consistent with the lack of substantial modification in microfractographic features [15]. Wei et al. [21] suggest that environmentally-assisted fatigue crack growth is controlled by one or several steps defined as follows: transport of active species to the crack tip, surface adsorption, dissociation of adsorbed molecules, hydrogen penetration and diffusion towards the site where the embrittling reaction takes place. In the following environmentally assisted fatigue crack growth enhancement in γ-alloys is analyzed according to this framework with a special attention paid to the influence of temperature on these mechanisms.

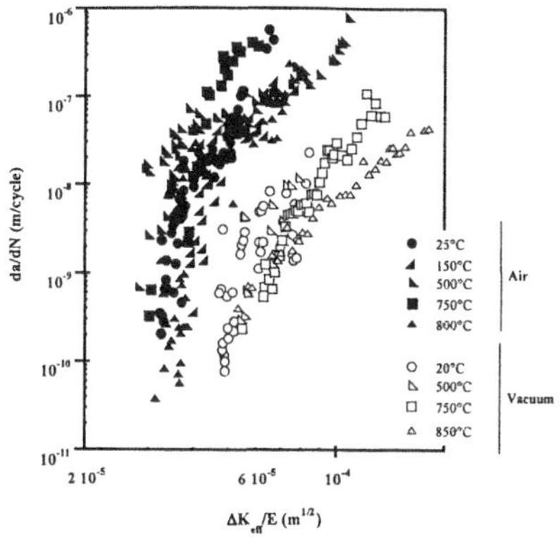

Figure 6: Environmental influence at different temperatures.

Identification of active species. The nature of the active species and the determination of the mechanisms involved in environmentally assisted propagation have first to be investigated. Numerous studies have shown that aluminides are prone to environmental embrittlement in presence of a moist atmosphere [10, 11, 22, 23]. This embrittlement, resulting in a loss of ductility, is suggested to be due to hydrogen produced by the dissociation of adsorbed water vapor molecules on surfaces and then dragged into the bulk material by mobile dislocations where the embrittling reaction occurs. However, this embrittling effect of water vapor can be partly or totally alleviated in presence of oxygen due to a competitive adsorption process between these two species. As oxygen adsorbs at a rate comparable with that of water vapor, it blocks adsorption sites which cannot be occupied by water vapor molecules and thereby limits the hydrogen production.

Ancillary testing under different atmospheres with intermediate water vapor content and different amount of oxygen has been carried out at room temperature and at 500°C in order to verify this assumption. The results are reported in Figure 7. It can be seen that all these conditions result in almost the same behavior in the near-threshold region. This behavior is intermediate between that obtained in ambient air and that observed in vacuum [24]. That means that water vapor controls the fatigue crack growth enhancement and does not interfere with oxygen at any temperature [25].

Surface reactions. This last conclusion is somewhat contradicting with the analysis proposed by Li and Liu [26] and based on surface reaction kinetics. Indeed, according to this analysis, the beneficial effect of oxygen due to the reduction in hydrogen production on surfaces and the subsequent limited embrittlement at room temperature described here above should be promoted at elevated temperature. As a consequence, if such a mechanism applies to FCG in TiAl alloys, the environmentally induced FCG enhancement should diminish at high temperatures. The data presented in Figure 6 show that this enhancement is nearly temperature independent. Furthermore the results obtained here at 500°C in an Argon/Oxygen mixture (Figure 7) demonstrate that oxygen does not prevent the crack growth enhancement due to the residual moisture content. Furthermore they also strongly suggest that oxygen, afetr adsorption and dissociation, does not embrittle the crack tip either. Indeed the enhancement obtained in the oxygen atmosphere is similar to that observed in argon. Since the oxygen content is similar to the content in ambient air, it is further suggested that even in ambient air oxygen would not significantly prevent water vapor assisted fatigue crack growth. Finally since the FCG enhancement in air is almost temperature independent, one can deduce that even at high temperatures up to 800°C oxygen has no effect. However it should be noticed that a different behavior is observed in iron aluminides [25] where oxygen does prevent such moisture-induced fatigue crack growth enhancement. The role of base compound, aluminum content and/or oxide layer has to be more deeply examined to get a deeper insight into these processes.

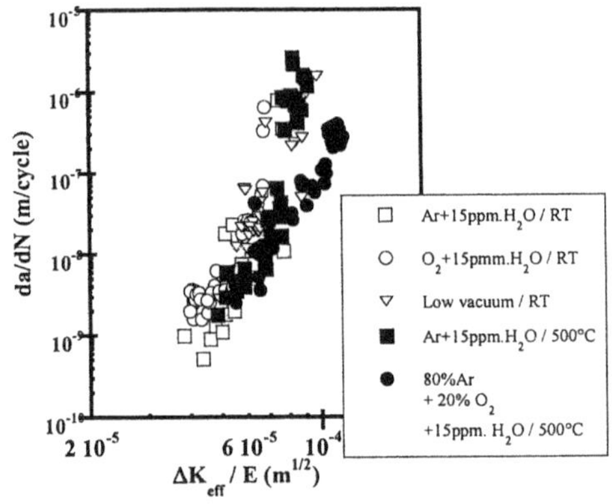

Figure 7: Fatigue crack growth behavior under different atmosphere containing water vapor and/oxygen at room temperature and 500°C.

Crack tip embrittlement. These results support a prevailing role of water vapor as active species and therefore a possible role of hydrogen-assisted fracture at the crack tip. However, the precise nature of the mechanisms operating at the crack tip still needs clarification. Indeed, the environmentally assisted fatigue crack propagation of conventional alloys has been attributed to two distinct mechanisms [27, 28]:

- a water vapor adsorption assisted regime: the adsorption of water vapor molecules induces an enhancement of the crack propagation by lessening the energy required to extend the crack [29].

- a subsequent hydrogen assisted regime due to hydrogen resulting from the surface dissociation of adsorbed water vapor molecules. This hydrogen is then presumably dragged into the strained material at the crack tip by mobile dislocations where it interacts with the fatigue damage [21].

It should be noticed that the second regime requires the attainment of a saturating adsorption on freshly created fracture surfaces. In addition, critical conditions depending on frequency, water vapor content and total pressure also determine the triggering of this regime. This hydrogen-assisted regime is typically observed under nitrogen containing traces of water vapor (up to 15 ppm.) for fatigue crack growth rates lower than 10^{-8} m/cycle [30]. However, the behavior observed under low vacuum conditions, i. e. roughly the same residual moisture content, does not show evidence of any hydrogen-assisted mechanism. By analogy the behavior exhibited in the present study on a γ-alloy in low vacuum and intermediate atmospheres (Figure 7) would be representative of the saturating adsorption-assisted regime. Therefore there is a reluctant enhancement observed in ambient air. This enhancement can be legitimately related to a hydrogen-assisted fracture at room temperature Furthermore the lack of modifications in da/dN curves or threshold values and the similarity of fracture surfaces [15] suggest that this same mechanism is also responsible for the enhancement observed at elevated temperature. This would mean that the temperature dependence of the different steps (adsorption, dissociative surface reaction, hydrogen transport) does not control the crack growth rate. This however raises several questions. Indeed it has been shown here above that oxygen does not interfere to limit hydrogen production. It comes out that the surface hydrogen production is enhanced by temperature but not quantitatively modified. Then the hydrogen transport mechanism is a concern since if at room temperature transport of hydrogen by moving dislocation is assumed to prevail, this mechanisms should become less efficient as temperature is increased. It could be replaced by lattice diffusion but one would then expect at least a transient effect on the FCG behavior which is not actually observed in the present investigations. Clearly these issues still need to be examined.

CONCLUSIONS

The influence of temperature on the fatigue crack growth behavior of a quaternary (2Mn-2Nb) γ-alloy has been investigated. The detailed analysis of the effects of temperature on the different mechanisms involved leads to the following conclusions:

• Increasing temperature only slightly improves the intrinsic fatigue crack growth resistance as observed in an inert environment, but only in the high crack growth rate regime in relationship with the enhanced fracture toughness around the brittle-to-ductile transition. Besides the near-threshold mechanisms do not seem extremely different from those involved in conventional ductile alloys.

• Closure effects are relevant under all the conditions investigated but the magnitude of crack tip shielding induced by closure is nearly temperature-independent whatever the environment. In particular the oxide-induced closure mechanism does not seem to prevail in air, even at high temperatures.

• Finally, a strong specific influence of environment has been highlighted both at room and elevated temperature. Water vapor has been shown to control this enhancement, independently of the presence of oxygen. Moreover, this environmental effect exhibits the same magnitude at the different temperatures investigated. As a consequence no "anomalous" temperature dependence was noticed on this alloy.

• The lack of influence of temperature on environmental fatigue crack growth enhancement is not fully consistent with an analysis of moisture-induced embrittlement of aluminides merely based on surface reaction kinetics. Clearly the identification of controlling mechanisms in environmentally assisted cracking of aluminides requires further investigations.

REFERENCES

1. Balsone, S. J., Wayne Jones, J. and Maxwell, D. C. (1994) In: *Fatigue crack growth in a cast gamma titanium aluminide between 25 and 954°C* W. O. Soboyejo, et al. (Eds), TMS, 307.
2. Chan, K. S. and Shih, D. S. (1998), *Metall Mater Trans A* 29 (1), 73.
3. Soboyejo, W. O. and Lou, K. (1994) In: *Micromechanisms of fatigue and fracture in gamma based titanium aluminides* W. o. Soboyejo, et al. (Eds), TMS, 341.
4. Soboyejo, W. O., Deffeyes, J. E. and Aswath, P. B. (1991), *Mater Sci Eng A-Struct Mater* 138 (1), 95.
5. Soboyejo, W. O., Aswath, P. B. and Mercer, C. (1995), *Scr Metall Mater* 33 (7), 1169.
6. Venkateswara Rao, K. T., Kim, Y. W. and Ritchie, R. O. (1995), *Scripta metall mater* 33 (3), 459.
7. McKelvey, A. L., Campbell, J. P., Venkateswara Ro, K. T. and Ritchie, R. O. (1996), *Fatigue '96*, Berlin, Germany, G. Lutjering and H. Nowack (Eds), Pergamon, Berlin, Germany., 1743.
8. McKelvey, A. L., Rao, K. T. V. and Ritchie, R. O. (2000), *Metall Mater Trans A* 31 (5), 1413.
9. McKelvey, A. L., Rao, K. T. V. and Ritchie, R. O. (1997), *Scripta Mater* 37 (11), 1797.

10. Stoloff, N. S. and Duquette, D. J. (1993), *Jom-J Min Metall Mater Soc* 30.
11. Liu, C. T. and Kim, Y. W. (1992), *Scr Metall Mater* 27 599.
12. Suresh, S. and Ritchie, R. O. (1983), *Fatigue Crack Growth Tresholds Concepts*, Philadelphia, Pennsylvania, D. L. Davidson and S. Suresh (Eds), The Metallurgical Society of AIME, Philadelphia, Pennsylvania., 227.
13. Suresh, S., Zamiski, Z. A. and Ritchie, R. O. (1981), *Metall. Trans.* 12A 1435.
14. Hénaff, G., Bittar, B., Mabru, C., Petit, J. and Bowen, P. (1996), *Materials Science & Engineering A* 219 212.
15. Mabru, C., Bertheau, D., Pautrot, S., Petit, J. and Hénaff, G. (1999), *Eng Fract Mech* 64 23.
16. Zhu, S. J., Peng, L. M., Moriya, T. and Mutoh, Y. (2000), *Mater Sci Eng A Struct Mater* 290 (1-2), 198.
17. Lesterlin, S., Sarrazinbaudoux, C. and Petit, J. (1996), *Rev Metall-Cah Inf Tech* 93 (9), 1135.
18. Lesterlin, S., Sarrazinbaudoux, C. and Petit, J. (1996), *Scripta Mater* 34 (4), 651.
19. Rosenberger, A. H., Worth, B. D. and Larsen, J. M. (1997), *Structural Intermetallics*, Seven Springs, Pa, M. V. Nathal, et al. (Eds), The Minerals, Metals & Materials Society, Seven Springs, Pa., 555.
20. Mabru, C. (1997),ENSMA - University of Poitiers (France) / The University of Birmingham (U. K.),
21. Wei, R. P. and Simmons, G. W. (1981), *Int. J. Fract.* 17 (2), 235.
22. Liu, C. T., Lee, E. H. and McKamey, C. G. (1989), *Scripta metall mater* 23 (6), 875.
23. Hénaff, G. and Tonneau, A. (2001), *Met Mater Trans A* 32A (March), 557.
24. Mabru, C., Hénaff, G. and Petit, J. (1997), *Intermetallics* 5 (5), 355.
25. Tonneau, A., Hénaff, G., Mabru, C. and Petit, J. (1998), *Scripta Mater.* 39 1503.
26. Li, J. C. M. and Liu, C. T. (1995), *Scr Metall Mater* 33 (4), 661.
27. Hénaff, G. and Petit, J. (1996), *Physicochemical mechanics of materials* 32 (2), 69.
28. Petit, J., Hénaff, G. and Sarrazin-Baudoux, C. (1997) In: *Gaseous Atmosphere Influence on Fatigue Crack Propagation* R. A. Smith (Eds), Kluwer Academic Publishers, 301.
29. Hénaff, G., Marchal, K. and Petit, J. (1995), *Acta Metall et Mater* 43 (8), 2931.
30. Petit, J., De Fouquet, J. and Hénaff, G. (1994) In: *Influence of ambient atmosphere on fatigue crack growth behaviour of metals* 2, Section VI on Influence of Environmental condition, A. Carpinteri (Eds), Elsevier, 1159.

Temperature-Fatigue Interaction
L. Rémy and J. Petit (Eds.)

GROWTH OF SHORT FATIGUE CRACKS FROM STRESS CONCENTRATIONS IN N18 SUPERALLOY

F. SANSOZ [(1,2)], B. BRETHES [(3)] and A. PINEAU [(1)]

(1) Centre des Matériaux, UMR 7633 CNRS, Ecole des Mines de Paris, B.P. 87, 91003 Evry cedex, France
(2) Currently at: Mechanics of Materials Laboratory, Department of Mechanical Engineering, University of Rhode Island, Kingston, RI 02881, USA
(3) SNECMA, Etablissement de Villaroche, 77550 Moissy Cramayel, France

ABSTRACT

DEN notched specimens containing a small semi-circular slot (0.1 mm) were made of a powder-metallurgy Ni base superalloy, alloy N18, in order to study the growth of short fatigue cracks from a stress concentration. Fatigue crack growth tests were conducted at 650°C with trapezoidal cycles 10s-300s-10s. Typical downtrends of crack growth rates were observed in these specimens during the crack propagation. Non-uniform stress and strain gradients at the notch root were calculated by FEM modelling using viscoplastic constitutive equations. The stress intensity factor was determined using these profiles and a weight-function method. To account for crack closure effects, a methodology was developed to calculate the effective stress intensity factor in the crack depth and at the free surface of notched specimens. It is shown that, for small crack lengths, in-depth opening ratios are significantly less pronounced in notched specimens than in unnotched specimens. Moreover, the crack closure effect determined at the free surface is higher than that calculated in-depth. The effect of a notch on this difference is addressed. Using these calculations, it is shown that the differences in crack growth rates observed between short and long cracks are no longer existent when crack closure effects are properly considered.

KEYWORDS

Short fatigue cracks, notch plasticity effects, crack closure, Finite Element calculations, 3D analytical predictions, powder metallurgy superalloy.

INTRODUCTION

Since powder metallurgy superalloys are used in the manufacturing of turbine disks for aeroengines, a clear understanding of the notch effects is required for a good assessment of defect tolerance at elevated temperature. One of these superalloys, N18 alloy, exhibits an excellent mechanical strength and good fatigue and creep resistances up to 650°C. However, during the processing route, a very small amount of inclusions are carried in the material. The size of the biggest inclusions is no more than 100 μm, but a small semi-elliptical crack could eventually be initiated under stress concentrations such as blade fixtures. Furthermore, due to high service temperatures, strongly non-uniform viscoplastic stress and strain fields are developed in the vicinity of these notches. The objective is, therefore, to take into account the notch plasticity effects in the growth behaviour of these semi-elliptical fatigue cracks.

In unnotched specimens, Pearson [1] showed differences in growth between physically small cracks (< 0.5 mm) and long cracks (> 0.5 mm). In order to correlate these differences, a number of authors developed the concept of intrinsic threshold [2]. In this approach, it is considered that the range of applied load is smaller than the stress intensity range, $\Delta K = K_{max} - K_{min}$, and is equal to $\Delta K_{eff} = K_{max} - K_{op}$, where K_{op} is the opening stress intensity factor calculated when the crack is fully opened [3]. Furthermore, in the case of short cracks, the determination of the effective stress intensity range ΔK_{eff} is strongly dependent of the crack length. On the other hand, Smith and Miller [4] investigated the behaviour of physically small cracks emanating from notches. Due to the notch plasticity effects, the load applied far from the notch can not be directly used and a local approach must be considered to determine the stresses within the notch. This approach was used successfully in several studies [5,6] in which crack closure effects were shown to significantly reduce the crack growth rates differences observed between short and long cracks. More recently Pommier et al. [7,8] have tested N18 alloy at 650°C. These authors showed that stress relaxation effects occurring at notch root can largely modify the effective stress intensity factor ΔK_{eff} when the crack is small in length (< 0.5 mm). Besides, on René 95 alloy, a methodology [9] was suggested to combine both the notch plasticity effects and the in-depth and the surface growth of penny-shaped cracks. However, the determination of the crack closure effect along the front of semi-elliptical cracks has not been fully investigated, in particular when creep-fatigue loading and notch plasticity effects are both considered. These objectives are partly achieved in this study by proposing a methodology to calculate the effective stress intensity range, ΔK_{eff}, in-depth and at the free surface of semi-elliptical cracks. The role of notch plasticity under creep-fatigue loading is addressed. This methodology is then applied to correlate the crack growth rates of short fatigue cracks measured on N18 alloy with Double-Edge-Notched specimens, specifically designed to study the effect of a stress gradient on the behaviour of small cracks.

MATERIAL AND EXPERIMENTS

Material and experimental procedure

N18 alloy is a Ni based superalloy. Its chemical composition is: Ni - 11.5% Cr - 15.7% Co - 6.5% Mo - 4.35% Al - 4.35% Ti - 0.5% Hf (weight %). All tests were performed on a "bulk" microstructure, which is obtained through a specific heat treatment procedure given elsewhere [10,11]. The monotonic and cyclic yield stresses at 650°C are 1050 MPa and 1150 MPa respectively. For a more comprehensive overview on this alloy, see reference [10].

Fatigue tests were carried out on Double-Edge-Notched (DEN) geometry containing two symmetrical U-shaped notches. The notch root radius is 2 mm and the reduced cross-section is 5 x 10 mm^2. A microstructural defect is simulated by a small semicircular EDM slot of 0.1 mm in depth located at the centre of the notch root on one side of the specimen. This machined defect is shown in Fig. 1.a. No precracking is made on the DEN specimens and the crack length on the free surface of the specimen is measured up to 1 mm from the initial semi-circular defect (0.1 mm). Tests were performed at 650°C with trapezoidal cycles 10s-300s-10s, maintaining a hold time of 5 min at maximum applied load. These specimens were tested with a constant nominal stress S_{max} varying from 600 MPa to 900 MPa, which represent respectively 0.5 and 0.8 times the monotonic yield strength (σ_0), and load ratios equal to 0, -0.5 or -1. A high-resolution optical system (Questar) was used to measure crack lengths at the free surface of specimen in the notch bottom. This technique has proved to be efficient to detect *in-situ* half-surface crack increments as small as 10 μm, while a conventional Potential Drop method is not enough sensitive to measure the length of very small cracks [11,12].

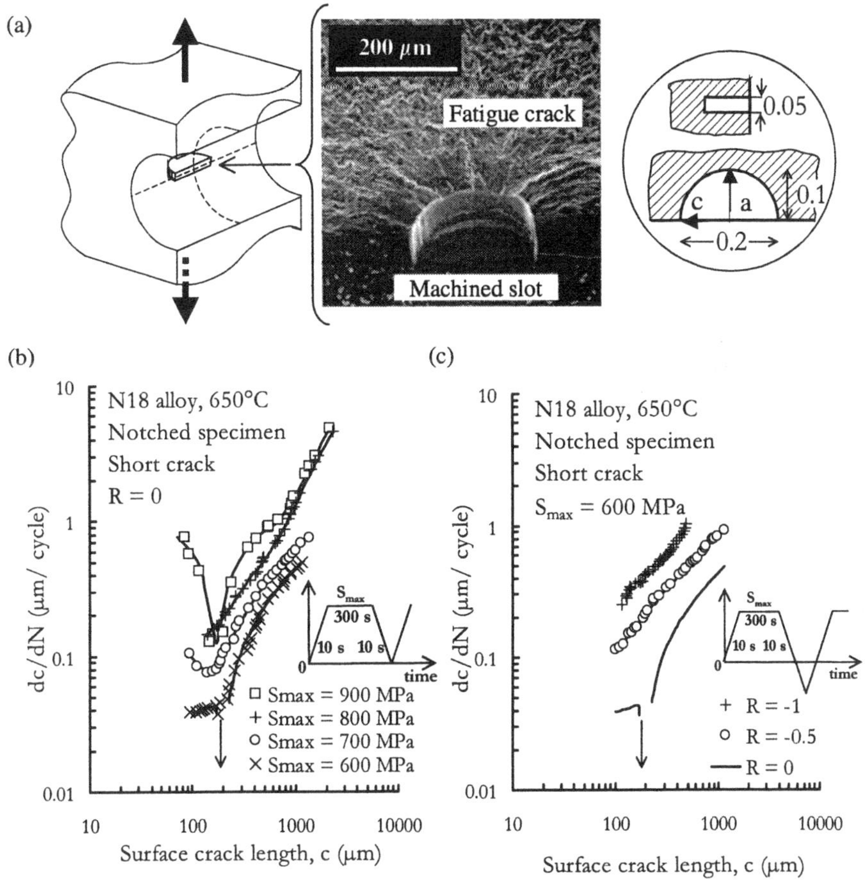

Fig. 1. Short crack growth rates measured in notched DEN specimens (N18 alloy at 650°C, cycles 10s-300s-10s): (a) semi-circular initial slot located at the center of the notch root. All dimensions in mm; (b) effect of maximum applied stress at R = 0; (c) effect of mean stress.

These results on short cracks were compared to the growth of long cracks measured on conventional unnotched specimens (KB2.5) containing a semicircular EDM defect of 0.3 mm in depth. On these specimens, a precracking was carried out at room temperature with a loading frequency of 10 Hz to obtain a semicircular crack with a depth of 0.5 mm. Fatigue crack growth tests were conduced at 650°C with trapezoidal cycles 10s-300s-10s and a load ratio $R = S_{min}/S_{max}$ of 0.1 or 0.3. The crack growth rate was measured on these specimens for crack lengths up to 2 mm in depth using a potential drop technique. Further details of the experimental procedure in notched and unnotched specimens are given elsewhere [12].

Results

Typical results measured on notched specimens with R = 0 are shown in Fig. 1.b. Two stages of crack propagation were observed with these tests. The first stage corresponds to decreasing rates when surface crack lengths are less than 200 μm. Then above this critical crack length, a steady state of crack propagation is observed with increasing rates. The downtrend of the crack growth rate curves or short crack effect is more pronounced for a low applied load S_{max} of 600 MPa. For this load, the arrow in Fig. 1.b represents a crack growth rate less than 10^{-8} m/cycle occurring during the crack propagation. Moreover it is observed that lowering the R ratio from 0 to –1 overcomes the down trend effect; see Fig. 1.c. These results strongly suggest that short crack effects are linked to crack closure effects.

MODELLING AND DISCUSSION

Stress and strain fields at notch root and ΔK calculations

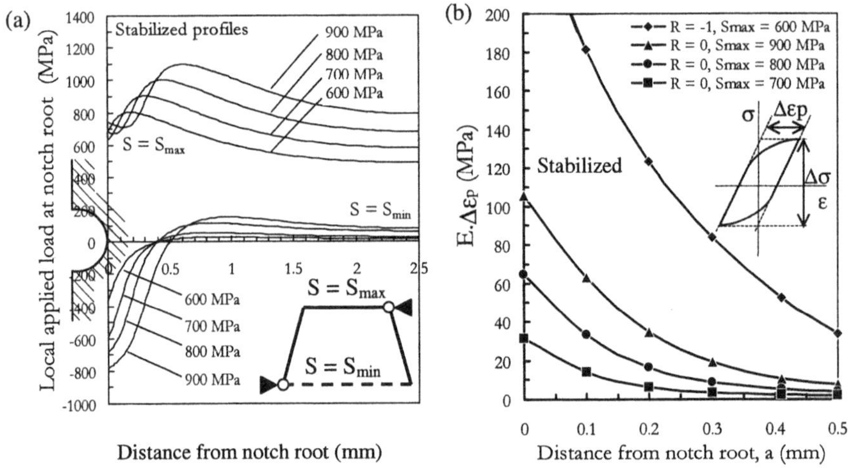

Fig. 2. Stabilized stress and strain profiles calculated at notch root with R = 0. Effect of applied loading: (a) on local stress at maximum load (S_{max}) and minimum load (S_{min}); (b) on pseudo-elastic stress calculated from plastic strain range.

The stress intensity range, ΔK, was calculated using the weight functions method introduced by Wang and Lambert [13,14], which was established for semi-elliptical cracks under non-uniform stress gradients. The local stress-strain field near the notch in the absence of a crack was calculated by Finite Element Method (FEM). The material behaviour was represented using an elasto-viscoplastic constitutive set of equations proposed by Lemaitre and Chaboche

[15]. The coefficients required for full identification of these equations, were identified using Low Cycle Fatigue tests performed at 650°C [11,12]. Detailed results of FEM calculations at notch root with cyclic loading are given in reference [12]. These FEM calculations showed that the tensile stress ahead of the notch progressively decreases to reach a stabilized condition, which was obtained after about 50 creep-fatigue cycles. Stabilized profiles at R = 0 with different applied loads are represented in Fig. 2.a. As expected, significant compressive stresses are noticed when the specimen is unloaded. This leads to an increase of the local applied stress range at notch root.

A simple correction to calculate ΔK accounting for the cyclic plasticity at notch root was used as proposed by Haigh and Skelton [16]. In this approach, the equivalent stress intensity range, ΔK^*, is calculated as:

$$\Delta K^* = (U.\Delta\sigma + E.\Delta\varepsilon_p).\sqrt{\pi a} \times F \tag{1}$$

where F is the LEFM geometry shape factor given by $F=\Delta K/(\Delta\sigma.\sqrt{\pi.a})$, Δσ is the total portion of the local stress range, U is the crack closure coefficient ($U=\Delta K_{eff}/\Delta K$), E is the Young's modulus and $\Delta\varepsilon_p$ is the plastic strain range in the vicinity of the notch. The pseudo-elastic stress, $E.\Delta\varepsilon_p$, calculated from stabilized profiles is represented in Fig. 2.b for different applied loads. For moderate applied stress ranges (<900 MPa), the strain range remains predominantly elastic. In this case, the calculated pseudo-elastic stress, $E.\Delta\varepsilon_p$, is less than 100 MPa. But for higher applied stress ranges (R = -1, S_{max} = 600 MPa), $E.\Delta\varepsilon_p$ can be larger than 200 MPa. However, the correction on ΔK given by equation (1) is less than 10% for this range when the crack depth is more than 90 μm. These calculations suggest that an elastic approach in terms of ΔK should apply in this situation, as indicated earlier [11]. Stress profiles at S_{max} and S_{min} similar to those shown in Fig. 2.a were approximated by a polynomial expression used in Wang and Lambert's results [13,14] to calculate both K_{max} and K_{min}. The results, which were used to compare the crack growth rates measured both on short cracks and on long cracks, are shown in the following.

Crack closure modelling and ΔK_{eff} calculations

In order to compare the crack growth rates of notched and unnotched specimens, the effective stress intensity factor, ΔK_{eff}, corresponding to these geometries must be determined first. In the present work, only plasticity-induced crack closure mechanisms will be considered. For unnotched specimens containing a semi-circular crack, closure effects are generally more pronounced near the surface than along the crack front within the specimen, due to the loss of constraint near the surface [17]. To investigate this effect on notched specimens, the current study utilises a simplified 2D FEM model to calculate the crack closure at the crack depth; see steps 1 and 2 in Fig. 3. Results of this model are then used in a 3D analytical simulation in order to calculate closure effects at the surface points of the crack; see steps 3 and 4 in Fig. 3. Both the 2D FEM model and the 3D analytical method will be explained in this section.

The 2D FEM model is based on a node-release technique similar to that proposed by Newman and Armen [18], which was used to simulate the growth of a fatigue crack in a 2D mesh. This technique was performed with the viscoplastic constitutive equations used previously to calculate the stress gradients in a uncracked geometry, as shown in Fig. 2.a. Plane strain conditions were assumed within the specimen. The mesh of a half-notched DEN specimen was compared to unnotched geometry as shown in Fig. 4.a. In both meshes, the propagation area, also shown in this figure, was refined with 8-nodes quadratic elements. The size of the smallest element in this area is 20 μm. Crack propagation is simulated from 0.095 mm to 1 mm measured from the free surface. In this analysis, the release of one node corresponds to three loading cycles (10s-300s-10s). Nodes are released at the minimum load level. The initial crack length, a_0, is equal to the depth of the machined EDM slot (0.10 mm). This model calculates the crack opening ratio S_{op}/S_{max} of a 2D crack propagating in a notched specimen subjected to different applied loads and R ratios. The profile of the crack opening ratio as obtained from this 2D model is assumed in this study, to be similar to that experienced by the in-depth point along a 3D semi-elliptical crack.

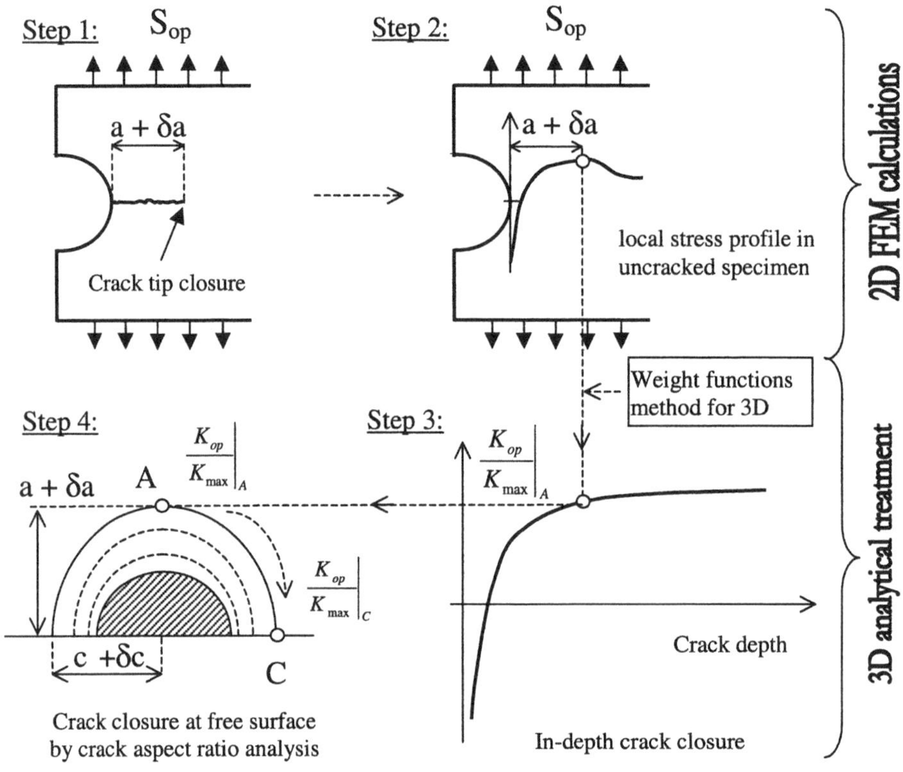

Fig. 3. Calculations of the opening stress intensity factor K_{op} for a semi-elliptical fatigue crack under non-uniform stress field, in-depth (A) and at free surface (C).

On the basis of this assumption, a procedure was established to determine the opening stress intensity factor, K_{op}, of a semi-elliptical crack under a stress gradient (steps 2 and 3 in Fig. 3). This procedure is described as follows. For a stress level equal to S_{op}, the stabilized stress profile at the notch root in a specimen without a crack is determined. The corresponding K_{op} is then calculated using Wang and Lambert's weight functions given at the deepest point of the crack front (point A). Figure 4.b shows the results of this procedure when S_{max} = 700 MPa and

R = 0 with a semi-circular crack front (a/c = 1). It is observed that the numerical results obtained on unnotched specimens are in good agreement with the experimental data for long cracks, as given in reference [10] ($K_{op} = 0.24\ K_{max}$). Furthermore, the opening ratios between unnotched and notched specimens are very similar when the crack is long (>200 μm). However, the crack closure effect is significantly less pronounced with notched specimens when the crack is very small (<200 μm). This result is in agreement with those of Pommier et al. [8] and Clung and Sehitoglu [19] who showed that crack opening levels are small in the notch root vicinity and, then, increase rapidly and stabilize out of the notch field.

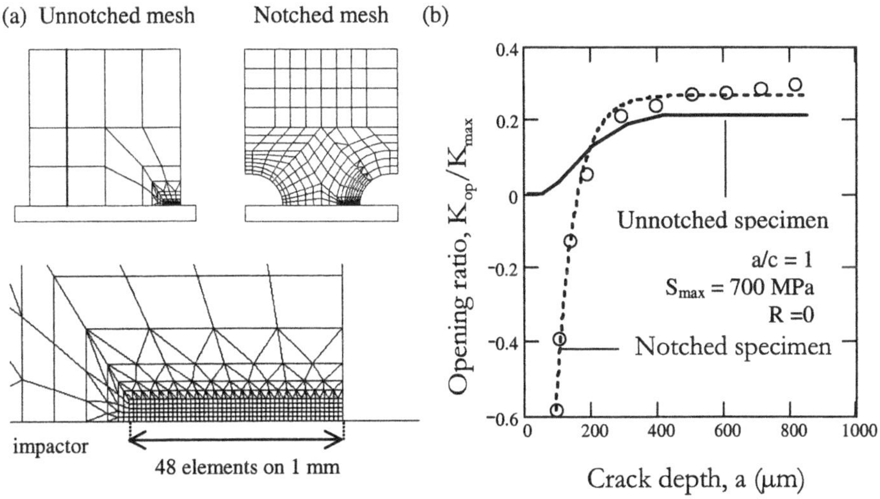

Fig. 4. 2D Finite Element calculations of crack closure: (a) geometry of unnotched and notched meshes used in the calculations; (b) final results K_{op} / K_{max} versus crack depth after analytical procedure.

The crack closure at the free surface (point C) was estimated from the knowledge of the opening ratios inside the specimens and using the method proposed by Jolles and Tortoriello [19]. In this technique, the crack growth increments at A and C, δa and δc respectively, are correlated as follows:

$$\delta a = \left(\frac{\Delta K_{eff,A}}{\Delta K_{eff,C}}\right)^n \times \delta c = \left(\frac{U_A \Delta K_A}{U_C \Delta K_C}\right)^n \times \delta c \qquad (2)$$

where n is the Paris's law exponent, U_A and U_C are the crack closure coefficients given at A and C, respectively. The method consists in the prediction of the crack aspect ratio, a/c, calculated iteratively from equation (2) imposing the shape of the initial defect. Figure 5 represents typical results obtained on DEN specimens tested with S_{max} = 900 MPa and R = 0. The analysis of the crack front shape obtained from interrupted tests as shown in this figure, indicates that the use of the ratio U_C/U_A = 1 is not a valid assumption. A correction to this approach was the use of a procedure introduced by Jolles and Tortoriello [19], which is applied to fit the ratio U_C/U_A using experimental measurements of the crack front shape, as shown in Fig. 5. The ratio used in this figure is fitted into the form:

$$\frac{U_C}{U_A} = 0.55 + (1 - 0.55) \cdot \exp\left[-2 \cdot \frac{a - a_0}{t}\right] \tag{3}$$

where t is the thickness of the specimen (t = 5 mm) and a_0 is the initial crack depth. This result shows that crack closure effects are more important along the free surface than inside the specimen. Jolles et Tortoriello [19] reached similar conclusions from observations on unnotched specimens ($U_C/U_A = 0.91$). It can, however, be noticed that the difference in crack closure along the crack front is less significant in unnotched specimens than in notched ones. The difference between these two geometries can be related to the increase of local stress range in the vicinity of the notch. Further studies to validate equation (3) for different applied loads are necessary. In the present work, the effective stress intensity factor in C, $\Delta K_{eff,C}$, was calculated using equation (3) for notched DEN specimens and $U_C/U_A = 0.9$ for unnotched specimens.

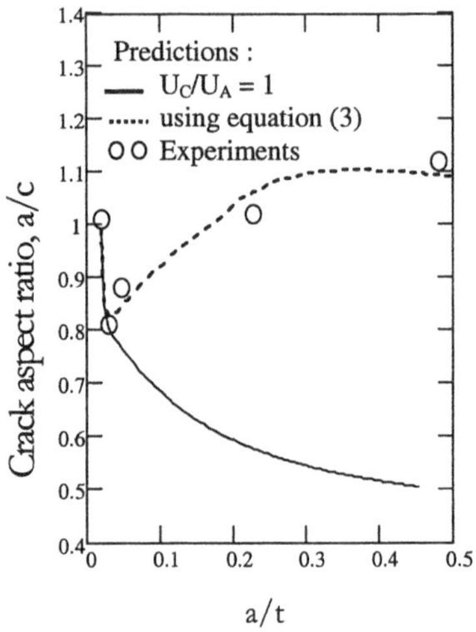

Fig. 5. Experimental examinations of crack front in notched specimens tested at 650°C with R = 0 and S_{max} = 900 MPa, and analytical predictions of crack aspect ratio.

Crack growth rate analysis

The crack growth rates obtained from notched specimens tested at 650°C are shown in Fig. 6. This figure includes also the results for long cracks obtained on conventional CT type specimens [10] and relatively long crack data obtained in this study using unnotched KB2.5 specimens. In Fig. 6.a, where no crack closure effect is considered, short and long cracks exhibit significant differences in crack growth behaviour at the free surface (point C), particularly at high ΔK values. It is also observed that a decreasing dc/dN pattern is present at low values of ΔK. In this case, to calculate ΔK, it was assumed that a/c = 1, which is in good agreement with the experimental results shown in Fig. 5, at least when half surface crack lengths reached approximately 1 mm. Similar results, which are presented elsewhere [11], were obtained for in-depth crack growth rates (point A). Figure 6.b shows the results obtained considering the crack closure estimate presented in this paper. In this figure, it is observed that the short crack growth rates effects found in notched specimens at lower values of ΔK no longer exist when closure effects are properly taken into account. This methodology seems, therefore, to be efficient to assess the significance in lifetime predictions of the very early stage of crack propagation.

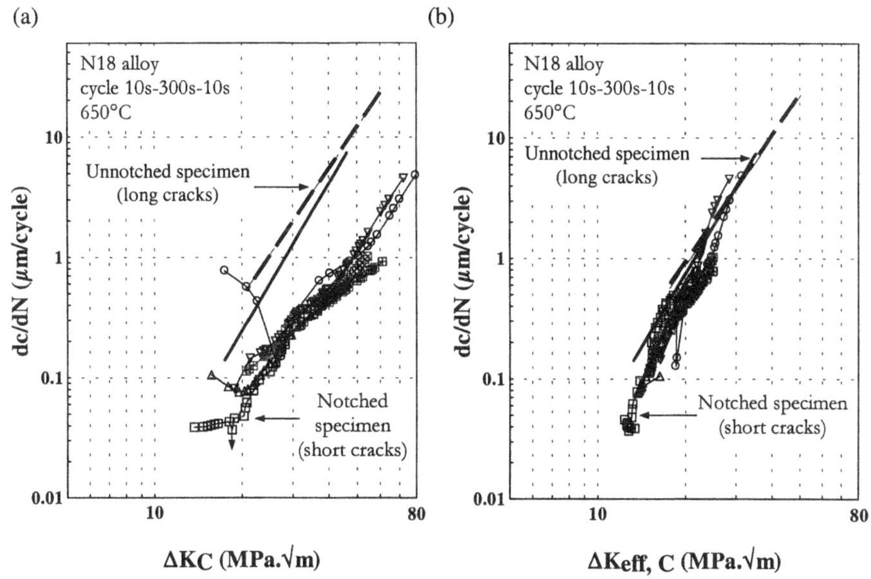

Long cracks / Unnotched specimen: KB2.5 ——— CT - - -

Short cracks / Notched specimen:

S_{max} =	0.5 σ_o	0.6 σ_o	0.7 σ_o	0.8 σ_o
R = 0	□	△	▽	○
R = -0.5	⊞			
R = -1	⊠			

Fig. 6. Comparison of fatigue crack growth rates measured in unnotched and notched specimens: (a) no crack closure is considered; (b) after 3D crack closure analysis.

CONCLUSIONS

1. The growth of short fatigue cracks was measured on N18 alloy at 650°C using DEN notched specimens. Typical downtrends of the crack growth rates were observed when R is equal to zero. This effect is no longer existent when negative R ratios are applied, keeping the same value for the maximum applied stress.

2. The stress intensity factor for a semi-elliptical crack and the crack opening were determined through FEM modelling using viscoplastic constitutive equations. A methodology was established in order to calculate the opening ratios within the specimen and at the free surface. It has been shown that crack closure effects calculated within the specimen are less pronounced in the vicinity of the notch root. However, the crack closure effect is more significant at the free surface than within the specimen. A phenomenological equation has been given to represent the change of crack closure along the crack in notched bodies. FEM calculations simulating the growth of a 3D crack from notches are necessary to extend these results.

3. Using the current methodology, it has been shown that the differences in growth between short and long cracks are simply interpreted by a mechanical effect related to crack closure effects. This methodology shows its ability to assess the significance of the early stage of propagation in lifetime predictions of turbine disks.

ACKNOWLEDGEMENTS: financial support from SNECMA is greatly acknowledged. Thanks also are due to Dr J.C. Lautridou from SNECMA and Prof. H. Ghonem from University of Rhode Island for many fruitful discussions and to Dr. J. Besson for his help in numerical calculations.

REFERENCES

1. S. Pearson (1975). *Engng Fracture Mech.* **7**, 235-247.
2. B. G. Journet, A. Lefrancois, and A. Pineau (1989). *Fatigue Fract. Engng Mater. Struct.* **12**, n°3, 237-246.
3. W. Elber (1971). ASTM STP 486, pp. 230-242.
4. R. A. Smith and K. J. Miller (1977). *Int. Journal of Mechanical Sciences* **19**, n°1, 11-22.
5. B. N. Leis (1985). *Engng Fracture Mech.* **22**, n°2, 279-293.
6. C. S. Shin and R. A. Smith (1988). *Engng Fracture Mech.* **29**, n°3, 301-315.
7. S. Pommier, C. Prioul, J. C. Lautridou and P. Bompard (1996). *Fatigue Fract. Engng Mater. Struct.* **19**, n°9, 1117-1128.
8. S. Pommier, C. Prioul and P. Bompard (1997). *Fatigue Fract. Engng Mater. Struct.* **20**, n°1, 93-107.
9. R. H. Van Stone, M. S. Gilbert, O. C. Gooden, and J. H. Laflen (1988). ASTM STP 969, T. A. Cruse Ed., Philadelphia, pp. 637-656.
10. A. Pineau (1997). In *Proc. of the Conf. "Engineering against fatigue"*. Eds J. H. beynon, M. W. Brown, R. A. Smith, T. C. Lindley and B. Tomkins. Sheffield (UK). 557-565
11. F. Sansoz, B. Brethes and A. Pineau (2001). To be published in *Fatigue Fract. Engng Mater. Struct.*
12. F. Sansoz (2000). *Ph.D. thesis*. Ecole des Mines de Paris
13. X. Wang and S. B. Lambert (1995). *Engng Fract. Mech.* **51**, n°4, 517-532.
14. X. Wang and S. B. Lambert (1997). *Engng Fract. Mech.* **57**, n°1, 13-24.
15. J. L. Chaboche and J. Lemaitre (1985). Mécanique des matériaux solides. Dunod, Bordas, Paris.
16. J. R. Haigh. and R. P. Skelton (1978). *Mater. Sci. Engng,* **36**, 133-137.
17. J. Z. Zhang and P. Bowen (1998). *Engng Fract. Mech.*, **60**, n°3, 341-360
18. J. C. Jr Newman and H. Jr Armen (1975). *AIAA journal*, **13**, n°8, 1017-1023
19. R. C. McClung and H. Sehitoglu (1989). *Engng Fract. Mech.*, **33**, n°2, 237-252
20. M. Jolles and V. Tortoriello (1983). ASTM STP 791, pp. I-297-I-307

Design and Structures

Temperature-Fatigue Interaction
L. Rémy and J. Petit (Eds.)

THERMO-MECHANICAL ANALYSIS OF AN AUTOMOTIVE DIESEL ENGINE EXHAUST MANIFOLD

K. HÖSCHLER*, J. BISCHOF**, W. KOSCHEL
Institute for Jet Propulsion and Turbomachinery, RWTH Aachen, D-52062 Aachen, Germany
* *current address: Rolls-Royce Deutschland Ltd & Co KG, D-15827 Dahlewitz, Germany*
** *current address: EADS Deutschland GmbH, D-81663 Munich, Germany*

ABSTRACT

The following paper presents a detailed description of a thermo-mechanical analysis method of an exhaust manifold for a four-stroke automotive Diesel engine, where the focus is on the physically correct description of the transient heat transfer on the hot gas side. For this purpose, a simple, but very effective method has been developed to calculate the quasi steady state heat transfer conditions and gas temperatures along the inner manifold gas path. The method is based on a coupled 1-dimensional exhaust gas flow and a 3-dimensional FE thermal analysis. The influence of the operational characteristic of a four-stroke engine on the local heat transfer is considered by appropriate correction factors.
The impact of the transient consideration on the stress-strain state of the component and therefore on the life is verified and explained via a dedicated example. The analysis results show very clearly the impact of the transient consideration on the stress distribution. The peak stresses do not occur at full load conditions, but at a transient state. Taking this transient peak stress with the corresponding strain range over the cycle at cyclic stabilized material conditions and the peak temperature, the life of the component can be analysed with the help of appropriate fatigue data.

KEYWORDS

Thermal Analysis, Mechanical Analysis, Manifold, Automotive Diesel Engine, Life Analysis

INTRODUCTION

Thermal and mechanically high loaded automotive engine components, like exhaust gas manifolds, necessitate a reliable design and analysis to fulfil the requirements with regard to weight, cost and life. An essential precondition for this accomplishment is a sufficiently precise transient thermal and mechanical analysis of the component using the Finite Element

Fig. 1: Exhaust Gas Manifold of a four cylinder turbocharged automotive engine

method.
Since one of the main loading for compact exhaust manifolds as depicted in Fig. 1, comes through the thermally induced strains, the priority must be on the correct description of the transient heat transfer on the exhaust gas side and on the outer surface of the component. The main difficulty is the setting-up of an adequate heat transfer model on the exhaust gas side describing the physically correct quasi-steady state heat transfer mechanism under consideration of the intermitting working process of a four-stroke automotive engine and possible superposition of individual cylinder exhaust gas flows for multi-cylinder engines. The model must be able to deliver the quasi-steady state local heat transfer coefficients and local exhaust gas temperatures along the gas channel for all transient stages of an engine loading cycle. For this requirement the temperature change of the exhaust gas along the manifold channel, especially during the acceleration and deceleration of an engine, needs to be considered by a coupled description of the exhaust gas flow and the temperature analysis.
The transient temperature development of the whole component may cause high local transient thermally induced strains, which are much higher than for steady state maximum loading conditions. For this reason, a transient stress and strain analysis is required as well, which should consider possible local non-linear effects of the material, which can be described by appropriate constitutive equations for cyclic material and creep behaviour. Based on the cyclic stabilized stress/strain hysteresis loop at the highest loaded location and the corresponding temperatures, the life of the component can be analysed with the help of appropriate fatigue data.

THERMAL ANALYSIS PROCEDURE

The thermal boundary conditions on the inside of the exhaust gas manifold and on the outside can be described by appropriate heat transfer correlations for forced and natural convection, radiation and heat conduction into adjacent components (e.g. cylinder head, turbocharger). The heat exchange between a gas and a wall can simply be described by

$$\dot{Q}_{conv} = A_S \, \alpha \, (T_G - T_W) \tag{1}$$

where $\dot{Q}_{conv}$ is the convective heat flow, A_S is the heat transferring surface, α denotes the heat transfer coefficient, T_G is the gas temperature and T_W the wall temperature. Since the heat transfer coefficient depends on the wall temperature for forced convection correlations in a pipe and vice versa, only a coupled temperature – flow analysis will lead to reasonable results.

General Procedure

For the correct analysis of the local heat transfer coefficient and corresponding gas temperature, all state variables describing the gas flow within the gas channels are necessary. This analysis can be performed by a simultaneous computation of the following one-dimensional equations:

1. Equation of continuity
2. Equation of momentum including gas friction and pressure loss due to the geometry, combining and dividing flows
3. Equation of energy including a heat flux into or out of the manifold walls
4. Equation of ideal gas

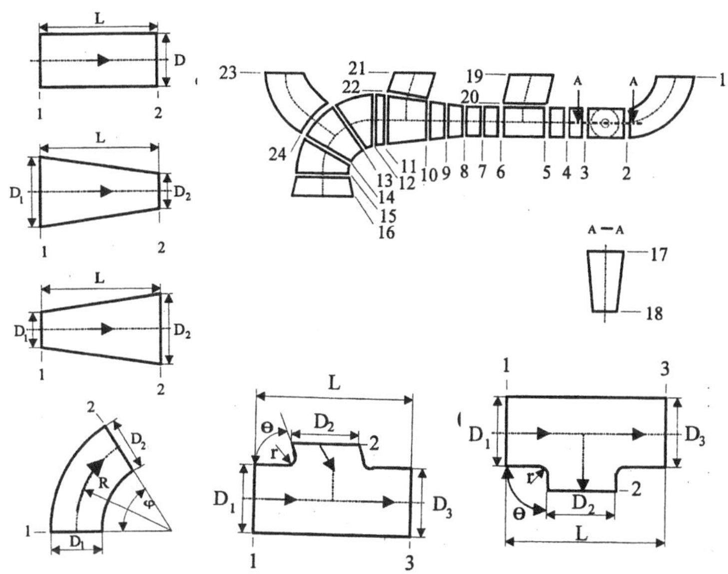

Fig. 2: Basic segments of the exhaust channels and position numbers

These equations are solved by an iterative step-by-step procedure for standard geometrical segments, where the values of the state variables at the exit of such a segment depend on the corresponding values at the inlet and the changes within the segment. To apply this procedure for the manifold displayed in Fig. 1, the exhaust gas channels have been divided into appropriate segments, which are shown in Fig. 2. The channel system consists of the following six elementary geometries:

- Straight channel
- Nozzle
- Diffuser
- Bend
- Flow combining T-segment
- Flow dividing T-segment

For the first four un-branched segments the equations are:
Equation of continuity:

$$\dot{m} = \rho_1 c_1 A_1 = \rho_2 c_2 A_2 = \text{const.} \tag{2}$$

Equation of momentum:

$$p_2 + \rho_2 c_2^2 = p_1 + \rho_1 c_1^2 - \tau_{1\to2} \tag{3}$$

where $\tau_{1\to2}$ characterises the pressure loss caused by gas friction and curvature.
Equation of energy:

$$\frac{c_2^2}{2} + c_{p,2} T_2 = \frac{c_1^2}{2} + c_{p,1} T_1 - q_{1\to2}, \quad \text{with } q_{1\to2} = \frac{\dot{Q}_{1\to2}}{\dot{m}} \quad \text{and } \dot{Q}_{1\to2} = A_S \alpha (\overline{T}_S - \overline{T}_{W,S}) \tag{4}$$

Equation of ideal gas:

$$\frac{p_i}{\rho_i} = R_G T_i \qquad i = 1,2 \tag{5}$$

For branched segments the first three equations are different and are given in the following where the indices are referencing the description in Fig. 2.

Equation of continuity:

$$\dot{m}_3 = \dot{m}_1 \pm \dot{m}_2 \tag{6}$$

Equation of momentum:

- T-segment with combining flow:
  From position 1 to 3:

$$p_3 + \rho_3 c_3^2 = p_1 + \rho_1 c_1^2 - \tau_{1\to3} - \zeta_{13}\frac{\rho_3}{2}c_3^2 \tag{7}$$

  From position 2 to 3:

$$p_3 + \rho_3 c_3^2 = p_2 + \rho_2 c_2^2 - \tau_{2\to3} - \zeta_{23}\frac{\rho_3}{2}c_3^2 \tag{8}$$

- T-segment with dividing flow
  From position 1 to 3:

$$p_3 + \rho_3 c_3^2 = p_1 + \rho_1 c_1^2 - \tau_{1\to3} - \zeta_{13}\frac{\rho_1}{2}c_1^2 \tag{9}$$

  From position 1 to 2:

$$p_2 + \rho_2 c_2^2 = p_1 + \rho_1 c_1^2 - \tau_{1\to2} - \zeta_{12}\frac{\rho_1}{2}c_1^2 \tag{10}$$

$\tau_{1\to2}$ characterises the pressure loss caused by gas friction and the curvature, ζ_{12} the pressure loss due to the dividing or combining flow.

Equation of energy for branched segments:

$$\dot{m}_3\left(\frac{c_3^2}{2} + c_{p,3}T_3\right) = \dot{m}_1\left(\frac{c_1^2}{2} + c_{p,1}T_1\right) \pm \dot{m}_2\left(\frac{c_2^2}{2} + c_{p,2}T_2\right) - \dot{Q}_{1\to3}\text{, and } \dot{Q}_{1\to3} = A_S\alpha(\overline{T} - \overline{T}_W) \tag{11}$$

For a T-segment with combining flow, the number of unknown coefficients corresponds to the number of available equations, so that all unknown state variables can be determined. In contrast, for the analysis of a T-segment with a dividing flow (e.g. for an exhaust gas re-circulation), two additional relations need to be known to determine the unknown coefficients. These relations are:

$$\dot{m}_2 = q\,\dot{m}_1, \qquad 0 \le q \le 1 \tag{12}$$

This describes the ratio of re-circulation mass flow, and

$$T_{t,2} \approx \frac{T_{t,1} + T_{t,3}}{2}. \tag{13}$$

This equation describes the approximate interdependence between the total temperatures at the inlet and exit cross-section.

As already described in the preceding equations, the pressure losses caused by friction, by the curvature of bent pipe segments and caused by combining or dividing the flows respectively need to be known to solve the set of basic equations. There are sufficient references in literature (e.g. Ref. 1,2) describing these pressure losses.

The determination of the balance of energy for the described geometries requires the calculation of the corresponding heat flow. According to equation (1), beside the local gas- and wall temperature, the local heat transfer coefficient needs to be known. Based on appropriate Nusselt-correlations for forced convection of turbulent flows in pipes the relevant heat transfer correlations can be calculated. For bent pipes an additional correction factor needs to be considered, which can be found in Ref. 3.

Furthermore, an additional heat flow to the inner walls of the exhaust gas channels due to radiation of the exhaust gas is characterized by:

$$\dot{Q}_{Rad} = \varepsilon_{G,W}\sigma A_S (T_G^4 - T_W^4) \tag{14}$$

where the equivalent emission coefficient can be calculated by the following approximation:

$$\varepsilon_{G,W} = \frac{1}{\frac{1}{\varepsilon_G} + \frac{1}{\varepsilon_W} - 1} \tag{15}$$

with ε_G and ε_W as the emission coefficients for the exhaust gas and the wall. According to Ref. 4, ε_G can be set to 0.8 to 0.99 for Diesel engines.

On the outer surfaces of the manifold, the heat fluxes comprise the heat transfer in or from adjacent components like the cylinder head and, as in this case, also the turbocharger. Further fluxes are described by forced and natural convection and radiation. In the present case, standard correlations from Ref. 5 have been used, as well as transient temperature measurements of the adjacent components.

The determination of the Reynolds- and Prandtl-number for the analyses of the pressure loss and the Nusslet correlations requires the calculation of the specific heat capacity, the dynamic viscosity and the heat conductivity of the exhaust gas. Since the values depend on the composition of the exhaust gas, the relative mass portions of the individual gas components (like O_2, N_2, CO_2, NO_2, CO, CH_4 and H_2O) have to be determined prior to the thermal analysis of the exhaust gas manifold. A good summary of the procedure to calculate these gas properties can be found in Ref. 2.

Consideration of the Working Cycle of the Engine

With the procedure described in the last paragraph the state variables of the exhaust gas along the channels of the exhaust gas manifold and the local heat transfer coefficients could be calculated. This procedure assumes a constant mass flow rate through all manifold gas admissions simultaneously, so that always steady state conditions are calculated as long as the boundary conditions (inlet gas temperature and pressure, pressure at the exit, number of revolutions of the engine) are kept constant.

This assumption is valid for the calculation of the gas temperature (for explanation see below), but not for the calculation of the local heat transfer coefficients, since these are significantly influenced by the working principle of a four-stroke engine. Fig. 3 shows the distribution of the gas temperature and the mass flow measured for a one-cylinder engine in the exhaust path (Ref. 6). It is obvious that the mass flow oscillated around zero as long as the exhaust valve of the cylinder is closed. When the exhaust valve opens (EVO), the mass flow increases rapidly,

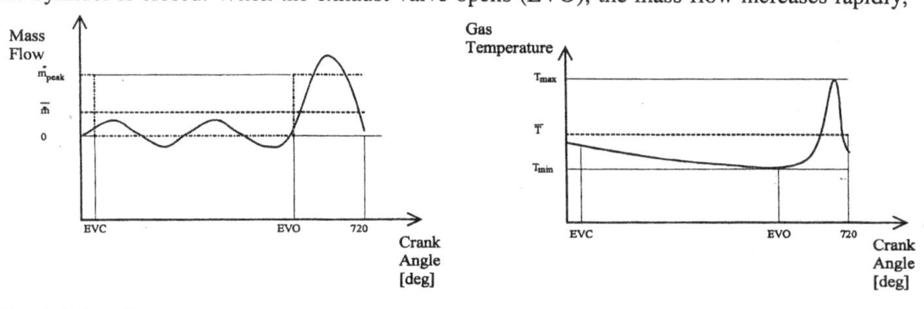

Fig. 3: Mass flow and exhaust gas temperature in the exhaust gas channel of a one-cylinder four-stroke engine

reaching a peak value and decreases then until the exhaust valve closes (EVC). A similar peak temperature behavior can be measured for the exhaust gas temperature.

During the time of one working cycle, the variation in exhaust gas temperature will only slightly influence the transient wall temperature distribution due to the short time nature. In contrast to this, the variation in mass flow has a large effect on the heat transfer into the manifold wall through the heat transfer coefficient. The higher the local velocity, the higher the Reynold number and therefore via the Nusselt number the higher the heat transfer coefficient. Since during the time when the exhaust valve is closed no mass flow occurs in the exhaust channel, it can simply be assumed, that during this time no convective heat transfer occurs (adiabatic condition), whereas only during the remaining time the heat transfer takes place. A calculation of the heat transfer coefficients based on the mean velocity over one working cycle would therefore not represent the right physics, whereas a calculation based on the peak velocity does (see Fig. 3). This peak velocity or the corresponding peak mass flow only depends on the opening time of the exhaust valve and the known mean mass flow.
For multi-cylinder engines it has to be considered that more than one cylinders exhaust gas mass flow can be within the exhaust manifold system at the same time. The sum of the mass flows is a function of the ignition order of the cylinders, the channel geometries and the time when the exhaust valves are open. For the considered manifold, a given ignition order and the closure time of the exhaust valves, the distribution of the peak mass flows at a specific locations is displayed in Fig. 4.
It can be seen that only the double peak mass flow is present for one working cycle in some areas of the exhaust gas channels. The described absolute peak mass flows, which are the base input for the correct heat transfer calculation, can then be related to the local mean mass flows and the mean exhaust gas temperatures as shown in the last paragraph. Since these mass flows depend only on the geometric and engine specific data, the ratio is independent of the engine speed.

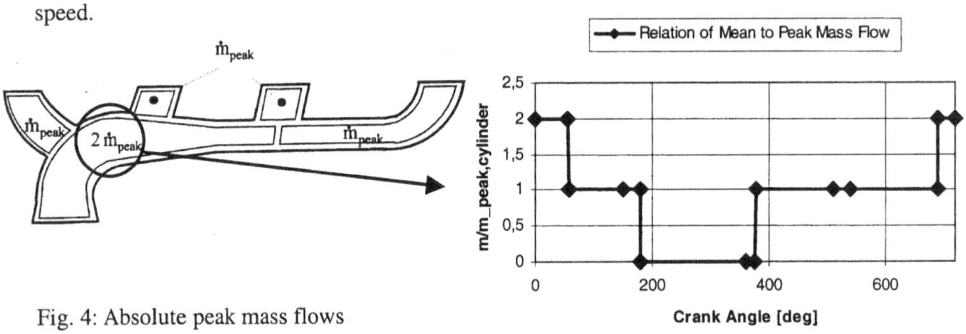

Fig. 4: Absolute peak mass flows

For the analysed manifold geometry the differences in mean and peak velocities and the resulting heat transfer coefficients at full load conditions are shown in Fig. 5. The charts emphasize the necessity for the consideration of the peak mass flows for the calculation of the heat transfer coefficients.
Summarizing these findings the following procedure can be applied for the calculation of the physically correct quasi steady-state heat transfer coefficients and gas temperatures along the exhaust gas channels:

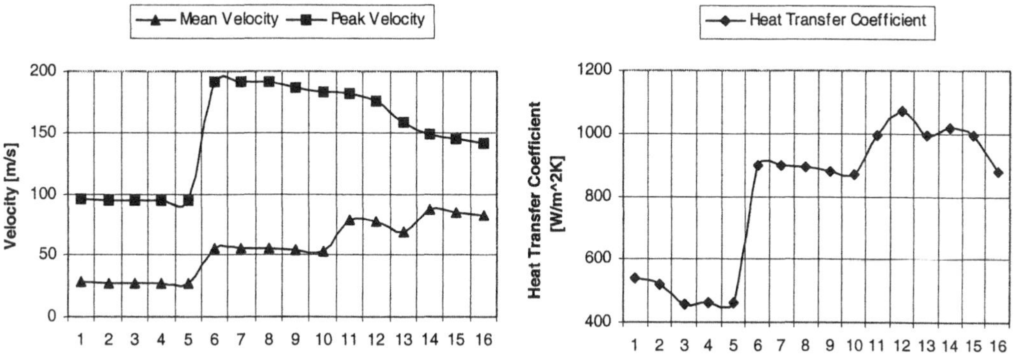

Fig. 5: Exhaust gas velocities and heat transfer coefficients for full load conditions

1. Calculation of the mean exhaust gas state variables along the gas channels assuming quasi-steady conditions at the inlet cross-sections.
2. Calculation of the heat transfer coefficients based on the ratio of absolute peak to mean exhaust gas velocities along the different segments of the exhaust gas manifold.

Transient Temperature Calculation for a Square Cycle

Following the procedure described in the last two sub-paragraphs, the exhaust gas manifold from Fig. 1 has been analysed with regard to the temperature distribution of the square cycle outlined in Fig. 6, where the speed of the crankshaft is displayed over the time. It can be seen that only during one third

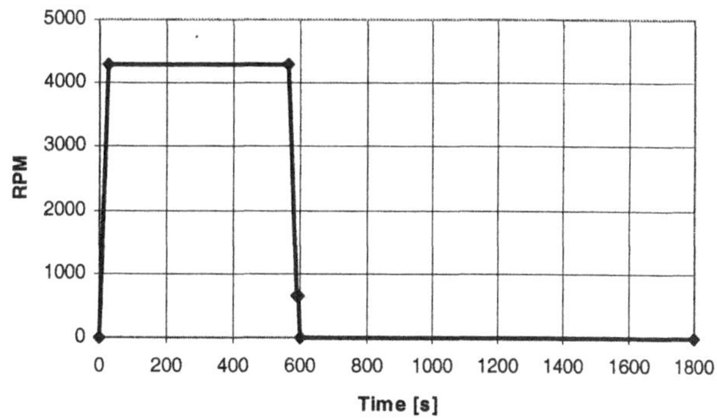

Fig. 6: Transient Square Cycle

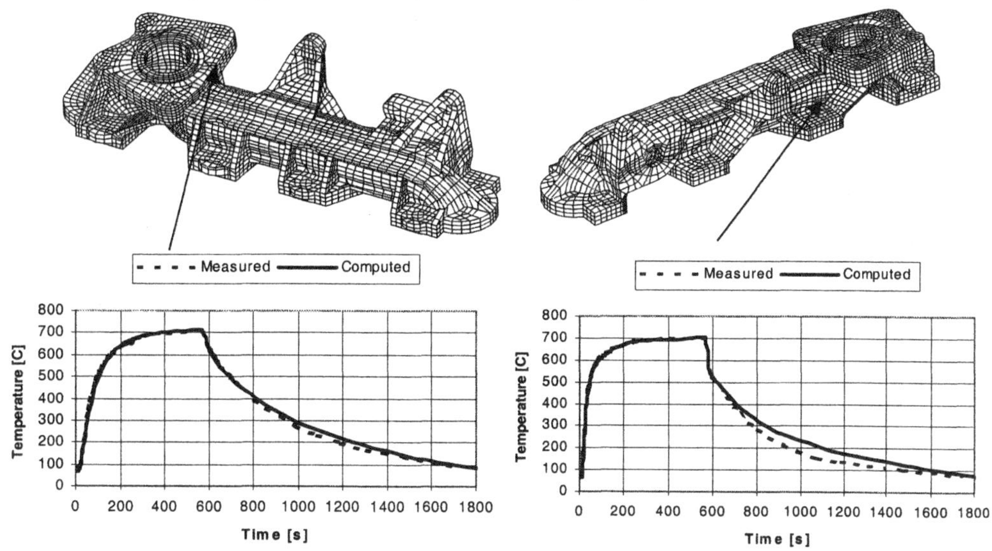

Fig. 7: Comparison of analysed transient temperature behaviour with measurements

of the whole time the engine is running on full load conditions. The thermal boundary conditions on the inside were calculated according to aforementioned procedure, whereas the boundary conditions on the outside of the component were only roughly adjusted so that the model temperatures match the measured ones at specific locations of the manifold surface. This confirms again that the inner heat transfer mainly influences the transient temperature behaviour of this component. Fig. 7 shows the excellent transient agreement between the measured and the calculated temperatures at two selected positions when the engine is running. The deviations during the cooling phase of the cycle can be accepted, since this does not influence significantly the following stress/strain analysis.
The good overall match of the analysed temperatures with the measured ones is displayed in Fig. 8, where the temperatures at full load conditions at specific locations are compared. The deviations are below two percent, which once again supports the developed procedure.

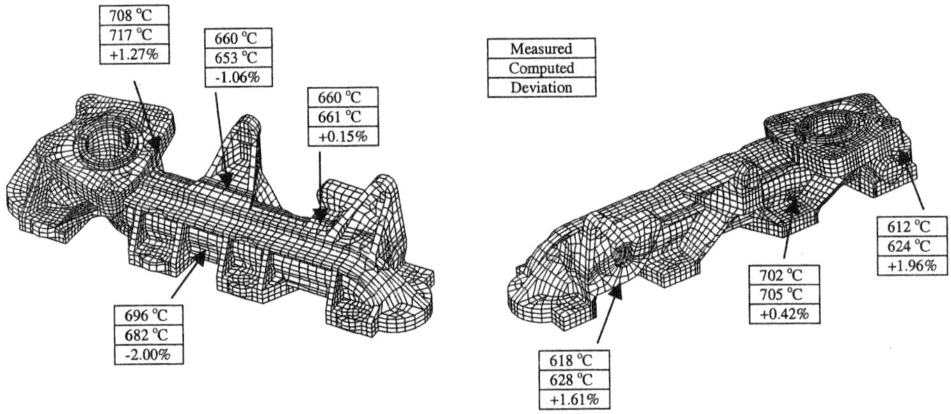

Fig. 8: Overall temperature deviations at full load conditions

MECHANICAL ANAYSIS PROCEDURE

The impact of a transient consideration on the mechanical behaviour of the fuel manifold is subject of the following chapter. Since the results are intended to be used for a following lifing analysis, the transient stress and strain behaviour of the component for the regarded cycle are required for stabilised material conditions. For this reason the manifold has been analysed transiently for the first two cycles to obtain a stabilized mechanical behaviour, assuming that the material does not show significant transient hardening effects.

Mechanical Boundary Conditions and Material Behaviour

Beside the thermally induced strains, the exhaust gas manifold is subjected to mechanical loadings due to the fixing screws, the differential gas pressure on the inner surfaces and forces due to the turbocharger and the exhaust gas re-circulation pipes. Under normal conditions, the influence of the external component load and the pressure differences can be neglected for this kind of solid component. Only the fixing screws may introduce a considerable mechanical load. To assure correct boundary conditions on the side of the cylinder head, the gasket

between these two components and its non-linear compressive behaviour has been modelled as well.
Plastic material data were in this case only available for monotonic tensile curves at several temperature levels. Based on this information only a simple isotropic hardening behaviour could be assumed which excludes transient hardening effects. This is sufficient to demonstrate the influence of the transient consideration on the stress-strain behaviour of the component, for reliable analyses, the cyclic transient material behaviour should be known and utilised.

Transient Mechanical Analysis

As already mentioned, the component has been analysed transiently for the first two cycles to obtain the stress and strain distribution at stabilized conditions.
The analysis showed a pronounced peak stress location near the flange to the exhaust turbocharger (Fig. 10), which occurs during the acceleration of the engine just a short time before the maximum speed is reached. A second peak can be seen, when the engine speed decelerates to idle. Fig. 9 summarises the temperature development at this location, the von Mises equivalent stress, the maximum principle stress and the resulting equivalent plastic strain. It is obvious that the peak stresses occurring during the acceleration and deceleration are significantly higher than at stabilized steady state conditions. A detailed view on the principle stress distribution is given by the explanation that during the acceleration phase, the temperature rapidly increases on the inside of the manifold channel, whereas the outer side is still relatively cold. This temperature difference causes a high tensile stress on the outer side and a high compressive stress on the channel side. During deceleration of the engine, the situation inverts. The inside temperature of the manifold decreases, whereas on the outer side

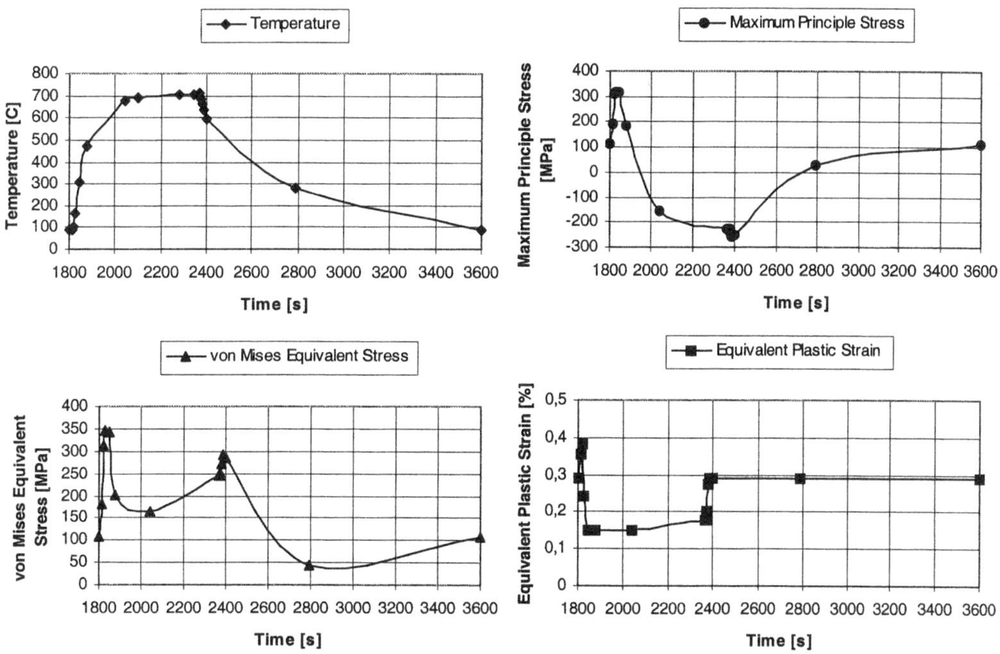

Fig. 9: Temperature, stresses and plastic strain development during the 2. cycle at the critical location

due to the bulk mass of the flange to the turbocharger the temperature decrease reacts slower. This causes a small compressive peak stress at the outer location and a tensile stress at the inner location.

Fig. 10 gives an impression about the location of the highest stress area and the introduced plastic strain range for the second cycle. Taking this transient peak stress, the corresponding strain range over the cycle at cyclic stabilized material conditions and the peak temperature, the life of the component can be analysed with the help of appropriate fatigue data.

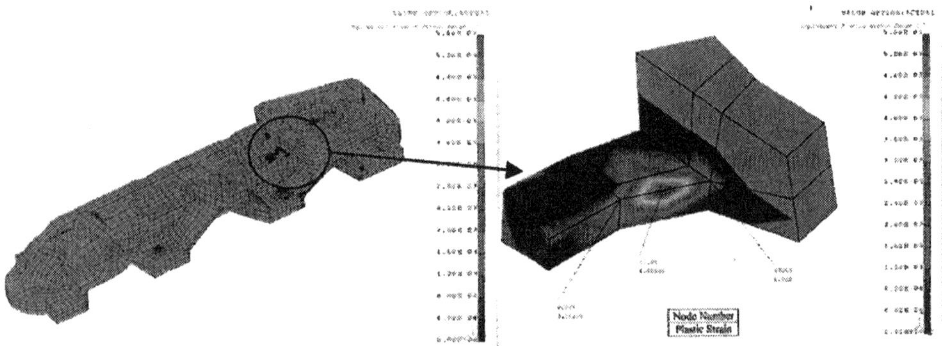

Fig. 10: Equivalent plastic strain range for stabilized conditions

The described method allows the correct identification of possible fatigue failure locations of exhaust manifold components. It supports the manual or automatic geometry optimisation of the component with regard to low cycle fatigue and therefore supports the intention of automotive companies to reduce development times by reducing the number of necessary tests.

ACKNOWLEDGEMENT

The authors would like to acknowledge RENAULT S.A. for their support and delivery of the measured data.

REFERENCES

1. Beitz, W., Küffner, K.-H. (Eds), (1986) *Dubbel, Taschenbau für den Maschinenbau* Springer Verlag
2. VDI-Wärmeatlas, (1999) 8. Edition, VDI-Verlag GmbH Düsseldorf
3. Hausen, H. (1976), *Wärmeübergang im Gegenstrom, Gleichstrom und Wechselstrom*, 3. Edition, Springer Verlag
4. Pflaum W., Mollenhauer K., (1977) *Wäremübergang in der Verbrennungskraft-maschine*, 2. Edition, Springer Verlag
5. Renz, U., (1984) *Grundlagen der Wärmeübertragung*, Institute for Heat Transfer & Climatology, RWTH Aachen
6. Meissner S., Sorensen S.C., (1986), *Computer Simulation of Intake and Exhaust Manifold Flow and Heat Transfer*, SAE Technical Paper Series No. 860242

THERMOMECHANICAL FATIGUE DESIGN OF ALUMINIUM COMPONENTS

VERGER, L.[1,2], CONSTANTINESCU, A.[1], CHARKALUK, E. [2]
[1] Laboratoire de Mécanique des Solides (CNRS UMR 7649)
Ecole Polytechnique, 91128 PALAISEAU - FRANCE
[2] P.S.A. Peugeot-Citroën - Direction des Plates-Formes, Techniques et Achats
18 Rue des Fauvelles, 92250 LA GARENNE COLOMBES - FRANCE

ABSTRACT

This paper presents a global approach of thermomechanical design in the automotive industry. The three steps of the method are the following: the loading definition, the modelling of the constitutive law, a failure criterion. Cracked area and lifetime prediction are described in the case of aluminium alloy cylinder heads submitted to transient thermal loadings. The main hypothesis and modelling choices are presented and results are illustrated by a series of computations on real 3D structures.

INTRODUCTION

Only a small number of components in automotive technology are submitted to thermomechanical loading cycles of a severity level capable of producing low-cycle fatigue. Most of these parts are related to the engine: cylinder-heads, exhaust manifolds, crank cases,... It is important to remark that the low-cycle fatigue problems in such engine components are related with the *start - operate - stop* cycles and not with the combustion cycles.
Therefore they involve large temperature changes. A previous study was dedicated to cast-iron exhaust manifolds [1]. This work showed an original global approach which conducted to a reliable prediction of damage location and of lifetime in spite of basic assumptions like the separation of the damage and the mechanical constitutive behaviour.

This paper presents a similar methodology applied to aluminium cylinder heads. The words *global approach* shall denote the following series of items:

- thermal analysis
- elasto-viscoplastic mechanical analysis
- fatigue analysis

The cylinder head is made of an aluminium alloy (in this case an A356 alloy) and is locally submitted to temperature cycles ranging from 20°C up to 260°C.
The mechanical behaviour of the aluminium alloy is plastic at low temperatures and highly viscoplastic at the upper temperatures. Moreover, aluminium alloys are also subject to thermal ageing effects for temperatures over 150°C.

Therefore different steps had to be defined in order to apply the global approach in this case:

- modelling of the ageing behaviour of the material
- modelling of the cyclic viscoplastic behaviour of the material
- determination of the optimised parameters of the models on the basis of isothermal and non-isothermal uniaxial tests
- determination of a low-cycle fatigue criterion on the basis of isothermal uniaxial tests and non-isothermal uniaxial and multiaxial tests
- computation of the Finite-Element Analysis with realistic boundary and loading conditions
- determination of the lifetime of the structure using the chosen fatigue criterion.

In order to keep the above methodology at a complexity level compatible with the constraints of the industrial design engineers, several assumptions regarding the *uncoupling* of the different phenomena have been made: uncoupling between thermal computation and mechanical computation, uncoupling between ageing evolution and mechanical behaviour and uncoupling between the constitutive law and the damage modelling. Another important hypothesis is the existence of a stabilised mechanical response of the structure.

A series of other difficulties like the identification of the temperature dependant parameters of the constitutive law with non-isothermal tests data could not be simplified [2].

The paper presents an overview of this approach and the specific details of the applications in the case of aluminium structures. The results are illustrated by a series of computations on real 3D structures.

MODELLING THE MATERIAL BEHAVIOUR

The studied automotive cylinder heads are made of an A356 aluminium alloy composed with 7% silicon and 0.3% magnesium (in mass percentage). In critical areas during the engine processing, the temperature of the material can reach an amount of 300°C.
At these temperatures, the material shows two main behaviours : in one hand a considerable thermal ageing resulting in the loss of at most 70% of its mechanical characteristics, in the other hand a highly viscoplastic mechanical behaviour as the temperature approaches 30% of the melting temperature.

Ageing behaviour

In order to stabilise the mechanical characteristics of the studied alloy at an upper level despite of the high temperature, the entire cylinder head is submitted to a typical T7 thermal treatment after casting : homogenisation at 540°C to dissolve the precipitates, water quenching, artificial ageing at 200°C during which the consistent Mg_2Si precipitates are formed.
The temperature influence on this original T7 alloy has been sized during an appropriate experimental program of conventional elastic force and hardness measurements on specimens submitted to constant temperature dwells (see Fig. 1).

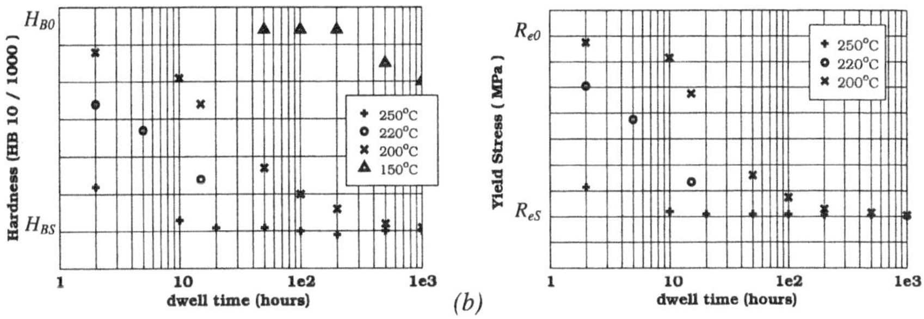

(a) *(b)*

FIG. 1. *Temperature ageing evaluation of a T7 A356 aluminium alloy [4] by conventional elastic force (a) and hardness (b) evolution*

A quite simple exponential function has been used to model the temperature and time-dependant ageing of the alloy :

$$H_B(t,\theta) = H_{BS} + \frac{(H_{B0} - H_{BS}) - \exp(0.95\ln(\theta) + 15.7\ln(t) - 86.4)}{1 + \exp(0.95\ln(\theta) + 15.7\ln(t) - 86.4)}$$

with H_{B0} the original hardness in Brinell scale, and H_{BS} the hardness of the fully aged material. This function has proved to be correctly fitting the experimental data on specimens [4] as well as on processed cylinder heads (see Fig. 2).

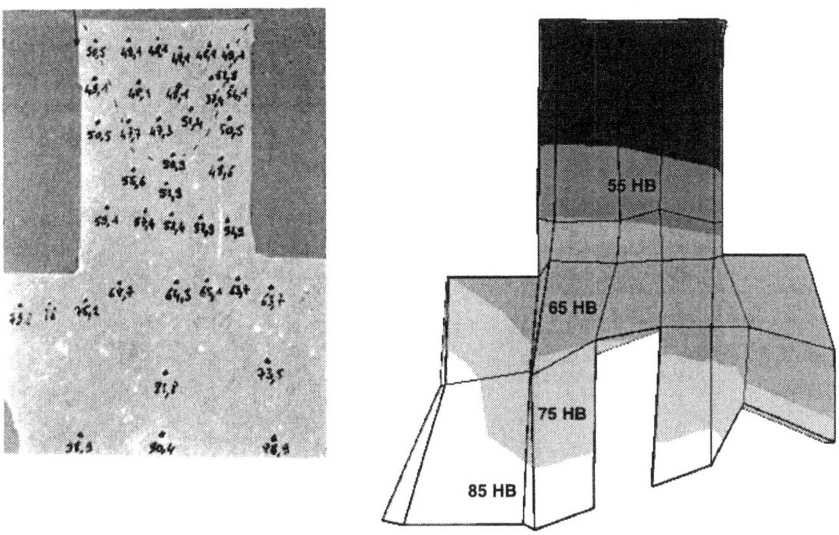

FIG. 2. *Comparison of hardness measurements (a) and calculations (b) on a section of intervalve space of a tested cylinder head*

Mechanical behaviour

Several tests have been performed on specimens in order to show the cyclic mechanical behaviour of A356 aluminium alloy (see Fig. 3). These are tension/relaxation/compression (TRC) cyclic tests at different temperatures in the range of 20°C to 300°C, performed at various strain rates : 10^{-3}/s, 5.10^{-4}/s and 10^{-4}/s.
A viscous behaviour has been shown by exponential stress/time relaxation response. Plastic yield stress and hardening are noticed at each temperature. A classical viscoplastic model with one single inelastic strain variable similar to those developed by Chaboche [5] has then been used with temperature-dependant parameters (see Fig. 4).

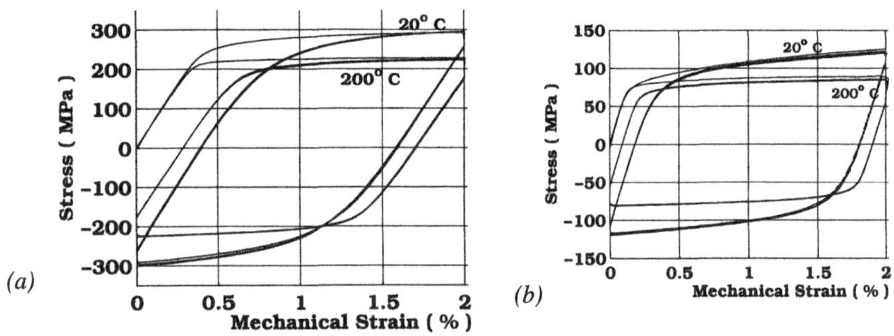

FIG. 3. *Response of T7 (a) and artificially aged (b) specimens to a cyclic tension-relaxation-compression test at several temperatures*

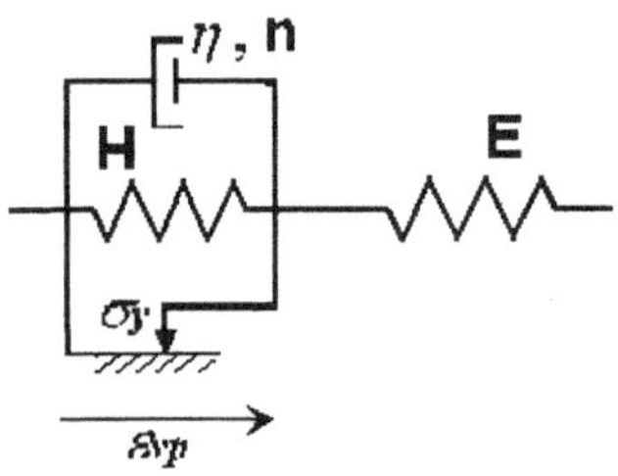

FIG. 4. *Schematic viscoplastic model with one single inelastic strain variable*

Uncoupled damage

As some classical low cycle fatigue (LCF) tests have been performed in the same range of temperatures, it has been shown that the A356 aluminium alloy undergoes a slight cyclic softening during almost the whole life of the specimens (see Fig. 5). The coupled modelling of such a cyclic damage with the viscoplastic behaviour would be very expensive in terms of identification tests and computation times. Furthermore, several studies have already shown that this coupling may not be necessary to perform reasonably predictive lifetime calculations [3,6].

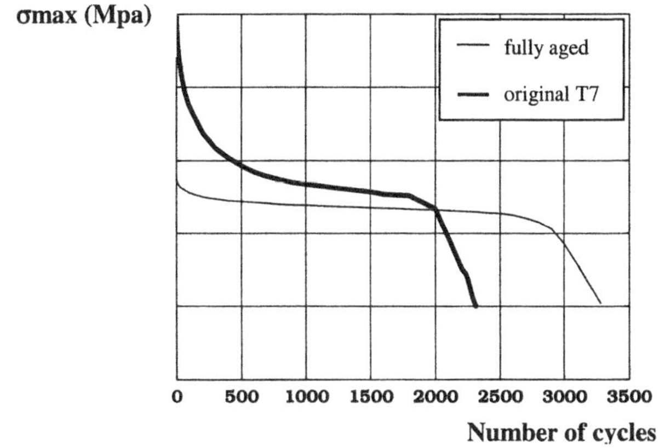

FIG. 5. *Maximum stress evolution during LCF tests at constant temperatures*

As a result, it has been chosen not to include any damage variable in the constitutive law used to represent the mechanical response of the studied material. The cyclic damage will then be calculated from the mechanical-variable results of a "stabilised cycle" numerical simulation (for more details, see the *Fatigue Life Assessments* part of the present document).

BEHAVIOUR IDENTIFICATION

The unified viscoplastic model chosen in this case is available over the whole range of temperature provided that its five parameters are computed as temperature-dependant functions.

The identification method is based on an optimal control approach [7]. The parameter functions are determined through the minimisation of a cost functional defined as:

$$J(p)=\frac{1}{2}\int\left[\sigma_{\exp}-\sigma_{calc}(p,t)\right]^2 dt$$

where the time-integral measures the difference between the experimental and the computed stresses ($\sigma_{\exp}$ and σ_{calc} respectively) over a complete experiment. In order to determine parameters from several experiments, the corresponding cost functionals can be added using weight factors. The minimisation itself is performed by a classical BFGS gradient descent algorithm.

According to the ageing of the alloy both measured and calculated on the structure (see Fig. 2), it has been chosen to characterise the material under two stable states: on one hand the original T7 state which makes up the " cool " part of the cylinder head, and on the other hand the fully aged state found in the " hot " part of the structure, which undergoes the thermomechanical fatigue phenomena. By the mean of a set of calculations on structure, the consequences of the lack of realistic ageing modelling in the transitional area on the fatigue predictions have been shown of the second order.

A particular effort had to be made on the determination of the experimental database to be used for the identification analysis [2]. Indeed the viscosity parameters η and **m** show an important coupling behaviour on the TRC tests simulation at constant temperature (see Fig. 6). Then, the use of non isothermal experimental data provides us with a useful additional information. Furthermore, the loading conditions applied on specimens for non-isothermal tests can be defined using the typical thermal and strain history registered on the engine component (see Fig. 7). As a result, the parameters identified from both isothermal TRC and non-isothermal "structure-like" measurements are likely to perform realistic behaviour simulations on structure calculations.

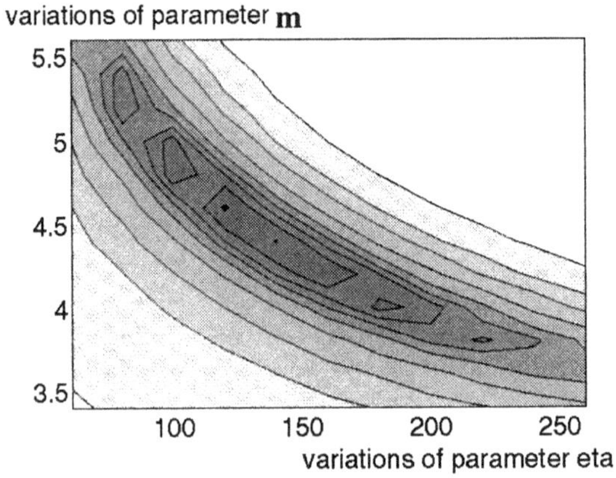

FIG. 6. *Isovalue of the cost functional showing the (η,**m**) coupling in the long shape of its minimum valley*

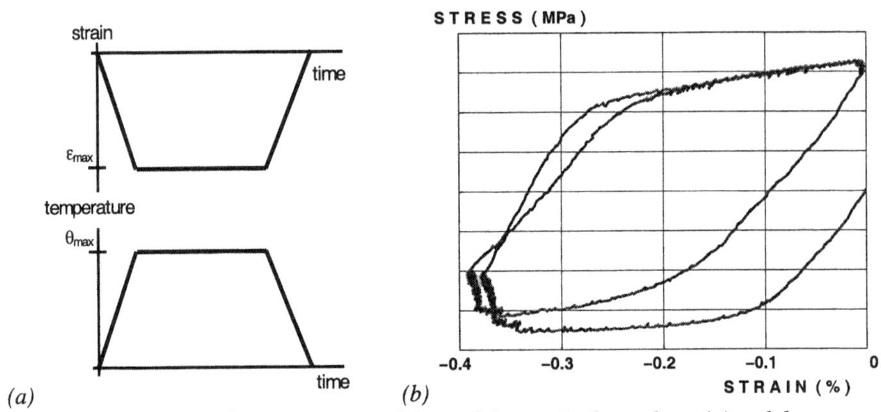

FIG. 7. *Strain and temperature conditions of the non isothermal test (a) and the stress/strain response of the viscoplastic model (b)*

FATIGUE ANALYSIS

The thermomechanical loading cycles considered for a cylinder head correspond to a start/stop cycle of the engine. This implies to calculate the whole assembly of cylinder blocks, gasket and cylinder head in order to perform the thermal and mechanical analysis. The heating conditions consist in an inflow of hot gases through the fire deck and the exhaust pipes of the cylinder head. Then a dwell time at maximum temperature is performed, followed by a cooling down due to engine shutting down and circulating pressurised cold water. Therefore three computations will determine the global behaviour of the structure:

- a combustion and fluid flow simulation supplying the heat exchange coefficients and the thermal fluxes,
- a transient thermal computation simulating heating and cooling down and supplying the temperature distribution (and consequently the ageing state distribution, see Fig. 8),
- a mechanical computation with the temperature field as the main load parameters.

The details of a similar approach applied on cast iron exhaust manifolds are available in [8].

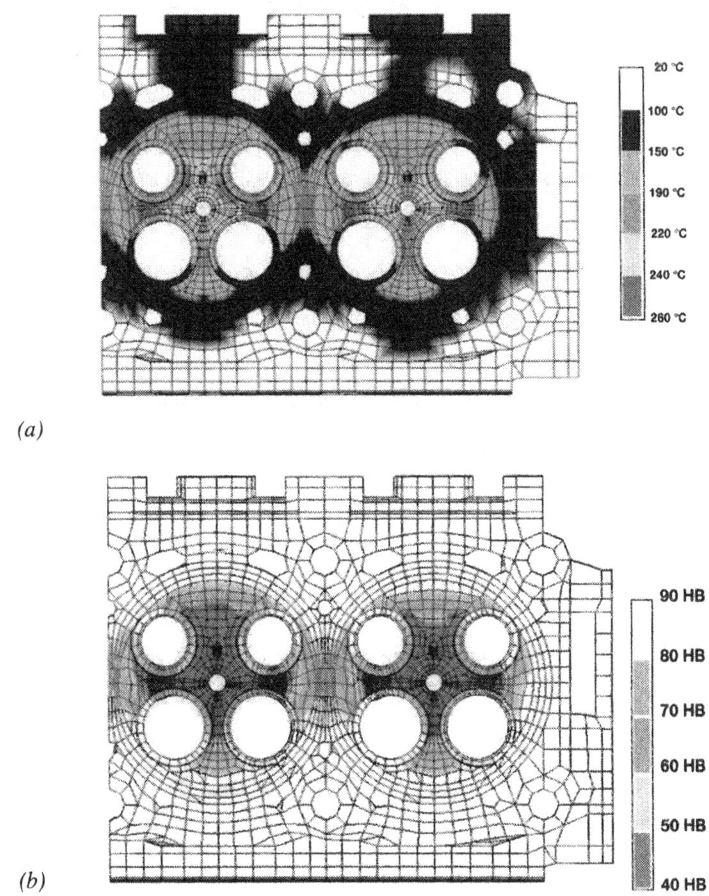

FIG. 8. *Maximum temperature field (a) and hardness distribution after 10000 cycles (b) computed on a cylinder head submitted to thermomechanical cycles*

Once the mechanical fields and a stable limit cycle are obtained, the fatigue analysis consists in applying a low cycle fatigue (LCF) criterion compatible with the anisothermal multiaxial context. In a preceding study, CHARKALUK et al. [1] have shown the inadequacy of a series of classical criteria in such a context. Usually based on the use of the plastic strain range (Manson-Coffin) or the maximal stress (SWT, Ostergren) of the cycle, these criteria have in common the major drawbacks of having no physical meaning under transient temperature conditions.

In this work as well as in the previous exhaust manifold study [8], it has been chosen to interpret the dissipated energy per cycle as the adequate damage indicator to be associated with a macroscopic crack initiation. The fatigue criterion becomes:

$$\Delta W \times N_f^{\beta} = c$$

where the dissipated energy is calculated over the stabilised cycle obtained from the mechanical analysis.

The choice of ΔW as a damage indicator proved to be consistent with numerous LCF results of the literature [9]. Moreover, ΔW is an intrinsic variable which takes into account both the multiaxial stress/strain state and the anisothermal effects.

In the case of the A356 alloy, the fatigue criterion has been established using classical isothermal LCF tests on fully artificially aged specimens at different temperatures between 20°C and 300°C under strain control. A series of TMF tests on clamped specimens were also conducted to assess the fatigue criterion. Figure 9 represents the comparison between experimental lifetime and calculated results for these different tests as well as for two real structure calculations. All the lifetime results stay within an acceptable ±2.5 error margin on lifetimes.

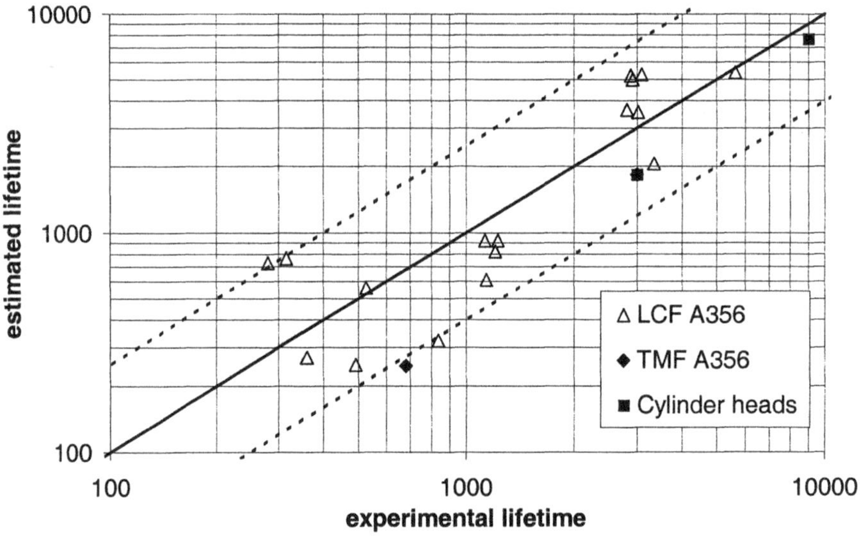

FIG. 9. *Comparison of the estimate lifetime and experimental lifetime for A356 alloy LCF and TMF tests and several structures*

The prediction of the crack location (see Fig. 10) as well as the lifetime has been proved to be fully satisfactory on complete structure calculations. It should be noticed that these results have been achieved on large FEM models of more than 800 000 degrees of freedom in computing times compatible with the constraints of the industrial design engineers.

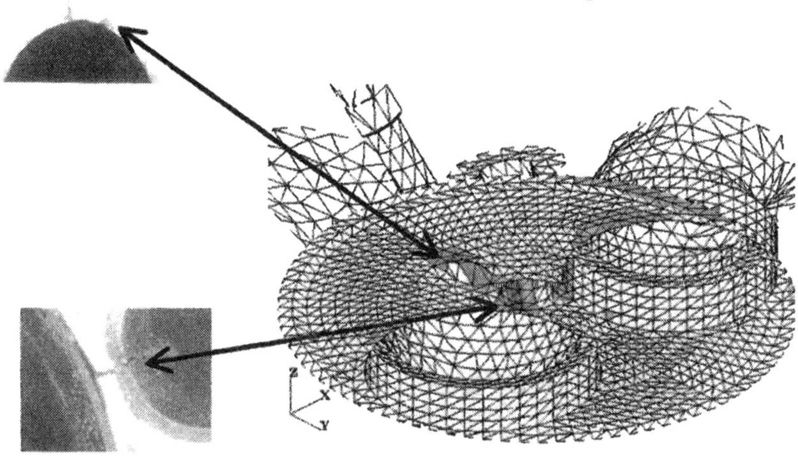

FIG. 10. *Crack initiation on a cylinder head prototype and distribution of the fatigue criterion on the FE model*

CONCLUSION

A complete design approach has been derived for structure undergoing thermomechanical fatigue. Formerly applied to cast iron exhaust manifolds, the method is developed here for the design of aluminium alloy cylinder heads. Based on the use of a simple elasto-viscoplastic law and an energetic failure criterion both suitable under multiaxial and non isothermal loading conditions, the design approach also calls for some important assumptions regarding the uncoupling of modelled phenomena which permit the computations of large structures. This design approach has been successfully applied on several industrial structures. It has shown a good correlation between prediction and experiment for the crack initiation location as well as for the durability which demonstrates the robustness of the approach. Moreover this method is compatible with an engine development schedule and permits to decrease the number of validating thermal shock tests on real components.

Acknowledgements

The authors would like to thank Mr K. Dang Van, LMS-Ecole Polytechnique, and Mr A. Bignonnet, PSA Peugeot Citroën, for fruitfull discussion.

REFERENCES

1. Charkaluk, E., Constantinescu, A., Bignonnet, A. and Dang Van, K. (1999) In: Dimensionnement en fatigue des Structures, SF2M 18th Journées de printemps
2. Verger L., Constantinescu A. and Charkaluk E. (2000) In: IUTAM Creep in Structures 2000, Murakami and Ohno (Eds). Kluwer.

3. Charkaluk, E. and Constantinescu, A. (2000) Materials at High Temperature, 17, N°3
4. Boussac, O. and Callais, T. (1998) Influence du Vieillissement Thermique sur les Caractéristiques Mécaniques à Température Ambiante de l'AS7G03 Y39, Rapport Interne PSA
5. Lemaitre, J. and Chaboche, J.-L. (1985) Mécanique des Matériaux Solides, DUNOD (Paris)
6. Sermage, J. P., and Lemaitre, J., and Desmorat, R. (2000) Fatigue Fract. Engng Mater. Struct., 23
7. Bourgeois, L. (1998) Contrôle Optimal et Problèmes Inverses en Plasticité, Ecole Polytechnique (Paris)
8. Lederer, G., Charkaluk, E., Verger, L. and Constantinescu, A. (2000) In : SAE Technical paper series, 2000-01-0789
9. Verger, L., Charkaluk, E. and Constantinescu, A. (2001) presented at the 6th International Conference on Biaxial/Multiaxial Fatigue and Fracture, Lisbonne

Temperature-Fatigue Interaction
L. Rémy and J. Petit (Eds.)

THERMOMECHANICAL FATIGUE IN THE AUTOMOTIVE INDUSTRY

A. BIGNONNET and E. CHARKALUK
P.S.A. Peugeot Citroën
Direction of Research and Automotive Innovation
chemin de la malmaison, 91570 Bièvres, France

ABSTRACT

The thermomechanical fatigue assessment in the automotive industry is discussed. The hypothesis and the principal aspects of the design strategy are presented.
There are the thermomechanical loading, the mechanical constitutive law of the material, the damage parameters and the fatigue strength criteria. The good understanding of these different steps allows to perform predictive calculations of automotive parts subjected to thermomechanical loading.

KEYWORDS

Thermomechanical fatigue, viscoplasticity, multiaxial fatigue, loading analysis, finite elements computations, dissipated energy

INTRODUCTION

Numerical simulation benefits these 20 past years of an increased performance of numerical tools and the development of new algorithms. Today this enables ever quicker calculations of structures. Nevertheless whilst academic work is quite active, few studies have been realised on structures submitted to multiaxial thermomechanical loading.
In the automotive context, engineers are faced with the design of components with complex thermomechanical loading. There are engine components like crankcase, piston and cylinder head, exhaust system and brake components. All of these parts are severely loaded and their computational fatigue design includes numerous difficulties leading to complex problem.
These difficulties come from four main steps of the design. That is to say: the identification of the thermomechanical loading, the mechanical constitutive behaviour of the material and its temperature dependence, the damage driving force identification and the fatigue criterion itself.

These four aspects are linked together; the loading and the material behaviour determine the mechanical response of the structure and from that the fatigue strength of the structure can be estimated through a damage indicator and a pertinent fatigue criterion (fig 1).

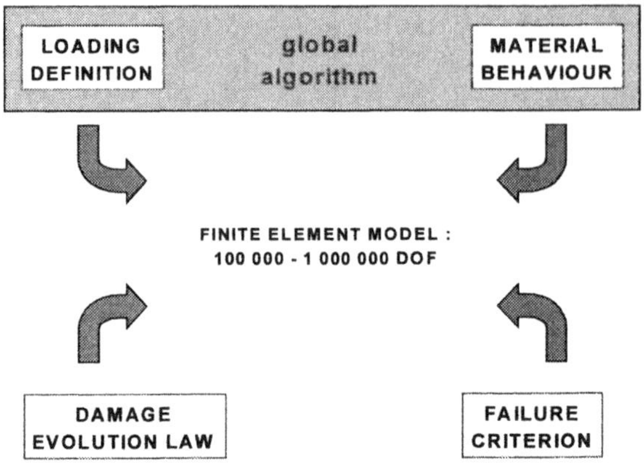

Fig. 1 : The four steps of fatigue assessment

For a design strategy, one should not concentrate separately on these four aspects of the problem. It should be a global approach to obtain an important robustness and to allow its integration in a development process [1] by the ability to answer to the asked question in the appropriate time. That is to say in our case: is the fatigue strength of the component guaranteed or not before the realisation of tooling for a prototype ?

In the following, we will put in light the hypotheses and the choices, which appears to be important or even essential with the objective to propose a robust fatigue design approach, ie from the description and the modelisation of the loading to the estimation of the fatigue life. The adaptation of the design strategy will be then illustrated on automotive components.

LOADING

Representation of in service loading

In the first step, it is important that the thermal loading, applied at the validation stage on engine testing and when calculating on virtual prototype, should be representative of the customer's uses from the point of view of the fatigue damage. An analysis method, which can be used, is the Stress-Strength Interference Analysis [2] which is frequently used for high cycle fatigue studies.
The main element of this method are: a counting method (for example rainflow counting) on the appropriate and accessible variable (for example forces, accelerations, temperatures), a cumulative damage law or equivalent damage method (for example Miner's rule), a risk analysis and the description of the fabrication scatter (material and process).
The extension of this method to the thermomechanical fatigue is particularly delicate due to the non-linearity (behaviour, loading ...) and supposes several simplifying hypothesis. Nevertheless, this can be reasonably done and it is used to define bench test to represent the customer's use with the desired severity. Figure 2 schematically represents the different steps of this work for a typical thermal loading of an exhaust pipe. The acceptance criteria is given with regard to an objective customer taking into account the fabrication scatter on a definite risk.

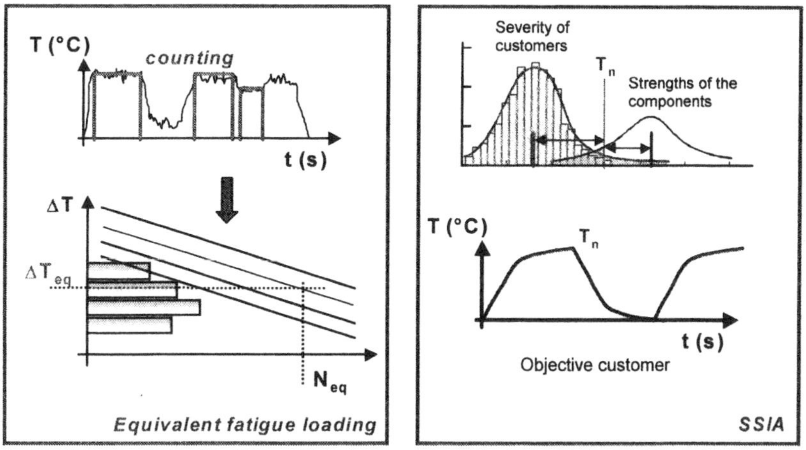

Fig.2 : Transformation of in service temperature analysis to an equivalent loading, with a known severity for tests or calculation, and decision strategy.

Simulation of the thermal loading

For engine components, the principal element of the loading is the thermal evolution (the mechanical loading is essentially linked to the bolt tightening between the different components and to the press fit of valve seats or valve guides in the case of the cylinder head; this different aspects can be represented rather easily). This is a dominant step of the approach since the accuracy of thermal data conditions at the first order the accuracy of the mechanical response of the structure and therefore the estimated fatigue strength.
The objective is to simulate numerically the thermal evolution of the components for operating engine conditions defined to be representative of the customer use.
The input data come from engine combustion simulation using dedicated numerical tools (KIVA, FIRE, FLUENT, ...). These data are used as thermal flux on the fire deck of the cylinder head on one hand, and are associated on the other hand to the gas flow in the exhaust conducts and pipes to determine the gas temperatures and therefore the internal exchange coefficient.
A transient thermal calculation done in both case enables the determination of the thermal loading which is used for the thermomechanical calculation [3].
The determination of the thermal loading is still today the major difficulty of a thermomechanical fatigue approach in the automotive industry (particularly for thermal exchange between the engine components: cylinder head, crankcase, piston, exhaust pipe ...) and will necessitate extensive work in the forthcoming years.

MECHANICAL BEHAVIOUR

The origins of failures encountered at high temperature on automotive components is linked to thermal loading. Three main aspects can be underlined:

- the temperature cycles between ambient and maximum temperature (250°C in the cylinder head, 800°C and above on exhaust pipe, or 700°C for brake disk) reach a domain where mechanical strength of material becomes quite low.
- high temperature dwell time favors creep (that is the case for engine crankcases).
- steep thermal gradients leads to inelastic deformations in confined zones.

In other respects failure analysis also shows:
- the cyclic aspect of the loading
- high and variable temperatures

Within these temperature ranges (0,3 to 0,5 Tm) the behaviour of the material is strongly non-linear with threshold. We can no longer speak of elasticity.

Material testing is performed on material / process representative specimens (similar microstructure and metallurgical quality as the actual component) at different strain rate with dwell time at imposed strain. The material behaviour shows a strain rate dependence of this threshold and a stress recovery during dwell time. As an example, the mechanical behaviour of a 356 Al Alloy is given on figure 3.
Testing at imposed stress show creep of material. These phenomena clearly demonstrate the role of viscosity and therefore simple plasticity is not able to represent our problem. We are faced to an elastoviscoplastic behaviour.

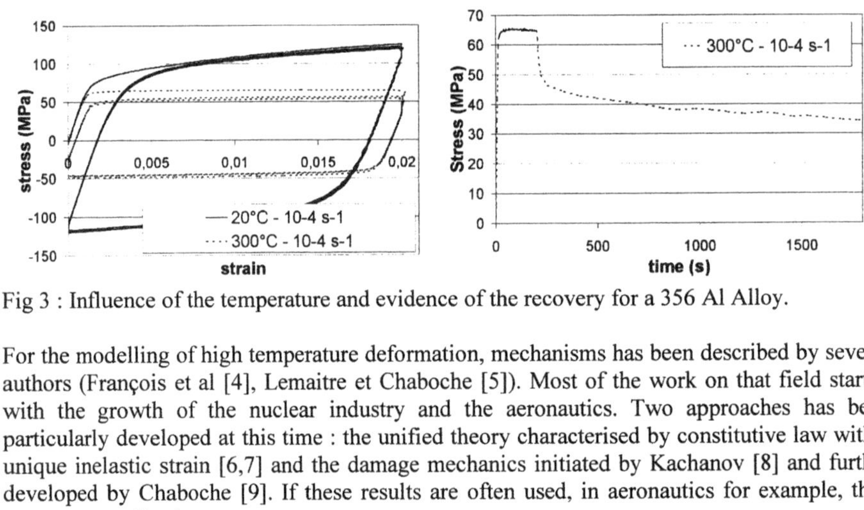

Fig 3 : Influence of the temperature and evidence of the recovery for a 356 Al Alloy.

For the modelling of high temperature deformation, mechanisms has been described by several authors (François et al [4], Lemaitre et Chaboche [5]). Most of the work on that field started with the growth of the nuclear industry and the aeronautics. Two approaches has been particularly developed at this time : the unified theory characterised by constitutive law with a unique inelastic strain [6,7] and the damage mechanics initiated by Kachanov [8] and further developed by Chaboche [9]. If these results are often used, in aeronautics for example, they present some disadvantages :

- the constitutive law can rapidly become complex to model the whole physical phenomena. The number of characteristic parameters become quite important and their eventual coupling makes difficult their identification.
- these mechanical models are not necessarily representative of the whole physical phenomena which can explain that they are often imperfect to model creep and recovery (François et al [4])
- coupling damage and behaviour leads to difficulties in parameters identification and numerical implementation, but overall conducts to lengthy calculation.

Based on these observations, one can conclude that to be appropriate to the automotive context, thermomechanical modelling should respect several constraints :

- the necessity to represent the behaviour of the material in the structure by simple constitutive equations. In the context, one should favour the representation of the mechanical behaviour without coupling with damage (fig 4).
  From experience most of the time this coupling is not necessary and does not improve the predictivity of the behaviour of the structure. Usually it is not necessary to describe the evolution of the structure cycle by cycle and particularly the early stage of the life of the structure. Therefore the identification of the parameters of the constitutive equation can be done on steady state cycles from material testing.
- the implementation in a finite element code using a numerical integration algorithm for the constitutive law stable and robust enough to authorised large integration steps.

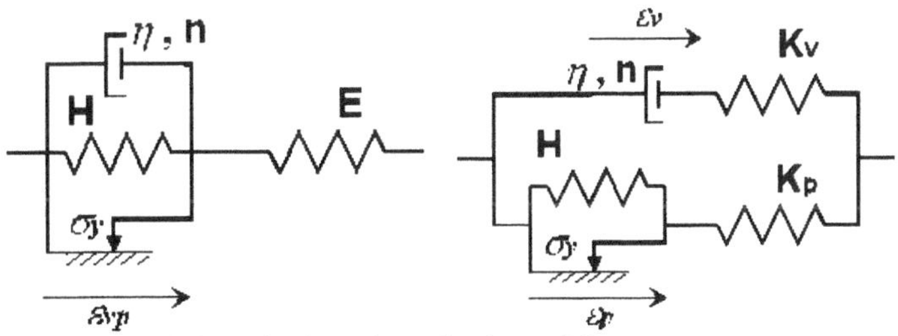

Fig 4 : Unified model and viscoelastic – elastoplastic model

DAMAGE PARAMETER AND FATIGUE CRITERIA

The final objective is to evaluate the fatigue strength, or life time of the structure, from the stabilised response of this structure. To that a criterion must be defined upon the mechanical variables linked to the stress and strain fields. The first studies about the low cycle fatigue criteria give some useful parametric laws. Nevertheless all the classical approaches like Manson – Coffin ([10], [11]), Strain Range Partitioning (SRP [12]) or the function defined by Smith – Watson – Topper (SWT [13]) initially expressed from uniaxial isothermal tests were not really adaptable to the thermomechanical fatigue of the complex structures. The difficulties came from the multiaxial character of the loading of such structures as well as the large temperature range see under operation. Actually the usual interpretation of low cycle isothermal fatigue tests with the Manson – Coffin law shows in one hand large differences in fatigue life for test with the same strain amplitude but at different temperature (as shown on figure 5 for cast iron) and in the other hand, it is not clear how it can be generalised to the multiaxial case. An interpretation in term of cumulative plastic strain as proposed by the Strain Range Partitioning method on a stress-strain isothermal curve is there somewhat hazardous.

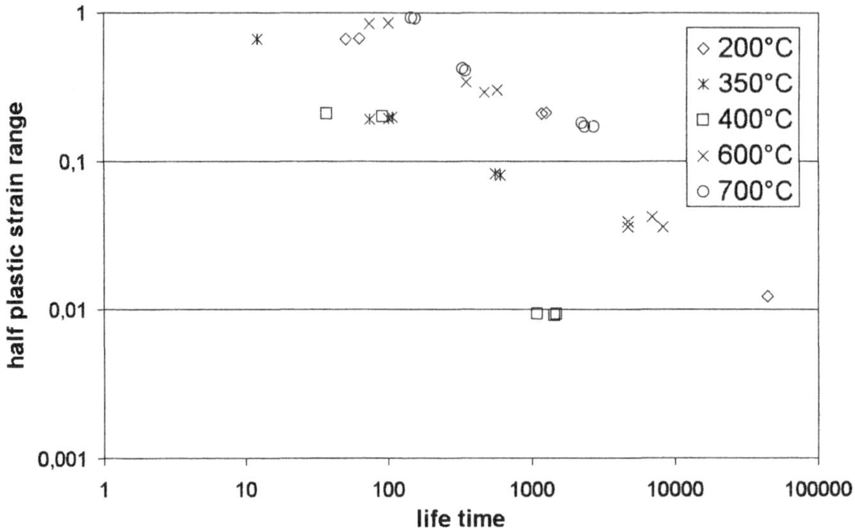

Fig 5 : Manson Coffin curve for cast iron at 200, 350, 400, 600 and 700°C

An interpretation of anisothermal test using the Smith Watson Topper function ($SWT = \sqrt{E \cdot \sigma_{max} \cdot \Delta\varepsilon}$) is difficult from the point where the choice or the definition of σ_{max} had no real sense and is in fact impossible in the anisothermal case.
Therefore the principal criticism for all these criteria is the impossibility of their generalisation to a complex loading.
It is also the case for the creep fatigue laws developed by Chaboche [14]. These laws has been generalised to anisothermal cases using a reduced stress σ / σ_u, where σ_u is the ultimate strength at the considered temperature, used to define the maximal stress or the mean stress independently of the temperature. This reduced variable is not justified from mechanical and physical point of view. For that reason Chaboche [14] propose to redefine the damage law in the anisothermal context.

A possible way to overcome these difficulties is the energetic approach. Skelton [15] describes the damaging process by the growth of crack through a process zone of constant size ρ at the crack tip. He considers the growth of the crack when a sufficient quantity of energy has been cumulated in this process zone. From that Charkaluk and Constantinescu [1] observe that the dissipated energy per cycle, at a steady state ΔW_S, can be representative of the fatigue behaviour of a material.
The interest of the dissipated energy per cycle comes from three main points :

- this energetic quantity can be generalised to multiaxial situation
- its determination is dependant of the temperature because it integrates the whole loading path and therefore allows calculations with anisothermal loading.
- ΔW_S is representative of the cyclic behaviour of materials.

The fatigue criterion can be simply expressed by a relationship :

$$\Delta W . N^{\beta} = C \tag{1}$$

PRACTICAL EXAMPLES

Exhaust Manifold

A practical application is given in the following example with the exhaust manifold of a 2.1 l turbocharged diesel engine. This is a silicon molybdenum (4 % wt Si, 0.8 % wt M_0) cast iron component.

Experimental conditions. The fatigue assessment and validation is made on a standard prototype durability cycle with a classical motor testing device. The equivalent damage history consists in a succession of full load at 4000 rpm and light load at the same regime. The transient period from full to light load and light to full load takes 30 s and the total cycle takes approximately 700 s. The external surface of manifold is monitored by regular visual inspection to detect crack apparition.

Numerical method for thermal computation. The aim of the thermal computations is the prediction of the temperature field distribution on the manifold from engine data, exhaust gas temperature and mass flow rate, which can already be estimated in the early phases of the project.
The loading cycle is clearly a transient one. The thermal computation can be made with two steps. A steady state temperature field distribution is first determined and afterwards the transient response can be computed under assumptions concerning the evolution of the boundary conditions and the evolution of thermal exchange coefficients.

Representation of the material behaviour. The constitutive law has to characterise the material from ambient to 800 °C. At 20 °C it presents an elastoplastic behaviour, above 700 °C an almost purely viscous one. The fact that one has to deal with a simple unifying law over the whole temperature range is dictated both by the F.E. transient analysis and the limited time devoted to the lifetime prediction during the development of new components. This conduct to the choice of an elastoviscoplastic (EVP) material model. A reasonably representative model of the high temperature EVP behaviour of the material is achieved with a viscoelastic – elastoplastic constitutive law with six temperature depending parameters or with the unified model with five temperature depending parameters (fig.4). These models and their parameters can be identified from isothermal cyclic tension relaxation recovery tests at several temperatures. An important point is to take care of the representativity of the test in the sense of aging and stabilisation of the behaviour corresponding to long term service conditions. Generally it is necessary to conduct some additional anisothermal test to get a unique solution for the parameters set.

Fatigue life estimation. The structural analysis is conducted on a Finite Element Model. The mesh of the manifold contain about 15,000 3D brick element. The analysis has been performed in two steps : first the manifold has been screwed on the cylinder head and second a series of thermal cycles have been computed. The imposed temperature distribution was a result of the previous transient thermal calculation.
The number of cycles has been chosen in order to obtain a stable cycle for the mechanical fields. The thermomechanical analysis were completed within approximately 6 hours of CPU time.

The results has been analysed in term of dissipated energy per cycle, as indicated in figure 6. The predictivity of the calculation is satisfactory as indicated by numerous comparative results between experiments and calculation on fig.7.

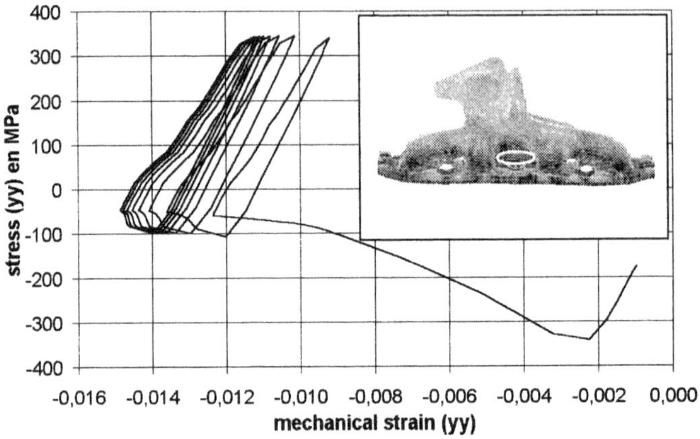

Figure 6 : Thermomechanical calculation and dissipated energy cartography

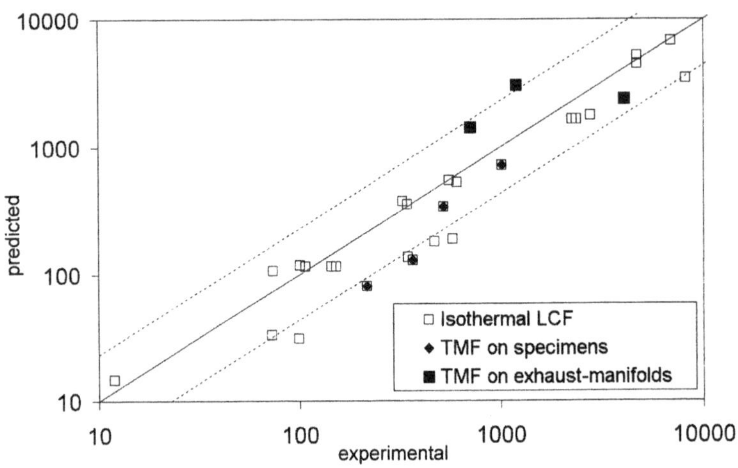

Figure 7 : Predicted versus experimental fatigue life for cast iron exhaust manifold.

Cylinder head

Cylinder heads are very complex components as well for the geometry as for the loading. They have to support severe fatigue loading which are: high cycle mechanical fatigue at 150 to 160°C in the upper part and low cycle thermomechanical fatigue closed to the combustion chamber (the fire deck). The example addressed the thermomechanical problem.

Whilst the cylinder head is much more complex than the exhaust pipe (numerical models rank around several hundred thousand elements and may reach one million of degrees of freedom), the numerical approach is the same.

Loading : The applied loading corresponds to an equivalent damage loading which represents whole life of the vehicle for a customer of identified severity (cf. figure 1).
The thermal loading came from the combustion. From the radial distribution of the thermal flow and the mean flow a transient thermal calculation is performed. The result is the temperature cartography at each node of the model and at each time instant.
The mechanical loading comes essentially from the tightening on the cylinder block and the press fitting of valve seats and valve guides.
The application of the thermal loading shows that the inter valve bridge is often above 200°C and goes up to 250°C.

Numerical simulation and fatigue strength prediction : Once the constitutive model chosen for the material behaviour (see previous chapter) and the numerical implementation realised, the cylinder head calculation can be done.
For such complex models, the sub modelling technique is recommended. It consists in a global elastic calculation of the structure and for the area where the behaviour is strongly non linear a sub model is extracted. The temperature and displacements at the boundary of the sub model are then applied to this sub model to perform the viscoplastic calculation.
This technique allows to limit the viscoplastic calculation only where it is necessary. This help to minimise the calculation time, which can be typically 2.10^5 s CPU.
The mechanical response of the structure is then determined and the indicator for fatigue strength can be extracted. In our case the dissipated energy per cycle at a stabilised state allows us to make the link between this mechanical state and the thermomechanical fatigue damage in a multiaxial and anisothermal context as shown in figure 8. The predictivity is rather good, within a factor of two in life cycles.

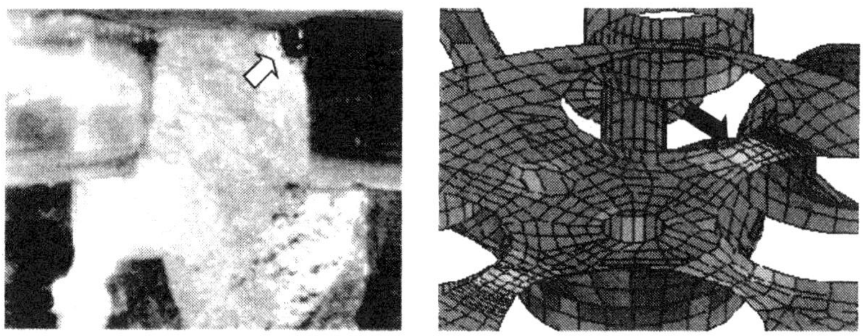

Figure 8 : Fatigue crack at inter valve bridge and numerical simulation.

Brake disc
Another automotive component which suffer thermal fatigue is the brake disc. The material of these component is usually a lamelar cast iron.

Loading : Under usual conditions when braking from maximum speed to 0 Km/h, under a pressure of 80 bars, the disc temperature goes from ambient to 450°C. Under exceptional situations much severe loading are recorded and the temperature can reach 700°C.
The physical phenomena analysis shows that the pressure on the pads associated with the disc rotation generates a thermal flow from friction, this flow is modified by convection and conduction. The thermal flows produce thermal expansion which then generates thermal stresses. These last ones will modify the mechanical loading. The system is therefore coupled.

Whilst the structure is circular, the thermal and mechanical response is not axisymetric.
At last, for the maximal temperatures (they can vary from 400° to 700°C), the material behaviour can be plastic or viscoplastic depending on the loading severity.

Damage observation : In case of exceptional loading and severe braking, the risk is a bowl failure; nevertheless this kind of situation is not expected in customer use.
Under "maximum" or "daily" braking we have to fight against radial cracking or honey comb cracking of the friction track.

Numerical simulation and fatigue assessment : An ideal simulation could be a coupled incremental calculation taking the contact into account in a 3D model with an incremental displacement of the loading.
This could bring somehow lengthy calculation incompatible with design office constraints. Design offices usually make some simplifying hypothesis which can be : axisymetry or no contact, coarse mesh either elasticity .... Most of those hypothesis are too strong and the results are therefore limited and do not allow fatigue strength prediction.
A different strategy for the numerical modelling allows to overcome this difficulty. The kinetic energy to dissipate is introduced in the thermal calculation through the modelling of the complete system : disc and pad device. A contact calculation indicates the surface concerned by the thermal exchanges and the friction. The transient thermal calculation, performed under the loading axis, consists in determining the temperature field in each point of the disc as well as the thermal stresses. Finally, knowing the contact surfaces and the thermal loading, a stationary thermomechanical calculation can be done. For example let's take a braking of 4 seconds from V max to 0 Km/h, the angular velocity decreases linearly and the disc makes 40 laps. The thermal simulation allows to obtain a stabilised thermal cycle after 2 or 3 brakings. For each lap, a stationary mechanical calculation is performed.

From this calculation of the mechanical fields, the fatigue strength can be estimated.
For the "daily" braking the temperature remains below 400°C, the mechanical response of the structure remains in the elastic domain. Therefore the high cycle fatigue strength is predicted using a multiaxial high cycle fatigue criterion, the Dang Van criterion in our case, based on material fatigue characteristics which are nearly constant from ambient to 400°C on this material.

For the severe braking, locally the mechanical response of the structure could reach the plastic domain and will tend to a plastic shake down. Therefore anisothermal low cycle fatigue criterion is used and the dissipated energy per cycle leads to reasonable estimation of the fatigue life.
Figure 9 give an example of results with comparison of a bowl failure in exceptional braking.

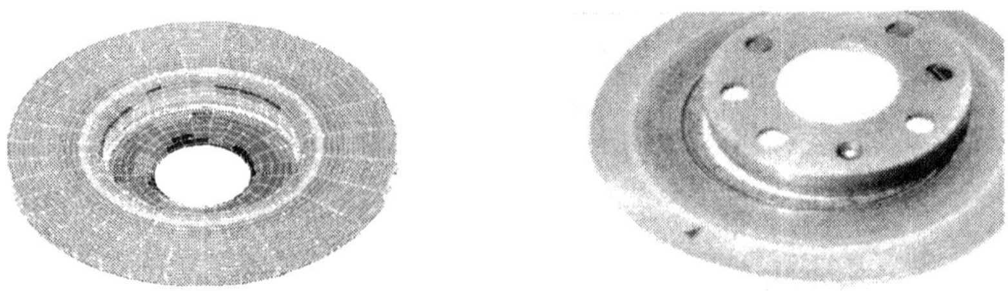

Figure 9 : bowl failure and dissipated energy cartography.

CONCLUSION

In the automotive context the objective of the fatigue assessment from a structural calculation in the development phase is to predict if the studied component or system is conform to the technical specification. The decision to take is to accept or to refuse the conception proposed for these components or systems.

For automotive components working at high temperature several difficulties are encountered :

- the necessity to be included in a development scheme, that means to obtain predictive calculation results in a short time.
- the multiaxiality of mechanical fields induced by the complexity of the structures and of the loading.
- the anisothermal character of the loading
- the elastoviscoplastic behaviour of the materials.

To address this problems a global design approach is necessary, including :

- the knowledge of the loading
- the modelling of the material behaviour and its implementation in a F.E code.
- the identification of the damage parameter and a failure or fatigue strength criterion.

Today the key point is to ensure the robustness of the approach in view of the decision of acceptance or refusal of the component from the calculation phase.

The two essential domains that have to be improved because they can lead to bad decision are :

- The simulation of the thermal loading. This parameter acts at the first order, an error of 10% on the temperature leads usually to a factor of 2 on the fatigue life.
- The choice of the viscoplastic model which affects significantly the calculated mechanical field and therefore the damage indicator.

AKNOWLEDGMENT

The authors thank M.L. Nguyen-Tajan and J.J. Thomas for their contribution to this review paper.

REFERENCES

[1] E. CHARKALUK and A. CONSTANTINESCU. Energetic approach in thermomechanical fatigue for silicon molybdenum cast-iron. *Materials at High Temperatures*, **17**, (3), pp. 373-380, 2000.

[2] J. J. THOMAS, G. PERROUD, A. BIGNONNET, and D. MONNET. Fatigue design and reliability in the automotive industry. In G. Marquis, editor, *Fatigue Design'98-3rd International Symposium on Fatigue Design*, pp. 1-11, 1998.

[3] G. LEDERER, E. CHARKALUK, L. VERGER, and A. CONSTANTINESCU. Numerical lifetime assessment of engine parts submitted to thermomechanical fatigue, application to exhaust manifold design. In *SAE Technical paper series*, 2000-01-0789, 2000.

[4] D. FRANCOIS, A. PINEAU, and A. ZAOUI. *Comportement mécanique des matériaux – vol II.* Hermès, 1994.

[5] J. LEMAITRE and J.L. CHABOCHE. *Mécanique des matériaux solides*. Dunod, 1985.

[6] H. SEHITOGLU, X. QING, T. SMITH, H.J. MAYER and J. A. ALLISON. Stress Strain Response of a Cast 319-T6 Aluminium under Thermomechanical Loading, *Met. Trans. A*, **31A**, January, pp. 139-151, 2000.

[7] E. NICOULEAU-BOURLES, N. EL-MAYAS, D. MASSINON and G. CAILLETAUD. Thermomechanical fatigue of aluminium alloys : experimental study and numerical simulation. Thermal Stresses 99, Third International Congress of Thermal Stresses, Cracovie, pp. 241-244, 1999.

[8] L.M. KACHANOV. Time on the rupture process under Creep conditions, *Izv. Akad. Nank.*, SSR, Otd. Tekh. Nank., n°8, pp. 26-31, 1958.

[9] J.L. CHABOCHE and C. STOLTZ. Détermination des durées de vie des aubes de turbines à gaz. *Revue Française de Mécanique*, **50-51**, pp.71-82, 1974.

[10] L.F. COFFIN. A study of the effects of cyclic thermal stresses on a ductile material. *Trans. ASME*, **53-A76**, pp. 931-950, 1953.

[11] S. S. MANSON. Behaviour of materials under conditions of thermal stresses. Technical Report TN 2933, N.A.C.A., 1953.

[12] G. R. HALFORD and S. S. MANSON. Life prediction of thermal-mechanical fatigue using strainrange partitioning. In *Thermal Fatigue of Materials and Components – ASTM STP 612*, 1976.

[13] K. N. SMITH, P. WATSON, and T. H. TOPPER. A stress-strain function for the fatigue of metals, *J. Mater.*, **5**, (4), pp. 767-778, 1970.

[14] J.L. CHABOCHE. Une loi différentielle d'endommagement de fatigue avec cumulation non linéaire. *Revue Française de Mécanique*, (50-51):71-82, 1974.

[15] R. P. SKELTON. Energy criteria for high temperature low cycle fatigue, *Mat. Sci. Tech.*, **7**, pp. 427-439, 1991.

Temperature-Fatigue Interaction
L. Rémy and J. Petit (Eds.)

STRUCTURAL CALCULATION AND LIFETIME-PREDICTION IN THERMOMECHANICAL FATIGUE OF ENGINE COMPONENTS

E. Nicouleau*, F. Feyel**, S. Quilici***,****, G. Cailletaud****
* Renault, Direction de la Mécanique, 67 rue des Bons-Raisins, 92508 Rueil-Malmaison,
** Onera-DMSE, 29 av. de la Div. Leclerc, 92322 Châtillon,
Armines, and * Centre des Matériaux de l'Ecole Nationale Supérieure des Mines de Paris, UMR CNRS 7633, BP 87, 91003 Evry France

ABSTRACT

The purpose of the paper is to show the successive steps of a numerical method for the prediction of crack initiation in industrial components submitted to thermo-mechanical fatigue. Finite element computations using realistic constitutive equations are first used to determine the stress and strain fields in the component. The constitutive equations account for cyclic viscoplasticity, creep, and aging effect. The industrial use of such theories requires robust integration techniques to reach the mechanical steady state. Since the real meshes have 10^5–10^6 degrees of freedom, parallel computers are also needed. Damage models with creep-fatigue interaction are finally applied to determine crack initiation. The example of aluminum cylinder heads is chosen to illustrate the demonstration.

KEYWORDS

Viscoplastic modeling, Aging effect, Aluminum alloys, Thermomechanical fatigue, F.E. structural calculation, Cycle skip, Parallel computations

INTRODUCTION

People in industry are well known to have complex components, severe loads and busy time schedules. This is why they need more and more powerful experimental and numerical tools to perform better and faster predictions. The current numerical power allows to compute models describing a part of the component by considering a reasonable description of the geometry. Several types of problems must be considered with the same component (stress, temperature, flow) with some exchanges between the different computations. The mechanical section of these procedure is usually achieved using just elastic behaviour, but it is now obvious that physical aspects of the deformation, involving viscoplasticity and aging are often needed for high temperature applications.

The first step of the prediction is the evaluation of a reliable stress and strain field. Good constitutive equations are then required. They must correctly represent (*i*) the yield domain, since many areas in the component will remain elastic, (*ii*) cyclic loading (Bauschinger effect) and the viscous effect, (*iii*) aging effects, depending on the thermomechanical history. Such a model has been developed for aluminum alloys [18, 19]. In a companion paper of the present conference, its mechanical response is compared to microstructural evolutions [13]. It has been shown that the final state of the material in each point of the component depends on the maximum temperature reached at this point, so that one will observe an *aging field* in the cylinder head, and that a simplified solution which would affect the same asymptotic properties to every point can induce errors in the evaluation of the stress redistribution during the simulation of the operation. Due to the complexity of the stress and temperature histories, it is decided to build a numerical framework to compute the whole life of the component. The various ingredients needed are:
- a robust integration technique with automatic time stepping; the design of the method must be such that any model evolution can be simply implemented into the finite element code;
- a technique for accelerating the convergence of the cycle by cycle calculation; nevertheless, since the material state continuously changes during the life of the specimen, the complete history must be kept in memory to apply the life prediction methods;
- a solution to accept very large meshes; the range 10^5–10^6 degrees of freedom (dof) is necessary to have fine elements in the area of interest together with a good geometrical representation of the total mechanical part, with realistic boundary conditions;
- the life prediction model; this model must take into account microstructural evolution, together with creep, fatigue, and creep–fatigue interaction. It is implemented in a post-processor which directly highlights the critical areas of the component, by integrating a damage variable on each Gauss point or node of the mesh.

The products of the study are finally a material library including various integration methods, a cycle skip technique to describe the total history with a reduced number of computations, and parallel computation for reaching a large number of dofs. The implementation and industrial use of all these methods is then a long term process. First results were produced in an academic framework [6]. The present paper gives an updated overview of the problem, with larger and more realistic computations.

CONSTITUTIVE EQUATIONS

The following constitutive equations are specially developed to account for the precipitation and coalescence phenomena in a copper bearing aluminum alloy, affecting the hardening mechanisms, then the mechanical properties. The models previously proposed to represent the behaviour of aluminum alloys were designed to represent hardening[16], or hardening and softening[7]. Arguments concerning the microstructure evolution will be found elsewhere [21, 13] In the present study, the model has to represent softening only. A very simple formulation is then chosen. Aging is represented by a scalar variable a, starting from zero, and tending to an asymptotic value a_∞, depending on temperature. An additional partition is assumed between elastic and viscoplastic strains. Plastic flow and hardening variables are defined as follows:

$$\text{yield function f}: f = J(\underset{\sim}{\sigma} - \underset{\sim}{\mathbf{X}}) - R - k \quad \text{with } J = (\text{ von Mises invariant}) \tag{1}$$

$$\text{viscoplastic flow: } \dot{\underset{\sim}{\varepsilon}}^p = \dot{p}\,(\partial f/\partial \underset{\sim}{\sigma}) = \dot{p}\,\underset{\sim}{\mathbf{n}} \quad \text{with} \quad \dot{p} = \langle f/K \rangle^n \tag{2}$$

$$\text{kinematic hardening: } \dot{\underset{\sim}{\alpha}}_i = \dot{p}\,\left(\underset{\sim}{\mathbf{n}} - (3D_i)/(2C_i)\,\underset{\sim}{\mathbf{X}}_i\right) \quad \text{and} \quad \underset{\sim}{\mathbf{X}}_i = (2/3)C_i\underset{\sim}{\alpha}_i \tag{3}$$

$$\text{isotropic hardening: } \dot{r} = \dot{p}\,(1 - (R/Q)) \quad \text{and } R = bQr \tag{4}$$

Aging is defined by the simple equation $\dot{a} = \langle (a_\infty - a)/\tau \rangle$, to represent the fact that the phenomenon can only increase the value of a. At initial state, there is no hardening, and the asymptotic value a_∞ is also equal to zero, provided the temperature is below the threshold which produces a microstructural evolution. The material parameters are *a priori* temperature and aging dependent. In fact, an explicit expression is chosen for $k(a, T)$, and two kinematic variables are introduced, only one of them depending on aging:

$$k = R_0(T) + R_0^*(T)(1 - a) \tag{5}$$

$$\underset{\sim}{\mathbf{X}}_1 = \frac{2}{3}C_1(T)\underset{\sim}{\alpha}_1 \quad ; \quad \underset{\sim}{\mathbf{X}}_2 = \frac{2}{3}C_2(T)(1 - a)\underset{\sim}{\alpha}_2 \tag{6}$$

NUMERICAL IMPLEMENTATION

INTEGRATION METHOD

The key point for a correct numerical implementation valid for non isothermal loadings is to integrate the state variables, and not the hardening variables [5]. The model is implemented in ZéBuLoN [2], using the user interface for the development of constitutive equations Z-front [3]. The model is then available for use in ZéBuLoN, but also in commercial codes, through Z-mat [11]. This library can be linked to well known codes like Abaqus, Marc, or Ansys, and allows the user to choose a predefined model, or to develop new models by using predefined building bricks like `elasticity`, `flow`, etc... This is made possible by a careful programming, which keeps the section concerning constitutive equations independent from the global system, as explained later. In both cases, the data file will have about no modification, if compared with a native model in the code.

The purpose of the integration box at this level is then only to start with the actual value of the state variables, the increment of external prescribed parameters (like temperature in a mechanical computation) and the strain increment, and to provide the increment of state variables, the updated stress value, and the jacobian matrix $\partial\underset{\sim}{\sigma}/\partial\underset{\sim}{\varepsilon}$ to evaluate the consistent stiffness matrix. In the present case, the variables are the elastic strain $\underset{\sim}{\varepsilon}^e$, the two kinematic state variables $\underset{\sim}{\alpha}_1$ and $\underset{\sim}{\alpha}_2$, the isotropic state variable r (or the accumulated viscoplastic strain p), and the aging variable a. Note that, the evolution of aging being independent of the mechanical variables, its history could also be obtained by a post-processing of the thermal computation, then aging would be considered as an "external parameter" for the mechanical computation. Due to the low amount of computation needed to compute this evolution, the direct computation of a during the mechanical calculation, which avoids an additional step in the numerical procedure, is the best solution. Two integration strategies can be chosen in the code: Runge–Kutta integration with automatic time step definition, and a θ–method. In the first case, only the explicit expression of the derivatives has to be introduced in the software: for the second one, a Newton method is used to solve the following implicit system in terms of $\Delta\underset{\sim}{\epsilon}^e$, $\Delta\alpha_1$, $\Delta\alpha_2$, Δp, Δa, for given values of $\Delta\underset{\sim}{\epsilon}$ (increment of mechanical strain) and Δt:

$$\underset{\sim}{\mathbf{R}}_e = \Delta\underset{\sim}{\epsilon}^e - \Delta\underset{\sim}{\epsilon} + \Delta p\, \underset{\sim}{\mathbf{n}} = 0 \tag{7}$$

$$\underset{\sim}{\mathbf{R}}_{a1} = \Delta\alpha_1 - (\underset{\sim}{\mathbf{n}} - D_1\underset{\sim}{\alpha}_1)\Delta p = 0 \tag{8}$$

$$\underset{\sim}{\mathbf{R}}_{a2} = \Delta\alpha_2 - (\underset{\sim}{\mathbf{n}} - D_2\underset{\sim}{\alpha}_2)\Delta p = 0 \tag{9}$$

$$R_v = K\,(\Delta p/\Delta t)^{1/n} - \langle f \rangle = 0 \tag{10}$$

$$R_a = \Delta a - \langle (a_\infty - a)/\tau \rangle\ \Delta t = 0 \tag{11}$$

In the previous equations, variables like α_i are expressed as $\alpha_i(t) + \theta\Delta\alpha_i$, and quantities like f and $\underset{\sim}{\mathbf{n}}$ are also expressed at the intermediate time $t + \theta\Delta t$, so that the final system may be quite complex. Note that time independent plasticity can be obtained as a limiting case provided $K = 0$ in equation 10.

CYCLE SKIP

This is an old heuristic method, used first at Onera, first for representative volume elements [20] then for finite element computations [14, 17]. It has been recently implemented in Abaqus and ZéBuLoN [12]. The method is based upon the fact that for cyclic calculations, besides the *small time scale* involved in time integration of material equations during each cycle, one can exhibit a *large time scale* corresponding to the state variable evolution from one cycle to the other. Recognizing that the state variable evolution along this large time scale is much smoother, it is then possible to drop the calculation of some intermediate cycles by using simple explicit extrapolation techniques. Denoting by T the cycle period, and choosing a particular instant τ in the small time scale local to each cycle, one can express the vector Y of state dependent variables in function of number of cycles N by $Y(N)$ and use a second order Taylor series expansion to compute $Y(N + \Delta N)$:

$$Y(N) = Y((N-1)T + \tau)\ ,\ with\ 0 < \tau < T \tag{12}$$

$$Y(N + \Delta N) = Y(N) + \Delta N Y'(N) + \frac{\Delta N^2}{2} Y''(N) \tag{13}$$

First and second derivatives $Y'(N)$ and $Y''(N)$ in (13) can then be obtained by choosing three subsequent cycles N_1, N_2, N_3 ($N_1 < N_2 < N_3$). The number of allowed skipped cycles ΔN is then calculated such as to minimize the second order term in (13) compared to the first order one. Knowing the value of ΔN, the state dependent variables are then extrapolated.

Using such an approach, the stress history can be known during the whole life of the component. Up to now, this input has not been used in the life prediction, which will be shown in example 1, but the opportunity is now available. Since the model used for the prediction of crack initiation has a damage variable [15], damage cumulation for a variable loading history is easy. On the other hand, the model is valid from LCF conditions to the HCF range.

PARALLEL COMPUTATIONS

Parallel computations started in engineering applications about 15 years ago. The first computers had a shared memory, and a small number of processors. On this type of machines, automatic parallelization algorithms have been tested. It was a sort of "small scale" parallelism, without any change of the general algorithms, the only modification concerned the parallel processing of the most critical loops. The machines were very specific and expensive. The perspective has now changed, with the "farms", or computers clusters, made of very simple machines, like workstations or PCs. In a cluster, the nodes are connected by a fast network, but they are independent, and have to cooperate through "message passing". As a consequence, the code must be rewritten, to explicitly account for these exchanges. Such a modification has been made in the finite element code ZéBuLoN. The work started three years ago [9, 10]. Some developments are still in progress, but nowadays the code has been tested on many configurations, and could be used on a every day basis to compute industrial components. The biggest calculation made up to now had more than 3 million degrees of freedom (dof). A 1.5 million dof mesh will be shown at the end of the paper.

The code uses a domain decomposition method, called FETI [8]. Such methods rely on a non-overlapping partition of the structure in a number of sub-domains. Classical finite element methods applied to computational mechanics involve two main stages:
- a *global stage* during which the linear tangent response of the structure is computed. This leads to the solution of a big sparse linear system $Kq = F$ where the unknowns q are the increments of nodal displacements, F the external forces and K the stiffness matrix;
- a *local stage* consisting of a loop over all integration points in the structure to compute the material response using the presiously presented constitutive equations. Starting from the strain increment deduced from the nodal displacement increments, the purpose of this step is to provide the increment of the state variables describing the material and the new stress field. As soon as all behaviour models are built upon the local state assumption, this local stage can be seen as the solution of N *independent* differential systems.

The use of domain decomposition methods will automatically ensure the parallelization of the local stage: each domain has to take care only of its own integration points. That means that the information concerning the behaviour (viscoplastic strain, hardening state variables) will be distributed on all the computers, and known only by the relevant node. On the other hand, the global stage requires some extra work: without any care, the global solution found by collecting the solution of all subdomains *is not* continuous across domain boundaries. A new step has then to be used in the process to compute the forces needed at the domain boundaries in order to ensure displacement continuity. An iterative conjugate gradient algorithm is used for that purpose. The size of the auxiliary problem at the boundaries depends on the number of nodes at the interfaces. For a given number of nodes in the subdomain, the interface has then to be as small as possible (compact shape) to save time and get a good convergence during the iterations. Usually, the time spent for "pasting" together the different domains remains small, if compared with the solution of the $Kq = F$ system, and with the integration procedure in each subdomain. All this work is, of course, done automatically without user interaction. ZéBuLoN contains everything needed for that purpose. MPI and PVM message passing libraries may be used at user convenience, and one should note that the parallel architecture of ZéBuLoN prevents the user from duplicating its input data file, even if no disk-connection exists between computational nodes.

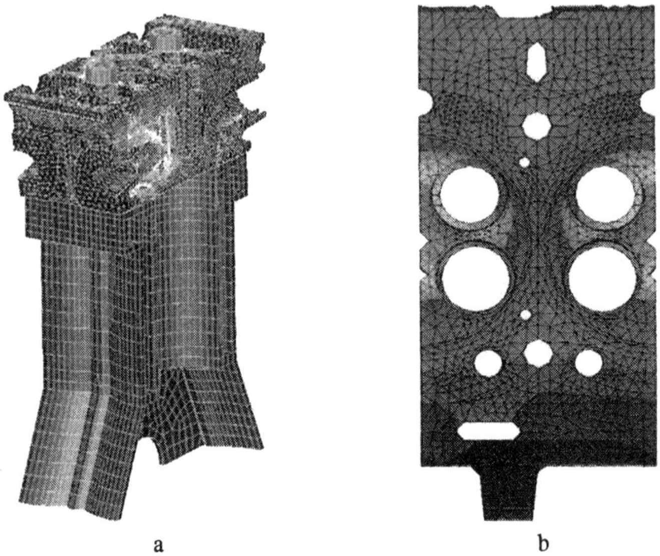

Figure 1: Data of the computations, (a) Cylinder head and block, (b) Temperature field

EXAMPLE 1

STRESS–STRAIN COMPUTATION

The cylinder head is one of the most critical components of an engine (and one of the main parts containing the camshaft driving system with valves, camshaft...). This component is submitted to mechanical loads (for instance valve seats insertion by shrink fitted, clamping, high pressure of gas combustion) and thermal fields involving high level of stresses in valve bridges. Sometimes, failure occurs by Thermomechanical fatigue (TMF). The present example describes a Diesel engine (direct injection) cylinder head computed with the FEM method using quadratic tetrahedral elements (220000 nodes) using Abaqus [1] and Z-mat [11]. Two half blocks of the engine are considered, as shown in Fig.1a.

The first step of our procedure is the thermal calculation. A hot/cold cycle is classically chosen to "severise" engine conditions by introducing high temperature gradients. Using a thermal FE calculation, a hot/cold cycle is computed, trying to get a good correlation with experimental measurements (thermocouples plugged in different location on a real engine). This virtual hot/cold cycle is setup for a constant regime, and for a given variation of engine load (from 5 to 100% of the load). The temperature history is then known on each point of the structure. Figure 1b shows a typical temperature field at maximum temperature. High temperatures can be observed, in the valve bridges, specially in the exhaust–exhaust area (VBEE, valve–bridge exhaust–exhaust), a little less in the intake–exhaust area (VBIE, valve–bridge, intake–exhaust). The temperature variation is more than 200°C for the hot zones, and not far from 100°C for water.

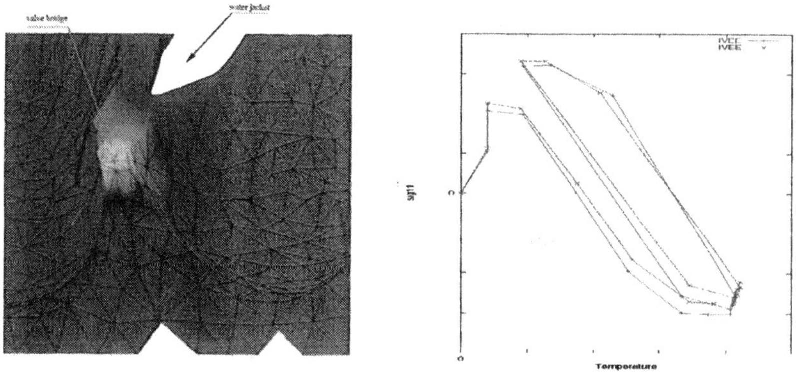

Figure 2: Cumulated plastic strain after 2 cycles

The objective of the mechanical calculation is to analyse the thermo-mechanical low cycle fatigue of the hot parts and the High Cycle Fatigue (HCF) of the cold ones. Complex boundary conditions are considered. The assembly, composed by the cylinder head, the cylinder head gasket and a simplified cylinder block is managed by the `CONTACT PAIR` card in `SMALL SLIDING` definition and the `GASKET` elements with the clamping controlled by the `PRE TENSION SECTION` surfaces. A non linear behaviour of the gasket is introduced. The boundary conditions are given by: (1) the border conditions to take into account adjacent blocks, (2) the assembly, (3) the temperature history.

The stress-strain fields are computed with the viscoplastic model and aging. After two cycles, the asymptotic conditions are obviously not reached, but it can be predicted that the final status will give an aging field, with higher value in the valve bridges, and practically no aging in the "cold" zones. Taking into account the full history is only possible using cycle skip. This was not made for this computation. Figure 2a shows the cumulated plastic strain reached after 2 cycles. Plasticity is well localised in the longitudinal bridges. The predominant plastic strain component is xx, the value obtained is large at the surface of the combustion chamber, but becomes negligible near the water jacket. During the history, tensile stress is first observed in the bridge areas, due to the fit of the valve seats. A tensile plastic strain is found in aluminum, at the interface aluminum–steel seats, specially for VBEE. Classical TMF loops are then observed, with compression at high temperature and tension at RT. Figure 2b shows the temperature–stress σ_{xx} history in the two valve bridges.

LIFETIME-PREDICTION

The input needed for life prediction are the stress, temperature, and aging history. A full calculation would use the total history, together with a progressive damage accumulation. At the present time, a simplified evaluation is made, using the aging values obtained at the end of the computation, and the stress and temperature variations during the second cycle. These variations are supposed to be periodic for the whole life. A detailed description of the model is given in classical textbooks (see for instance [15]). A brief summary of the equations needed for the

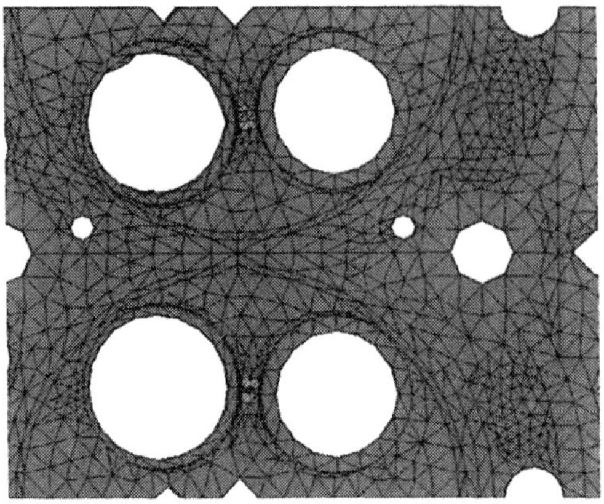

a.

N_C computed on a period P :

$C = \frac{1}{N_C} = \frac{1}{\kappa+1} \int_P \left(\frac{\chi(\underset{\sim}{\sigma})}{A}\right)^r dt$

A, κ, r are material coefficients, $\chi(\underset{\sim}{\sigma})$ = linear combination of von Mises, eigen stress and hydrostatic pressure

N_F computed for one cycle :

$F = \frac{1}{N_F} = (\beta \mp 1)(1-\alpha)\left(\frac{\Delta J^*/2}{M^*}\right)^\beta$

with $1 - \alpha = \gamma \left\langle \frac{\Delta J^*/2 - \sigma_l^*}{1 - S_{eq}^*} \right\rangle$

$\gamma, \beta, \sigma_l^*, M^*$ are material coefficients, ΔJ^* and S_{eq}^* are stress amplitude and maximum equivalent stress normalized by the actual (aging dependent) ultimate stress

b.

Figure 3: Prediction of the number of cycles to crack initiation. (a) Life contour, (b) Model

computations is made in Fig.3b. On one hand, the contribution of creep damage is defined by the integration of an equivalent stress, on the other hand, fatigue damage comes from a stress dependent relation for the modelling of Woehler's curve. Creep–fatigue damage interaction is obtained by calculating first the number of cycles to failure in fatigue N_F and in creep N_R, then by applying a nonlinear interaction rule. As shown in [4], the numerical procedure is simply a cycle-by-cycle computation, which just needs the values C and F defined in Fig.3b:

$$(1 - D_0)^{\kappa+1} - (1 - D_1)^{\kappa+1} = C \qquad (14)$$

$$\text{then } \left(1 - (1 - D_1)^{\beta+1}\right)^{1-\alpha} - \left(1 - (1 - D_2)^{\beta+1}\right)^{1-\alpha} = F \qquad (15)$$

$$\text{then } D_0 = D_2 ... \qquad (16)$$

One can see that, according to the numerical values of the two coefficients κ and α, the model will deliver a more or less nonlinear response, since they govern the nonlinearity of damage evolution. As a matter of fact, creep damage was very low for the conditions of the test, but could become predominant in the "real life". The location of the critical zone and the number of cycles to crack initiation are in good agreement with the experimental data.

EXAMPLE 2

A large 3D linear computation is presented in this section in order to evaluate the state of the art in a cheap cluster environment, which could be easily available for a small group of engineers (less than 10^5 Euros). The studied structure is a "model" cylinder head. It is called "model" because it is built using a smaller structure which is symetrised and duplicated in order to increase the number of dofs, thus leading to a weird structure. But this is a good image of the real industrial structure to be computed. It involves 1.6 million degrees of freedom (357420 quadratic tetrahedral elements). The structure is clamped at one end and a unity displacement is applied at the other end. The boundary conditions are set to represent the behaviour of the component at high temperature, under nonuniform thermal field previously defined by thermal computations, at maximum regime. For the present case, the only result wished is CPU time, in order to demonstrate the ability of the computational procedure to deal with such a class of problems. The material behaviour is then just isotropic elasticity.

This mesh is automatically split into 32 subdomains using a tool developed at Onera named SplitMesh [22]. Parallel computation is done on a PentiumIII cluster installed at Centre des Matériaux de l'Ecole des Mines de Paris. The cluster consists of 48 nodes, but only 32 are used for this computation. Each node contains a PentiumIII running at 800 MHz, 768 Ram megabytes, and 16 gigabytc of IDE-disk. All nodes are linked together by a Fast Ethernet switch. This computation requires about 300 megabytes of memory in order to store and factorise the global stiffness matrix of all domains. The elastic solution lasted 11 wall clock minutes with 3 minutes to read the file and load the problem data.

This type of result is very encouraging, since it demonstrates that non–linear computations are now possible for this class of problems. It would take about one day to make two or three cycles, so that one can expect to cover the full life of the component in about one week, using the cycle skip technique. Of course, it is still important to reduce these numbers to be able to introduce material nonlinearities in the design process, but the present calculation procedure can now be applied for smaller structures.

CONCLUDING REMARKS

This paper describes the various elements needed for the life prediction of structures or components working at high temperature. Clearly, the numerical computations to be performed will involve more and more internal variables to have a good model of the material behaviour, and more and more degrees of freedom to correctly represent complex geometries. These two trends are taken into account by the development of robust integration techniques, cycle skip and parallel computations. It is shown that the threshold of 10^6 degrees of freedom can now be reached, but not for routine computations, with non linear material behaviour. The future improvements on CPU time will come from computer, specially through an increase of the number of nodes. The present numerical methods have been currently tested for about 100 nodes, they need to be adapted for 1000-node machines.

An other direction for the research is the influence of the cracks in components. Many attempts were made in the past, in the field of continuum damage mechanics. More recent developments are made with interface elements. This type of study must be connected with metallurgical mechanisms, the final aim being to get a prediction of the behaviour of the component taking into account the transition between initial damage, or short crack development leading to the macroscopic initiation and the real crack propagation. Even if in most of the cases the crack initiation must be avoided, it could be important to check if the subsequent propagation is catastrophic or remains stable.

References

[1] HKS Abaqus-5.8 user manual. 2000.

[2] J. Besson and R. Foerch. *Revue Européenne des Éléments Finis*, 7(5):535–566, 1998.

[3] J. Besson, R. Leriche, R. Foerch, and G. Cailletaud. *Revue Européenne des Éléments Finis*, 7(5):567–588, 1998.

[4] G. Cailletaud and J.-L. Chaboche. Lifetime predictions in 304 stainless steel by damage approach. In *ASME PVP COnference*, Orlando (USA), 1982.

[5] G. Cailletaud and J.-L. Chaboche. *Comput. Methods Appl. Mech. Engrg*, 133:125–155, 1996.

[6] G. Cailletaud, C. Depoid, D. Massinon, and E. Nicouleau-Bourles. In G.A. Maugin, R. Drouot, and F. Sidoroff, editors, *CONTINUUM THERMOMECHANICS: The Art and Science of Modelling Material behaviour"*. Kluwer Academic Publishers, 2000.

[7] Chaboche, J.-L., El-Mayas, N., and Paulmier, P. *C. R. Acad. Sci. Paris*, 320(II):9–16, 1995.

[8] C. Fahrat and F.X. Roux. *Computational Mechanics Advances*, 2(1):1–101, 1994.

[9] F. Feyel. PhD thesis, École Nationale Supérieure des Mines de Paris, 1998.

[10] F. Feyel, G. Cailletaud, and F.-X. Roux. *Revue Européenne des Éléments Finis*, 7(1-2-3):55–72, 1998.

[11] R. Foerch, F. Azzouz, S. Quilici, and G. Cailletaud. In *Abaqus Users Meeting, Chester (UK)*, pages 213–227, may 1999.

[12] R. Foerch, V. Gros, V. Mounoury, S. Quilici, and G. Cailletaud. In *Abaqus Users Meeting, Newport (USA)*, May 2000.

[13] I. Guillot, B. Barlas, G. Cailletaud, M. Clavel, and D. Massinon. In *SF2M-ESIS Conference on Temperature–Fatigue Interaction*, may 29–31 2001.

[14] S. Kruch. In Owen et al., editor, *4th Int. Conf. on "Computational Plasticity"*, 1992.

[15] J. Lemaitre and J.-L. Chaboche. *Mechanics of Solid Materials*. Cambridge University Press, Cambridge, U.K., 1990.

[16] D. Marquis. PhD thesis, Univ. Paris VI, 1989.

[17] K. Nesnas and K. Saanouni. *Revue Européenne des Éléments Finis*, 9:865–891, 2000.

[18] E. Nicouleau-Bourles. PhD thesis, École Nationale Supérieure des Mines de Paris, 1999.

[19] E. Nicouleau-Bourles, N. El-Mayas, D. Massinon, and G. Cailletaud. In J.J. Skrypek and R.B. Hetnarski, editors, *Thermal Stresses '99*, pages 241–244. Cracow, june 1999.

[20] S. Savalle and J.-P. Culié. *La Recherche Aérospatiale*, 5:263–278, 1978.

[21] Smith, T.J., Maier, H.J., Sehitoglu, H., Fleury, E., and Allison, J. *Metall. and Mat. Trans.*, 30A:133–146, 1999.

[22] Splitmesh user manual. 2000.

Temperature-Fatigue Interaction
L. Rémy and J. Petit (Eds.)

THERMO-MECHANICAL FATIGUE LIFE ANALYSIS ON MULTIPERFORATED COMPONENTS

P. KANOUTE*, D. PACOU*, D. POIRIER*, F. GALLERNEAU*, J.-M. CARDONA**
*Office National d'Etudes et de Recherches Aérospatiales,
29 avenue de la Division Leclerc 92322 Chatillon, France.
**Centre des Matériaux URM CNRS 7633,Ecole des Mines de Paris,
BP.87 91003 Evry, France.

ABSTRACT

This paper describes thermomechanical fatigue tests performed on multiperforated superalloys specimens for turbine blade applications. Following a short description of the specific device designed to produce a thermal gradient in the thickness of wall specimens, the experimental results are presented together with the corresponding calculations. A 3D finite element analysis is needed to determine the stress state of the specimen. Specifically, recent models developed at ONERA to predict the lifetime of superalloy components, that can account for the anisotropy material, oxidation, creep and fatigue interaction effects, are applied to predict the fatigue life of the test specimens.

KEYWORDS

Thermal gradient; single crystal superalloys; fatigue life prediction; finite element analysis.

INTRODUCTION

Many complex phenomena take place in structural components working at high temperature under cyclic conditions. The strong requirements in design procedures (safety, reliability), involve the development of sophisticated life-time prediction techniques, taking into account the cyclic viscoplastic behaviour, crack initiation under creep, oxidation and fatigue conditions. Single crystal nickel based alloys are widely used as blade materials in aero gas turbine because of their excellent resistance to high temperature. The increase of both creep strength and thermal fatigue resistance is respectively due to the elimination of grain boundaries and due to the achievement of a reduced Young's modulus in <001> orientation that is along the axis of the blade. Tailoring of their chemical composition towards improving the high temperature strength has resulted in a reduced oxidation resistance. Coatings such as aluminides formed by a diffusion process are generally applied in order to provide blades with adequate protection against environmental degradation.

The mechanical behaviour of this kind of anisotropic material, such as single crystal superalloys, is today well known. Several constitutive models have been proposed [1,2] and applied with success for the design of single crystal blades. Thermomechanical fatigue tests have been developed for turbine blade alloys over the past 20 years to validate constitutive equations and damage models under complex loading situations. These experiments are generally performed on volume element simultaneously submitted to controlled load (or displacement) and temperature. This kind of test however may not be sufficiently representative of the behaviour that can be obtained in cooled turbine blades where a thermal gradient is generated in a very thin wall. The major cause of failure in current single crystal blades of aero gas turbines is thermally induced stresses, which result from thermal strains over the blade thickness caused by temperature gradients during heating and cooling cycles.

A specific device has been designed in order to check the capability of lifetime prediction methods, recently developed at ONERA, for thin walled turbine blades. This experimental facility has the capability of generating a thermal gradient up to 200°C for a maximum external temperature of 1050°C, in a smooth tubular specimen of 2 mm thick. In order to understand the impact of the presence of perforations in the inlet and outlet regions of the blades on their lifetime, the experimental device has been applied to mutiperforated tubes made of Nickel-Base superalloy single crystal. The description of the experimental facilities and the experimental results are given in the first section. In the second section are presented the results of the 3D finite element analysis performed on multiperfored tubes in order to determine the stress state at the stabilised cycle. The last section gives the application of a phenomenological fatigue-creep-oxidation interaction model, stress based and recently developed for anisotropic material at ONERA, to predict the fatigue life of each test.

EXPERIMENTAL DETAILS

THE EXPERIMENTAL SET-UP

In complex highly cooled turbine blades, thermal stresses may become at least as important as the centrifugal stress inducing the creep of the material during the stabilized regime of the engine. These thermomechanical stresses are particularly generated during the start-up and the shut-down of the engine. The thermomechanical fatigue tests generally used over the last 20 years provide accurate information on material response but are inadequate to reproduce the very high thermal gradient during heating and cooling cycles. To simulate the thermal gradient present in the blade, an experimental facility has been developed which allows to cool the internal wall of hollow cylindrical fatigue specimen, during the heating of the external wall. The difficulty is to obtain the highest possible thermal gradient but with an acceptable gage section temperature distribution. The heating is obtained with a coil-heating fixture connected to a 12 kW audio frequency induction heating unit. One advantage in experiments involving thermal cycling is the time required for heating and cooling which can be kept very short. Another advantage using induction unit in this application is the localized nature of the heating of the specimen external surface. The thermal gradient is then generated by cooling the internal surface with a constant pulsed airflow at room temperature as shown in the figures 1 and 2.

A good distribution of the temperature is obtained by means of an alumina sleeve put close to the internal surface, which induces a cool airflow. A sonic neck system above the specimen has been chosen to have a constant and controlled air delivery that can be checked by pressure measurements at the specimen inlet and outlet.

The initiation of a macroscopic crack, typically 1 mm long is detected in the specimen by the potential drop technique [3].

Fig. 1. Thin wall thermal gradient experimental device.

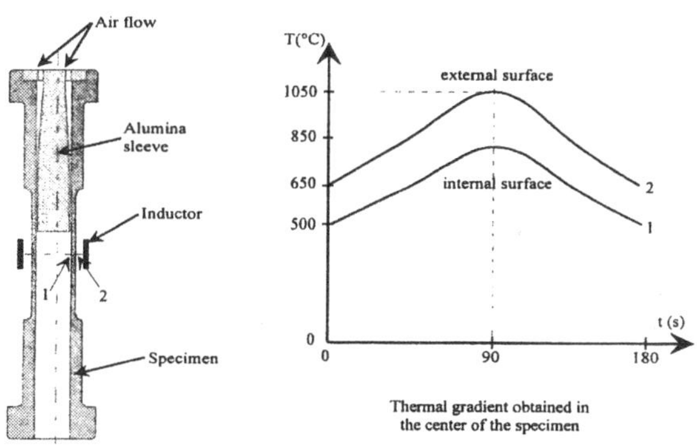

Fig. 2. Schema of the heated/cooled specimen.

THE TEST SPECIMENS

The aim of these experimental tests is to simulate the thermomechanical load produced in the inlet and the outlet regions of the blades, in order to study the influence of the presence of multiperforations on the lifetime. Therefore, the experimental specimens were pierced following two types of patterns.

All the specimens are made of AM1 single crystal superalloy and oriented along the <001> crystallographic orientation. The perforations are performed with a laser, close to the <010> orientation. Industrial C_1A coating has been deposited on the external surface of the specimen, as it is made on the real blade, to protect the alloy against oxidation but also against corrosion and erosion phenomena.

EXPERIMENTAL RESULTS

Firstly, preliminary tests have been realized in order to verify if a specific thermal gradient along the circumferential, the axial and the radial directions is generated by the presence of the holes. Experimental specimens with four isolated holes are considered. The tests are performed with an induction-heating unit of 28 mm of height, centered on the holes. The tests have been performed at several temperature, from 650°C in the region just outside the hole from 1050°C in the region of the specimen at the highest temperature. The temperature measurements performed with thermocouples and infrared camera have shown the absence of circumferential thermal gradient around the hole, as shown in the figure 3. On the other hand, an axial thermal gradient of around 50°C/mm has been noticed on both sides of the inductor.

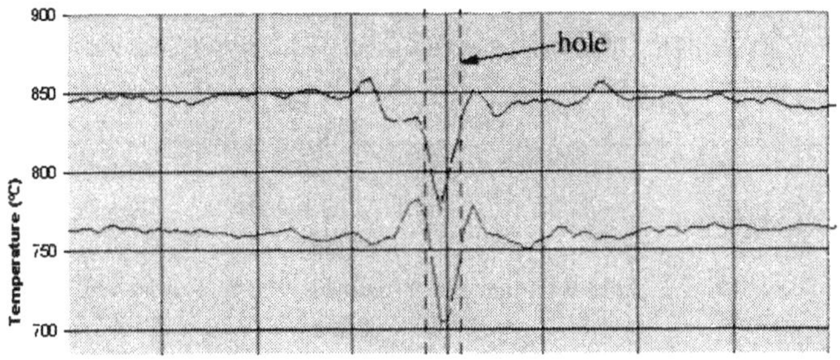

Fig. 3. Circumferential thermal gradient in the region just outside a hole.

These observations are in perfect correlation with results obtained on turbine blades by SNECMA. Only an axial thermal gradient and a wall thermal gradient were noticed. In order to simulate at best the operation conditions of the turbine blades, we tried to recreate these thermal gradients in the experiments.

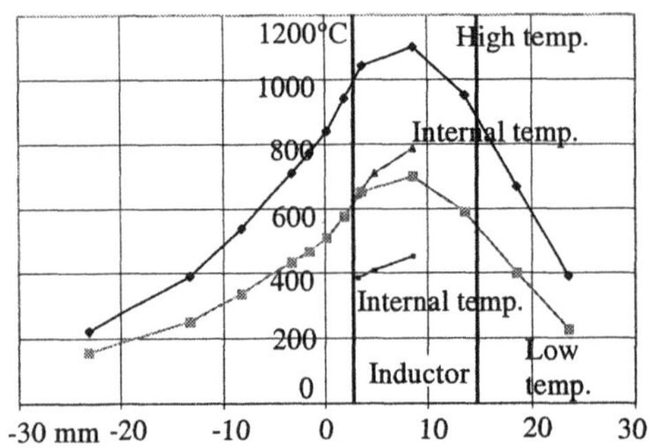

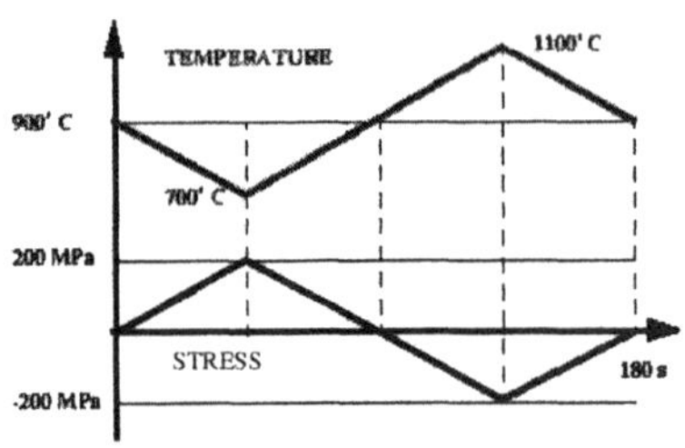

Fig. 4. Axial thermal gradient obtained on the specimen and the corresponding thermomechanical load.

Many thermocouples have been welded on the specimen on the gage circumference of inner and outer surfaces to verify a good homogeneity of the temperature fields. Several tests have been performed to optimise for the experimental specimens the axial thermal gradient and the temperature near the hole at the highest temperature. The best compromise has been obtained by modifying the inductor (12 mm of height and 2 whorls instead of 3) and by shifting it upstream the airflow which allows to rise axial and radial thermal gradients of around 250°C near the region at the highest temperature, as shown in the figure 4.

Several thermomechanical cycles have been considered for the tests. It has been noticed that the crack initiation occurs either in the hot region, which is characteristic of a usual fatigue-creep damage mechanism, or at the perforation at the highest temperature. Observations of rupture zones in the regions of the perforations, by using optical and scanning electron microscopes, show that many cracks occur at the perforations perpendicularly to the axial load, as it can be seen in Figure 5. The observation of the breaking surface reveals that the cracks grew only at the coated external surface, in intergranular and transgranular manner.

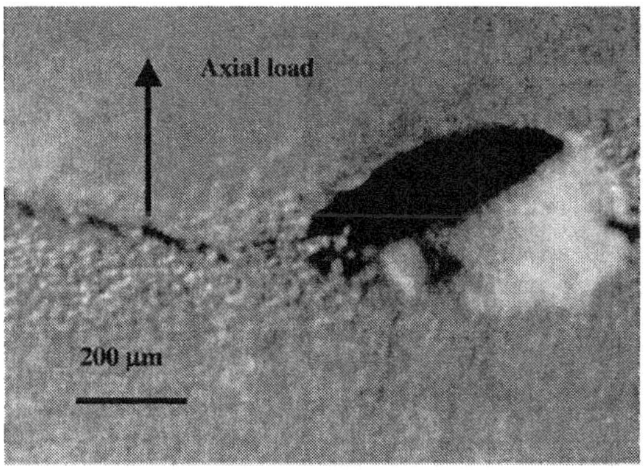

Fig. 5. Micrograph of a crack initiated around a hole (optical microscope).

All these test results have been systematically compared to life predictions as part of the doctorate of J.-M. Cardona [4]. A finite element analysis of the specimen was first carried out by using crystallographic constitutive equations, from which the main results are presented in the following section.

FINITE ELEMENT ANALYSIS

The determination of the stress-strain in the multiperforated specimens requires a 3D finite element analysis. The modeling of single crystal superalloy behaviour has been widely studied these last years. It has been shown that either crystallographic or macroscopic models could predict accurately their cyclic viscoplastic behaviour. The model used in this work is the crystalline model [2] developed at the School of Mines of Paris. This anisotropic model uses a multicriteria approach, for which the resolved shear stress on each system τ_s is computed from the macroscopic stress tensor and the orientation tensor $\tilde{m}$ of the slip system and one yield criterion is defined for each system. It assumes that the only mechanism of plastic deformation is the crystallographic slip. The viscoplastic strain rate is the result of the contribution of viscoplastic $\dot{\gamma}^s$ shear on the N slip systems. The constitutive equations are then written on each slip system, with the definition of two scalar variables for each slip system, a kinematic hardening variable and an isotropic one, which is the accumulated slip on each system. Cross hardening may be also introduced in the expression of the isotropic hardening by means of an interaction matrix. However, no latent hardening has been experimentally observed for anisotropic superalloys. The only isotropic hardening is self-hardening, which is present at low temperatures (under 760°C). On the other hand, cubic slip has been observed even at low temperatures. Two families of material constants are then introduced, characterizing the octahedral slip and the cube slip. It can be noticed that a link has been established between this crystallographic model and the macroscopic one developed at ONERA allowing to deduce the material constants of the macroscopic model from those of the crystallographic one and conversely [5]. Figure 6 shows that the two models give exactly the same prediction for tension loading in the <001> and the <111> crystallographic orientations and close predictions in the <101> direction.

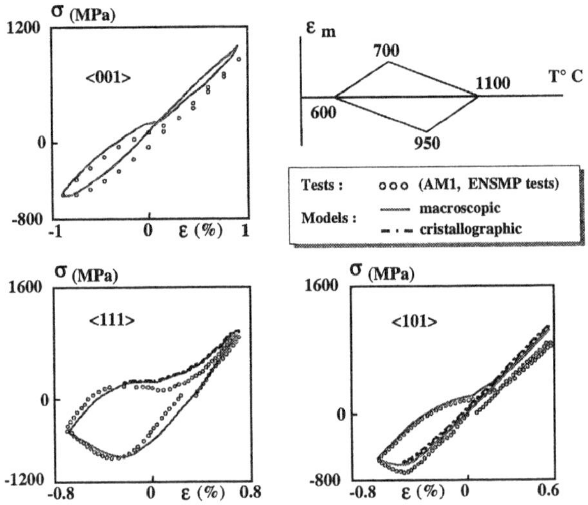

Fig. 6.Validation of the constitutive equations: prediction of thermomechanical fatigue loops at the stabilised cycle for a specimen representative of a volume element of the material.

The cubic anisotropy material leads to a three dimensions calculation in the case of a multiperforated specimen. In order to limit the size of degree of freedom in the calculation, the mesh includes one layer of quadratic element in the thickness of the specimen and the size of the elements just outside the hole is around 0.2 mm. In figure 7 the thermal field applied at maximum temperature during the heating to the test specimen is illustrated. The next figure shows the von Mises equivalent stress versus the deformation in the region just outside the hole. The strong stress relaxation with time and the need to simulate a great number of cycles (around 30) for stabilisation can be noticed. The comparison between the simulations and the experimental data at the stabilised rate shows the fairly good accuracy of the model to predict the stress-strain state of the specimens, as illustrated in the figure 9.

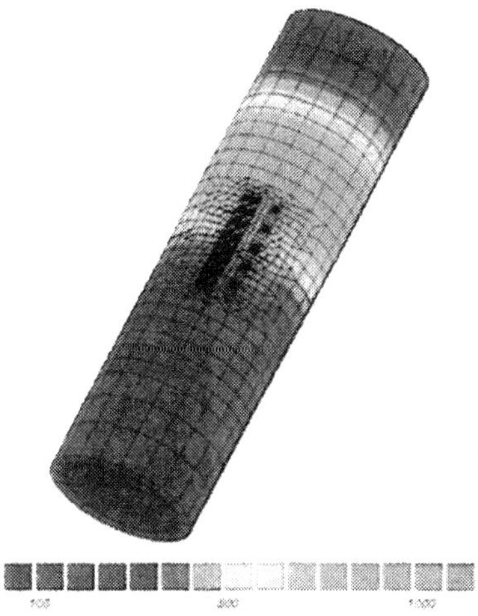

Fig. 7. Thermal field applied to the specimen.

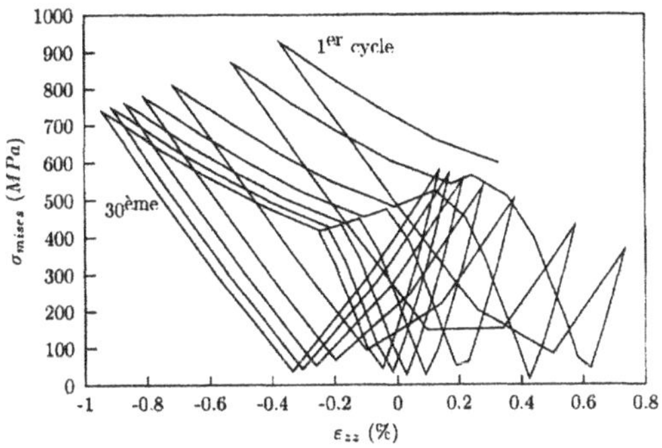

Fig. 8. Evolution of the stress-strain state with cycles.

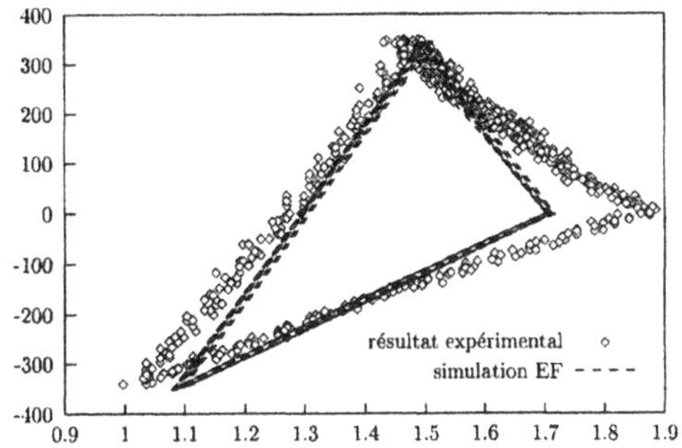

Fig. 9. Stress-strain state at the stabilised cycle.

FATIGUE LIFE PREDICTION

Previous microscopic observations of the rupture zones and breaking surfaces of superalloys have shown that the failure process can be divided into two damage processes [6,7]: a microinitiation phase and a micropropagation phase. In the first initiation, microcracks appear at the surface of the specimen, due to coating degradation, and grow very slowly during this period. The second phase is characterised by the development of few microcracks that propagate more quickly than the others in the depth of the material. One of those cracks forms then the main macroscopic crack that rapidly leads to the rupture of the specimen. In addition to these damaging effects due to the fatigue processes, damages due to creep and oxidation may also interact with fatigue damage. Several observations have effectively shown the prominent part played by the coating on the fatigue strength of the superalloy at low temperatures. The phenomenological fatigue-creep-oxidation interaction model proposed by ONERA is based on the differentiation of these two damage processes. This continuous damage model is a stress-based model that assumes that the progressive deterioration processes can be described by scalar damage variables.

This model has been detailed in previous publications [8,9,10] in its uniaxial and multiaxial forms. We will then recall here only its main characteristics. Four damage variables are introduced in this creep-fatigue-oxidation interaction model. D_I is related with the surface crack microinitiation, D_p with the micropropagation. D_p increases only if D_I has reached the initial value of 1. Interaction effects are introduced between fatigue and oxidation, described by the variable D_{ox}, only during the microinitiation phase. The last variable, D_c, describes creep damage, that can be developed during the microinitiation phase, but interacts with the fatigue damage during the micropropagation phase. The originality of the model lies in the way of writing the evolution equations in anisothermal conditions. Using reduced stresses $S_I = \sigma / \sigma_{uI}(T)$ or $S_p = \sigma / \sigma_{uP}(T)$ as parameters describing the thermomechanical fatigue cycle, where $\sigma_{uI}(T)$ and $\sigma_{uP}(T)$ are respectively the ultimate stresses in microinitiation and micropropagation and which are temperature independents, the fatigue laws can then be temperature independent [11]. Only the creep and the oxidation phenomena, which are time dependent and thermally activated, are described by temperature dependent laws.

The fourth damage growth laws due to microinitiation (eqn.(1)) and micropropagation (eqn.(2)) in fatigue, to creep (eqn.(3)) and to oxidation (eqn.(4) and (5)) are expressed by the following relations:

$$\left\{\begin{array}{l} dD_I = \dfrac{1}{C}\left\langle \dfrac{S_{II} - S_{lI}(\bar{S}_H)(1-D_{ox})}{(1-D_{ox}) - S_{eq\max}} \right\rangle^b dN \\ dD_P = \left[1-(1-D_P)^{\beta+1}\right]^{1-\alpha} \left\langle \dfrac{S_{II} - S_{lP}(\bar{S}_H)}{1-S_{eq\max}} \right\rangle \left(\dfrac{S_{II}}{M^*(\bar{S}_H)(1-D_P)} \right)^{\beta} dN \\ dD_c = \left(\dfrac{X^c(\sigma(t))}{A(T(t))} \right)^{r(T(t))} \left[1-D_c\right]^{-k(T(t))} dt \\ dD_{ox} = \dfrac{1}{2} D_{ox}^{-1} \left(\dfrac{K^*}{e_o} \right) \left[1 + \dfrac{\langle X_{ox}(S) - S_{l\,ox}(\bar{S}_H) \rangle}{B} \right]^{2m} dt \\ (k^*)^2 = \displaystyle\int_0^{\Delta t} \left[k_o \exp\left(-\dfrac{Q}{RT(t)} \right) \right]^2 dt \end{array}\right\}$$

The indices I and P stand respectively for initiation and propagation. A_{II} is the octahedral shear amplitude, S_{PH} or S_{IH} the mean hydrostatic pressure as defined by Sines [12]. S_{eq} is the Hill equivalent stress. Limit stresses S_l are defined by $S_{lI} = S_{lIo}\,(1-h_1\,\text{Mean}(tr\,S_p))$, $S_{lP} = S_{lPo}(1-h_2\text{Mean}(tr\,S_p))$, and $S_{lox} = S_{loxo}(1-h_3\text{Mean}(tr\,S_p)) + \text{Mean}(trS_p)$. Similar relation is used for the parameter $M^* = M_o^*(1-h_4\text{Mean}(tr\,S_p))$. For damage creep law, Hayhurst's multiaxial criterion [13] has been extended to take into account the material anisotropy of the single crystal.

The proposed formulation makes the model attractive for its identification on a large scale of temperatures. Pure fatigue tests performed at one temperature are thus sufficient to identify all the parameters of microiniation and micropropagation laws. Pure creep tests performed at several temperatures are nevertheless required to identify the creep damage law. Material anisotropy effects are taken into account by four-order tensor introduced in A_{II} and S_{eq} equations. For cubic anisotropy, two independent material constants are required. The model has been completely identified by F. Gallerneau [14] for the AM1 single crystal superalloy in the <001>, <110> and the <111> crystallographic orientations.

This model, already used with success for thin wall smooth tubes [15] has been applied to predict the fatigue life of the multiperforated test specimens. The fatigue-creep-oxidation interaction model is applied as post-treatment of the finite element calculation. We have reported in the figure 10 the experimental and the predicted lifetimes to the initiation in the specimens of a macroscopic crack, typically 1 mm long. We can notice that the model provides fairly good predictions of the lifetime of the tested specimens. The potential method is difficult to apply on multiperforated specimens to estimate the fatigue life to initiation. For the longest life range, the fatigue life to initiation is certainly overestimated.

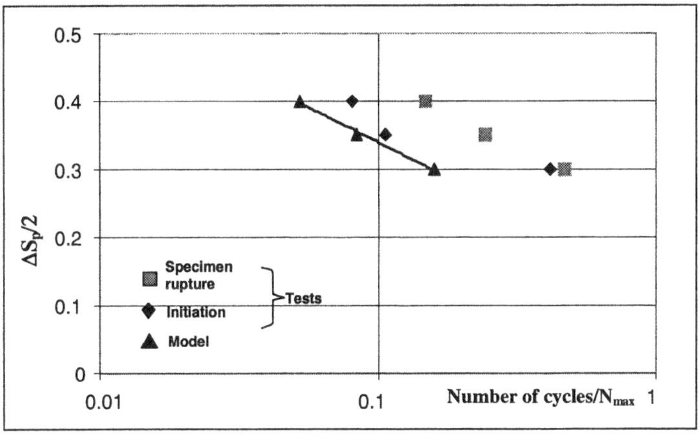

Fig. 10. Comparison of experimental and calculated lifetimes.

CONCLUSION

In this paper fatigue test results obtained at ONERA on a coated single crystal superalloy for turbine blades are presented. These experimental tests have been performed on multiperforated thin wall thermal gradient tubes which can be considered to represent structural components of turbine blades. An experimental set-up, developed at ONERA, has then been applied allowing to reproduce in laboratory a specific thermal gradient in the region just outside the holes of a thin walled specimens. Life predictions of the set of tests have been performed systematically at the School of Mines of Paris. A finite element analysis of the specimen is made first by using a crystallographic viscoplastic model. Then, a fatigue-creep-oxidation interaction damage model is applied as post-treatment of the finite element calculation. The comparisons of the lifetime predictions with respect to the experimental one show a fairly good correlation. These results allow to conclude that this experimental device is efficient to validate our life prediction method under complex thermomechanical loads in real structural components of turbine blades.

Acknowledgements- Support for this work by SNECMA is gratefully acknowledged.

REFERENCES

1. Nouailhas, D., Culié, J.-P., Development and application of a model for single crystal superalloy, Fredd A.D and Walker K.P Ed., *High Temperature Constitutive Modeling – Theory and Application*, ASME, MD-26, AMD-121, New York, 151.
2. Méric, L., Poubanne, P., and Cailletaud, G., (1991), Single crystal modeling for structural calculations: Part I - Model presentation, *Journal of Engn. Mat. and Techn.*, 113, 162.
3. Pollicella, H., Baudin, G., Cailletaud, G., (1985), *Mesure de longueur de fissure, déformation et endommagement par une technique de potentiel électrique*, 60th Meeting of the Structures and materials Panel AGARD Specialists Meetings, San Antonio (Texas) USA, 22-26 avril.
4. Cardona, J.M., (2000), Comportement et durée de vie des pièces multiperforées: application aux aubes de turbine, Thesis, School of Mines of Paris.
5. Nouailhas D., Cailletaud G., (1992). Comparaison de divers critères anisotropes pour monocristaux cubiques à face centrée (CFC), Note aux Comptes rendus de l'Académie des Sciences de Paris, t.315, série II, 1573.
6. François, D., Pineau, A., and Zaoui, A.,(1992). *Comportement mécanique des matériaux*, Hermès Ed.
7. Klesnil, M. and Lukas, P., (1980). *Fatigue of metallic materials*, Elsevier Ed.
8. Gallerneau, F., (1995), Etude et modélisation de l'endommagement d'un superalliage monocristallin revêtu pour aube de turbine, Thesis, School of Mines of Paris.
9. Gallerneau, F., Nouailhas, D., and Chaboche J.-L., (1995), *Etude et modélisation de l'endommagement en fatigue d'un superalliage monocristallin revêtu*, Journées de Printemps, Fatigue et traitement de surface, Paris.
10. Gallerneau, F., Nouailhas, D. and Chaboche, J.-L., (1996), *A fatigue damage model including interaction effects with oxidation and creep damages*, FATIGUE'96, Berlin.
11. Gallerneau, F., Nouailhas, D., and Chaboche, J.-L., (1996). Fatigue damage behavior of a coated single crystal superalloy, Proc. Of ECF'11, *Mechanisms and Mechanics of Damage and Failure*, Petit, J., Ed, pp. 1275-1280.
12. Sines, G., (1959), Behavior of metals under complex static and alternating stresses, *Metal. Fatigue*, 145-169.
13. Hayhurst, D.R., (1972). Creep rupture under multiaxial state of stress, *J. Mech. Solids*, vol. 20 n° 6, 381-390.
14. Gallerneau, F., (1999). *Modelling of anisotropy effects of a single crystal superalloy on its fatigue-creep resistance*, ICAF'99, Seattle, July 12-16.
15. Chaboche, J.-L , Culié, J.-P, Gallerneau, F., Nouhailhas, D., Pacou, D., Poirier, D. (1997). *Thin wall thermal gradient: experimental study, F.E. analysis and fatigue life prediction*, The 5th International Conference on Biaxial/Multiaxial Fatigue and Fracture, Cracow (Poland), September 8-12.

Temperature-Fatigue Interaction
L. Rémy and J. Petit (Eds.)

MECHANICAL ANALYSIS OF AN AERO-ENGINE COMBUSTOR UNDER OPERATION CONDITIONS USING A UNIFIED CONSTITUTIVE MATERIAL MODEL FOR DEFORMATION SIMULATION

U.MÜLLER, K.HÖSCHLER, M.GERENDÁS, H.-J.BAUER, U.SCHOTH
Combustion Department/Thermo-Mechanics; Methods/Analysis and Simulation Technologies
Rolls-Royce Deutschland Ltd & Co KG
Eschenweg 11, D-15827 Dahlewitz, Germany

ABSTRACT

The operation environment of an aero-engine combustor leads to extreme temperatures and temperature gradients causing high strains during a flight cycle, especially under transient conditions between different power settings. Non-linear effects in terms of material deformation and creep-fatigue interaction become significant. The paper presents the thermal- and mechanical analysis of the combustion chamber of the Rolls-Royce BR700 aero-engine manufactured in a nickel-base alloy. The results are considered for use in a life assessment for the structure under operating conditions. The investigation is based on a thermo-mechanical 2D finite element analysis of the whole combustor module for a complete flight cycle considering boundary conditions as functions of environment and performance parameters. This analysis generates temperature distributions and mechanical boundary conditions for a 3D sector model of the combustor. A detailed viscoplastic analysis is then carried out for the most highly loaded area. This is performed under consideration of an CHABOCHE-type unified constitutive material model using material parameters adapted to uniaxial specimen test data in the Brite EuRam programme CPLIFE. The analysis is carried out using the finite element code ABAQUS in combination with an user-defined material subroutine (UMAT).

KEYWORDS

Combustion chamber, thermal analysis, viscoplasticity, unified constitutive material model, Chaboche, nickel-base alloy, fatigue

INTRODUCTION

The rapidly growing market of frequently flying regional jets requires reliable operating aero-engines which can cope with the high number of short distance flights. The BR715 jet engine, manufactured by Rolls-Royce was certified in August 1998 and went into revenue service on the Boeing B717-200 in September 1999. The high number of flights (up to 13 per day) of this regional airplane can only be guaranteed by reliably available engines. To fulfil this demand, the engine had to be excessively tested during ground and flight tests. Due to increasingly dynamic market conditions in a highly competitive business environment, airplane and hence engine manufacturers are required to react fast on new developments in customer requirements. This leads to extremely tight development programmes from market requirement identification and programme launch to certification and entry into service of the final product.

This tremendous increase in efficiency of design and development without compromises in safety and reliability of the final product can only be achieved by an intensive utilisation of state of the art analysis methods to prepare and support the engine test activities. The finite element method has proven to be a viable tool for the engineer to accomplish this task. Whereas in most other areas of the engine modern numeric analysis methods for stress analysis are well established, the evaluation of strength, durability and structural integrity of combustion chambers is still very much bound to engineering experience. This is, besides other problems, mainly due to the fact that combustors operate at extremely high temperatures. The material behaviour at these temperatures can not be simulated anymore by linear-elastic or "simple" elasto-plastic analysis. Combined plastic, creep and viscous effects lead to a relatively fast stress redistribution and overall residual deformations of the component.

The fast improvement of computational power brings the application of numerically expensive unified constitutive material simulations within reach for the description of practical problems. The authors are involved in the Brite/EuRam research programme CPLIFE which investigates the material behaviour and lifing methods for components operating under creep-plastic loading conditions in materials typical for high temperature applications. The paper shows how knowledge generated in this research programme will be incorporated in the design process of an engine combustor, and which problems still have to be solved.

GENERAL MODELLING PROCEDURE FOR COMBUSTOR COMPONENTS

Thermal Analysis of Combustor Assembly Model

To get the transient 3D temperature distribution in the combustor liner, which is the basis for the stress analysis and lifing procedure, a comprehensive modelling process must be completed. First a 2D model of the complete combustor module is developed, which includes inner and outer casings, compressor outlet guide vanes and turbine nozzle guide vanes. A data base provides all the necessary material data used in the analysis. Engine performance data and the specific flight cycle under investigation are fed into the calculation during the analysis. An example for the development of the compressor delivery pressure during the flight cycle is given in Fig.1.

To model specific parts of the combustor in a finite element analysis, several boundary conditions must be known at this location, i.e. the coolant and hot gas temperature on both sides of the combustor liner, the radiative heat flux to the wall and the local pressure drop across the wall. To have access to this information during the complete flight cycle, a transient 1D network representation of the air flow around and through the combustor is coupled within the FE analysis model.

Based on the knowledge about the combustor design and the flow structure, appropriate heat transfer correlations are applied to each surface. This complex setup is then matched with thermal paint results and with thermocouple measurements taken during engine testing.

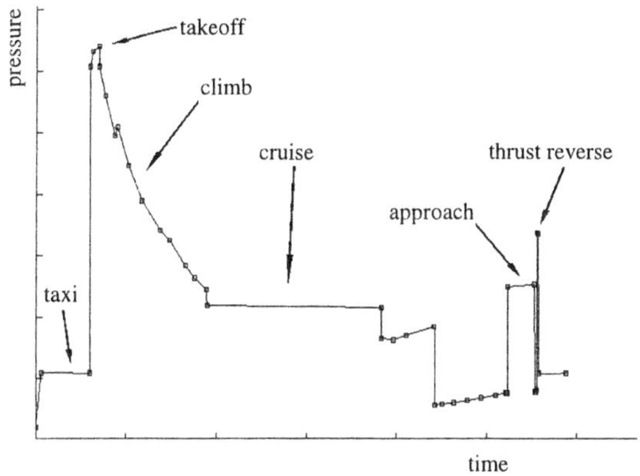

Fig.1: Development of Compressor Delivery Pressure during Flight Cycle

The result at a specific time instant of a transient 2D simulation is displayed in Fig.2. Shortly after an increase in compressor delivery temperature and pressure, the casing flange, where also the combustor is mounted, is still colder than the casing wall. The combustor with a relatively cold combustor head and hotter combustor liners is visible, as well as the relatively hot turbine vanes. This calibrated model is then used to derive the thermo-mechanical boundary conditions for the 3D model using a linear-elastic calculation.

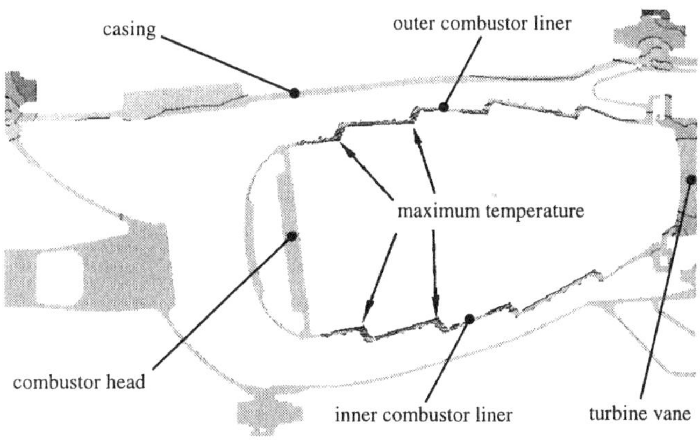

Fig.2: 2D Model of Combustor Module with Temperature Contours

Mechanical Analysis of Combustor Assembly Model

The finite element analysis package used by Rolls-Royce allows a combined thermo-mechanical analysis. After the thermal analysis has reached a satisfactory matching with experience or test results, all required mechanical boundary conditions are applied. That includes:

- Pressures from the air system on all free surfaces of the module.

- Carcass loads since the combustion chamber outer casing lies in the main structural load path of the engine. These loads are generated by a whole engine model which analyses the entire spectrum of engine manoeuvre conditions in its effect on the deformation behaviour of the engine as well as the structural loads on single components.
- Aerodynamic loads from the compressor outlet guide vane and the turbine nozzle guide vane.
- Interferences to simulate any specific mechanical conditions which describe the build condition of the module.

The thermo-mechanical analysis was run with a linear-elastic material formulation for a typical flight cycle (Fig.1). The result was used to identify the areas with the most severe loading throughout the cycle. For the example presented herein, the area of the first outer bay cooling ring was identified as the area with the highest local stress and also the highest strain variation over the cycle combined with a severe temperature level for large periods of the cycle (see Fig.3). The maximum linear-elastic stress level occurs during takeoff conditions.

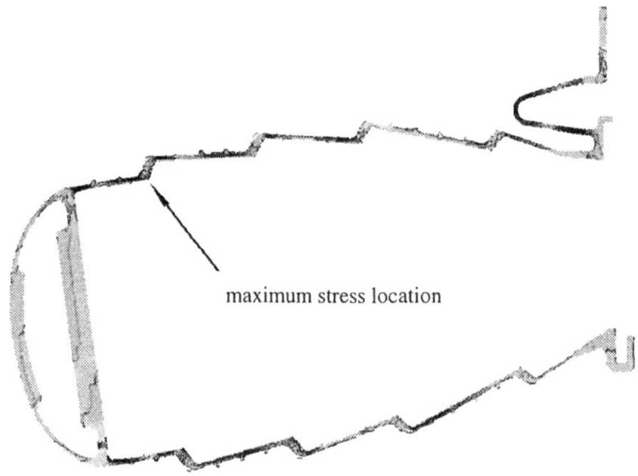

Fig.3: Maximum Stress Location in Combustor from Global Linear-elastic Analysis

Due to the simplified thermal and mechanical representation of the cooling holes in this rotational-symmetric model the results are only taken as a possibility to identify the most critical cooling ring of the combustor. High stresses are visible at the aft edge of the first combustor liner bay and at the edge of the cooling holes. The 3-dimensional nature of the structure in the area of concern requires a more detailed 3D finite element model analysis to determine the true stress and strain levels.

3D FINITE ELEMENT SIMULATION OF COMBUSTOR LINER

Finite Element Model

The thermal 3D finite element analysis was carried out in an in-house finite analysis system for its better adaptation to the special needs of the simulation of aero-engine components whereas for the mechanical analysis the commercial analysis package ABAQUS version 5.8 was selected because of its capability of relatively easy incorporation of user defined material models. Since the planned visco-plastic analysis is numerically very expensive, the model

was made as small as possible. Hence all possible geometric symmetry properties of the structure were considered.
The first outer cooling ring of the outer combustor liner has circumferentially staggered rows of cooling holes. The cooling ring is considered as the critical feature. To minimize the influence of the interfaces, where boundary conditions of the 2D analysis will be introduced, on the area of interest, a part of the first and second bay of the outer combustion liner was incorporated into the 3D model.

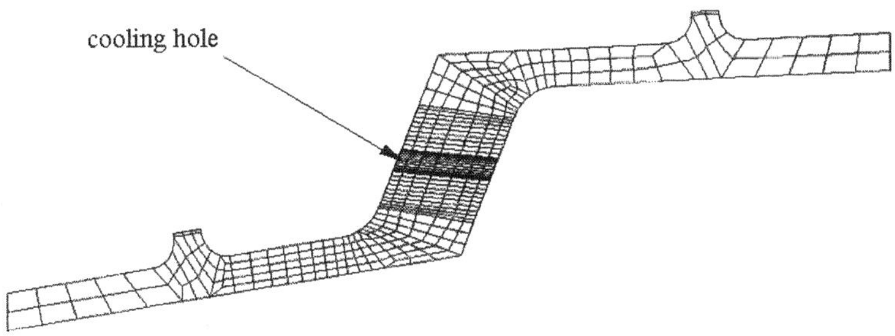

Fig.4: 3D Finite Element Mesh

The finite element mesh is shown in Fig.4. In total, about 2800 20-node brick elements are used. Due to the geometrical sector symmetry of the component in the area of interest in combination of the planar symmetry of each periodic sector, only a sector covering one half hole of each row was modelled. The cyclic/planar symmetry is guaranteed by the prevention of normal displacements for the sector cut faces in circumferential direction.

3D FE Thermal Analysis

The 3D model of the cooling ring segment is embedded into the 2D model of the combustor module to get the appropriate time-dependant boundary conditions on the model interfaces. Again, all necessary thermal boundary conditions and heat transfer correlations are applied to the respective surfaces based on the knowledge about the flow structure. The result of such a transient 3D analysis is displayed in Fig.5. During takeoff the liner material is locally heated up to relatively high temperatures at the lower corner of the cooling ring.

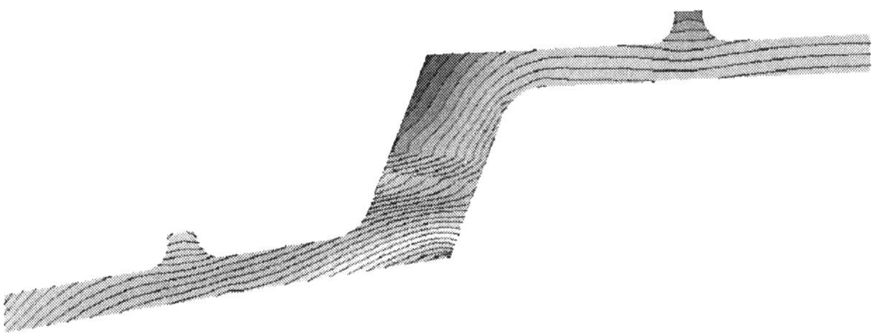

Fig.5: 3D Model of Cooling Ring Segment with Temperature Contours

3D FE Linear-Elastic Mechanical Analysis

The mechanical 3D finite element analysis was carried out with mechanical boundary conditions from the 2D thermo-mechanical analysis and temperature fields from the thermal 3D analysis. A summary of all applied boundary conditions is shown in Fig.6.

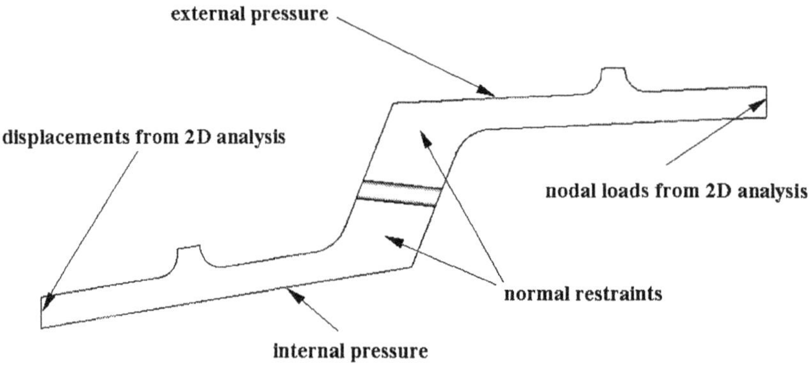

Fig.6: Mechanical Boundary Conditions

A mixed force/displacement approach for the simulation of the global model influence on the local 3D model has been selected. This minimizes disturbing effects from deviations of the local stiffnesses in the 2D and 3D simulation and also prevents creep effects from unrealistically reducing the carcass load influence on the local stress conditions. A displacement field is applied at the front cut surface simulating the axial and radial displacements at that location derived from the 2D analysis. On the aft cut surface nodal loads are applied which were derived from the associated nodes of the 2D analysis. Appropriate pressure loads are applied to all other free surfaces. All boundary conditions vary accordingly throughout the flight cycle.

CHABOCHE-TYPE UNIFIED CONSTITUTIVE MODEL FOR SIMULATION OF COMBUSTOR MATERIAL BEHAVIOUR

In the research programme CPLIFE [1] different approaches for unified constitutive material laws are investigated for the simulation of the material behaviour of nickel-base alloys used in components under environmental conditions typical for combustors and turbines. The approach followed by Rolls-Royce is based on a proposal from Chaboche which initial simple version was first published in 1977 [2]. The model, subject to various further developments and refinements, is able to predict the transient stress and strain history of a material point exposed to complex mechanical loading at isothermal conditions and even varying temperatures.
The very extensive material test programme carried out in CPLIFE and the associated material parameter optimisation and test simulation activities was the basis for Rolls-Royce to use the following set of evolution equations

$$\underline{\dot{\varepsilon}} = \underline{\dot{\varepsilon}}^{el} + \underline{\dot{\varepsilon}}^{in} + \underline{\dot{\varepsilon}}^{th} \tag{1}$$

$$\underline{\dot{\varepsilon}}^{el} = \frac{1+\nu}{E}\underline{\dot{\sigma}} - \frac{\nu}{E}Tr(\underline{\dot{\sigma}})\underline{\delta} \tag{2}$$

$$\dot{\varepsilon}^{th} = \alpha(\dot{T} - \dot{T}_0)\underline{\delta} \tag{3}$$

$$\underline{\dot{\varepsilon}}^{in} = \frac{3}{2}\left\langle \frac{J(\underline{\sigma} - \underline{X}) - \alpha_R R - k}{K_0} \right\rangle^n \frac{\underline{\sigma}' - \underline{X}'}{J(\underline{\sigma} - \underline{X})} \tag{4}$$

$$\underline{\dot{X}}^i = \frac{2}{3} c^i a^i \underline{\dot{\varepsilon}}^{in} - c^i \Phi(p) \left(J(\underline{X}^i) / (a^i / \Phi(p)) \right)^{\kappa^i} \underline{X}^i \dot{p} - \beta^i \left| J(\underline{X}^i) \right|^{r^i - 1} \underline{X}^i \tag{5}$$

$$\Phi(p) = \Phi_s + (1 - \Phi_s) e^{-bp} \tag{6}$$

$$\underline{X} = \sum_{i=1}^{2} \underline{X}^i \tag{7}$$

$$\dot{R} = b(Q - R)\dot{p} \tag{8}$$

with	$\underline{\mathbf{X}}$	-	Kinematic hardening variable
	R	-	Isotropic hardening variable
	Φ	-	Influence of isotropic on kinematic hardening,
and the conventions	δ	-	Kronecker symbol
	Tr($\underline{\mathbf{F}}$)	-	Trace of tensor $\underline{\mathbf{F}}$
	J($\underline{\mathbf{F}}$)	-	2nd invariant of tensor $\underline{\mathbf{F}}$
	$\underline{\mathbf{F}}$'	-	Deviator of tensor $\underline{\mathbf{F}}$
	< >	-	MacCauley brackets.
The parameters are	E	-	Youngs's Modulus
	ν	-	Poisson Ratio
	a_i	-	Saturation value of kinematic hardening variables
	c_i	-	Kinematic hardening exponents
	β_i	-	Coefficients of kinematic hardening recovery
	r_i	-	Kinematic hardening recovery exponents
	Φ_s	-	Coefficient of effect of isotropic on kinematic hardening variable
	b	-	Isotropic hardening exponent
	κ_i	-	Exponents of dynamic recovery on kinematic hardening
	α_R	-	Coefficient of isotropic hardening
	K_0	-	Overstress parameter
	k	-	Strain rate dependent initial yield stress
	n	-	Strain rate sensitivity parameter.

For varying temperature, the equations for kinematic and isotropic hardening have to be extended by

$$\underline{\dot{X}}^i = \left[\underline{\dot{X}}^i\right]_{\dot{T}=0} + \frac{1}{C^i}\frac{\partial C^i}{\partial T}\underline{X}^i \dot{T}, C^i = c^i a^i \tag{9}$$

$$\dot{R} = \left[\dot{R}\right]_{\dot{T}=0} + \left(\frac{1}{b}\frac{\partial b}{\partial T} + \frac{1}{Q}\frac{\partial Q}{\partial T} \right) R\dot{T} \tag{10}$$

This model concentrates on the modelling of the monotonic and transient cyclic behaviour as well as the creep behaviour with the non-linear kinematic and simple isotropic formulation. For better representation of mean stress relaxation under strain controlled loading, the approach incorporates an extended kinematic hardening equation proposed by Ohno and Wang [3],[4]. The material simulation was converted into a FORTRAN programme and then incorporated in a user-defined material subroutine UMAT for use in ABAQUS.

ANALYSIS AND RESULTS

2D Analysis Results

The stress level and distribution in the module is caused, as previously mentioned, by mechanical as well as thermal effects. The temperature and associated stress distribution for takeoff conditions in the area of the first outer cooling ring are shown in Fig.7.

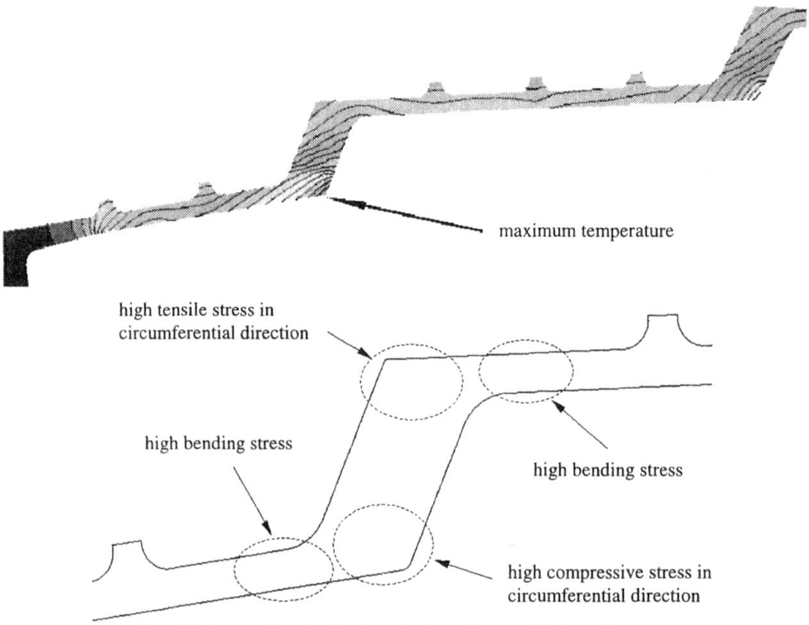

Fig.7: Results of Linear-elastic 2D Analysis at End of Takeoff

The maximum temperature occurs at the inner aft edge of the cooling ring. A relatively strong temperature gradient develops from the inner to the outer edge of the cooling ring. As a consequence, high compressive stresses in the circumferential direction occur at the aft inner edge combined with relatively high tensile stresses at the outer surface. A pronounced temperature gradient also exists from the centre to the end of the bay segments. This leads to relatively high shell bending stresses in the combustor liner skin. Since the cooling holes are introduced in a region of very intensive thermal and mechanical loading, a 3D investigation is inevitable to get a correct understanding of the actual local stress and strain conditions in this area.

3D Analysis Results

A comparison of the results of the linear-elastic analysis with the analysis featuring an unified constitutive material law is given in Fig.8. The graphic shows the von Mises stress in the cooling ring at full power conditions which induces the most severe stresses in the linear-elastic analysis.

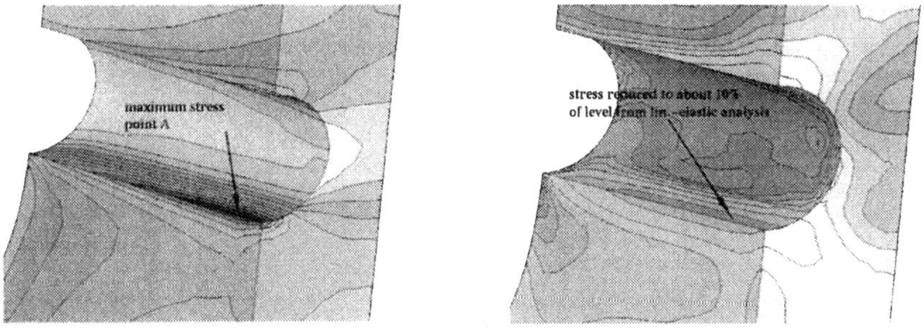

Linear-elastic Analysis **Visco-plastic Analysis, First Cycle**

Fig.8: Von Mises Stress at Full Power Conditions in Typical Cooling Hole

As expected, the maximum stress in the linear-elastic analysis occurs at the inner edge of the most inner cooling hole and is mainly dominated by the circumferential stress. Due to the exposure to high temperature in this area, significant relaxation effects lead to a massive redistribution process of the stresses already in the first cycle. The node with the highest stress in the linear-elastic analysis has after experiencing the exposure of takeoff conditions for the associated period only less than 10% of that level. The bending stress in the transition from cooling ring to the second combustor bay are now more dominant since the temperature in this area is significantly lower and therefore non-linear material responses are less pronounced. In Fig.9 the variation of the circumferential stress at point A (see Fig.8) during a complete flight cycle is shown. The non-linear material simulation leads to a significant mean stress shift with a change in sign for the non-linear solution and the development of high tensile stresses in those parts of the flight cycle that have lower temperatures.

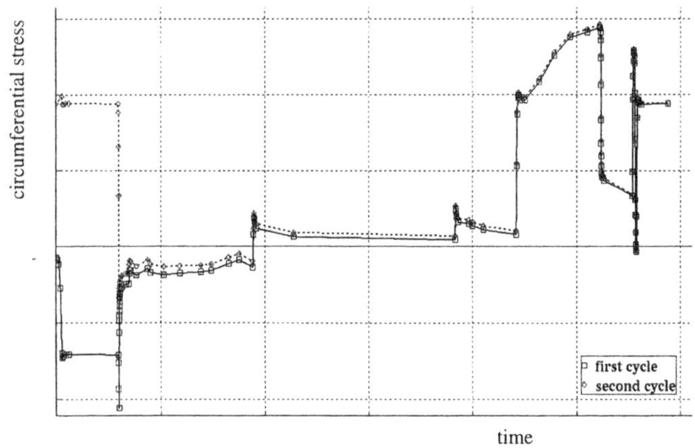

Fig.9: Circumferential Stress versus Time at Point A

SUMMARY AND OUTLOOK

Life assessment approaches which are based on linear-elastic stress analyses tend to significantly underpredict the life of structures like combustion chambers since they neglect the supporting effects of the surrounding structures when massive redistribution processes take place. The aim of the project is to generate more realistic input data for those lifing methods which consider mean stress relaxation or the interaction of creep- and cyclic fatigue. As an example, Fig.10 shows the stress-strain hysteresis of point A for a cycle later in the engine life which can now be used in life analysis.

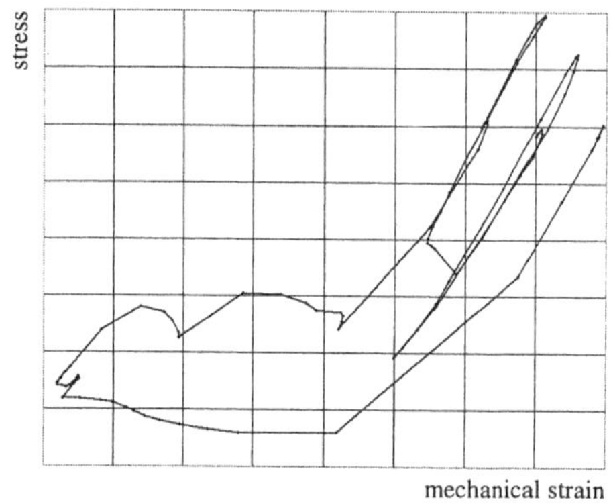

Fig.10: Stress versus Mechanical Strain at Point A for Higher Number Cycle

Although the analysis method described above is seen as an important contribution on the way to an effective method for life evaluation of hot components, a number of obstacles still have to be overcome. One major problem for extensive use in the design and development process is the very long computation time, even for relatively small models like the combustor cooling ring segment presented above. The life target for combustion chambers of modern civil aero-engines is significantly above 10000 flight cycles. Since a stabilised cycle is of interest for lifing, effective cycle skip algorithms need to be applied to generate results in an reasonable timeframe.

ACKNOWLEDGEMENT

The authors would like to acknowledge the financial support for the CPLIFE programme by the European Community under the Industrial & Materials Technologies Programme Brite-EuRam III. Many thanks also to all the members of the CPLIFE working group for their very supportive co-operation.

REFERENCES

1. Brite/EuRam project BE97-4034, Lifing methods for components operating under creep-plastic loading conditions (CPLIFE)
2. Chaboche, J.-L. (1977), *Bulletin de l'academic polonaise des sciences, Serie des sciences techniques, Vol. XXV, No. 1*, Viscoplastic equations for the description of cyclic and isotropic behaviour of metals
3. Ohno, N. (1997), *Transactions of the 14th International Conference on Structural Mechanics in Reactor Technology (SMiRT 14), Lyon, France, Aug. 17-22*, Current state of the art in constitutive modelling for ratchetting
4. Ohno, N. (1998), *Int. J. Mech. Sci, Vol. 40*, Constitutive Modelling of cyclic plasticity with emphasis on ratchetting

Temperature-Fatigue Interaction
L. Rémy and J. Petit (Eds.)

LIFETIME PREDICTION ON STAINLESS STEEL COMPONENTS UNDER THERMAL FATIGUE LOAD

P.O.SANTACREU
Usinor Recherche et Développement, Centre de Recherche d'Isbergues, Ugine S.A., F-62330 Isbergues, France

ABSTRACT

Thermal fatigue of austenitic and ferritic stainless steel grades has been experimentally and numerically investigated. A special test has been developed to determine the thermal fatigue resistance of clamped V-shape specimen. Examination of the failed specimens indicated that cracks could be mainly attributed to out-of-phase thermal fatigue in case of ferritic grades and to in-phase thermal fatigue in case of austenitic grades. A numerical method is proposed for the design and the lifetime prediction of components under thermal fatigue load. Thus, the viscoplastic strain amplitude is used as the crack initiation criterion for ferritic stainless steels. Due to a coupling with oxidation and creep during the in-phase thermal fatigue of austenitic grades, the phasing between the thermal and the mechanical loads has to be taken into account in the criterion. The hydrostatic pressure at the maximal temperature can be proposed as a such phasing factor.

KEYWORDS

Thermal Fatigue, Stainless Steel, Exhaust, Damage, Life Prediction.

INTRODUCTION

Background

Our study deals with the development of numerical lifetime assessment tools dedicated to the design of stainless steel automotive parts operating at high temperature, focusing on the fatigue design of exhaust manifolds submitted to severe thermal loads (figure 1.). Exhaust suppliers test the manifold on engine dynamometers under cyclic conditions which are generally specified by auto makers.

Today, exhaust gas temperature can be as high as 950°C. Hence cyclic thermal stresses and plastic strains are generated in the more clamped areas and may lead to the failure of the component. Generally, the part has to pass approximately 1500 cycles to be considered valid for production and so the design needs to be optimised in that aim. In an effort to reduce both the number of costly motor bench tests and development time of a part, simulation tools have to be proposed. Those tools consist in a mechanical behaviour model for high temperatures and in a damage model under non isothermal mechanical loads.

Objectives

In aim to promote the use of stainless steel in exhaust manifold application, studies were undertaken by Ugine-Usinor to develop high temperature stainless steel grades, provide high temperature mechanical properties and propose methods for fatigue design of such compounds. A collaboration with nCode was also engaged to develop a thermomechanical fatigue (TMF) post processing software which includes different existing fatigue criteria. In future, the study of the coupling between creep and oxidation appears to be an interesting way to improve both the understanding of material TMF resistance and its modelling.

MATERIALS AND EXPERIMENTS

Materials

The studied materials are stainless steel grades commonly used for exhaust manifold application in form of bent and hydroformed tubes or deep-drawn sheets. The considered thickness for sheet is below 2 mm. Three types of grades are distinguished :

- stabilised ferritic grades containing from 11 to 18% Cr, like EN1.4512 (AISI409) or EN1.4509 (AISI441) are characterised by low ultimate tensile strengths at temperature above 800°C (around 50 MPa), a thermal expansion coefficient around 12.10^{-6}/°C and a good cyclic oxidation resistance up to 950°C. Creep resistance of ferritic grades can be significantly improved by an intergranular precipitation of stable niobium intermetallic compounds;
- austenitic grades containing around 18% Cr and 10% Ni, like EN1.4301 (AISI304) or EN1.4541 (AISI321) are characterised by higher ultimate tensile strengths (around twice those of ferritic grades) but higher thermal expansion coefficient around 20.10^{-6}/°C leading to a very poor cyclic oxidation resistance ;
- austenitic refractory grades containing 20% Cr and 12% Ni at least, like EN1.4828 (~AISI309S) whose properties are close to austenitic grades but with a better oxidation behaviour.

Antoni et al. presented in detailed a comparison between cyclic oxidation properties of stainless steel in ref. [1].

Thermal fatigue testing

Method. A special test has been developed to determine the thermal fatigue resistance of steel sheet specimens. The testing rig and the experimental procedure are described in references [2] and [3]. This test permits to impose thermal cycle on a clamped V-shaped specimen by alternate resistance heating and air cooling (figure 2). It has been also adapted to the case of welded specimen [4]. The thermal fatigue life of a specimen is expressed as the number of cycles to failure. For a given grade, the fatigue life depends on the maximal and minimal temperature of the cycle, holding time at the maximal temperature and specimen thickness. The advantage of this test is that it is both simple for classing the stainless steel grades and representative of the thermal fatigue process occurring in an exhaust manifold, and so aiming a study of the damage mechanisms.

Experimental results. Some results obtained on the different stainless steel grades for 250°C-900°C cyclic conditions and 2 mm-thick specimen are displayed on figure 3. We notice :

- EN 1.4541 (AISI 321) and 1.4301 (AISI 304) austenitic grades exhibit a poor thermal fatigue resistance compared to the ferritic grades EN 1.4512 (AISI 409), F14Nb (14%Cr Nb-stabilised) and EN1.4509 (AISI441) ;
- EN 1.4509 (441) offers the best thermal fatigue resistance, even compared to the refractory grade EN1.4828 (~AISI309) which is more sensible to the detrimental effect of the holding time at the maximal temperature.

Oku et al. investigated also the thermal fatigue resistance of ferritic stainless steels and found same difference between ferritic and austenitic grades [5]. In fact, microstructural observations performed on specimens revealed that the fatigue crack propagation occurs in intrados of the specimen in the case of ferritic grades and in extrados of the specimen in the case of austenitic grades. The difference between the thermal expansion coefficient of ferritic grade and those of austenitic grades is not sufficient to explain by itself the difference between the thermal fatigues lives and crack locations. Finally, it has to be noticed that these results differ significantly from results obtained in isothermal conditions – low-cycle or high-cycle fatigue – where resistances follow generally the high temperature tensile strength.

NUMERICAL MODELLING

Modelling procedure and assumptions

The general approach for a lifetime prediction of a component using finite element analysis (FEA) includes the three following steps :

- geometrical modelling and meshing of the part ;
- thermal and mechanical simulations (uncoupled)
- finally, lifetime assessment by the post-processing of computed local values, like the equivalent strain or stress.

The challenge of the thermal analysis is to reproduce at best the real thermal field : for example, the hot area due to the convergence of exhaust gas flow and the gradient between the motor flange and the body of the manifold. Usually, gradient through the thickness is not reproduced. Generally, thinning and residual stresses brought by the forming process are not taken into account in FEA. Concerning the modelling of complete component, meshing using 3D-spatial shell elements implies that some meshing rules have to be proposed for areas containing small curvatures or weld seams. Studies are undertaken in Usinor dealing with the transfer of the thickness and residual strain fields between explicit software used for forming simulation and implicit one used for the fatigue design. In the same way, the modelling of weld seam and small curvature using shell elements is actually studied on the basis of the developed thermal fatigue test.

Constitutive law for material and identification procedure

Because of the involved temperatures, an elasto viscoplastic behaviour description has to be preferred to a solely elastoplastic one. In fact, the viscoplastic behaviour of a metal subjected to cyclic loading at high temperature is well-described using a non linear kinematic hardening model coupled with a Norton law; like the model proposed by J.L.Chaboche [6]. All the parameters were assumed to depend only on temperature and are identified using the stress-strain curves derived from low-cycle fatigue tests performed in isothermal conditions - from room temperature to 950°C - and for different strain amplitudes and rates.

In our identification procedure no relaxation tests were performed; so our set of parameters did not allow to simulate a long period creep process (strain rate below 10^{-4} s^{-1}). Because the stabilised strain-stress loop was chosen for the parameter identification, we supposed that the material reached a saturated cyclic hardening state. Watanabe et al., in ref. [7], have preferred the first half cycle which appears to closer describe a softened material especially when a recovery process occurs during a long period at high temperature. The difference is mainly significant at low temperature. It is clear that a complete coupled metallurgical behaviour will be a significant improvement for the model but identification and implementation in FE code are substantially more complex.

Application to the thermal fatigue specimen

ABAQUS [8] was used as solver for both thermal and mechanical analysis of the different experiments where thickness, maximal and minimal temperatures, holding time and grade were varied. Only a quarter of the specimen was meshed using 8-nodes 3D finite elements (figure 4). First, the thermal analysis was done to fit precisely the experimental measurements by thermocouples : only the four first cycles are simulated.

A UMAT procedure was necessary to perform the thermomechanical analysis using the elasto-viscoplastic Chaboche model : so we used the Z-ABA software [9]. Different experimental conditions were simulated. Figure 5 shows a comparison between the experimental and calculated clamping force which is considered as a satisfying result in regard of our assumptions. Also, figure 5 evidences an accommodation process just after the half-first cycle and thereafter the clamping force - or stress- reaches a stable loop (also

evidenced by Watanabe in [7]). The interest for the modelling of few cycles rather than modelling only one heating is also demonstrated.
Figure 6 shows two typical features of the thermal fatigue process of a component :

- an in-phase mode at extrados of the specimen implying creep at the maximal temperature under tension ;
- an out-of-phase mode at intrados of the specimen implying creep at the maximal temperature under compression.

These two modes would determine the main damage mechanisms involving in a stainless steel part under thermal fatigue.

DAMAGE EVALUATION AND POST-PROCESSING

Thermal fatigue damage process

Failures of exhaust parts are often attributed to an out-of-phase thermal fatigue process due to compressive strains that occur at high temperature [7]. Many of these examinations are performed on ferritic stainless steel grades but a such conclusion can not be generally applied to austenitic grades where in-phase thermal fatigue process appears more detrimental : tensile state of stress occurring at high temperature. In fact, a crack tip oxidation coupled with fatigue is the main damage mechanism of the austenitic grades (figure 7). Intergranular cavitation is also observed due to creep damage.

Thermal fatigue criterion for ferritic grades

Taira, in ref. [10], formulated a Manson-Coffin like criterion which relates the equivalent viscoplastic strain amplitude $\Delta\varepsilon_{vp}$ accumulated during each cycle to the number of cycles to failure N in form,

$$K\ (\Delta\varepsilon_{vp})^n = N \quad (1)$$

where the constants K and n may depend on the holding time and the maximal temperature of the cycle. In fact, a mean value of $\Delta\varepsilon_{vp}$ is computed for cycles 2 to 4. Evolution of $\Delta\varepsilon_{vp}$ as function of time is shown on figure 8 where the increment at the intrados is greater than at the extrados. The failure is naturally related to this local quantity. Relation (1) has been identified on EN1.4509 (AISI441) grade for different maximal and minimal temperature, respectively from 850°C to 950°C and from 100°C to 250°C. Model and experimental results are displayed on figure 9. The holding time plays both on viscoplastic strain – through creep - and on parameter K, but it is not still included in the criterion.

Particular case of austenitic grades

Equation (1) can not predict the thermal fatigue failure of austenitic specimen because the viscoplastic strain amplitude is always greater in intrados than in extrados even though cracks occur at extrados (figure 10). The conclusion would be the same using other criteria based on the strain, the stress or the dissipated energy. In case of austenitic grade a phasing factor able to distinguish in-phase or out-of phase mode appears necessary to establish a thermal fatigue law. As it is shown on figures 10 and 11, the hydrostatic pressure is a good local quantity to describe the phasing between the thermal load and the mechanical load. A tensile state of stress ($p>0$) at high temperature leads to the opening of the cracks and therefore the oxidation penetration. Cracks propagate rapidly in in-phase mode. Unfortunately, no such relation is still identified and it is one of our managed aims.

CONCLUSION

Thermal fatigue resistance

The thermal fatigue resistance of different stainlesss steel grades was studied by means of a specific test. Further, the developed test appears as a useful mean to study damage process and to identify or validate damage criteria. So, two typical features of the thermal fatigue are simulated :

- an in-phase mode implying creep at the maximal temperature under tension which appears to be the most detrimental mode for austenitic grades ;
- an out-of phase mode implying creep at the maximal temperature under compression which appear to be the most detrimental mode for ferritic grades.

Results evidenced that the ferritic grade EN 1.4509 (AISI441) offers the best resistance compared to the austenic grades which are more sensible to the detrimental effect of holding time at high temperature due to a creep and oxidation coupling with fatigue.

In the out-of-phase damage mode of ferritic grades, the viscoplastic strain amplitude was used as the crack initiation criterion using a non isothermal Manson-Coffin law. Concerning the in-phase damage mode of austenitic grades, the phasing between the thermal load and the mechanical load has to be taken into account in the criterion. Finally, the general approach for a lifetime prediction of real component is presented but some difficulties have still to be solved for application ; particularly meshing rules have to define for small curvature and weld seam with 3D-shell elements.

AKNOWLEDGEMENT

The author wishes to acknowledge the valuable inputs of C.Simon and O.Cleizergues and thank I.Evenepoel, H.Sassoulas (now at CEA), B.Proult and F.Moser (Ugine-Savoie-Imphy) for the performing of the finite elements analysis and the experiments.

REFERENCES

1. Antoni , L., Herbelin, J.-M., (1999), in EFC Working Party Report on Cyclic Oxydation of High temperature Materials : Mechanisms, Testing Methods, Characterisation and Lifetime Estimation M.Schütze, W.J. Quadakkers Eds, Publication N°27 in European Federation of corrosion series. Inst. of Materials p.187.
2. Sassoulas , H., Santacreu, P.-O, (1999), 18ème Journée de Printemps de la SF2M-Dimensionnement en Fatigue des Structures : Démarches et Outils, Paris, 2-3 Juin, p. 161
3. Santacreu, P.-O. et al, (1999), Thermal Stress'99, Cracow, Poland, June 13-17, p.245.
4. Renaudot, N. et al., (2000), SAE Technical paper series N°2000-01-0314 SAE 2000 World Congress Detroit Michigan March 6-9.
5. Oku, M. et al, (1992), in Nisshin Steel Technical Report, 66, p37.
6. Lemaitre, J, Chaboche, J.-L.,(1985), Mécanique des Matériaux Solides, Dunod Eds., Paris.
7. Watanabe,Y., et al, (1998), SAE Technical paper series 980841, SAE Int. Congress, Detroit Michigan, February 23-26.
8. Abaqus, (1998), Hibitt, Karlsson and Sorensen, Inc.
9. Transvalor, Northwest numerics Inc., (1999) Z-Set /Z-Aba version 8 manuel.
10 Taira, S, (1973), in Fatigue at elevated Temperatures, ASTM STP 520, p. 80.

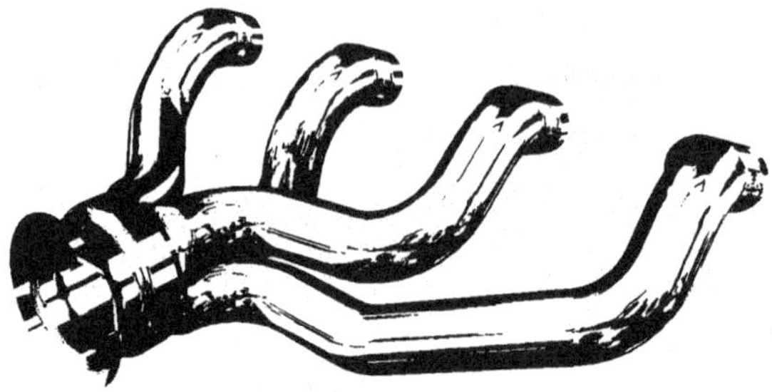

Fig. 1. A 4-cylinders stainless steel exhaust manifold *(with courtesy of Faurecia)*

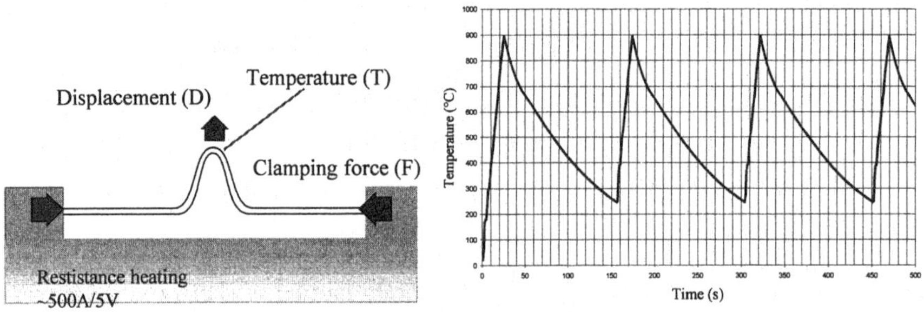

Fig. 2. Principle of the thermal fatigue test developed and example of thermal cycle.

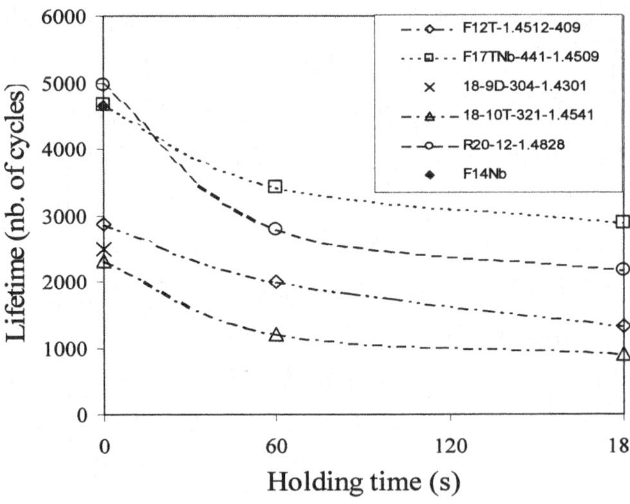

Fig. 3. Thermal fatigue life of the different stainless steel grades as function of the hold time at the higher temperature : 2 mm-thick specimen, cycle 250°C ↔900°C under air.

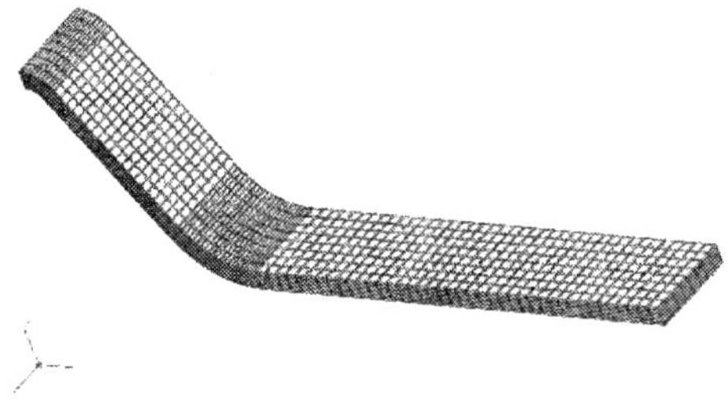

Fig. 4. 3D meshing of a ¼ of the thermal fatigue specimen.

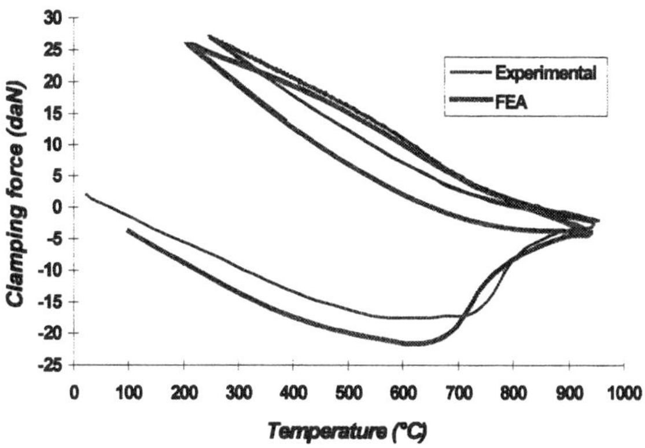

Fig. 5. comparison between experimental and calculated clamping force for EN1.4509 (AISI 441) 250°C↔ 900°C no holding time.

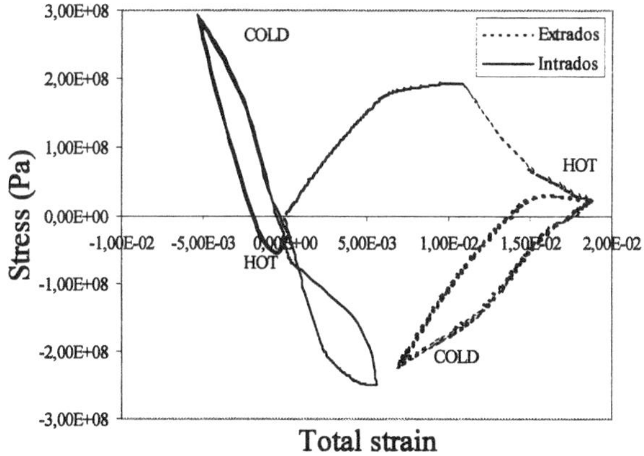

Fig. 6. Total strain versus stress curve at intrados and extrados of the specimen - for EN1.4509 (AISI 441) 250°C↔ 900°C no hold time.

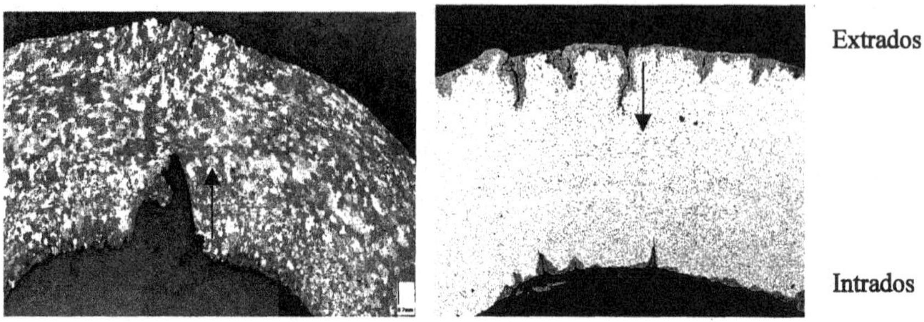

Fig. 7. SEM observations of a failed ferritic specimen (left) and austenitic specimen (right): EN1.4509/AISI441 and EN1.4541/AISI 321 for 250°C↔950°C cycle.

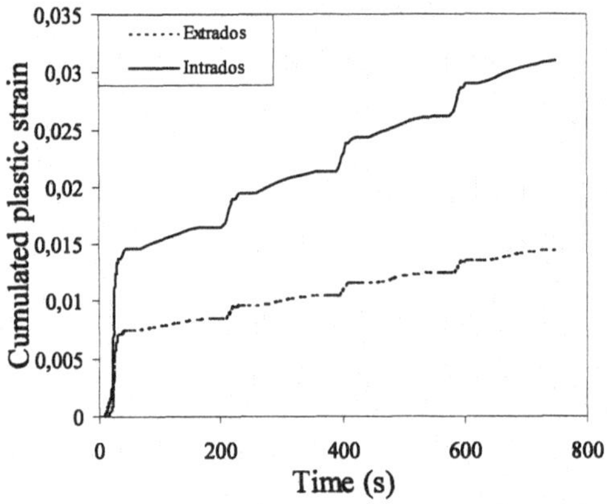

Fig. 8. Cumulated viscoplastic strain versus time at intrados and extrados – EN 1.4509 (AISI 441) 250°C↔ 900°C no holding time.

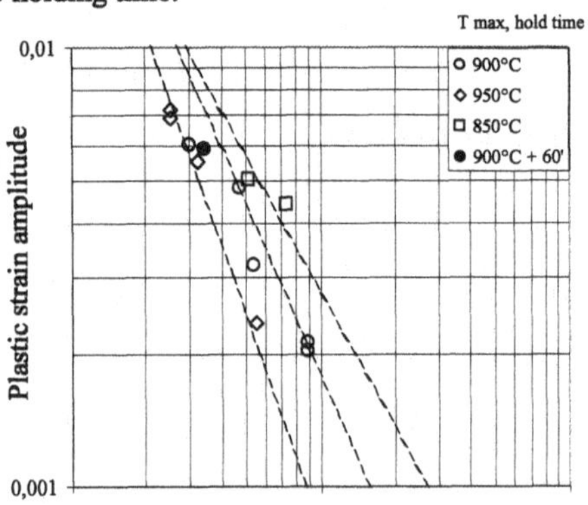

Fig. 9. Viscoplastic strain amplitude versus fatigue life for EN1.4509(AISI441) grade for three different maximal temperatures : model (dot line) and experimental results (symbols)

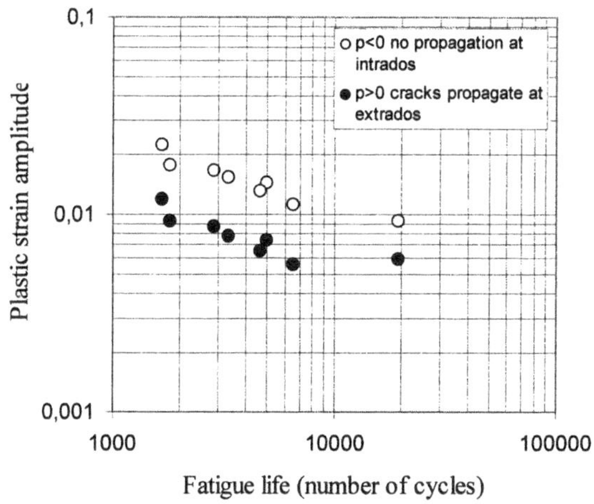

Fig. 10. Viscoplastic strain amplitude versus fatigue life for EN1.4828(~AISI309S) austenitic refractory grade: black dots indicate crack propagation.

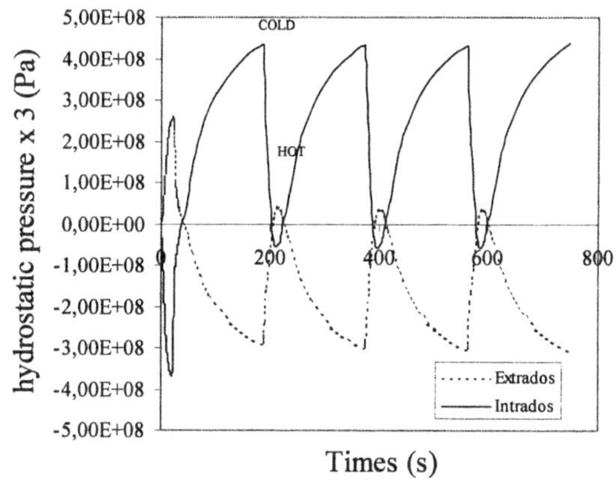

Fig. 11. Hydrostatic pressure (trace of stress tensor) versus time at intrados and extrados of the thermal fatigue specimen.

Temperature-Fatigue Interaction
L. Rémy and J. Petit (Eds.)

ISOTHERMAL AND THERMO-MECHANICAL FATIGUE LIFE MODELLING OF COMPONENTS AND STRUCTURES AT ELEVATED TEMPERATURE

X.B. LIN, P.R.G. ANDERSON, V. OGAREVIC and M. BENNEBACH
nCode International Limited, 230 Woodbourn Road, Sheffield S9 3LQ, UK

ABSTRACT

Elevated temperature fatigue-life projects, funded by leading automotive manufacturers and component suppliers, are being carried out at nCode International Limited. These projects aim to develop an integrated durability solution for engineering components at elevated temperature, and involve material characterisation, software development of numerical damage models, software validation and verification. It is expected that the procedure developed can eventually allow industry to use and compare various models to predict the life of components under high temperature conditions and to achieve "right-first-time" design. The integrated durability procedure is explained in this paper, followed by an overview of some damage models at elevated temperature, including creep, fatigue and creep-fatigue interaction models. Some details of software development and material testing are also described.

KEYWORDS

Creep, fatigue, isothermal fatigue, thermo-mechanical fatigue, creep-fatigue interaction, thermal stress analysis, visco-plastic analysis, FEA, CAE

INTRODUCTION

It has been generally recognised that lifetime predictions of components at the design stage are key to the achievement of reduction in development times and improvements in product quality. The traditional "design-it, break-it, redesign-it..." loop is too expensive and slow for global competitiveness. Great efforts have been made to develop numerical models to predict creep, fatigue and creep-fatigue lives at elevated temperature. However, it seems that the design of components under high temperature environments is still based on some simple rules or previous experience in each particular industry.

A couple of projects, funded by several leading automotive manufacturers and component suppliers, are being carried out at nCode International. The projects aim to develop an integrated durability procedure, which would enable automotive manufacturers to use various well-developed models to predict the life of their high temperature components, such as exhaust manifolds, exhaust systems and other hot engine components. The projects involve material characterisation, software implementation of some damage models that have been chosen, software connection to commercial finite element (FE) solvers, together with software validation and verification.

In this paper, a brief description of the integrated high-temperature durability procedure is given, followed by an overview of several well-known damage models related to the prediction of fatigue, creep and fatigue-creep interaction lives under isothermal or thermo-mechanical fatigue conditions. The models include the traditional low cycle fatigue, Larson-Miller creep, Chaboche's non-linear creep, fatigue and creep-fatigue damage evolution, Halford's total strain partitioning and Taira's equivalent temperature procedures. Software development and material testing are also explained.

INTEGRATED DURABILITY PROCEDURE

Life predictions for components at an early design stage are essential in the automotive industry, because automotive manufacturers increasingly require reductions in the time and cost for bringing new designs to production while assuring a high reliability level of the vehicle. The need to predict the durability of components under high temperature conditions, such as exhaust manifolds has recently become urgent. It is well known that the stress, strain and life predictions at high temperature are much more difficult than at room temperature, since materials exhibit time dependent behaviour, and damage mechanisms also become complex due to the mixture of creep, fatigue and oxidation that occur under these conditions. However, great efforts have been made in the research community to develop various damage models for high temperature use. It has now become possible to use these models for engineering applications.

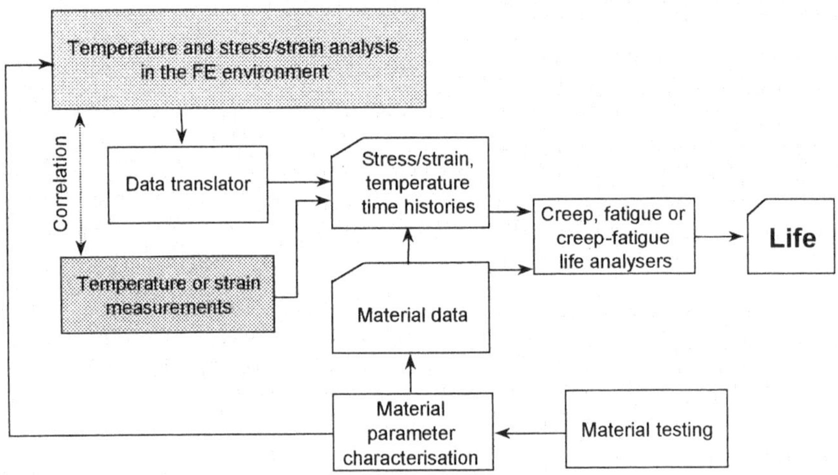

Fig. 1. An integrated durability procedure for use under high temperature conditions.

An integrated high temperature durability procedure for industrial applications is shown in Fig 1, in which it can be seen that life predictions under high temperature conditions generally require temperature, stress, strain and material inputs. Temperature at some locations can be measured in rig or in-service tests, and its distribution over the entire component can also be modelled by performing a thermal analysis in the finite element environment, together with some necessary computational fluid dynamics (CFD) calculations. Strain may be measured in high temperature environments, but specialised techniques and instruments are usually needed. Direct stress/strain modelling by FE has recently become possible for a given temperature and mechanical loading cycle. In some commercial FE software packages, such as ABAQUS and ANSYS, visco-plasticity models are available. Using these models, the appropriate stress or strain time histories can be obtained for life calculations over the entire component. Correlation between measured and calculated temperature or strains can also be performed, as shown in Fig.1, to make sure that good modelling has been achieved. For a successful life prediction, material parameters in both damage models and stress-strain relations must be determined from relevant material tests, although these tests are usually both expensive and time consuming. This procedure obviously enables identification of the life-critical locations and provides a basis for design optimisation. Equally obviously, the benefit of using such a procedure would be enormous.

The projects being carried out at nCode International intend to establish the integrated procedure described above. The procedure consists of the following technical tasks:

- Identifying of some well-recognised damage models
- Developing software to implement these damage models and to link them with FEA solvers
- Testing appropriate materials
- Determining material parameters for all damage models in the software
- Validating the software
- Verifying the software predictions for the target automotive components

Some details of damage models, software development and material testing are given below.

DAMAGE MODELS

A very brief description is given below of several damage models dealing with the prediction of fatigue-creep lives at elevated temperature.

Low-cycle local strain fatigue model

The local strain approach is a classic fatigue model [1-3], which has been used to predict the low-cycle fatigue crack initiation life at ambient temperature. This model is recommended in some industrial standards and great success has been achieved in engineering practice. The model requires both total strain-life and cyclic stress-stain curves, which can be obtained from small specimen strain-controlled fatigue tests. The Neuber notch correction may be used to convert the elastic to elastic-plastic stress/strain, and fatigue cycles contained in a strain time history are counted by the well-known rainflow method.

This model may still be applicable to high temperature conditions, if the temperature is above the room temperature but not so high as to cause significant creep effects in the material. However, the material cyclic stress-strain and strain-life curves must be obtained from strain-controlled fatigue testing at the relevant elevated temperature rather than room temperature.

Larson-Miller creep model

It has been recognised that most materials fail due to creep at high temperatures. Many methods have been developed to predict the creep rupture time. Among them the Larson-Miller parameter method [4] is well known. It is an extrapolation method, which uses the material data obtained from short rupture life measurements to predict the long creep life. In some conditions, such as operating at very high temperature, undergoing low frequency cycling or long hold times, the damage due to fatigue is very limited and a creep model can give a reasonable estimate of the lifetime.

The Larson-Miller creep damage model can be expressed as follows,

$$PLM = T(\log t_r + C) \tag{1}$$

In the above equation, T is the absolute temperature; t_r is the rupture time; C is a material constant; and PLM is the Larson-Miller Parameter which can be expressed in terms of stress,

$$PLM = A_0 + A_1 \log\sigma + A_2 (\log\sigma)^2 + \ldots \tag{2}$$

Material constants C, A_0, A_1 and A_2 can be determined from the constant stress (load) creep rupture test at several different temperature levels. Fig. 2 illustrates a typical Larson-Miller curve re-plotted from the work of Yamauchi *et al.* [5].

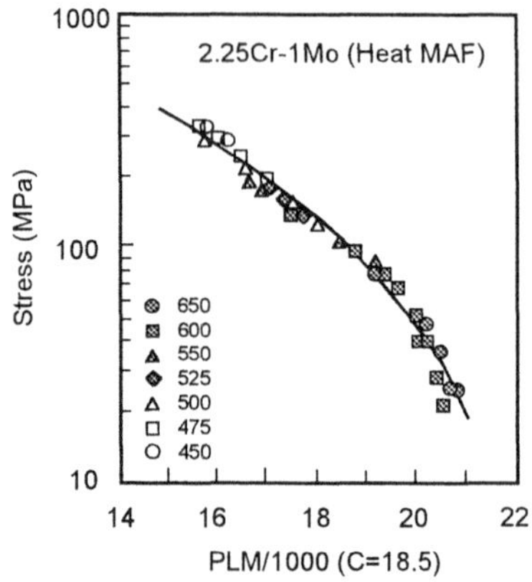

Fig. 2. An illustration of the Larson-Miller Parameter curve

It is worth noting that under conditions of varying stress or temperature a damage summation rule is needed to use the Larson-Miller equation to estimate the rupture time. The following equation is a widely-used life-fraction rule [6],

$$\sum \frac{t_i}{t_{ri}} = 1 \tag{3}$$

where t_i and t_{ri} are the time spent and time to rupture under condition i, respectively.

Chaboche's fatigue model

The non-linear continuous fatigue damage models proposed by Chaboche et al. [7-9] can be used to describe the progressive deterioration processes before macroscopic crack initiation in both room and high temperature situations. They have been successfully used to predict the fatigue life, in different materials, for various loading conditions, such as two stress level fatigue tests, block loading programs, and strain-controlled fatigue tests. One of these models can be written as follows,

$$dD = \left[1-(1-D)^{\beta+1}\right]^{\alpha}\left[\frac{\sigma_{\max}-\bar{\sigma}}{M(1-D)}\right]^{\beta} dN \tag{4}$$

where D is the damage variable; $\sigma_{\max}$ and $\bar{\sigma}$ are maximum and mean stresses during a cycle, respectively; and

$$M = M_0(1-b\bar{\sigma}) \tag{5}$$

$$\alpha = 1 - a\left\langle\frac{\sigma_{\max}-\sigma_l}{\sigma_u-\sigma_{\max}}\right\rangle \tag{6}$$

$$\sigma_l = \bar{\sigma} + \sigma_{l0}(1-b\bar{\sigma}) \tag{7}$$

The material parameters contained in the above damage equation can be determined mainly from stress-controlled fatigue testing together with others derived from conventional data. σ_u is the ultimate stress strength; σ_{l0} is the fatigue limit; and b is used to describe the effect of mean stress on the fatigue limit. Exponent β is obtained from the S-N curve under fully reversed conditions, by plotting $\sigma_{\max}$ as a function of $N_f(\sigma_{\max} - \sigma_{l0})/(\sigma_u - \sigma_{\max})$. Co-efficient $M_0 a^{-1/\beta}$ is obtained from one point of the S-N curve. The independent values of a and M_0 have no importance if only fatigue damage is being considered.

The above model is cycle dependent, but the damage evolution rate depends on the current damage level. A difference from the traditional Miner's damage rule, which always over-estimates the fatigue life for block program loading conditions [9], is that this model is able to describe the effect of cycle sequence. The fatigue life can be estimated by the integration of equation (4) from D=0 to 1. Cycle-by-cycle integration is required for the variable stress condition. Fig. 3 shows the S-N curves for steels A201 and A517 together with life predictions made by the Chaboche continuous damage model. It can be seen that the model covers both low and high cycle fatigue regions, and has taken into account the mean stress effect.

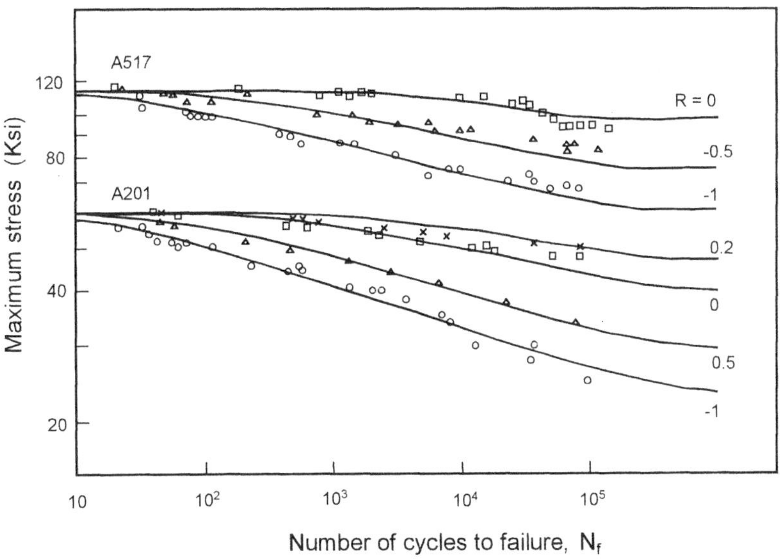

Fig. 3. S-N curves for A201 and A517 steels together with the predictions made using the Chaboche fatigue model [9] (R is the stress ratio defined by $R = \sigma_{\min}/\sigma_{\max}$).

Chaboche's creep model

Based on Rabotnov-Kachanov's work [10,11], Chaboche and Lemaitre [8,12] proposed a creep damage equation, which is expressed as follows,

$$dD = \left(\frac{\sigma}{A}\right)^r (1-D)^{-k} dt \tag{8}$$

where A, k and r are temperature dependent material parameters. They can be determined from isothermal creep rupture tests. In general, k is larger than r. By integrating the above creep damage equation for a constant stress and temperature case, we can obtain the following damage evolution equation

$$D = 1 - \left(1 - \frac{t}{t_r}\right)^{\frac{1}{k+1}} \tag{9}$$

where the rupture time t_r is

$$t_r = \frac{1}{k+1}\left(\frac{\sigma}{A}\right)^{-r} \tag{10}$$

The creep life for a variable stress and temperature case can be estimated by using a numerical integration scheme. Failure is considered to occur when the damage parameter D reaches 1

Chaboche's creep-fatigue interaction model

It has been assumed by Chaboche *et al.* [13-15] that the creep and fatigue damage, which is related to intercrystalline and transcrystalline micro-cracks respectively, can be added together to give the total creep-fatigue damage, that is.

$$dD = \left(\frac{\sigma}{A}\right)^r (1-D)^{-k} dt + \left[1-(1-D)^{\beta+1}\right]^{\alpha} \left[\frac{\sigma_{\max} - \bar{\sigma}}{M(1-D)}\right]^{\beta} dN \tag{11}$$

The means for determination of the material parameters in the creep and fatigue damage equations have been indicated previously. However, in order to describe the creep-fatigue interaction properly, some creep-fatigue interaction tests are also required to determine the parameter a, (Eqn. 6) unlike the case for pure fatigue.

In general, the above differential equation has to be integrated by a numerical increment method to calculate the creep-fatigue life. The damage due to creep and due to fatigue may be independently accumulated during each stress cycle. Satisfactory predictions [13,14,16] have been made by this creep-fatigue interaction model for various tests, such as stress or strain controlled visco-plastic fatigue tests, visco-plastic tests at two temperature levels, creep tests after fatigue and thermo-mechanical fatigue tests.

Halford's total strain partitioning model

Halford's TS-SRP (Total Strain Range Partitioning) model [17-19] is one of the Halford strain-based damage models, which had been developed initially for isothermal creep-fatigue life predictions. The model introduces the "failure" and "flow" concepts. The failure tests are used to determine the "failure" behaviour of the material, whilst the "flow" tests are designed to characterise the stress-strain behaviour (that is partitioning inelastic strains into PP, PC or CP type strains). The mathematical equations of this model are:

$$\Delta\varepsilon_t = \Delta\varepsilon_{el} + \Delta\varepsilon_{in} = B(N_{f0})^b + C'(N_{f0})^c \tag{12}$$

where

$$B = K_{ij}(C')^n \tag{13}$$

$$C' = \left[\sum F_{ij}(C_{ij})^{1/c}\right]^c \tag{14}$$

The subscripts ij denote the type of cycle, i.e. ij=PP, PC, CP or CC. Equation (14) is derived from an interaction damage rule and four generic SRP inelastic strain-range versus cyclic- life relations for a theoretical zero mean stress condition. The intercepts B and C' are time dependent, and are determined by the flow or stress-strain-time response of the material for a particular duty cycle. Fig. 4 illustrates equation (12), in which we can see that the model has assumed that the elastic and inelastic failure lines for isothermal creep-fatigue cycles are parallel to the corresponding failure lines of pure fatigue (PP cycles). This is based on some experimental results for several alloys. It should be indicated that the above formulae have also been extended to thermo-mechanical fatigue modelling [20,21]. In this case, bithermal, instead of isothermal, fatigue tests are required to determine the material parameters contained in the above equations. The maximum and minimum temperatures used in bithermal fatigue tests should be similar to the temperature range for the particular duty cycle under investigation.

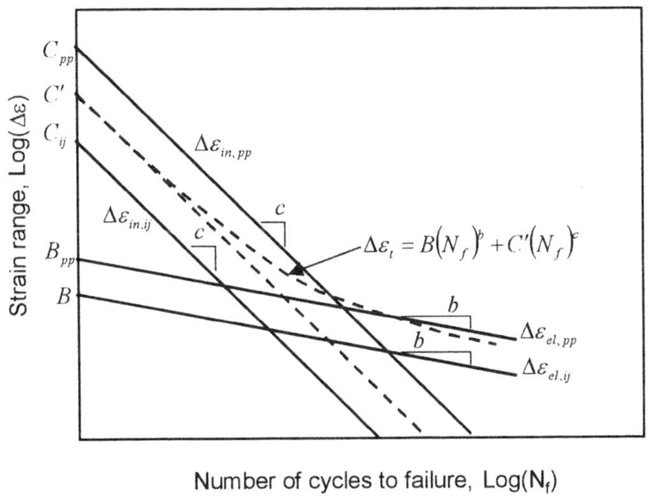

Fig. 4. Relation between total strain-range and life for creep-fatigue cycles.

For creep-life predictions, the TS-SRP model only requires total mechanical strain range, creep hold time and cycle type (PC, CP or CC), and, of course, basic material parameters.

Taira's model

This model [22] can be used to correlate both thermal fatigue and low cycle fatigue. Taira's damage equation is as follows:

$$\lambda(T)\left(\Delta\varepsilon_p\right)^n N_f = C_1 \tag{15}$$

where C_1 is a material constant independent of the temperature, and the effect of temperature is included in the damage factor $\lambda(T)$, which can be determined from test results on low cycle fatigue at several temperature levels. The thermal fatigue life can be predicted using an equivalent temperature in the above equation. The equivalent temperature is calculated on the assumption that the equivalent temperature level is that at which a test of mechanically induced low cycle fatigue gives the same number of cycles to failure as a test under the same plastic strain range in thermal fatigue. Please note that the rate of cycling in thermal and low-cycle fatigue tests should be about the same within the cycle and there should be no hold times.

This model is basically a traditional plastic strain based approach. Care should be taken to have a similar strain rate between testing cycles, from which the material parameters are determined, and the duty cycle under investigation.

SOFTWARE DEVELOPMENT

The software is designed with the intention of maximising the applications of a variety of isothermal fatigue, creep, thermo-mechanical creep-fatigue damage models as well as their combinations. Possible combinations of mechanical load and temperature for each damage model included in the projects are identified and implemented. Three categories, namely "Pure Creep", "Pure Fatigue", and Creep-fatigue interaction", are further divided to enable users to select different approaches for their applications.

Implementation of the damage models mentioned above is comparatively straightforward in most cases. However, a great effort has to be made to generalise the Chaboche models in order to deal with different possible combinations of stress and temperature cases, such as block loading programs, isothermal cycling, thermo-mechanical cycling, and even random stress and temperature cases. Interpolation of material parameters to an arbitrary temperature is made for both Chaboche creep and fatigue models. A rainflow cycle counting scheme that is able to record the cycle order is employed for the non-linear fatigue or creep-fatigue damage evolution.

Appropriate stress or strain, such as measured strains and stresses/strains obtained from a visco-plastic analysis for a given temperature history, are required by the numerical models for life predictions. It may be possible to use elastic stresses or strains from a thermal stress analysis, if they are thought to be close to those from a visco-plastic analysis. However, care should be taken if doing so.

Connection to finite element solvers will be developed to allow users to predict creep, fatigue or creep-fatigue lives for an entire component. Temperature and stress analyses are required before the life can be predicted. However, it is worth indicating that it can be very time-consuming to perform a visco-plastic finite element analysis for a large model. It is difficult to do such analysis for practical loading time histories. In practice, a duty cycle is usually selected, and the analysis is only performed until the stress-strain loop is stabilised. Engineering judgment is needed to select a typical duty cycle.

MATERIAL TESTING

A comprehensive material test program is being carried out to produce supporting material data required for the various damage models. Several common automotive materials for manifolds, cylinder heads and exhaust systems, such as austenitic stainless steels, cast aluminiums and cast irons have been included in the program.

Standard smooth cylindrical specimens with diameter 7 mm at the test gauge section are used for tests. Test temperatures are selected based on practical operating conditions for relevant components and experience of material behaviour. Different temperature levels or ranges are used for the different material types. Various test types are required to produce the material parameters for the variety of numerical damage models. It is worth indicating that some tests can be used for more than one model, and, as a result, advantage can be taken to reduce the number of specimens in the test program. Individual test types are briefly summarised as follows:

- Tensile tests at different strain rates at several temperatures. Basic material properties, such as proof strength, UTS and Young's modulus, can be obtained from this type of test.
- Creep tests with constant stress (load) at several temperatures. They are basically required by both the Larson-Millar and Chaboche creep models.

- Isothermal fast stress-controlled fatigue tests with different stress ratio at several temperatures. Chaboche's fatigue damage equation is based on this type of S-N test.
- Bithermal fatigue tests, including "failure" and "flow" tests. Halford's TS-SRP model will be used for thermo-mechanical fatigue life predictions in the present project. Therefore, the bithermal fatigue tests are performed, as indicated previously.
- Isothermal low cycle fatigue tests at several temperatures. This type of strain-controlled fatigue tests is basically required for Taira's model.
- Creep-fatigue interaction tests with different loading frequency or hold time. To use Chaboche's fatigue and creep damage equations for creep-fatigue life modelling, these interaction tests are necessary.
- Thermo-mechanical fatigue tests with a specific temperature range. This type of test can be used to verify various damage models in this project, and also to obtain further material behaviour information useful for some models.

The physical properties of the materials, such as thermal expansion, specific heat, thermal diffusivity and thermal conductivity from ambient temperature up to specified high temperature, have also been measured. A material database, including all test information and results, will be established for use with the software.

CONCLUDING REMARKS

The integrated durability procedure described in this paper can be used to make life predictions and to improve the design methodologies currently adopted in industry. The software tools developed are of practical use for a variety of components in different industries, such as aero engines, heat exchangers, nuclear pressure vessels and turbine blades, although automotive components are targeted at present. However, one should be aware that some difficulties still exist. Visco-plastic stress/strain analysis is not realistic for a temperature or mechanical loading history measured in the field, which is often both long and complex. Some simplified methods may need to be developed. The effect of oxidation and how to handle it is still not clear, although some models have taken it into account by virtue of the way in which the relevant materials parameters are measured. The multiaxial problem must also be dealt with in the future. More research is needed in all these areas .

REFERENCES

1. Basquin, O.H. (1910) *Am. Soc. Test. Mater. Proc.* **10**, 625
2. Coffin, L.F, Jr.. (1954) *Trans. ASME* **76**, 931.
3. Manson, S.S. (1953) In: *Heat Transfer Symposium*, pp. 9-75.
4. Larson, F.R. and Miller, J. (1952) *Trans. ASME* **74**, 765.
5. Yamauchi, M. *et al.* (1996) In: *6th Int. Conf. on Creep and Fatigue, IMechE*, pp. 511-519.
6. Robinson, E.L. (1938) *Trans. ASME* **160**, 253.
7. Chaboche, J.L. (1974) In: *Rev. Fr. Méc. No. 50-51*, pp. 71-82.
8. Chaboche, J.L. (1981) *Nuclear Engng Design* **64**, 233.
9. Chaboche, J.L. and Lense, P.M. (1988) *Fatigue Fracture Engng Mater. Struct.* **11**, 1.
10. Kachanov, L.M. (1958) *Lzv. Akad. Nauk. SSR, Otd Tekh. Nauk*, No.8, pp.26-31.
11. Rabotnov, Y.N. (1969) *North Hollan Publishing Company, Amsterdam. London.*
12. Lemaitre, J. and Chaboche, J.L. (1978) *J. de Meca. App. 2*, 317.
13. Chaboche, J.L., Policella, H. and Kaczmarek, H. (1977) *Applicability of the SRP Method and Creep-Fatigue Damage Approach to the LCHTF Life Prediction of IN-100 Alloy.* ONERA.

14. Chaboche, J. L., Policella and H., Savalle, S. (1976) *Application of the Continuous Damage Approach to the Prediction of High Temperature Low-Cycle Fatigue.* ONERA.
15. Cailletaud, G. and Levaillant, C. (1984) *Nuclear Engng Design* **83**, 279.
16. Chaboche, J.L (1980) In: *Symp. On Low-cycle Fatigue and Life Prediction.* Firminy, France.
17. Halford, G.R. and Saltsman, J.F. (1983) In: *ASME International Conference on Advances in Life Prediction Methods*, D.A. Woodford and J.R. Whitehead, eds., pp. 17-26.
18. Saltsman, J.F. and Halford, G.R. (1987) In: *Symposium on Low Cycle Fatigue – Directions for the Future*, ASTM, STP 942.
19. Saltsman, J.F. and Halford, G.R. (1988) *NASA Technical Paper 2779*, NASA.
20. Halford, G.R. *et al.* (1992) In: *Advances in Fatigue Lifetime Predictive Techniques*, ASTM STP 1122, pp. 107-119.
21. Halford, G.R. *et al.* (1992) In: *Advances in Fatigue Lifetime Predictive Techniques*, ASTM STP 1122, pp. 120-142.
22. Taira, S. (1973) In: *Fatigue at Elevated Temperatures, ASTM STP 520*, pp. 80-101.

AUTHOR INDEX

KEYWORD INDEX

Kolibri 360 GmbH
Ferdinand-Jühlke-Straße 7
99095 Erfurt, Deutschland
produktsicherheit@kolibri360.de